AVID

READER

PRESS

**ALSO BY
ADAM HIGGINBOTHAM**

*Midnight in Chernobyl:
The Untold Story of the World's
Greatest Nuclear Disaster*

Challenger

A TRUE STORY OF HEROISM AND DISASTER ON THE EDGE OF SPACE

Adam Higginbotham

AVID READER PRESS

New York London Toronto Sydney New Delhi

Avid Reader Press
An Imprint of Simon & Schuster, LLC
1230 Avenue of the Americas
New York, NY 10020

"Whitey on the Moon" by Gil Scott-Heron used with permission of Brouhaha
Music Inc. Rumal Rackley, administrator.

First Avid Reader Press hardcover edition May 2024

AVID READER PRESS and colophon are trademarks
of Simon & Schuster, LLC

Simon & Schuster: Celebrating 100 Years of Publishing in 2024

For information about special discounts for bulk purchases,
please contact Simon & Schuster Special Sales at 1-866-506-1949
or business@simonandschuster.com.

The Simon & Schuster Speakers Bureau can bring authors to your
live event. For more information or to book an event, contact the
Simon & Schuster Speakers Bureau at 1-866-248-3049 or visit our
website at www.simonspeakers.com.

Interior design by Lewelin Polanco
Diagrams by Alexis Seabrook

Manufactured in the United States of America

1 3 5 7 9 10 8 6 4 2

Library of Congress Cataloging-in-Publication Data has been applied for.

ISBN 978-1-9821-7661-7
ISBN 978-1-9821-7663-1 (ebook)

For Isla

CONTENTS

CAST OF CHARACTERS

NASA Headquarters, Washington, DC

Robert Frosch
Administrator: head of NASA, 1977–1981

James Beggs
Administrator: head of NASA, 1981–1986

William Graham
Deputy Administrator

Jesse Moore
Associate Administrator for Spaceflight: overall head of the Space Shuttle program, 1984–1986

Mike Weeks
Deputy Associate Administrator for Spaceflight, Technical

Johnson Space Center, Houston, Texas

Christopher Kraft, Jr.
Center Director, 1972–1982

George Abbey
Director of Flight Operations

Gene Kranz
Deputy Director of Flight Operations

John Young
Chief of the Astronaut Office

Arnold Aldrich
Space Shuttle Program Manager

Maxime Faget
Director of Engineering and Development

Tom Moser
Head of Structural Design, 1972–1982

Dorothy "Dottie" Lee
Subsystems Manager of Aerothermodynamics

Jay Greene
Flight Director, Mission Control

Jenny Howard
Booster Systems Engineer, Mission Control

Steve Nesbitt
Public Affairs Officer, Johnson Space Center; Chief Commentator, Mission Control, 1986

Marshall Space Flight Center, Huntsville, Alabama

Wernher von Braun
Center Director, 1960–1970

William Lucas
Center Director, 1974–1986

George Hardy
Deputy Director of Science and Engineering

Judson Lovingood
Associate Director of Engineering and Propulsion

Stanley Reinartz
Manager, Shuttle Projects Office

Larry Mulloy
Project Manager, Space Shuttle solid rocket boosters

Kennedy Space Center, Merritt Island, Florida

Gene Thomas
Launch Director

Cecil Houston
Resident Manager for the Marshall Space Flight Center

Johnny Corlew
Quality Assurance Inspector, pad closeout crew

Charlie Stevenson
Leader of the Ice Team, engineering department

Morton Thiokol, Wasatch Division, Utah

Jerry Mason
General Manager and Senior Vice President

Cal Wiggins
Deputy General Manager and Vice President

Bob Lund
Vice President of Engineering

Joe Kilminster
Vice President of Space Booster Programs

Allan McDonald
Director of the Space Shuttle Solid Rocket Motor Project

Arnie Thompson
Supervisor of Structures Design for Solid Rocket Motor Cases

Roger Boisjoly
Senior Scientist, Structural Mechanics, Space Shuttle Solid Rocket Motor Project

Bob Ebeling
Manager, Final Assembly, Space Shuttle Solid Rocket Motor Project

Astronauts

Joe Allen

Jim Bagian

Guion Bluford

Charles Bolden, Jr.

Manley "Sonny" Carter, Jr.

Richard "Dick" Covey

Bob Crippen

Anna Fisher

Robert "Hoot" Gibson

Frederick Gregory

Henry "Hank" Hartsfield

Frederick "Rick" Hauck

Bruce McCandless II

Ron McNair

Story Musgrave

George "Pinky" Nelson

Ellison Onizuka

Judy Resnik

Sally Ride

Francis "Dick" Scobee

Rhea Seddon

Mike Smith

James "Ox" van Hoften

The Space Participants

Jake Garn

Greg Jarvis

Christa McAuliffe

Barbara Morgan

Bill Nelson

The Rogers Commission

William Rogers
Chairman

Neil Armstrong
Vice Chairman

Dr. Richard Feynman
Nobel Prize–winning physicist

General Donald Kutyna, U.S. Air Force
Former head of Pentagon shuttle operations

Titles in January 1986 unless otherwise noted.

PROLOGUE

Flight Control Room Two
Johnson Space Center, Houston
January 28, 1986, 8:30 a.m.

The coffee, as usual, was terrible: bitter and thin, the color of tea; almost certainly undrinkable. He filled a cup anyway, returned to his console, and plugged in his headset. It promised to be a long morning.

Steve Nesbitt had arrived at his office early, checking for the latest weather updates from the Cape before taking the short walk out past the duck ponds to Building 30, and up in the elevator to the third floor of Mission Control. But from what he'd already seen on TV, there was no way they would launch today: it was freezing down in Florida, and there were two-foot icicles hanging from the gantry. Space Shuttle mission 51-L seemed sure to face yet another delay.

Nesbitt had been with NASA public affairs for just over five years, and was there for the triumph of the first Space Shuttle launch in 1981— helping to respond to a clamor of press and media inquiries from all over the world. Since then, he had become chief commentator for Mission Control, and delivered the live commentary from Houston on almost every one of the twenty-four shuttle flights. But he was still nervous.

Responsibility for translating the bewildering patois of engineering jargon and acronyms spoken by NASA engineers and astronauts into language the public could understand began with the launch countdown commentary that boomed from the loudspeakers at Cape Canaveral. After that—once the count reached zero and the spacecraft left the ground—everything that happened was on Nesbitt's watch. There was no script, and he knew his words went out live to anyone watching a launch on television—either on the three national networks, on the recently launched cable channel CNN, or over NASA's own dedicated satellite feed; he relied instead on his Ascent Events List, which mapped a series of milestones the shuttle would pass on its way to orbit, from the slow roll it would execute as it roared away from the launchpad to the moment its main engines cut out, at the edge of space.

The hushed environment of the Flight Control Room had been devised to concentrate the minds of each of the flight controllers on their own tasks, and only recently had a TV set been installed near the Flight Director's console, to display images of the shuttle in flight. Nesbitt rarely had time to look at that, as he focused attention on the console in front of him. Here, he had access to real-time information about the spacecraft: on his headset, he could listen in to dozens of audio "loops" connecting groups of NASA engineers and flight controllers on the internal communications network; and on a pair of black-and-white monitors, he could see telemetry data transmitted back to Earth from the shuttle, columns of numbers updated every second describing any one of thousands of technical parameters of its performance in flight.

With more than seventy feeds to choose from, Nesbitt had his regular preferences: "Flight Ops Procedures," which included data on the shuttle's engine performance, and the "Trajectory" display, which showed its speed, altitude, and downrange distance. Even with all this at his fingertips, Nesbitt found the live commentary nerve-racking, and practiced often. He took his duty of public service seriously, and hated it when other commentators took flight with flowery language, like Hollywood PR guys. He wanted to play it straight.

And yet, suffering from the effects of a cold he'd picked up the day before, even as the final countdown began, Nesbitt would have welcomed another launch delay: his throat was sore, and he wasn't certain he could talk through the whole ascent without his voice straining or cracking. He waited in silence for his cue: for the shuttle engines and the giant solid rockets to light; for his counterpart at the Cape to announce that *Challenger* had cleared the tower.

It was almost exactly 11:38 in the morning when Nesbitt saw the numbers on his screen start to move, and a few seconds later keyed his mic to speak:

"Good roll program confirmed. *Challenger* now heading downrange."

At the console position next to him, the flight surgeon—a Navy doctor in full uniform—had her eyes on the big TV set across the room. It was a perfect launch. *Challenger* was less than half a minute into flight when Nesbitt gave his next update.

"Engines beginning throttling down, now at 94 percent," he said. "Normal throttle for most of the flight is 104 percent. We'll throttle down to 65 percent shortly."

The flight surgeon watched the shuttle climb higher into the cloudless sky over the Atlantic; Nesbitt kept his gaze on the monitors. "Velocity 2,257 feet per second," he said. "Altitude 4.3 nautical miles, downrange distance three nautical miles." The numbers all looked good; at sixty-eight seconds, he reported the next key moment on the list in front of him. "Engines are throttling up. Three engines now at 104 percent."

Ten feet away, down on the next row of consoles, astronaut Richard Covey confirmed the change with the shuttle commander: "*Challenger*, go at throttle up."

"Roger, go at throttle up."

The spacecraft was one minute and ten seconds into flight.

Four seconds later, Nesbitt heard a loud crackle in his headphones.

Beside him, the surgeon saw *Challenger* abruptly obscured by a ball of orange and white flame.

"What was that?" she said.

But Nesbitt was staring at his monitors.

"One minute fifteen seconds. Velocity 2,900 feet per second," he said. "Altitude nine nautical miles. Downrange distance seven nautical miles."

Then Nesbitt looked up, and followed the surgeon's gaze toward the

TV set. Something terrible had happened. There was no sign of *Challenger*, just the expanding fireball where it had once been—and the exhaust trails of the shuttle's two booster rockets, twisting in opposite directions across the sky. His console was no help: the data streams had frozen. Around him, the other flight controllers sat stunned, faces slack with shock. No one said a word.

Nesbitt knew he had to speak, but he had no information to explain what he was witnessing. His mind raced. He thought of his responsibility to the public, and to the astronauts' families. He thought, suddenly, of the attempt on Ronald Reagan's life nearly five years before: in the confusion that followed, CBS news anchor Dan Rather had announced that White House press secretary James Brady had been killed—only to discover that Brady, despite the bullet in his head, remained very much alive. Nesbitt didn't want to make a mistake like that.

A few moments of quiet extended into half a minute. An agonizing silence enveloped the NASA commentary loop; an eternity of dead air. On the TV screen, the cloud drifted in the wind; fragments of debris fluttered toward the ocean. The Flight Director polled his team in vain for answers.

It was forty-one seconds before Steve Nesbitt spoke again.

"Flight controllers here looking very carefully at the situation," he said, his voice flat and impassive. "Obviously a major malfunction."

PART ONE

The Last Man on the Moon

CHAPTER ONE

FIRE ON PAD 34

Martha Chaffee was in her kitchen, making hot dogs for the kids' dinner, when she saw Michael Collins at the front door; in that moment she realized she was a widow. The astronauts and their families formed a small, tight group in the neighborhood—"Togethersville," they called it—and that evening a few of the wives were already gathering in the living room of the Chaffees' yellow-brick house on Barbuda Lane. Someone had mentioned an accident, but at first there seemed no reason for Martha to worry: the cars pulling up on the street outside could be part of the routine Friday-evening exodus from work. Besides, although Martha's husband, Roger, was away, training down at Cape Canaveral, he wasn't even scheduled to fly. Yet now Collins was standing on her doorstep, solemn and alone. There could be only one explanation.

"I know, Mike," she said. "But you've got to tell me."

She turned, and Collins followed her silently down the narrow hallway. In the den where her two children were watching the TV, someone reached over to turn it off before the news bulletins began. It was January 27, 1967.

At thirty-one, Roger Chaffee was the youngest and most inexperienced of the three astronauts assigned to the crew of Apollo 1—NASA's first manned test flight in the new rocket program intended to eventually land men on the moon. Dark-haired and good-looking, Chaffee had been an astronaut for a little more than three years, and had yet to fly in space. His rookie status was exaggerated by his baby face, his slight stature, and his archetypal straight-arrow background: a Boy Scout whose obsession with flight started early, building model planes with his barnstorming pilot father; a punctilious Navy aviator with a degree in aeronautical engineering from Purdue, where he had met Martha, the college homecoming queen; a devoted husband who drew up the blueprints for their home in Houston, and liked to relax by building his own rifles and making ammunition, which he carried on weekend hunting trips. The mission commander, veteran astronaut Gus Grissom—the second American in space, a short, plainspoken forty-year-old test pilot with a reputation for drinking and skirt-chasing—told reporters that Chaffee was "a great boy"; over the months they spent training together, the younger man showed his admiration for his commander by adopting parts of Grissom's distinctive body language and salty vocabulary, an affectation so incongruous that Chaffee's friends often teased him about it. The final member of the crew was the senior pilot, Ed White, another veteran astronaut and a lithe athlete whose image had been stamped on the consciousness of the nation when he became the first American to make a space walk, leaving the capsule of Gemini IV to drift high over the Pacific Ocean, tethered only by a gilded umbilical cord feeding him oxygen.

By the time Chaffee, White, and Grissom learned of their assignment to fly Apollo 1—in which they planned to spend up to fourteen days in Earth orbit, giving them time to check out the many complex new systems developed to take later missions to the moon and back—the space race with the Soviet Union was entering a white-knuckle phase. At the moment in May 1961 when President Kennedy had committed the country to land an American on the moon before the end of the decade, the recently established National Aeronautics and Space Administration had not yet managed to put a human being into Earth orbit, but instead suffered one humiliation after another at the hands of the Soviets. Cosmonaut Yuri Gagarin had become the first man in space just the month before.

But over the next five years, a growing team of scientists, engineers, and technicians, engaged in a project of experimental engineering on a scale nearly unprecedented in history, achieved a series of developmental leaps that enabled NASA to close the gap with the Soviet program: in February 1962 John Glenn became the first American in orbit, helping to make him and the other six astronauts of the Mercury program into national heroes; in June 1965, Ed White embarked on his spectacular space walk, and later that year the crews of Gemini VI and Gemini VII made the first successful orbital rendezvous. In June 1966, Surveyor 1, a three-legged robot spacecraft, made the first American controlled landing on the lunar surface and transmitted more than eleven thousand photographs back to Mission Control in Houston. The prospect of eventually following Surveyor with a manned mission no longer seemed entirely out of reach. Even so, the obstacles to sending men to walk on the moon and then returning them safely to Earth remained daunting. Although Kennedy's final deadline to do so lay just three years away, the technology necessary to accomplish the task—the massive Saturn V rockets, the Apollo capsules that would carry the three-man crew across 250,000 miles of deep space to their destination, and the delicate, insect-like lander intended to deliver two of them to the lunar surface—was still being designed and built. And the process was proving fraught with problems.

As the summer of 1966 drew to a close, Chaffee, Grissom, and White flew to California to attend a conference with senior NASA engineers and North American Aviation, the contractor responsible for building their spacecraft—and one of thousands of private corporations hired by the government to design and manufacture the hardware for the lunar program. The meeting was held in a low fieldstone building amid the sprawling complex of hangars and industrial spaces at the company's Downey plant, ten miles east of Los Angeles International Airport. Formally known as the Customer Acceptance Readiness Review, the conference was intended to be the final step before North American shipped the crew's Apollo capsule to Florida for handover to NASA. Work had been slow, and the contractor's final testing of the capsule wasn't complete. But Joe Shea, who ran the Apollo spacecraft program, told everyone that he intended to go through with the review anyway. Shea—arrogant, charismatic, and a gifted systems engineer—had been hired years earlier as an enforcer to help integrate the numerous moving parts of the moon program, someone who could both handle the demands of its management and understand the finest details of its experimental technology. With his compulsion for puns, his trademark red socks, and his relentless dedication to the job, Shea had won the admiration of his staff and built a public profile as a spokesman for NASA that would soon approach that of the German rocket scientist Wernher von Braun, father of the US space program.

The list of enduring problems with Chaffee, Grissom, and White's capsule, known as Spacecraft 012, was long, and the crew had become intimately familiar with its deficiencies during previous visits to the Downey plant. Many of the astronauts were not merely skilled pilots, but technicians with expertise and advanced qualifications in aeronautical engineering and celestial mechanics. From the very beginning, they recognized that they were placing their lives in the hands of government contractors who were often inventing the technology of space travel as they went along. When asked what he was thinking about when preparing for launch aboard his Mercury-Redstone rocket, Alan Shepard, the first American in space, had infamously replied, "The fact that every part of this ship was built by the low bidder."

Now all of the Apollo astronauts were involved in helping devise design changes and revisions to the equipment that would carry them toward the moon. Gus Grissom rarely let Spacecraft 012 out of his sight: during all of 1966, Grissom's son Mark reckoned he spent no more than eighteen days at home with his family. And, in private, the Apollo 1 commander was deeply concerned by what he saw at Downey: poor design and shoddy workmanship plagued the capsule. To the astronauts, half the staff at the California plant seemed diligent and careful—but the other half were clueless and incompetent. Roger Chaffee challenged the North American engineers over the faults he found, and sketched out possible solutions on the factory floor. But some of the most serious failings of the spacecraft were persistent and intractable. The environmental control system—which supplied oxygen and maintained the temperature inside the capsule—was prone to leaks of ethylene glycol solution, and had already caught fire in one test. The thick bundles of electrical wiring that snaked through the interior of the spacecraft were badly installed, and their insulation often frayed, increasing the risk of sparks and short circuits in the cabin. The main hatch was a clumsy, three-layer design that opened inward into the cramped cockpit, and—unlike those on the earlier Mercury and Gemini capsules—could not be jettisoned in an emergency by explosive charges. Instead, it required Ed White, lying prone in the middle of the three crew couches, to reach behind his head and use a ratchet handle to operate a mechanism securing the six latches holding it shut. Even under ideal circumstances, it would take three men between forty and seventy seconds to open the hatches and get out of the spacecraft. Almost without exception, the Apollo astronauts detested the hatch design, and had lobbied to have it replaced with one that could be quickly and easily opened from the inside. Joe Shea refused. "Too much money and not enough time," he told them.

Leading the Acceptance Readiness Review meeting for Spacecraft 012—surrounded at the conference table by the crew, the North American executives, and colleagues from NASA including spacecraft designer Max Faget, Director of Flight Operations Chris Kraft, and more than a dozen officials from Houston—Shea was upbeat and lighthearted. Talk was punctuated by laughter and wisecracks. The major faults with the spacecraft were not on the agenda, and the discussion focused mainly on a succession of minor glitches or issues that had been discussed before. Nevertheless, the meeting wound on for six hours. At one point late in the day, their attention turned to the question of flammable materials in the cockpit. Since the beginning of the space program, astronauts had used nylon netting and strips of Velcro glued to the walls of the spacecraft to corral checklists and pieces of equipment that might otherwise fall beneath their couches during tests, or float around in zero gravity. And the cabin of Spacecraft 012 was no exception: it was festooned with Velcro installed at the request of Grissom, White, and Chaffee. But all of it was flammable. This was a long-standing concern for the NASA engineers not just because of the risks of a fire breaking out in space—they still had no idea how flames might behave in zero gravity, and attempts to develop a workable fire extinguisher for use in orbit had come to nothing—but because the cockpit atmosphere was composed of pure oxygen, pressurized at launch to more than sixteen pounds per square inch.

Pure oxygen—rather than a mixture of oxygen and nitrogen similar to air—had been used in space capsules by NASA since the Mercury days, for both engineering and medical reasons. The designers regarded the equipment necessary to create a mixed-gas environment as too heavy and complex to carry into space, and flight surgeons feared that an astronaut breathing nitrogen and oxygen might suffer decompression sickness—the "bends"—if the capsule lost cabin pressure in orbit. And although in pure oxygen, fire spreads more quickly and burns more fiercely than in air, at first the risks seemed manageable. Back-of-the-envelope calculations suggested that the one-man Mercury capsules were so small that any fire starting inside would burn up the available oxygen supply in a few seconds and then extinguish itself. But the three-man Apollo cabin was almost six times larger than the Mercury capsule, with a corresponding increase in the danger of fire. So, at first, the contractors at North American planned to use a mixed-gas atmosphere. However, NASA's then head of the spacecraft program disagreed, insisting once again on pure oxygen, leading to a heated meeting in which he confronted his opposite number at North American. The two managers began shouting at each other until, at last, the NASA official ended the argument. "You are the contractor," he said. "You do as you're told. Period."

The agency understood the potential consequences of this decision. In July 1963, an internal NASA document reported that, in tests of the pure-oxygen atmosphere, "[i]t has been observed that a number of otherwise nonflammable materials, even human skin, will burst into flame." But in the complex web of engineering compromises necessary to build a capsule light and practical enough to carry three men to the moon, on an almost impossible deadline, but without unduly endangering the lives of the crew, this was deemed an "acceptable risk." The contingency plan outlined in the 1963 document was simple: "Fires in the spacecraft must be precluded at all costs."

The regulations governing flammable materials in the cockpit of Apollo 1 were strict: none were permitted within four inches of any potential source of ignition, including the approximately fifteen miles of problematic wiring, 640 switches, indicators, computer controls, and circuit breakers that filled the capsule. So when the Readiness Review meeting in Downey considered the question, Joe Shea engaged in a few minutes of back-and-forth with the North American engineers before repeating the rules and issuing an instruction: "Walk through the goddamned spacecraft," he said—and clean up all the Velcro and anything else that might feed a potential fire. A minor issue, easily addressed.

Finally, as the review wound down, Gus Grissom asked for the floor, and from a large envelope pulled two prints of a photograph. It showed him, Chaffee, and White sitting at a small table on which rested a scale model of the Apollo capsule. Facing it, the three astronauts had bowed their heads and pressed their hands together, as if in prayer, offering supplication to a higher power. Grissom handed one print as a souvenir to the executive in charge of North American's Space Division. "We have one for Joe Shea also," Grissom said. "Joe advised us to practice our backup procedures religiously, so here we are practicing."

Laughter filled the room. Grissom handed the picture down the table to Shea, who saw that the three astronauts had each signed the print, and added a personal inscription: "It isn't that we don't trust you, Joe, but this time we've decided to go over your head."

At the outset of 1967, Joe Shea was on the verge of becoming a celebrity. He was already in demand as a public speaker, and hobnobbing with *CBS Evening News* anchor Walter Cronkite. Now the editors of *Time* magazine were preparing a cover story on him timed to coincide with the launch of the first manned Apollo mission. But the spacecraft was still far from ready. It had arrived in Cape Canaveral from the North American plant in Downey on schedule, but half finished: accompanied by spare parts and uninstalled hardware, and a roster of tests that should have been conducted in California, but remained incomplete. The environmental control system continued to cause problems, threatening a major launch delay. The service module, which contained the spacecraft's main engine, was damaged during testing. And the Velcro and nylon netting added to the cabin by the astronauts had not been removed as Shea had instructed; instead, there was more of it than ever in the cockpit.

In early October, Shea had received a letter from Hilliard Paige, a senior executive at the Missile and Space Division of General Electric, advising NASA on spacecraft safety issues. Paige had recently witnessed a combustion test conducted by one of his staff on samples of Velcro in a pure-oxygen environment. He had watched, aghast, as the material ignited in a flash and was abruptly consumed by flames. The technician told Paige that he had tried bringing the issue to NASA, but had found it hard to get their attention. Paige made his fears explicit: "I do not think it technically prudent to be unduly influenced by the ground and flight success history of Mercury and Gemini under a 100 percent oxygen environment," he wrote. "The first fire in a spacecraft may well be fatal."

In response, Shea delegated the issue to a deputy, who, held up by the thicket of other problems with the program, took seven weeks to respond, but finally assured his boss that newly completed risk assessments indicated they had little to worry about. Shea sent Paige a letter reassuring the GE executive that NASA engineers were working on the issue, and enclosed a copy of the new assessment. But in a handwritten postscript Shea revealed his own concerns. "The problem is sticky," he wrote. "We think we have enough margin to keep fire from starting—if one ever does, we do have problems."

At the same time, Gus Grissom had become increasingly angry about the condition of Spacecraft 012—and the flight simulator on which he and the crew were supposed to be training—as the launch date neared. He worried that both his crew and the technicians were distracted, in too much of a hurry to get off the ground, and that fundamental safety issues were being neglected. And it was growing obvious to almost everyone involved that the Apollo 1 vehicle was simply not in a state to fly: if Spacecraft 012 were a horse, one former NASA official observed, it would have been shot. Yet Grissom felt his warnings were going unheard. The NASA managers and astronauts alike were in the grip of what they would later recognize as "'Go' Fever"—the desperate drive to push on toward a launch and keep to the schedule, regardless of the problems, in the belief that if they just kept going they could fix all the faults along the way. This seat-of-the-pants approach had worked so far, with the triumphs of the Mercury and Gemini programs, which— despite some close calls—had progressed without a single serious accident. And now, with the deadline for a successful moon landing so close, "'Go' Fever" was more virulent than ever. When Grissom's friend and fellow astronaut John Young asked the mission commander why he didn't complain more stridently about the poor wiring inside the Apollo capsule, Grissom was blunt: "If I say anything about it, they'll fire me."

At the end of November 1966, the persistent problems with the spacecraft pushed back the expected launch date of Apollo 1 into the new year, and NASA announced that the schedule of subsequent flights was sliding back; one mission was canceled altogether. But in public, Grissom stifled his worries about Apollo's first manned flight. In a round of end-of-year TV interviews intended by NASA to help build public support for the space program, he projected confidence that the mission was on track; only when answering a question about what would constitute a successful flight did he hint at his true feelings: "As far as we're concerned, it's success if all three of us get back," he said. Most of those listening assumed he was joking.

With Christmas approaching, the three men returned to their families in Houston. Roger Chaffee learned that the ingenious illuminated display he had created in his front yard—topped with Santa and his reindeer galloping over the roof of the house—had won first prize in the Nassau Bay Garden Club competition for holiday decorations. On Christmas Day, he gave Martha a special Apollo pin he and his crewmates had designed to commemorate their flight. The astronauts had intended to take the pins on their mission and hand one over to each of their wives as a surprise on their return from space. But the three men had grown impatient, and made them Christmas gifts instead. By the new year, they had devised new mementos to carry into orbit for their wives: three gold charms, tiny replicas of the Apollo 1 spacecraft, each made unique by the setting of a single diamond representing the position the astronauts would occupy in the capsule: Grissom on the left; White in the center; and Chaffee on the right.

::::::::::::::::::

It was shortly before one o'clock in the afternoon on January 27 when the three astronauts stepped out of the elevator onto the scarlet-painted gantry of Pad 34 for a practice countdown, the full dress rehearsal of the launch of Apollo 1 known as the "plugs-out test." Almost two hundred feet below them, the newly completed access roads of the sprawling Cape Canaveral Launch Complex stretched back toward the domed concrete Launch Control Blockhouse and, beyond that, to the ocean. The Atlantic glittered turquoise in the hard Florida light.

NASA officials had chosen the site as their gateway to the moon less than six years earlier, and since then the Kennedy Space Center had risen from the wilderness with prodigious speed. While the launch gantries were isolated on the Cape that gave the area its nickname—a narrow finger of sandy beaches and windblown scrub extending south into the ocean for thirty-four miles—most of the Kennedy facilities lay to the west, between the Banana and Indian Rivers, on the salt marshes of Merritt Island. The buildings from which the Apollo astronauts and their spacecraft would prepare to leave on the lunar voyage—the Operations and Checkout Building, with its crew dormitory and suiting-up room; the Vehicle Assembly Building in which the Saturn rockets would be stacked together; and the angular, modernist Launch Control Center in its shadow—had been constructed amid eighty-eight thousand acres of mosquito-infested wetland and alligator-filled lagoons. Remote and inhospitable, the location had been selected with care by agency

engineers: from the coastal launchpads, their spacecraft would fly east
out over the Atlantic, away from populated areas; and its proximity to
the equator meant that the rotation of the Earth would give the massive
rockets a boost to escape gravity. But nature had surrendered only reluctantly to the advance of the Space Age: it took the Army Corps of Engineers three years and stupendous quantities of insecticide to render the
area "biologically unfit" for the mosquito; the foundations of the monumental Vehicle Assembly Building, which was large enough to contain
the Great Pyramid of Cheops, were anchored by thousands of steel tubes
driven into bedrock 160 feet beneath the surface of the island, to stabi-
lize the boxlike structure in high winds. And the subtropical weather
could be unpredictable, bringing violent thunderstorms in the summer,
occasional hurricanes, and banks of fog that shut down launchpads and
runways with little warning in the winter. But today, the weather was
perfect for the test: clear, warm, and bright, with unlimited visibility.

Sealed in white nylon pressure suits, Grissom, White, and Chaffee
breathed pure oxygen from the environmental control packs they carried in their hands like luggage, connected to their helmets by flexible
hoses. As they had suited up that morning, the astronauts had seemed
tense and reserved. Grissom was preoccupied with failures in the spacecraft communications system, and had suggested that Joe Shea join
them inside the capsule for the duration of the test so he could experience for himself how bad things had become. But the North American
Aviation technicians had proved unable to add an extra audio headset
inside Spacecraft 012, and there was no way Shea was prepared to spend
the entire afternoon crouched in the cramped space at Grissom's feet
without being able to hear what was going on. So he planned to fly back
to Houston that afternoon instead, promising to go over the issue with
Grissom later, in the simulator. Having completed one more interview
with a reporter working on the *Time* story, Shea headed to the Melbourne airport.

The plugs-out test was a demanding and complex rehearsal, conducted on a huge scale. It would involve not just the astronauts and the pad
support staff at Cape Canaveral, but the coordinated work of thousands

of engineers and contractors behind the consoles there and almost nine hundred miles away, back at Mission Control in Houston. A full simulation of the launch of Apollo 1—from the initial powering up of its computers and guidance systems to the moment it reached orbit—the test was designed to check that each one of the systems of the spacecraft and its booster worked as intended. It would be as close to the experience of the real mission as possible without actually igniting the engines. In the weeks preceding the simulation, the booster—the giant Saturn 1B rocket—had been rolled slowly out to the launchpad, and the spacecraft lifted on top of it by crane. Once the astronauts were inside the capsule, the countdown would proceed through the moment their vehicle began running under its own power and, a few minutes later, as the clock reached zero, the pad crew would pull out the final umbilical lines connecting Apollo 1 to the gantry, just as the automatic systems would do at the instant of launch. Then, at the moment the rocket would clear the tower, Mission Control would take over to continue the simulated flight to orbit. To conclude the simulation, Gus Grissom planned to practice an emergency escape from the capsule: there was so much still needed to do if they were to meet the February launch date that the commander had to squeeze in as much preparation as possible. Despite its complexity, few at NASA regarded the test as dangerous. The Saturn booster would not be fueled with any of its volatile propellant, and the same routine had been a landmark in the progress of all previous manned launches—and had always passed off without incident. The fire rescue team that would be stationed near the pad for the real launch would not be part of the simulation.

It took more than an hour for the pad crew to seal the awkward hatches of Spacecraft 012, and, eventually, the technicians had to pound one of them closed. Inside, the cabin formed a truncated cube within the conical command module. The astronauts' adjustable couches were crammed into an eight-foot square, hemmed in on three sides by instrument panels and bulkhead equipment bays. The canvas couches allowed the men to lie on their backs for launch, gazing upward at the banks of switches and dials within easy reach above them, but which created a low, angled ceiling that made it impossible to stand; the hatch lay behind their heads. At around 2:45 p.m., the pad team reported that Grissom, White, and Chaffee were locked inside the capsule. The atmosphere was purged with pure oxygen and increased to a pressure of 16.7 pounds per square inch, forcing the inner hatch against its sills like the cork in a champagne bottle.

Almost immediately, things began to go wrong. There were more problems with the environmental control system, triggering the master alarm, and, just as Grissom had feared, radio communication between the capsule and the ground controllers continued to break down. As the astronauts ran though their checklists and threw switches, they struggled to hear the controllers' instructions. Hours went by, and the three men began to troubleshoot the radio interference, but the countdown continued even as the problems grew worse. At one point, local air traffic control chatter began to bleed into the exchanges the three men were having with the ground. At 6:20 p.m., with just ten minutes to go before the simulated launch time, the controllers put the countdown on hold, and Grissom began swapping out pieces of equipment to isolate the fault. By 6:30 p.m., the astronauts had spent more than five hours in the malfunctioning capsule, and launch control was preparing to resume the countdown. But Grissom was seething. "How are we going to get to the moon," he asked over the static, "if we can't talk between three buildings?"

From the seat beside him, Ed White cut in to tell him the controllers couldn't hear a thing he was saying.

"Jee-*sus* Christ!" Grissom said in exasperation.

For the next thirty seconds, there was silence. From inside the spacecraft, the instruments detected signs of motion, and scratching noises came from Grissom's open microphone as he apparently moved around on his couch. Then the telemetry registered a sudden voltage surge within the capsule, and the sensors attached to Ed White's chest recorded a rise in his pulse and respiration rates. At a few seconds after 6:31 p.m., ground controllers in Cape Canaveral and Houston were startled by a single, clipped exclamation coming over the VHF channel named Black-3. Grissom, or perhaps Chaffee, shouted, "Hey!"

Two seconds later, Chaffee—whose responsibility it was to maintain contact with the ground in an emergency—said, in a calm, disciplined voice, "We've got a fire in the cockpit."

A few miles away, inside a control room on Merritt Island, one of the North American engineers monitoring the test looked up from his notes. He turned to the man at the next console.

"Did he say, 'Fire'?" he asked. "What the hell are they talking about?"

Inside the launchpad blockhouse, an RCA technician watching the bank of monitors carrying TV images of the test heard Chaffee's voice, and turned to the feed from Camera 24. Trained on the White Room— the clean space on Level A8 of the launch gantry enclosing the side of the spacecraft—Camera 24 could be zoomed in to show a partial view through the cockpit window of the capsule. At first he saw only a bright glow inside, but then made out flames flickering across the porthole, and saw Ed White's hands reaching above his head, toward the hatch dog bolts.

Only a few seconds had passed since Chaffee's first report, but now every one of the men listening to channel Black-3—in the blockhouse, on Merritt Island, all the way over in Houston—heard the rookie astronaut again, the words garbled, but unmistakably agitated.

"We've got a bad fire—let's get out . . ." Chaffee said. "We're burning up!"

On his TV monitor, the RCA technician saw a flurry of movement inside Spacecraft 012. Ed White's arms reached out, then drew back, and reached up again. Then a second pair of arms joined White's as the blaze spread from inside the far left of the cockpit toward the porthole, before the technician's view was obscured by flames.

Channel Black-3 carried a brief, terrible sound that sounded to some like a scream of agony. Then silence. Eighteen seconds had passed since the first message of alarm.

Up on Level A8 of the launch gantry, the pad leader was standing at his desk on the swing arm, just twelve feet away from the spacecraft, when he heard the crew's first shout over his communications box. He made out only the word "fire," but that was enough: "Get them out of there!" he shouted at his chief technician. Then, as he turned back to the box to notify the blockhouse below of the emergency, he heard a loud whoosh, and a sheet of flame leapt over his head, singeing the papers on his desk. The chief technician, already heading toward the capsule, was thrown back against the gantry door by the explosion, and showered with fire and debris. Smoke and flames billowed through the swing arm walkway, driving the men back toward the tower elevator, where they found gas masks, and returned with three colleagues to fight the blaze, using the only fire extinguisher they could find.

It took the five men some five minutes, crawling in relay through the heat and choking smoke so thick they could barely see, to prize the heavy hatches off Spacecraft 012. The inside of the cabin was smoldering and blackened, lit from within by amber warning lamps still glowing on the instrument panels. Gus Grissom and Ed White lay at the foot of the hatch, their pressure suits fused together by the fire. Roger Chaffee remained strapped to his couch, where he had kept communications open until the end. Melted into the Teflon surface of the cockpit window was a single handprint, outlined in soot.

The pad leader retreated from the stricken spacecraft and picked up the headset connecting him to the blockhouse below. He thought hard about what to report, unwilling to reveal the truth over an open channel.

"I can't describe what I see," he said.

CHAPTER TWO

WHITEY ON THE MOON

The call came in over the red phone on Deke Slayton's desk at the Manned Spacecraft Center soon after 5:30 p.m. Houston time. Michael Collins, the senior astronaut present, watched Slayton's assistant snatch the receiver from its cradle and listen impassively for what seemed a long time. Then he hung up.

"Fire in the spacecraft," he said softly.

A dazed silence fell over the room. Everyone knew which spacecraft he meant, and the inevitable consequences of a fire within it. Then the red phone rang once more, bringing further details—and Collins recognized at once that someone had to get word to the families before the press reached them. He called the nearby Astronaut Office, almost deserted at that hour, and arranged to have two astronauts head out to the homes of Ed White and Gus Grissom. But no one could be found to take the news to Martha Chaffee. It had to be a close friend—and an astronaut—and Collins realized with a sinking feeling that the duty fell to him: the Chaffees lived just three doors down from him on Barbuda Lane.

It was only a mile and a half from the Space Center to the yellow-brick house, but Collins drove it slowly, dreading what came next.

Later that night, all three national TV networks interrupted their regular schedules with bulletins about the deaths of the crew of Apollo 1, and President Johnson issued a statement: "Three valiant young men have given their lives in the nation's service," it read. "We mourn this great loss and our hearts go out to their families." World leaders—including the Secretary-General of the United Nations, Queen Elizabeth II, and Pope Paul—sent condolences. Along the two-lane highway leading from Cape Canaveral to the nearest town, Cocoa Beach, flags flew at half-staff. The mayor of Houston announced that the city would enter an official state of mourning.

The Apollo launchpad fire was the most lethal accident in the short history of the US space program, and the nation reeled in shock: the disaster pulled back the curtain on the technicolor adventure that to the public had begun to seem routine. Four days after the fire, on the morning of January 31, TV viewers across the country watched live as a caisson drawn by six black horses bore Gus Grissom's flag-draped coffin to a burial plot at the crest of a hill in Arlington National Cemetery. That afternoon, the body of Roger Chaffee was lowered into the same frost-covered ground in the plot beside his mission commander. A detail of Navy enlisted men folded the flag that had covered Chaffee's casket into a tight triangle for presentation to his widow, who waited at the graveside in a black pillbox hat and sunglasses. President Johnson murmured words of consolation to Martha, and then bent to shake hands with Sheryl Chaffee, eight, and her five-year-old brother, Stephen; a trio of Navy Phantom jets roared overhead in the "missing man" formation. Escorting the shattered family was astronaut Gene Cernan, Martha's next-door neighbor, her husband's closest friend, and fellow Apollo lunar pilot. As a three-shot volley of rifle fire echoed over the silent hillside, Cernan wondered if the accident that had killed his colleagues meant the end of the moon program, too.

Back at Cape Canaveral, an investigation of the fire had begun almost at once, as stunned managers from NASA and its contractors converged on the scene aboard planes from Houston and Washington, DC. Joseph Shea flew down from Ellington Field on the night of the accident, and moved into a room inside the astronauts' quarters in the Operations and Checkout Building on Merritt Island. There, he listened for the first time to the tape recording of the astronauts' final moments, while nearby, teams of technicians began a forensic examination of the spacecraft in which they had died. NASA chief James Webb, determined to protect his agency from outside scrutiny, made a handshake deal with President Johnson to allow them to investigate themselves. Shea insisted on taking the lead in the inquiry. He knew he wouldn't be able to sleep until he discovered what went wrong. He began working back-to-back shifts, sixteen hours a day, in search of the answers and, like many of his colleagues, sought refuge from the grief and the strain in Seconal and scotch.

At the end of a ten-week probe, NASA's board of inquiry delivered a report that stretched to six parts and more than three thousand pages. Their findings revealed shocking incompetence, describing "many deficiencies in design and engineering, manufacture and quality control" of the Apollo spacecraft. They found that NASA personnel "failed to give adequate attention to certain mundane but equally vital questions of crew safety." In retrospect, the problems were glaring: most obviously, the cabin atmosphere of pure pressurized oxygen—and the hatch that made escape in an emergency impossible. But the capsule had also been filled with so much combustible material—seventy pounds in all—that by the time of the accident it resembled a bomb waiting to go off: there were five thousand square inches of Velcro alone, in a spacecraft supposed to contain no more than five hundred. North American engineers had scheduled the walk-through to remove flammable material from the capsule—the same one ordered by Shea in August the previous year—for January 29, two days after the fatal test. And the slipshod installation of the pipework and wiring at the North American factory in Downey, which had created hundreds of potential points of ignition for a blaze, plagued each spacecraft it built. When engineers at the Cape conducted an inspection of the latest Apollo capsule, recently delivered from the contractors, they uncovered more than 1,400 wiring faults. Shea came to conduct his own examination of the new vehicle, and left the building with tears in his eyes.

The intensity of the fire, which had reached temperatures high enough to melt aluminum, had incinerated so much of the cabin of Spacecraft 012 that it was impossible to prove conclusively how it had begun. But the evidence suggested that it started with a short circuit caused by worn insulation on a wire in the cable bundles tangled near Gus Grissom's feet, close to a leak in the pipework carrying ethylene glycol around the troubled environmental control system. The spark set light to a piece of nylon netting, which melted and glowed for a few moments before abruptly flashing into a sheet of flame that swept the cabin with explosive speed. Grissom, White, and Chaffee had died not from burns sustained in the fire but from asphyxiation—within seventeen seconds of the blaze starting, the intense heat melted their oxygen hoses, and ventilation fans sucked smoke and toxic gas into their pressure suits. The three men were overcome so quickly that they had no chance of escape.

The fire brought much of the work on the Apollo program to a halt overnight, and the potential need to overhaul the spacecraft design made the chances of landing men on the moon before 1970 seem slimmer than ever. Yet from the outset, NASA chiefs made clear that they had no intention of allowing the deaths to divert the agency from its goal—indeed, that only in reaching the moon would the country honor the three astronauts' sacrifice. Even so, once the agency published the final report of the inquiry, Congress launched its own investigation of the accident. Administrator Webb and other NASA chiefs were hauled to the Capitol to testify in a series of hearings in which the culture, competence, and budget of the Apollo program and the agency itself came under attack. Senator Walter Mondale of Minnesota, already a vocal NASA critic, suggested the Apollo engineers were guilty of "criminal negligence." Future secretary of defense Donald Rumsfeld, then a thirty-four-year-old congressman from Illinois, decried the idea that NASA had been allowed to investigate itself; the head of the House subcommittee overseeing the agency described the report as "a broad indictment of NASA and North American and the whole program." Administrator Webb was the first witness. He acknowledged the program's failings, but insisted that his astronauts and engineers alike assured him that if the problems were addressed, they would be confident to fly Apollo rockets into lunar orbit. In often passionate testimony, he declared that space exploration represented a zenith of human aspiration: the moon should remain a shared goal for all Americans.

"If any man in this room asks for whom the Apollo bell tolls," Webb said, "it tolls for him and me, as well as for Grissom, White, and Chaffee. It tolls for every astronaut test pilot who will lose his life in the space-simulated vacuum of a test chamber or the real vacuum of space."

It took more than a year for the Apollo program to regain momentum after the disaster, and it was not until August 1968 that engineers delivered another spacecraft to Pad 34, ready to take men into orbit. By then the capsule had been rebuilt and stripped of flammable material, with a one-piece hatch that could be opened in three seconds, new wiring, armored plumbing, and modifications to accommodate a mixed-gas atmosphere; the astronauts would now wear fireproof pressure suits woven from glass fiber and coated in Teflon. The management of both NASA and North American was overhauled, and the agency created a separate bureaucracy to impose more stringent safety standards on the spacecraft. But some of the most profound changes were in the spirit and attitude of the engineers who would go on to take Apollo to the moon. Shaken out of the complacency and arrogance that had set in during the successes of the Mercury and Gemini missions, the technicians embraced a new earnestness, and a shared focus on the hard work necessary to take men to the lunar surface before the end of the decade. But for the young engineers at NASA, the idea of the program as a stirring adventure, and the early exhilaration of constantly improvising solutions to previously unimagined technical challenges, was gone. The theoretical recognition that things could possibly go wrong was supplanted by the cold realization that they *would*. Those with backgrounds in the missile industry and in systems engineering were hit especially hard by the disaster—they were the ones who had believed that all risks could be removed from spaceflight; accidents should never happen. Those who came from the world of flight testing, like the majority of the astronauts themselves, found it easier to move on after the accident. These men recognized the realities of the work they were doing, and were prepared to live with the consequences.

"You lose crew," one said later. "Pilots die flying experimental aircraft."

Joseph Shea, whose bosses came to believe was so driven and traumatized by the deaths on Pad 34 that he was at risk of a nervous breakdown, remained in charge of the Apollo spacecraft program only until the eve of the first congressional hearings. By the time James Webb began his testimony, Shea had been unwittingly lined up to take the fall for the accident: prevented from appearing at the hearings himself, sidelined into a sinecure in the administration's DC headquarters, and replaced as head of the program. In the weeks that followed, Shea found himself with little to do and spent his days alone in his new office reading the newspapers, or killing time walking the galleries of the District museums. He resigned from NASA just four months later, embittered by the way he had been treated, but tortured by his role in the deaths of Grissom, Chaffee, and White. For years afterward, Shea kept the signed photo the three men had given him—*It isn't that we don't trust you, Joe*—displayed prominently by the front door of his house, where he had to pass it every day. Often, he stayed up late into the night thinking about the accident and what he could have done differently. Sometimes, he admitted to wishing that he had joined the three astronauts inside Spacecraft 012 that January afternoon, taking his place at Gus Grissom's feet before the hatch was hammered shut for the last time.

::::::::::::::::::

Not quite six years after the fire, on December 14, 1972, Gene Cernan stood beside the Apollo 17 lunar rover and gazed across the silver-gray desolation of the Taurus-Littrow valley, the craters and the low mountains around him silhouetted against the black horizon of deep space. He positioned the rover's camera to perform its final duty—capturing TV images of the departure of his spacecraft, the lunar lander *Challenger*— and then, seized by an urge he would later find hard to explain, traced the initials of his daughter Teresa Dawn's name in the fine dust at his feet.

Even in his cumbersome pressure suit, in one-sixth gravity the walk back to the lander took only a few minutes. At the foot of the ladder, he delivered the short speech he had memorized for the occasion. "I'd like to just say what I believe history will record: That America's challenge of today has forged man's destiny of tomorrow. And, as we leave the moon at Taurus-Littrow, we leave as we came and, God willing, as we shall return: with peace and hope for all mankind."

As he climbed the ladder, he paused to look over his shoulder at the busy pattern of boot prints on the surface. He knew he would never be coming back. And the remaining missions of the lunar exploration program—until recently planned to include three further landings— had been canceled. Cernan, thirty-eight years old, a Navy pilot who had grown up milking the cows on the Wisconsin farm where his grandparents lived without electricity, would be the last of only a dozen men to walk on the moon.

Inside the cabin, Cernan and mission geologist Harrison "Jack" Schmitt completed the final checks to rendezvous *Challenger* with the command module orbiting above them. Cernan's finger moved to the switch that would fire the ascent engine.

"OK, Jack," he said. "Let's get this mother out of here."

Just as they had when Neil Armstrong and Buzz Aldrin became the first human beings to set foot on the lunar surface some three years earlier, US TV networks carried live pictures of the Apollo 17 mission. This time, however, viewers called the CBS switchboard in frustration: the coverage had made them miss the latest developments in the hot hospital drama, *Medical Center*.

For all the steep sacrifice, magnificent spectacle, and superhuman achievement of the Apollo program, it had not taken long for the

American people to grow tired of watching men walk on the moon. In the years NASA technicians were working to recover from the deaths on Pad 34, the technological optimism and Cold War certainties that once seemed to define the decade had faded, and the moon project was often eclipsed by strife at home and abroad. The nation had been drawn more inextricably into the horrors of Vietnam, and the struggle over civil rights had intensified on the streets of cities throughout the United States; the costly endeavors of the space program increasingly seemed wasteful and quixotic. The headlines of 1968 were dominated by the catastrophes of the Tet Offensive, the assassinations of Robert Kennedy and Martin Luther King Jr., and the nationwide riots that followed; at their headquarters in Washington, DC, single-minded NASA engineers continued to plot their way to the moon, while five blocks away an angry mob looted a department store.

It was not until December of that year, when the crew of Apollo 8 became the first men to reach lunar orbit, that the project had seized the public imagination once again. On Christmas Eve, the astronauts beamed back images of a distant, fragile Earth hanging in the dark void of space, and recited from Genesis. The verses crackled across the downlink to Mission Control, where rocket engineers wept at their consoles, and around the world a billion people—the largest broadcast audience in history—listened in. And when Neil Armstrong stepped from the footpad of the lunar lander into the Sea of Tranquility on July 20, 1969, an estimated 600 million people were watching live on television, and all three US networks dedicated their airtime to the events for thirty-six continuous hours. Two months later, Vice President Spiro Agnew laid out his timetable to land an American on Mars in 1986.

But even as President Richard Nixon spoke by telephone to congratulate the astronauts on the lunar surface, NASA was undergoing what would prove the first of many cutbacks. By decade's end, the extraordinary success of the project had been matched only by its exorbitant cost: at its peak, NASA had some four hundred thousand men and women at work on Apollo, and the price of the program's support facilities alone was $2.2 billion; the technology and materials of the lunar lander were so exotic that each one cost fifteen times its weight in gold. In total the project would cost the country an astonishing $28 billion— the equivalent to a third of all US military spending for 1969, at the height of the Vietnam War.

And as one ultimately successful mission to the moon's surface followed another, a series of equipment failures and narrow escapes made the engineers in charge of the program increasingly nervous. However routine it began to seem for the American public, the architects of Apollo knew that spaceflight had always been experimental, and sending men so far, into such a hostile environment, with so many opportunities for failure, was fraught with lethal hazards.

Although there had been no more fatal accidents since the launchpad fire that had almost ended the program before it started, there had been enough close calls to make it clear that death was never far away. After the aborted lunar voyage of Apollo 13 in 1970, in which three astronauts survived a catastrophic explosion in outer space only through a combination of good fortune and last-minute improvisation, the engineers began to fear that their luck would soon run out. If they pushed on to the planned end of the program, the chances were good that at least one Apollo mission would never return from the moon.

And with each subsequent lunar landing, public interest in the enterprise began to wane. Television viewers discovered that one area of the moon's dusty volcanic surface looked much like another. Audiences for the landings declined, and the networks scaled back their coverage. A growing public disillusionment with technology of all kinds, no matter what its provenance or intent, became intertwined with the conviction that the exploits of a group of homogeneously square-jawed and uniformly white pilots and scientists on a rock a quarter of a million miles away in space were doing nothing for ordinary Americans facing rising deprivation and economic hardship back on Earth. Even on the eve of Armstrong and Aldrin's departure for the moon, NASA Administrator Thomas Paine had been confronted by a march led by the civil rights leader Reverend Ralph Abernathy, who stood in the drizzle at Cape Canaveral at the head of a column of four wagons led by tethered mules, to protest at the inhuman squandering of federal funds. Abernathy told Paine that the money spent launching the gleaming rocket looming in the distance could have been better spent addressing the abject poverty of Black Americans. "If it were possible for us not to push that button and solve the problems you are talking about, we would not push that button," the Administrator replied. Instead, Paine suggested that the moon shot could inspire people to come together to solve the nation's wider problems: "I want you to hitch your wagon to our rocket and tell the people the NASA program is an example of what this country can do," he said. The following day, Paine invited ten of the protesters to watch the launch from the VIP area, beside senators, generals, and President Johnson. But the glaring inequalities of the program remained. The following year, Gil Scott-Heron expressed the way many Black Americans felt, with scornful irony, in his poem "Whitey on the Moon":

Was all that money I made las' year (for Whitey on the moon?)
How come there ain't no money here? (Hmm! Whitey's on the moon)
Y'know, I jus' 'bout had my fill (of Whitey on the moon.)

Nor were the astronauts themselves any longer perceived as the wholesome supermen who had once reliably lit up the covers of *Life*. In July 1972, news emerged that the crew of Apollo 15 had been reprimanded by NASA for smuggling hundreds of specially stamped envelopes to the moon and back as part of a scheme to sell them privately on their return and use the money to set up trust funds for their children. Separately, fifteen of the twenty-four Apollo astronauts were reported to have been charging for autographs, at $5 a time. Nixon's Justice Department opened an investigation into what it called the "commercialization of space." Humiliated NASA chiefs ensured that none of the three Apollo 15 crew would ever go into orbit again, but the myth of the perfect astronaut was shattered forever.

By the time Gene Cernan lifted off from Cape Kennedy at the beginning of December that year, public engagement with the space program had fallen so far that ABC, NBC, and CBS had no plans to screen any coverage of the mission at all; later, there would be persistent rumors that NASA had paid the TV networks to carry even the footage of Cernan taking what would prove to be twentieth-century man's last steps on the lunar surface.

Back in Houston, the visionary architects of Apollo had made swashbuckling plans for the next stage of man's exploration of the universe—including a permanent base on the moon, an orbiting space station accommodating a crew of one hundred, a reusable spacecraft to shuttle astronauts up to it from Earth, and a pair of massive nuclear-powered rockets to take a dozen men on a yearslong expedition to Mars. But the bottomless funding and the political will that had flowed from John Kennedy's decade-old challenge, and the lunar milestone sanctified by his death, had been consigned to the past. The agency faced drastic budget cuts.

A moon base was out of the question, the Soviets already had a rudimentary space station, and landing on Mars carried too many echoes of Apollo. Nixon, facing rising inflation and unemployment—and determined to nullify or destroy as many of the projects of Kennedy, his former nemesis, as he could—even considered ending manned spaceflight altogether. Yet to do so was regarded by his advisers as political suicide; the public may have perceived the Apollo program as extravagant and costly, but they still expected to see American astronauts in space, which remained a key battleground of Cold War technology. In August 1971, Caspar Weinberger, then head of the White House's Office of Management and Budget, wrote a letter to Nixon explaining that it was also essential to maintain the United States' dwindling prestige in the eyes of the world. An end to manned spaceflight, Weinberger wrote, would "be confirming . . . a belief that I fear is gaining credence at home and abroad: that our best years are behind us."

In January 1972, the new Administrator of the National Aeronautics and Space Administration, James Fletcher, arrived in Southern California for a midmorning appointment at Nixon's oceanfront mansion in San

Clemente. An unseasonably frigid wind rose over the cliffs as he and his deputy hurried along the walkways of the Western White House toward the President's office. They carried briefing documents and a white plastic model of the new vehicle that they hoped would ensure the future of manned spaceflight, and of NASA itself. The meeting lasted a little more than half an hour. Nixon toyed with the model, and mused on the possibilities of using the vehicle to dump nuclear waste in space; he asked Fletcher and his deputy if the vehicle would be worth the investment. "Even if it were not," the President told them, "we would have to do it anyway, because space flight is here to stay. Men are flying in space now and will continue to fly in space, and we'd best be a part of it."

Nixon's formal statement blessing the project had already been printed and stacked, ready for distribution to reporters waiting in a nearby hotel. "I have decided today," it read, "that the United States should proceed at once with the development of an entirely new type of space transportation system, designed to help transform the space frontier of the seventies into familiar territory, easily accessible for human endeavor in the eighties and nineties. It will revolutionize transportation into near space by routinizing it. It will take the astronomical costs out of astronautics."

Early drafts of the statement had referred to this new craft as the Space Clipper. NASA had put forward a number of other suggestions, including Astroplane, Skylark, Starlighter, and others—Hermes and Pegasus—which drew on the same myths of antiquity as the Mercury, Gemini, and Apollo programs. Presidential adviser William Safire's ideas would summon the spirit of maritime adventure and manifest destiny; his proposals included naming the first of the fleet *Yankee Clipper*. But Nixon rejected them all. Instead, this new project, an ungainly compromise shaped by the demands of congressional funding, Pentagon specifications, and political expedience, would carry the name it had been given since the earliest days of its development: the Space Shuttle.

CHAPTER THREE

THE SPACEPLANE

A t around 10:00 a.m. on April 1, 1969, twenty of the leading engineers at NASA's Manned Spacecraft Center in Houston received phone calls summoning them to a meeting—just as many of them were making the final preparations to launch Apollo 10 on the last rehearsal mission before landing on the moon. Instructed not to tell anyone where they were going, they convened on the third floor of Building 36, in a filthy room cluttered with cardboard boxes and surplus furniture. None of them knew why they were there; some suspected it was all part of an elaborate April Fool's prank. After a short wait, Maxime Faget—the center's diminutive Director of Engineering and Development—arrived, clutching an airline garment bag in his hand. Leaping onto a desk, Faget pulled a balsa-wood-and-paper model from the bag, and raised it above his head.

At forty-seven, Max Faget was regarded by many of the engineers he worked with as something approaching a genius. Inspired and intuitive, his imaginative leaps in spacecraft design, sketched out not at a drafting table but on ordinary drugstore graph paper, upset colleagues accustomed to the laborious precision of the engineering process—one liked to say that his name was an acronym for Flat-Ass Guess Every Time. But he confounded them by almost always turning out to be correct.

Five feet six, blunt, and opinionated, Faget could flush with anger in defense of his ideas, but recovered quickly and—despite his ego and his hatred of higher authority—was widely liked. He said that when he was absorbed in a problem he would pass into an almost trancelike state, oblivious to his surroundings, for hours, days, or even weeks at a time; he was always thinking of something new. One colleague, who had known Faget since the beginning of his career, later observed, "The United States could run for the next hundred years on the ideas Max had while he was shaving this morning."

Born in British Honduras, where his father, a ship's doctor, had briefly found a job with the colonial government, Faget came from an old Louisiana family with roots in France. The Fagets led a peripatetic life, moving from Central America to Seattle, San Francisco, New Orleans, and Norfolk, Virginia. Max and his older brother were enthusiastic sailors and builders of model airplanes; he graduated from Louisiana State University with a degree in mechanical and aeronautical engineering. After serving on submarines in the Pacific at the end of World War II, Faget was still in his mid-twenties when he was hired by a small government agency, the National Advisory Committee for Aeronautics—or NACA—in 1946. At the NACA laboratory in Langley Field, Virginia, he worked on a research program exploring rocket-powered hypersonic flight—more than five times the speed of sound—and became known for his improvised aerodynamic experiments; an engineer arrived at the workshop one day to find him tossing models made from paper plates off the balcony above the shop floor, in an attempt to make them fly.

It was at Langley that Faget and his colleague Caldwell Johnson developed the concept for the capsule that would take the first astronauts into space during the Mercury program; at the time the problem the engineers faced was to devise a rudimentary spacecraft that could carry a single human being, be light enough to be launched on the top of one of the US Army missiles available for the purpose, yet able to protect the astronaut from the intense aerodynamic heating the vehicle would experience on the way back into Earth's atmosphere. Previous aerodynamic research into developing intercontinental ballistic missiles to deliver nuclear weapons to distant targets—cities inside the USSR or China, for example—had shown that the kind of traditional rocket design depicted in science fiction or comic books, a streamlined cylinder tapering to a needlelike nose, would be of little use in reality.

Traveling the suborbital route necessary to reach halfway around the Earth, these ballistic missiles were intended to fly along a parabolic trajectory, like that of an artillery shell on a battlefield—yet tracing an arc thousands of miles long. Hurled by powerful rocket motors up and out of Earth's atmosphere, the missile would not reach its apogee until it was hundreds of miles out in the airless vacuum of space, before once again being seized by gravity and, at last, turning nose-down to plunge back toward the ground. But as it reentered the upper reaches of the atmosphere, the needlelike warhead of the missile would strike the air molecules it encountered with increasing speed, creating friction, and causing its skin to heat up; as the atmosphere grew more dense, the accelerating warhead encountered more air molecules, and the friction and heating intensified, reaching temperatures of up to 12,000 degrees Fahrenheit: heat so intense that the warhead would catch fire, melt, and disintegrate, breaking apart high in the stratosphere. The warhead, and the nuclear weapon it contained, would fall harmlessly to Earth in fragments, a million-dollar meteorite of fused and flaming metal.

Using wind tunnel experiments and mathematical modeling, Faget's fellow engineers at the Ames Research Center in California established that—counterintuitively—the most effective design for ballistic missile warheads was a blunt, curved shape. This would fall more slowly as it reentered the atmosphere, and generate a shock wave that would insulate the warhead itself from the destructive energy of friction, enabling it to remain intact until it reached its target.

It was this work that inspired Faget to conceive his simple but brilliant design for what would become the Mercury capsule: a squat cone that would be launched into space pointing upward on top of a rocket, but, when ready to return, would maintain its attitude to present its blunt end to the Earth as it reentered the atmosphere, both slowing its descent and presenting the broadest possible surface to the thickening air below.

A heavy heat shield covered the base of the capsule, sheathed in a thick layer of resin that charred and boiled away in the fierce temperatures of reentry, helping to safely dissipate the heat and protect the astronaut within. When NACA and its staff were absorbed into the newly created National Aeronautics and Space Administration in October 1958, Faget already had his design for the United States' first spacecraft ready to go.

::::::::::::::::::

As the end of the next decade approached, Max Faget had become the principal creative force behind the development of the spacecraft that would take men to the moon and back. But by the time the first Apollo capsule reached Earth orbit, much of his work was complete, and the engineers who worked for him in Houston would soon need a new task to keep them occupied.

Early in 1969, Faget was having lunch with his boss, Manned Spacecraft Center Director Robert Gilruth, who mentioned that he had recently attended a US Air Force presentation about the potential for a reusable winged spacecraft—one that could be launched into orbit, but then return to Earth to repeat the voyage again. "Max," Gilruth said, "these guys are talking about a crazy thing. Why don't you look at it and see what it's about?" Faget agreed to consider the problem, and soon came up with an initial concept. In the garage beside his home in Dickinson, a few miles southeast of the center, he drew on his boyhood experience to build a model of the design, with which to demonstrate the aerodynamic characteristics of this revolutionary new vehicle.

The prototype Space Shuttle that Faget pulled from the garment bag in Building 36 was three feet long and made from balsa wood, glue, and translucent brown paper. It looked ordinary enough: shaped like an aircraft, with a snub-nosed fuselage, short straight wings, and a horizontal tailplane with two vertical fins. As the assembled engineers looked on, Faget threw it twice across the room: first, it flew arrow-straight, just like a conventional aircraft; before he launched it a second time, he tilted the nose of the model up toward the ceiling at an angle of sixty degrees. Yet now his creation maintained this attitude throughout its flight path, presenting its entire underside to the path of onrushing air, as if falling, *horizontally*, through space.

The engineers didn't need Faget to explain the meaning of his demonstration: this was an elegant aerodynamic design for a spaceplane, one that would present its broad underside to diffuse the extreme heat of reentry, just like his teardrop-shaped capsule, and then tilt nose-down to glide through Earth's atmosphere like a conventional plane.

"We are going to build the next-generation spacecraft," Faget announced.

The monumental feats of the Apollo program may have required a workforce of hundreds of thousands scattered across the United States, and an investment of money and ingenuity unmatched since the effort to develop the atomic bomb; even so, the technological challenges of devising the spacecraft that took men to the moon would be dwarfed by those of building the Space Shuttle.

The scope of exploration, and distances, involved in the vehicles' respective journeys were scarcely comparable: Armstrong and Aldrin's trip to the lunar surface required them to break entirely free of Earth's gravity and embark on an eight-day round trip through more than 900,000 miles of outer space; the Space Shuttle would be required merely to travel into low Earth orbit—between 190 and 330 miles above sea level—where it would circle the planet for up to a week before returning home. In some ways, it was as if the sixteenth-century explorer Ferdinand Magellan had proposed to follow up the first circumnavigation of the world by rowing across Lisbon harbor and back.

But by the time Armstrong and Aldrin set down in the Sea of Tranquility in the summer of 1969, the principles of the rocket engines that took them there had been in use for more than forty years, much of their equipment was tailor-made for their mission, and all of it was designed to be used just once. The engines themselves were essentially destroyed by the processes of combustion, twisted and immolated by the high-pressure gases they briefly channeled into thrust, and designed only to survive their few minutes of use before tumbling into the ocean or disintegrating high in Earth's atmosphere.

At each successive step of their journey, the astronauts discarded another part of their vehicle, from the first stage of their Saturn V rocket crashing into the Atlantic to the segments of the lunar lander discarded on the moon. Even the ablative heat shield that protected their conical twelve-foot capsule from the fire of reentry was gradually destroyed as it approached its destination, the dense layer of resin sloughing off and vaporizing as it heated in the thickening air. At liftoff, their fully loaded rocket weighed more than three thousand tons; the remainder that splashed down in the ocean at the end of the ride weighed just seven.

However, no one had ever built a vehicle that could travel into orbit, then not only return to Earth entirely intact, but prove so robust it could be refitted and refueled to repeat the journey again and again. This would require a true spacefaring vessel, and in 1969 such a vehicle remained the stuff of science fiction, like orbiting space stations and bases on the moon. Intended to blast off like a rocket, go into orbit like a spacecraft, and land like an airplane, almost every element of the proposed new ship would have to survive the full range of the extraordinary forces exerted on man and machine by spaceflight. These would begin with liftoff, where the acoustic shock of its rocket engines screaming in unison could reach 167 decibels—powerful enough to kill a human being. As it ascended and accelerated, the shuttle would be buffeted by wind resistance and drag equivalent to several times its weight on the ground, until it reached the point known by the engineers as "Max Q"—the point at which the airframe risks being torn to pieces by the forces of maximum aerodynamic stress. On reaching the vacuum of space, it would be plunged into temperatures of minus 250 degrees Fahrenheit—the "cold soak," low enough to embrittle metal to the point that it can shatter like hard candy. On reentry, the skin of the craft, enveloped by ionizing air and blazing plasma, would heat from cold soak to 2,700 degrees Fahrenheit—high enough to melt steel—and reach a velocity of up to twenty-five times the speed of sound before slowing to a few hundred miles an hour to land on its own undercarriage under the control of its pilots.

Not only would the delicate mechanisms of the Space Shuttle have to be engineered to endure forces of such extremity and violence, but its design and construction would also be governed by a further new parameter, one of which NASA had no existing experience: a limited budget.

:::::::::::::::::

Regardless of the obstacles, the idea of a rocket-powered spaceplane was not new. In the 1930s, Austrian Nazi engineer Eugen Sänger detailed proposals for the *Silbervogel*, or Silverbird, a piloted "intercontinental long-range glider" theoretically capable of circling the globe in a matter of hours. Although Sänger conceived the *Silbervogel* as a civilian space glider, once the Second World War began, he offered it up as a candidate to fulfill Hitler's *Amerikabomber* plan. Launched from the ground on a rocket-driven sled, the flat-bottomed spacecraft would use its own rocket motor to reach an altitude of one hundred miles, crossing the Atlantic at 6,600 miles per hour, to bring the war to the United States by dropping bombs on New York City. Skipping over the upper atmosphere like a flat stone skimming across a pond—allowing it to radiate excess heat from its fuselage during each skip, in what Sänger termed "dynamic soaring"—the rocket plane would fly on, traversing the continent from east to west, and land in the Pacific territory of Imperial Japan. There it would refuel and rearm, making it possible to bomb Los Angeles on the way home to Germany. But despite years of theoretical studies, Sänger had difficulty persuading the Reich Ministry of Aviation that the idea was practical, and, mercifully, the *Silbervogel* was never built.

As the Third Reich staggered toward annihilation in the closing months of the war, those German engineers whose ingenuity had come to deadly fruition with the V-2 rocket bomb—which killed thousands of civilians in England and Belgium—began looking to barter their knowledge for lenient treatment at the hands of the Allies. Among them were Wernher von Braun, who would become the architect of the Saturn V moon rocket program—and his boss, Major General Walter Dornberger. After the war, both von Braun and Dornberger were among the 120 German rocket engineers brought to the United States as part of Operation Paperclip, a secret US government scheme to exploit the Nazis' scientific expertise in the emerging confrontation with the Soviet Union.

Dornberger once complained that Germany had been defeated

because his subordinates on the V-2 program were "more interested in the possibilities of space travel than with victory in war," but now the vanquished major general found a new life among his former foes, in Buffalo, New York. In 1950, he started work as an engineer with the Bell Aircraft Corporation, where he became a vocal advocate of spacecraft development. The Cold War was shifting into high gear, and the US Air Force was looking for ways to quickly and reliably drop nuclear weapons on cities behind the Iron Curtain. Dornberger suggested reviving Sänger's work on the *Silbervogel*, and even visited the inventor and his wife—by now living in Paris—to encourage them to come to the United States. But Sänger, who had already been the subject of a botched Soviet kidnap attempt led by Joseph Stalin's son Vasily, turned him down. So, with US Air Force funding, Dornberger set to work on his own plans for a new rocket plane, using the same principles as the Nazi transoceanic bomber from decades before.

In October 1957, the USSR startled the West with the announcement of the successful launch of Sputnik, the world's first artificial satellite; less than a week later, the Air Force merged Walter Dornberger's research into a classified program to build a sophisticated reusable spaceplane, known as Dyna-Soar, and later the X-20. Dyna-Soar was a flat-bottomed, blunt-nosed craft that would blast off vertically from Cape Canaveral atop a Titan missile that carried it into space, and then skip and glide over the Earth's atmosphere, or use its wings to soar even higher, into orbit. Engineers predicted that, by the early 1970s, a fleet of X-20s—traveling at the speed of an intercontinental ballistic missile, but guided by a human pilot—would be ready to intercept and destroy Soviet satellites and conduct rescue missions in space, spy on hostile territory, or deliver atomic bombs to targets deep inside the Soviet Union with little warning and terrifying accuracy. The Air Force spent years and hundreds of millions of dollars—more than $4 billion at twenty-first-century prices—on research and development work intended to overcome the many obstacles to flying such a complex machine.

At the time the Dyna-Soar program began, NASA did not yet exist; the United States had not even launched an unmanned satellite into orbit, and scientists and engineers understood little about the environment in the vacuum of space, or the technology that would be required to survive there. They feared the potential effects of weightlessness on spacecraft components and on the bodies of pilots; in the absence of aerodynamic data about reentry, they were concerned about how they might maintain control of a vehicle traveling at more than five times the speed of sound—the still largely unexplored realm of hypersonic flight—and the integrity of its structure as it encountered temperatures that they believed could reach as high as 4,500 degrees Fahrenheit.

The theoretical solutions the Dyna-Soar engineers devised for these challenges lay at the limits of existing technology. Instead of the heavy ablative heat shield later used on Max Faget's space capsules to resist the temperatures generated during reentry, they proposed that this new vehicle would have a nose cone of hexagonal graphite tiles, and a skin of overlapping shingles forged from esoteric metal alloys—molybdenum, niobium, and René 41; the pilot and equipment inside would be protected from the heat by a water-cooled aluminum cocoon. They hoped that these precautions would endure for six journeys to and from space before having to be replaced.

Intended to glide back to Earth to land like an ordinary plane under the control of a pilot, Dyna-Soar was designed with a tricycle undercarriage of metal skids and wire brushes—because rubber tires were expected to melt and catch fire before the spaceplane reached the runway. In March 1962, the Air Force announced a team of six test pilots to fly their first missions into space, and began construction of a prototype. Six months later, they unveiled a full-scale plywood mock-up of the Dyna-Soar—a faintly bat-like black V-shaped aircraft with a rounded nose and broad delta wings swept sharply up at each end. Presented at the annual Air Force Association conference in Las Vegas, the X-20 caused a sensation: *Reader's Digest* described this futuristic "space glider" as one of the most significant steps in aviation since the Wright brothers' first flight.

But in the years since Walter Dornberger had first turned his attention to bombing the Soviets from orbit, both the official aims and the public expectations of US spaceflight had changed profoundly. Yuri Gagarin had already become the first man in space, with the launch of Vostok 1 in April 1961, and the focus of US space exploration had shifted to keep pace with the USSR. The small capsules Faget had designed were carrying American astronauts into orbit on giant military rockets, and swinging safely back to Earth beneath braking parachutes; these might not be reusable, but they were cheaper and simpler than the experimental spaceplane, and did not require the use of exotic materials, or so dramatically test the boundaries of existing aerodynamic expertise.

More importantly, both Presidents Eisenhower and Kennedy had publicly committed the United States to the goals of exploring space for the betterment of humanity. In 1958, NASA had been established by the National Aeronautics and Space Act, which stated that "it is the policy of the United States that activities in space should be devoted to peaceful purposes for the benefit of all mankind."

Such high-minded sentiments were clearly at odds with designs for a hypersonic bomber intended to swoop down from orbit and unleash thermonuclear destruction from Moscow to Vladivostok—even before the United Nations ratified an international treaty banning states from placing atomic weapons in orbit, on the moon, or anywhere in outer space. Scientists and Air Force generals had sought to reorient the goals of Dyna-Soar to emphasize its potentially crucial role as a pure research project—designed to develop the technology of a reusable spaceplane— but to no avail. In December 1963, Defense Secretary Robert McNamara canceled the Dyna-Soar before it had ever made it out of the workshop, despite all the money and time sunk into its development.

Nevertheless, the Air Force remained determined to lay claim to its own dedicated space program, which would allow it to send military astronauts to conduct secret "black" operations in orbit. At the same time he made public the cancellation of Dyna-Soar, McNamara announced a new project, the Manned Orbiting Laboratory—or MOL—which the Pentagon would develop alongside NASA's Gemini and Apollo programs. Newspaper readers were allowed to infer that the MOL would be an altruistic enterprise, dedicated to scientific experiments in orbit. "The MOL is conceived as a literal laboratory in space, which would enable us to study man's adaptability and limitations over prolonged periods," the director of the program said at the time.

In reality, the laboratory was intended as a top secret manned spy satellite, crewed by Air Force and Navy astronauts with access to the most sensitive intelligence products of the CIA, and trained to shoot high-definition photographs of the USSR from orbit using a massive and sophisticated camera code-named DORIAN. On November 12, 1965, the Pentagon announced the names of eight men who would train at the Air Force Aerospace Research Pilot School to fly aboard the new spacecraft. That day, the military astronauts posed agreeably for press photographs—but within twenty-four hours, they had been fully briefed on the true nature of their mission, and vanished from public view, swallowed within the netherworld of classified Department of Defense operations.

The MOL program marked the beginning of a sometimes bitter struggle that would endure for decades, between the well-defined civilian and scientific openness of NASA and the Pentagon's clandestine hunger to extend the Cold War beyond Earth's atmosphere. But in the meantime, the defense agencies and NASA had begun collaborating on yet another project, one that would pioneer much of the technology for the spacecraft of the future.

:::::::::::::::::::

In October 1954, in a bid to expand their practical knowledge of hypersonic flight, two separate panels of US government experts resolved to build a new kind of experimental plane: a piloted research aircraft "capable of achieving speeds of the order of Mach number 7 and altitudes of several hundred thousand feet." Such speed was difficult to imagine: the bullet from a high-velocity rifle travels at around 3,000 feet per second; the aircraft the engineers were proposing would carry its pilot at velocities more than twice as fast, or 5,300 miles per hour. The proposed altitudes were higher than any aircraft had ever flown—above 100,000 feet, beyond what test pilots call the "sensible atmosphere"—where few air molecules persist, on the fringes of space. Existing knowledge of such extremes was limited to theoretical calculation and scale-model tests in wind tunnels, yet the expert panelists determined that the new aircraft be built as quickly as possible using only available technology.

The new aircraft, the X-15, made its first flight less than five years later, part of a crash development program backed by the US Navy and Air Force and operated by NASA. Three of the vehicles were built by the engineers of North American Aviation, each one a single-seat rocket plane with stub wings and a dart-shaped tail, resembling nothing so much as a missile with a cockpit at one end. Carried aloft beneath the wing of a B-52 bomber, the X-15 was dropped at a launch altitude of thirty-five thousand feet. The pilots fired their rocket engines for just ninety seconds, climbing steeply and accelerating to hypersonic speed, rising in a parabolic arc to the edge of space. They watched the curvature of the Earth expand beneath them, the sky darken from blue to black, and the pages of their cockpit flight plans float apart in zero gravity.

As the aircraft left behind the sensible atmosphere, its aerodynamic control surfaces—the rudder and ailerons—became useless, and the pilots had to employ a newly devised system of hydrogen peroxide thrusters, firing jets of gas from the nose and wings, to adjust its attitude. From the apogee of its ballistic trajectory, the X-15 then began a long fall back to Earth as a hypersonic glider, before coming in to land on dry lake beds in the high desert of California and Nevada. The skin of the vehicle was fabricated from a dense nickel alloy—Inconel X—necessary to cope with the extreme temperatures created by friction as it returned to the atmosphere.

Each pilot was outfitted in a custom-made, oxygen-fed pressure suit designed to protect him from exposure to the deadly high-altitude environment in the case of cabin depressurization or the need to eject. The X-15 ejection seat, a massive device equipped with automatic clamps designed to pin its occupant's arms and legs safely in place as he was shot free of the crippled aircraft by a powerful rocket motor, was supposed to work at speeds of up to Mach 4 and altitudes of 120,000 feet. But few of the pilots believed that.

Over the course of nine years, the X-15 was flown almost two hundred times by a dozen different men—including the thirty-year-old future astronaut Neil Armstrong. Their missions pioneered what one scientist called efforts to separate "the real from the imagined problems" of flying on the frontier of space. Before the first Mercury capsules were launched, the brief journeys of the X-15 gave NASA confidence that manned space-craft could leave Earth's atmosphere and return to it without losing control or burning up on reentry. And the initial goals of the rocket plane's experimental program, to fly to hypersonic speeds and record altitudes, were quickly exceeded. In a pair of flights in the summer of 1963, veteran test pilot Joe Walker took the X-15 beyond 330,000 feet, over the internationally recognized boundary of space—the Karman Line, sixty-two miles up—unofficially becoming the first man in history to travel into space twice. But almost from the beginning, the X-15 pilots' exploits were eclipsed by the glamour and pyrotechnics of the Mercury and Gemini spacecraft. While John Glenn and Alan Shepard were celebrated with ticker-tape parades and appeared on the covers of *Life*, few people outside the aerospace industry knew or cared about the X-15 or those who flew it. Yet the lack of attention had its benefits: ignored by the media, NASA's high desert rocket plane experiments proceeded with little oversight or interference from politicians or bureaucrats in Washington, DC. "We produced research data," one of the X-15 pilots wrote later, "instead of headlines."

Keen to extend the length of the program and keep its expensive aircraft flying beyond its initial aims, NASA soon began using the three X-15s as platforms for scientific experiments, mounting equipment in wingtip pods that the pilots opened in the upper reaches of the atmosphere, to gather data on esoteric high-altitude phenomena. Their exploration expanded knowledge of both aerospace engineering and medicine: flight surgeons discovered that the pilots' heart rates during X-15 flights were more than double those recorded on tests of other aircraft, due to the seemingly inexorably rising strain they felt while confined in their cockpits, anticipating each launch. The pressure suits they wore provided the basis for those used to outfit the Mercury astronauts. But testing the limits of existing technology in such an unforgiving environment was fraught with risk, and the pilots' experimental role often required them to push their aircraft far into the unknown. In an aircraft traveling at more than a mile a second, the smallest mechanical fault—or the tiniest miscalculation—could escalate at bewildering speed.

CHAPTER FOUR

THE MOST COMPLICATED MACHINE IN HISTORY

By the time Air Force major Mike Adams strapped into his X-15 on the morning of November 15, 1967, the experimental aircraft had been flying for eight years with only a single serious accident and a handful of major emergencies. The engineers and pilots at Edwards Air Force Base considered the rocket plane "mature," its flights so frequent that they saw them as routine.

Earlier generations of experimental high desert aviation had been conducted in secrecy, and witnessed only by a small handful of technicians and military personnel. Ten years before, the rate of attrition among test pilots had been so high that almost every week a new replacement seemed to arrive at Edwards following a crash. In the event of an accident, wives and relatives had been notified only after the fact. But now death no longer seemed to linger so close at hand, and Adams's wife, Freida, had made sure to be present for each one of his flights aboard the X-15. It was a little frightening, but exciting—like *Flash Gordon*—and now she found herself getting caught up in the moment as she watched him walk across the arid runway to climb into the cockpit of the sleek, black rocket plane for the one hundred ninety-first flight of the program.

Stoic and reserved, Adams, thirty-seven, had wanted to become an astronaut since graduating from test pilot school, and had narrowly missed being selected as part of the group by then preparing to go to the moon. Although later chosen as one of the eight military pilots training to fly in the Pentagon's Manned Orbiting Laboratory, he grew impatient with lengthening delays in the program, and dropped out to volunteer for the X-15 research project; it seemed like the next best thing to going into orbit. In October 1966, Adams had become the twelfth and final pilot selected to fly the experimental plane, and less than a year later had completed half a dozen missions in the cockpit. By November 1967, he still had his mind set on space.

:::::::::::::::::

As her husband reached launch altitude, fell away from beneath the wing of the B-52, and ignited the engine of his spaceplane, Freida Adams was standing in the control room at Edwards with Mike's mother, listening in as his reports came over the radio. With an abrupt punch of acceleration that pinned Major Adams back in his seat, within half a minute the X-15 was traveling at 2,000 feet per second and ascending rapidly through the stratosphere. In the control room, Adams's fellow test pilot Pete Knight called out his altitude: 83,000 feet, then 110,000 feet; at 150,000 feet, Adams shut down the engine, but the X-15 kept ascending, as the designers intended, toward the blackness of space. Just over a minute later, Knight called a new reading: 261,000 feet, or 49 miles above sea level.

Part of the purpose of Adams's flight was to conduct a series of atmospheric experiments, including one to collect micrometeorites, and use an electromechanical probe to gather data on solar radiation. Yet when activated—and unbeknownst to either Adams or the team monitoring the flight from the ground—the probe began creating electrical interference, which disabled the computer and automatic control systems of the X-15. By the time Adams reached the planned apogee of his trajectory, the plane's computer was repeatedly shutting down and restarting, and the nose of the aircraft was drifting from its intended flight path. Under normal circumstances this might have been a minor problem, easily corrected. But Adams—distracted by the malfunctioning computer, almost certainly disoriented by his severe and undiagnosed vertigo—misread his instruments and compensated in the wrong direction. As it began its descent into the atmosphere, the nose of the X-15 continued to wander from its flight path, rotating until the aircraft was falling back to Earth first sideways, then backward—and kept turning, whirling through one complete revolution after another; Adams had lost control. At an altitude of around forty-three miles, and traveling at five times the speed of sound, he radioed to the ground.

"I'm in a spin, Pete."

This was a previously unheard-of phenomenon. Uncontrolled aerodynamic spins were a well-known and potentially deadly part of conventional airplane flight, and one for which pilots had developed practiced recovery techniques. But no one had ever conducted wind tunnel studies or experiments on their potential in research aircraft like the X-15. There were no known ways to recover from one. Indeed, to Pete Knight and the experienced engineers manning the ground control station, the idea that a hypersonic plane could enter a spin in flight seemed impossible. Perhaps unable—or unwilling—to comprehend the situation, Knight didn't immediately acknowledge Adams's message. Instead, he told him to terminate his experiment and turn on the cockpit camera.

"I'm in a spin," Adams repeated.

"Say again."

"I'm in a spin."

"Say again."

For a few seconds, no one in the control room spoke. But the intercom remained silent, and Freida Adams felt the atmosphere in the room curdle; then she knew. A technician took her by the arm and led her away.

From his console, Pete Knight could see that the X-15 was falling fast. He began calling the altitude once more.

"OK, Mike, you're coming through about 135 now. . . . Let's get it straightened out. . . . Coming up to 80,000, Mike."

There was no reply.

Plunging earthward at more than three thousand feet per second, the X-15 at last stopped spinning—through some combination of Adams's struggles with the aircraft, its own inherent stability, and the actions of its automatic attitude control system. But as the X-15 continued hurtling toward the ground, this same automated system took hold of the aircraft, rolling it from side to side, oscillating with increasing violence. Within seconds, each motion reached the equivalent of 13 g's—an invisible fist wrenching the plane with thirteen times the normal force of gravity. The X-15's steel fuselage buckled, and the aircraft began to come apart in the air. When it reached 62,000 feet, and was still traveling at Mach 4, ground control lost all telemetry from the rocket plane. For almost another minute, Pete Knight continued trying to raise Adams over the radio. He was still trying as the first fragments of the plane hit the ground. Then a report came in from a chase plane pilot.

"Pete, I got dust on the lake down there."

"What lake?"

The wreckage of Mike Adams's X-15 was scattered over an area of the Mojave Desert twelve miles long and two miles wide. The central part of the fuselage came down among low hills a few miles south of China Lake Naval Weapons Center, near the town of Johannesburg, California. An emergency team arrived by helicopter within minutes. They found Major Adams's body, still in the cockpit.

Afterward, NASA conducted its own investigation into the crash and published a three-hundred-page report, finding evidence of equipment malfunction and oversights rooted in complacency. The experimental probe that had failed during Adams's ascent employed an electric drive that had never been tested for use at high altitude—but had been flown before, so technicians assumed it would be okay to fly again. Although, years earlier, Adams had participated in a centrifuge test that had revealed he suffered from vertigo so acute that it might incapacitate him in flight for minutes at a time, the information was never shared with the medical staff at Edwards.

The lack of public interest in the X-15 extended even to the death of one of the program's pilots. In contrast to the way the tragic fire on the launchpad at Cape Canaveral had transfixed the world just ten months earlier, Mike Adams's crash received scant attention: the *New York Times* printed the news on page 14; the *Sacramento Bee* carried a front-page report—below the fold, reported as the death of a local man.

But among the anomalies of Major Adams's last flight was the duration of his rocket engine burn, which had continued for four seconds longer than planned. Instead of the projected height of 250,000 feet—around 47 miles above sea level—Adams had ascended an extra 16,000 feet, or just over 3 miles, and across what the US Air Force then regarded as the official boundary of space. In January 1968, Freida Adams drove from her new home in Louisiana to Barksdale Air Force Base, where she accepted the award of her husband's silver astronaut wings, distinguishing Adams as the first American to die in spaceflight. The remains of the X-15 that killed him were buried at an unmarked site in the desert.

::::::::::::::::::::

The rocket plane program—already winding down before Adams died—continued for just one more year. Of the three rocket planes in the fleet, by then only one remained airworthy, and support for the program in the Air Force, NASA, and Congress had dwindled. Fear of another fatal accident only hastened its end, and on December 20, 1968, the aircraft's final mission was canceled.

Yet the X-15 had far surpassed its initial goals. And, despite the shadow cast over its record by Adams's death, it came to be regarded as the most successful flight research project in history. The aircraft proved so far ahead of its time that some of the speed and altitude records set by its pilots would remain unbroken for more than fifty years; its journeys beyond the reach of Earth's atmosphere made it the world's first operational spaceplane. Although some of the experiments flown on the X-15 went on to be used as prototypes for the Apollo moon missions, much of the experimental technology it proved in flight would lie dormant for decades. But the most important discoveries made by the pilots flying the rocket plane were among the most obvious: with the X-15, NASA had established practical design principles for a winged spacecraft that could return to Earth from orbit.

::::::::::::::::::

At the moment the last of the X-15s was finally grounded, the crew of Apollo 8 were preparing for their audacious Christmas trip into lunar orbit. Senior figures at NASA were becoming confident they could land men on the moon before the end of the following year and—belatedly— their thoughts had turned to what might come afterward. Yet, even as they anticipated their greatest triumph, there was little agreement about what their next steps should be.

NASA Administrator Jim Webb, a wily Washington, DC, insider who had led the agency since the inception of the Apollo program, had maintained an unswerving focus on fulfilling President Kennedy's end-of-the-decade promise for every moment of his more than seven years on the job. But, in part because he feared distracting public attention— and congressional funding—from the race to beat the Soviet Union to the moon, for most of that time Webb had refused to commit the agency to serious planning for any project beyond that goal. When asked to an-swer questions about NASA's long-term intentions in a letter from Pres-ident Johnson written soon after his election in 1964, Webb had taken a year to deliver a full reply, and then did so with a report of elaborately contrived bureaucratic circumlocution.

But by 1968, with much of the costly hardware necessary to take the Apollo astronauts to the moon designed, built, delivered, and paid for, Webb's dedicated evasiveness had also left NASA without a dynamic purpose for the future. In the absence of a new goal—and just as the financial cost of the war in Vietnam was escalating ruinously—Johnson and Congress began cutting the agency's budget. They slashed NASA's overall funding for 1969 by a quarter, and the army of staff at the agen-cy's contractors—North American, Grumman, Rocketdyne, and the rest of the aerospace companies who had worked on the moon shot—was cut almost in half, from 377,000 people to 186,000.

In part due to the pork barrel politics that had seen congressmen jostling to bring a part of the lunar bonanza into their districts, NASA facilities had ended up being constructed at sites scattered across the country in a network of semiautonomous "centers": it was no coincidence that the Manned Spacecraft Center in Houston, soon to be renamed in Johnson's honor, had found a home not only in the President's home state, but close to the districts of the handful of local politicians with seats on key congressional committees. Other NASA centers— including the launch and assembly facilities at Cape Canaveral, the Langley aeronautical research center in Virginia, and the rocket engine test site in southern Mississippi—each brought national attention, and jobs, to the regions around them in the boom years of the space race. But when Congress shut off the spigot of Apollo funding, those same communities felt the consequences.

In Huntsville, Alabama, Wernher von Braun had built his own fiefdom of rocketry at the Marshall Space Flight Center, among 1,800 acres of lush forest and rich ocher dirt carved from the middle of the US Army's sprawling Redstone Arsenal. From his hilltop office in Building 4200 he presided over a complex of workshops, laboratories, and test stands where a team of more than seven thousand men and women developed the family of powerful rockets that had taken every astronaut since Alan Shepard into space. And while the glamour and the news cameras went to Mission Control in Houston and the launchpads at the Cape, it was from Marshall that the technology emerged to bring the moon within America's grasp—and von Braun who received the lion's share of NASA's funding. Under his direction, the center had expanded to include its own post office, day care, and internal taxi service; armies of contractors flooded the area, bringing new roads, schools, and a jet airport. In less than twenty years the population of the small southern town nearby—until recently known as the Watercress Capital of the World—grew nearly tenfold, and became so identified with its influx of German scientists that local wits referred to it as "Hunsville."

But once the last of the Apollo hardware had shipped and the layoffs began, the bubble burst: in Huntsville, unemployment rose, restaurants fell quiet, and apartments emptied out in a city that had only recently seen waiting lists for motel rooms. The city was a bellwether of what was to come elsewhere: within a few years, the post-lunar blight would arrive in Houston, and strike the beachfront towns around Cape Canaveral. And if NASA was to save itself from wasting away to nothing, its leaders had to come up with a new idea, one as simple and bold as Apollo, with which to seize hold of the nation's imagination again.

Fortunately, Thomas Paine, who replaced an exhausted and disappointed Jim Webb as NASA Administrator in October 1968, had just such an idea. Paine, forty-six, was quite unlike his predecessor: a Washington, DC, neophyte who had little enthusiasm for space travel before arriving at NASA to serve as Webb's deputy; his name had been chosen from a list of company executives interested in taking any senior position in the federal government. The son of a US Navy commodore, and a veteran who had served on a submarine in the Pacific during World War II, Paine had spent much of his career working at General Electric. He was an exuberant and visionary technocrat who continued to see the world through a maritime lens. If a meeting ran too long, he might fill the time doodling a sketch of his boat, the USS *Pompon*, steaming across the surface of the ocean.

In 1969, with President Nixon newly elected to the White House, Paine pressed the administration to support a new phase of exploration for NASA: even more extravagant than Apollo, embracing the United States' destiny in the stars. The moon landing, Paine declared, "started a movement that will never end, a new outward movement in which man will go to the planets, first to explore, and then to occupy and utilize them."

At the center of these plans was a space station built in Earth orbit, the staging post from which to establish permanent bases on the moon and to launch teams of astronauts on a two-year mission to visit Venus and land on Mars. The Martian explorers would embark from Earth in 1981, propelled by nuclear rocket engines, which were already at an advanced state of development, at a test site specially constructed by scientists from the Los Alamos National Laboratory at Jackass Flats, Nevada. As the space station and the moon bases took shape, the astronauts would operate a three-part "Space Transportation System": including a "space tug" to ferry them down to the lunar surface, a nuclear-powered transport craft; and the keystone of the arrangement, a reusable vehicle that could launch from the ground like a rocket and land like an airplane, to move astronauts and equipment between the ground and Earth orbit.

But Paine, for all his vision and ebullience, had badly misread the political climate. Nixon had little interest in the space program, and was alert to the danger that NASA might ride a wave of global euphoria over the moon landing into an ever-more spendthrift future. And, even amid the patriotic enthusiasm surrounding the lunar triumph, the American public had scant appetite for sending men to Mars, an exploit that one congressman told the press could cost as much as $200 billion. In a Gallup poll taken immediately after the Apollo 11 mission, 53 percent of respondents said they opposed a Mars landing; a *Newsweek* survey revealed that more than half of the public wanted the President to spend less on space exploration. When Armstrong, Aldrin, and Collins were feted at a celebrity-packed gala dinner in Los Angeles less than a month after returning from their flight, and hosted by Nixon himself, protesters hung a banner from the office building opposite the venue that simply read: FUCK MARS.

As the President made it clear that he wouldn't pay for any further interplanetary adventures, Paine abandoned plans for bases on the moon, the nuclear-powered transport, and the space tug. He focused instead on winning support for funding a space station—and the reusable vehicle necessary to build it and resupply it once complete. This craft would be a "space shuttle" with which to realize a long-held dream of cheap, routine access to Earth orbit, an orbital delivery vehicle operated along the same budgetary principles as a commercial airline: a vehicle that erstwhile Nazi general Walter Dornberger described as "an economical space plane capable of putting a fresh egg, every morning, on the table of every crew member of a space station circling the globe."

Dottie Lee got the call one Friday morning in early 1969. When she arrived in his office, her boss in the Structures and Mechanics Division at the Manned Spacecraft Center in Houston was brief. "Monday, you're to report to Building 32. You don't tell anybody where you're going, or what you're going to do," he said. He handed her a document: "Read this."

Lee, forty-two years old, a formidable technician with coiffed flamered hair and a taste for vodka martinis, was one of the few female engineers working in Houston, and perhaps the only one supervising a team of men. A math graduate who had once imagined becoming a high school geometry teacher, she had instead been hired straight from college by NACA to work as a "human computer," making the complex aerodynamic calculations necessary for the early days of hypersonic flight. Lee was eight years into her career there, and had been promoted to Senior Mathematician, when she was asked to fill in as a temporary replacement for Max Faget's secretary, who was leaving on her honeymoon. Lee didn't know how to type, and spent most of her time at her desk in Faget's office calculating a triple integral. At the end of two weeks, Faget asked her to work for him full-time, as a project engineer. She learned on the job, first on the design of experimental rocket flights at NACA and, after the creation of NASA, in the Apollo program, specializing in the esoteric discipline of aerothermodynamics. By 1969, she was supervising a group of contractors at work on the heat shield protecting the Apollo command module on its return to Earth from the moon.

After the meeting with her division chief, Lee had to explain to the contractors that she had been reassigned, but gave them no details. "I'm going to be out of pocket for a little while," she said. "I cannot say anything of where I am." The document Lee had read outlined the principles of a reusable spacecraft that could fly into orbit and then land on a runway; the project was secret. On Monday morning she arrived at Building 32, a windowless hangar-like space in the northeast corner of the Houston campus, to discover a guard at the door with a clipboard in his hand. Only those whose names were on his list were allowed to enter.

Inside, a clandestine team of draftsmen and engineers representing each of the major disciplines of spacecraft construction—including aerodynamics, propulsion, structures, and thermal protection—set to work on the design of the Space Shuttle, under the supervision of Max Faget. The balsa-wood model that Faget had built in his garage was at the heart of these initial plans—for a fully reusable spaceplane system made up of two separate winged vehicles, each flown by its own crew, but piggybacked together for launch. The first was a booster craft, as large as a Boeing 747, which would carry the second—a lighter vehicle the size of smaller airliner—on its back to an altitude of fifty miles. There, the booster would release the smaller craft—the orbiter, the straight-winged spaceplane of Faget's design—which would carry on under its own power into space. The booster would then fly back to land on an airstrip at the launch site, ready to start again. The orbiter would carry out its mission in space before reentering the atmosphere and landing on an airstrip like a conventional airplane.

At first, the new spacecraft was planned on an even more ambitious scale than Apollo, with technology fit to carry American astronauts into the twenty-first century. Tom Paine called for a $14 billion project that would be ready to fly by 1975, powered by fourteen massive cryogenic engines and large enough to lift the components of a space station into orbit.

But long before President Nixon agreed to sign off on the creation of NASA's next-generation spacecraft, the agency's seamlessly futuristic aspirations were dashed once again on the unyielding economic realities of the 1970s. When funding for the combination of a space station and shuttle received congressional approval only with the narrowest of margins, Paine recognized that, by asking for both, he risked getting neither. So he shunted plans for the space station off into the distant future, on the understanding that the shuttle was the only remaining element of the once-complex Space Transportation System that could exist without any of the others. And if—in the absence of a space station, or moon bases, or the staging of interplanetary rocket missions—the proposed shuttle no longer had any specific destination to reach, then surely plenty of reasons could be found for it to visit low Earth orbit.

Even then, Congress and the White House balked at the extravagant price tag for the shuttle. Instead of $14 billion, Congress and the hard-headed bureaucrats of Nixon's new Office of Management and Budget agreed to allocate just $5.5 billion for development of the new vehicle. Based on NASA's existing calculations, this amount was far too little money to create an experimental machine that would require the research and development of so much untried technology. But the NASA chiefs—who had come to believe that the future of US manned space-flight, and the agency itself, depended on committing to an ambitious new spacecraft—agreed, regardless. It was the first of many fatal compromises.

Despite the budget cut, NASA continued to pitch the shuttle as a panacea for all of the nation's future orbital transport, a space truck that could be robust enough for routine operation, with quick turnaround times making it available to launch every two weeks. Most important, the shuttle would be so cheap to operate that it would soon pay for itself: in contrast to the Saturn V rockets that were taking astronauts to the moon at the staggering price of $185 million per launch, the reusable shuttle might cost a mere $350,000 each time. But these numbers only made sense if the new vehicle could fly almost as frequently as a commercial airliner, spreading the huge cost of its development and construction over a great number of launches.

To make this argument, NASA commissioned a cost benefit analysis from Mathematica Inc., a consultancy cofounded by one of the original developers of game theory. The Mathematica study, which would soon become infamous for its fantastical accounting, predicted that the shuttle would indeed provide the United States with a cheap alternative to expendable rockets—so long as it made at least 736 flights between 1978 and 1990, or fifty-seven missions a year; more than one a week.

At the launch rate imagined by Mathematica, the new workhorse spacecraft could not only provide for every one of NASA's own orbital needs, but also help the government turn a profit by taking on commercial satellite customers, and fly military missions, replacing all of the expendable rockets that the Pentagon used for launching its spy satellites. Nixon's accountants examined the projections and spluttered in disbelief. "They start at a number that strains credibility," wrote one, "and go up from there."

To support its case and make the numbers more plausible, NASA sought support from the Air Force—and Pentagon backing proved decisive in finally winning presidential endorsement for the shuttle. But in exchange, NASA had allowed the Air Force to set two specifications for the orbiter that would profoundly complicate its design.

::::::::::::::::::

The White House had recently canceled the "black" Manned Orbiting Laboratory program, partly due to the fruits of work by the National Reconnaissance Office—a three-letter government agency so secret that it did not yet officially exist. Staffed by officers from the Air Force and the CIA, the NRO had no headquarters building, but operated from behind an unmarked door on the fourth floor of the Pentagon. The new agency had overseen a series of classified reconnaissance satellites of steeply increasing sophistication that rendered obsolete the idea of placing human spies with cameras in orbit.

Code-named Hexagon, the NRO satellites could capture images of objects on Earth as small as two feet across; with such fine definition, analysts could count the number of people sitting on a picnic blanket in Gorky Park—or of ground-to-air missiles on a military base in Kazakhstan. But the satellites, which the Pentagon had been sending into space on modified Titan missiles, carried cameras loaded with sixty *miles* of film, were each the size of a Greyhound bus, and weighed fifteen tons. To carry this enormous payload into space, the new shuttle was designed with a cargo bay sixty feet long and fifteen feet wide; to accommodate Hexagon and similar, even more powerful, spy satellites already in development, NASA's new space truck would be required to lift up to twenty-two tons into polar orbit.

The Air Force also insisted that the shuttle be able to glide a thousand miles east or west after it reentered the atmosphere. This extensive "cross-range" ability would serve several purposes of interest to the Pentagon: it could avoid being forced down in communist territory in the event of an emergency in space; it would enable it to take off and land from its launch site after a single orbit, making it possible to undertake swift once-around-the-earth reconnaissance missions that would be back on the ground less than two hours after launch; and it would also make it possible to conduct offensive operations in space, including missions to snatch Soviet spacecraft from orbit and return to Earth before the owners of the kidnapped satellites had time to respond.

Yet such long-distance glides were beyond the capacity of Max Faget's ingeniously lightweight and straight-winged design. As the secret drawings coming off the drafting tables in Building 32 multiplied—first into dozens, then scores, of variations—Dottie Lee and the other engineers began work on a design that would meet Air Force expectations. What finally emerged was a big triangular-winged vehicle the size of a DC-3 airliner; something that began to look a lot like an enlarged version of the old Dyna-Soar spaceplane. This delta-wing shuttle had the range that the Air Force wanted, but was so much heavier than Faget's original concept that it necessitated saving weight elsewhere: it meant discarding the air-breathing jet engines the designers once considered for the orbiter, which would have allowed it to fly under its own power once it returned to Earth's atmosphere. Now the orbiter would have to glide down to its landing strip, plummeting to Earth at the speed and angles of a fighter jet, but approaching the runway with total precision to execute a perfect landing at the first attempt.

Other key parts of the original concept would soon be abandoned, too—including the rocket-powered escape system necessary to save the crew if the spacecraft faced imminent destruction, especially during launch. Another of Faget's innovations, some version of this system had been built into every previous NASA manned spacecraft since the beginning of the program; but now weight—and cost—meant that it had to go.

Faget—who had designed the Apollo capsule in which Grissom, White, and Chaffee had been incinerated in January 1967, and served on the subsequent board of inquiry into the fire—was well aware of the cost of faulty design. He opposed the new configuration of the nascent

spacecraft and persisted in producing variants of his own original concept long after drawings of the Pentagon's choice had been sent out to
contractors for production studies. The pugnacious engineer was a
powerful figure within NASA, and answered directly to the head of the
Manned Spacecraft Center. But when in November 1970 he wrote an
internal memo attempting to assert control over the specifications for
the shuttle, he was cut off by a relatively junior Air Force officer; it was
clear where the power lay. "We've made a pact with the devil," Faget later
told a friend in the astronaut corps.

Another serious compromise in the design of the new spacecraft
was yet to come. By 1972, congressional budget restrictions meant that
Faget and the other engineers were forced to abandon their plan for an
orbiter carried by a fully reusable piloted booster craft, complete with
its own internal fuel tanks. Instead they drew up proposals for an only
partially reusable and unmanned booster stage: constructed around a
separate external tank to carry all the fuel for the vehicle's main engines,
and two powerful strap-on rockets that would provide the majority of
the thrust to reach orbit. The expendable fuel tank, once emptied, would
be jettisoned at the edge of space, falling back into the atmosphere and
disintegrating before raining into the Indian Ocean in pieces. The exhausted strap-on boosters could descend on parachutes for recovery at
sea, refurbishment, and reuse on future missions.

At first, the designers proposed that these boosters use liquid-fueled
rocket engines like all previous manned spacecraft: these were not only
powerful but could be throttled up and down, shut off and started up
again, as required. But such engines were also expensive, and their delicate fuel tanks and plumbing might be too easily damaged to survive the
impact of an ocean splashdown and immersion in seawater. So the technicians turned instead to solid rocket technology. Giant segmented metal
cylinders packed with a rubbery compound of volatile fuel, solid rockets
were not unlike massive fireworks: once lit, they could not be throttled
or shut down, but continued firing until they burned out. As a result,
they could not be flight-tested before launch—and had never before
been used for manned missions; veteran engineers at NASA, including

Wernher von Braun, believed they were too dangerous ever to be used to carry humans. And although Max Faget and the other designers favored the liquid option, they were overruled from above: years of experience using the solid rockets to launch missiles and carry unmanned missions into space had shown they were simple, cheap, and apparently reliable.

So it was that the designs at last left the drawing board and, with Nixon's approval, by the end of summer 1972 NASA had awarded four main contracts for each element of the Space Transportation System. The orbiter would be built in California by the major contractor on the Apollo program—and builders of the ill-fated Apollo 1 capsule—the recently renamed conglomerate North American Rockwell; under the project direction of NASA's propulsion experts at the Marshall Space Flight Center, Rockwell's subsidiary Rocketdyne would handle the development of the vehicle's liquid-fueled main rocket engines. Marshall would also take overall responsibility for the giant external fuel tank, fabricated by defense contractor Martin Marietta at the Michoud Assembly Facility in New Orleans; and the manufacture of the motors for the solid rocket boosters, by the Thiokol Chemical Corporation, at its sprawling Wasatch facility in the deserts of northern Utah.

Structural assembly of the first in a fleet of four orbiters—to be named *Constitution* in honor of the country's impending bicentennial celebrations—began at Rockwell's Air Force Plant 42 in June 1974. In the meantime, NASA set out to assemble the new class of astronauts who would fly the revolutionary spacecraft: an intake of pilots and engineers unlike any other in the agency's history.

CHAPTER FIVE

THE FUTURE
BLACK SPACEMAN

March 24, 1976

Standing near the center of one hundred acres of landscaped grounds reclaimed from a humid parcel of cow pasture and swamp ten miles southwest of downtown Houston, the main administration building of the Johnson Space Center was a brutalist concrete box overlooking a small chain of torpid ornamental ponds and a parking lot. Known, with NASA's utilitarian bluntness, as Building 1, the office block and the campus surrounding it had been formed in prefabricated sections and thrown together over the course of two years with the same sense of speed and expedience as the manned space program itself.

Ever since its completion in 1964, the Space Center had been claimed proudly by the residents of Houston, but was not a part of the city at all: it was a forty-minute drive away from the center of town in Clear Lake, a remote exurb of strip malls and housing subdivisions, bordered by the Gulf Freeway and the brackish estuarine inlet for which it had been named. From the windows on the top floor of Building 1, the derricks and the vent stacks five miles away on the Houston Ship Channel loomed from a petrochemical haze, a ghostly blue forest on the near horizon.

And it was here, on a Wednesday morning in the final week of March 1976, that George Abbey convened a new meeting of NASA's Astronaut

Selection Board: the first since Armstrong and Aldrin had walked on the moon. Together, the twelve members of the board planned to recruit the largest single intake of new astronauts the agency had ever seen.

As he took his seat at the massive oval table in the Center Director's conference room, Abbey had been Director of Flight Operations in Houston—overseeing Mission Control, astronaut training, crew selection and their flight assignments—for barely three months. A forty-three-year-old graduate of the US Naval Academy in Annapolis, a trained pilot, and himself an unsuccessful applicant to fly in space, Abbey had been attached to NASA since the early days of the Apollo program, and ascended inconspicuously from the lowest rungs of the agency's Byzantine bureaucracy. Over the course of more than a decade in Houston, he had built a reputation as a workaholic technocrat, a true believer pledged to the cause of manned spaceflight. On call twenty-four hours a day, Abbey had five young children at home, but rarely saw them: after leaving his office, he often lingered into the small hours among those who manned the consoles at Mission Control, or in the taverns and hotel bars scattered along the two-lane highway specially built to serve the center, NASA Road 1.

As part of the team who had successfully brought home the crew of Apollo 13, in 1970 he had been awarded the Presidential Medal of Freedom. But unlike other, highly visible administration characters like Flight Director Gene Kranz—instantly recognizable by his severe blond flattop and fancy vests—Abbey remained unknown to the general public, a doughy middle-aged functionary in a blazer and khakis. Secretive, inscrutable, and Machiavellian, he combined a prodigious memory with a subtle knack for reading human behavior, a calculated reluctance to commit his instructions to paper, and a gift for gathering intelligence from those around him and using it well: the Thomas Cromwell of the Johnson Space Center.

Now Abbey and his boss, Center Director Chris Kraft, were under instructions to transform the elite astronaut corps into a body that at last began to reflect the diversity of the United States. They wanted to ensure that this new class of astronauts, the first who would be trained to fly the Space Shuttle in its routine operations, and the first new group to be admitted to NASA in more than ten years, would include both women and minorities. It would be the first serious effort to do so since the agency chose the Mercury Seven from a group of thirty-two volunteers in 1959. While the initial guidelines for what was then called "Project Astronaut" had been broad enough to include anyone with physical endurance and a demonstrable spirit of adventure, regardless of their gender or race, the leaders of the fledgling agency disregarded this rubric and insisted that successful applicants all be chosen from a field of military test pilots: specifically those with thousands of hours of experience flying high-performance jets. This stipulation had inevitably shrunk the pool of potential candidates to a small, uniform group composed almost entirely of trim, churchgoing white men with ready smiles, pretty wives, and a shared hatred of communism.

But by the time the Astronaut Selection Board gathered around the conference table in 1976, the world was finally changing. In 1972, the Equal Employment Opportunity Act had made it illegal for employers to discriminate on the basis of race or gender, and NASA had hired a Black woman to run its equal opportunity programs. In September that year, Administrator James Fletcher asked his senior management to develop a plan for the next selection of astronauts, "with full consideration being given to minority groups and women." In 1975, President Gerald Ford signed Public Law 94-106, which stipulated that all US military academies admit female trainees. Now, for the first time in their history, the Selection Board included a Black man and a woman. Their intention was to recruit a total of forty new astronauts to crew the Space Shuttle— twenty pilots, and twenty in the freshly created category of "Mission Specialists," who were not trained aviators but scientists, engineers, and physicians.

Not everyone on the Board was enthusiastic about the plans. Deke Slayton, one of George Abbey's predecessors as Director of Flight Operations, had been a member of the original Mercury Seven, and at one point scheduled to be the second American in space. But before he could fly, Slayton had been grounded by an irregular heart rhythm, and moved into management as the first Chief of the Astronaut Office. Placed in charge of all crew selection and flight assignment for the Apollo missions, Slayton came to embody the arrogance and contempt with which many astronauts treated the managers and flight surgeons who policed their access to space; he also occasionally covered for the missteps and recklessness of his military-trained colleagues. After Gene Cernan had crashed a NASA helicopter while buzzing women sunbathing on a beach near Cape Canaveral in 1971, Slayton concealed the true cause of the accident from Center Director Chris Kraft—to ensure that his good friend Geno did not lose his place as commander of the final mission to the moon. When it came time to replace Slayton, Chris Kraft had chosen Abbey for the job in part to wrestle control of the Astronaut Office back into the hands of the agency's senior managers.

As the board meeting began, Slayton—now a grizzled fifty-two, his dark military high-and-tight growing out into a more fashionable steel-gray mop that would soon fall about his ears—questioned whether it would be possible to find any suitable astronaut candidates that did not meet the old military criteria. Kraft told him that the skills required of Mission Specialists aboard the Space Shuttle meant their ranks would be filled from a broader pool of candidates than test pilots alone could provide. Furthermore, he planned to have NASA actively solicit applications for this new class of astronauts from among women and minority groups.

That was too much for Slayton, who rose to his feet. "I want no part of this," he said. He walked out, and never returned.

::::::::::::::::::::

Change had been a long time coming to NASA. Pressure to diversify the astronaut corps had existed almost from the moment the names of the Mercury Seven were announced to the public, and years before any human being had flown in space. In September 1959, Dr. Randy Lovelace, whose New Mexico research foundation had provided medical screening for the Mercury candidates, heard news that a female cosmonaut had begun training in the Soviet Union; on his own initiative, he set out to examine how women's suitability for spaceflight compared with that of men.

In search of an experimental subject, Lovelace approached twenty-eight-year-old Geraldyn "Jerrie" Cobb—an experienced pilot who had first flown at the age of twelve, gone on to set numerous altitude, speed, and distance flying records, and had recently been named Woman of the Year by the Women's National Aeronautic Association. In February 1960, Lovelace began subjecting Cobb to the same gruesome physiological and psychological tests endured by the Mercury astronauts—eighty-seven in all, including one that required swallowing three feet of rubber hose to test levels of stomach acids, another in which ice water was shot into her inner ear to induce vertigo, more than nine hours in a sensory deprivation tank, and furious centrifuge experiments designed to assess resilience to high g-forces. Cobb passed every one, and in August that year Lovelace presented his findings at the Space and Naval Medicine Congress in Stockholm, Sweden, where he announced that women's physical size and biology might also give them advantages for orbital flight that men lacked. His conclusions made headlines: *Life* magazine carried pictures of Cobb during testing, and *Time* referred to her as "Astronautrix Cobb"; in interviews she said she was keen to go into space as soon as she could. "I'd want to do it even if I didn't come back," she reportedly told a friend. "God has always been my pilot and I'm not at all afraid, just eager."

As Cobb began making public appearances, Lovelace expanded his cohort of test subjects. With the advice and financial assistance of Jacqueline Cochran—a veteran pilot and businesswoman who had directed the Women's Airforce Service Pilots during World War II, become the first woman to break the sound barrier, and was well-connected at NASA—a further twelve women fliers were subjected to two phases of astronaut-style testing. All of the women demonstrated that they were just as physically and psychologically qualified for spaceflight as men. In correspondence, Cobb began referring to her group as the FLATs: First Lady Astronaut Trainees.

Yet the truth was that the agency had no intention of training female astronauts—and some at NASA were embarrassed by what they saw as a publicity stunt. The press continued to regard the prospect of women in space as a diverting novelty, with stories about "Astrodolls," "Astronettes," and "Space Gals." But as Cobb's public renown grew, the papers began to describe her as a trainee astronaut, leading the public to believe that the agency had opened the space program to women. Even as the agency disavowed any involvement in Lovelace's program, and his tests remained almost entirely privately funded, NASA received an increasing number of inquiries from female volunteers to fly in space.

In June 1961—a month after Alan Shepard had become the first American in space—then NASA Administrator James Webb introduced Jerrie Cobb at a banquet. He praised her interest in the space program and her success in the three phases of medical tests, and announced that he was officially enlisting her as a consultant to the agency. Cobb's appointment did more to improve NASA's public image than it did to further the cause of female astronauts; the following month, shortly before a battery of further medical trials was scheduled to take place at the Navy's base in Pensacola, the tests were canceled. To spend the time and money involved in administering the experiments, the Navy had needed formal notification that NASA required data on the women pilots—and the space agency declined to provide it. Soon afterward, Cobb received a letter from NASA's Assistant Administrator for Public Affairs confirming that her consultancy was little more than window dressing: "Despite the manifest interest in your proposal from audiences who hear you speak," he wrote, "I am afraid that at the present we cannot undertake an additional program training women to be astronauts."

Undeterred, Cobb wrote to Vice President Johnson, in his capacity as Chairman of the National Aeronautics and Space Council, asking him to use his influence to press the case for women in space. Another of the FLATs, pilot and activist Jane B. Hart—whose husband was a Democratic senator—sent letters about the issue to every member of the Senate and House Committee on Space, and in March 1962 the two women sat down with Johnson in his office in Washington, DC. They warned the Vice President that the USSR had already begun training women cosmonauts, and urged him to start a crash program to put a female astronaut in space within the year. "A woman can do anything a man can do up there," Hart told reporters outside, "and besides, she weighs less and requires less food." But the meeting came to nothing; Johnson told Cobb and Hart that he was powerless to intervene: the matter was up to NASA. Beforehand, the Vice President's executive assistant had drafted a letter to the head of the space agency, which she suggested Johnson show to the two women to offer them some encouragement—and also win him some good publicity. The letter, seeking clarification about the criteria for NASA's astronaut selection, fell short of calling for action, but implicitly questioned the agency's failure to consider putting women into space: "I'm sure you agree that sex should not be a reason for disqualifying a candidate for orbital flight," it read. But the mercurial Vice President refused to show the draft document to Cobb and Hart. While he remained gracious with the two women during their visit, after they had left Johnson took a pen and scrawled, "Let's stop this now!" in inch-high letters across the signature line of the letter. It was never sent, but instead buried in White House files.

Three months later, Cobb and Jackie Cochran appeared before the Congressional Committee on Science and Astronautics, in what one New York lawmaker described as an attempt to permanently settle "this problem about women astronauts." In her opening statement, Cobb testified to their shared professional experience with male pilots, and explained that they had no intention of joining a battle of the sexes.

"We seek only a place in our nation's space future without discrimination," she said. "We ask as citizens of this nation to be allowed to participate with seriousness and sincerity in the making of history now, as women have in the past. There were women on the *Mayflower* and on the first wagon trains west, working alongside the men to forge new trails to new vistas. We ask that opportunity in the pioneering of space."

But this effort, too, went nowhere: the members of the committee listened to the women speak and then recommended that NASA continue to select astronauts as it had done to date, "strictly on a technical and professional basis." In his testimony, astronaut John Glenn, the recently minted national hero who had become the first American to orbit the Earth just a few months earlier, spoke for the status quo: "The men go off and fight the wars and fly the airplanes and come back and help design and build and test them. The fact that women are not in this field is a fact of our social order," he said.

The committee concluded that NASA's existing standards of astronaut selection were not only reasonable but expedient. With the deadline already set to land men on the moon by the end of the decade, the agency must meet that goal before concerning itself with lesser priorities. "Maybe we should ask the good ladies to be patient and let us get this thing accomplished first," said Ohio Democrat Walter H. Moeller, "and then go after training women astronauts."

Meanwhile, NASA announced that it would be accepting applications for a second class of astronauts in the spring of 1962; for the first time the process was opened to civilians. Among those who applied were six women. Of this half dozen, NASA management later said that some had not reached the necessary educational standards, and one was too old; all lacked the necessary experience as test pilots in high-performance jet aircraft. Every one of them was disqualified. When the new group of astronauts was presented to the press in Houston that September, it proved once again—to no one's great surprise—to be composed entirely of men.

Nine months later, on June 16, 1963, the Soviet authorities announced their latest propaganda triumph: twenty-six-year-old cosmonaut Valentina Tereshkova, a former textile worker and amateur parachutist with a recent trade school qualification as a "cotton spinning technologist," had become the first woman in space. In a three-day journey aboard her Vostok 6 capsule, Tereshkova had orbited the earth forty-eight times—her single spaceflight longer than the total of all those made by US astronauts to date.

Even as the drama of Jerrie Cobb's struggle with NASA was playing out in the newspapers and in Congress, the Kennedy administration had also been quietly attempting to bring racial integration to the astronaut corps. In September 1961, Edward R. Murrow, the pioneer broadcaster and head of the US Information Agency—effectively the propaganda arm of the State Department, with a remit to sell American principles and policy at home and abroad—had written a letter to NASA Administrator James Webb. Murrow, like others in Washington, DC, had watched Yuri Gagarin embark on a global tour as an ambassador for communism, on which he made stops in many of the countries of the world just emerging from colonial rule—including Liberia, Ghana, Cyprus, and India. Murrow recognized that America's astronauts would be useful tools in spreading the good word of democracy abroad, particularly as the fight for civil rights within the United States reached a critical inflection point. Murrow wrote:

Dear Jim,

Why don't we put the first non-white man in space?
If your boys were to enroll and train a qualified Negro and then fly him in whatever vehicle is available, we could retell our whole space effort to the whole non-white world, which is most of it.

As ever,
Yours,
Edward R. Murrow

Webb, by that time preparing for John Glenn's first flight into orbit and already embroiled in the debate about female astronauts, sent Murrow a polite brush-off. Despite the obvious political benefits that NASA might gain from selecting a Black candidate for spaceflight, Webb explained that it had so far chosen astronauts on strictly technical grounds; to adopt any wider criteria "would be inconsistent with our agency's policies."

But when that year NASA and the Air Force jointly announced a list of fifty test pilots who would be candidates for future missions in space, the White House was already taking an interest. President Kennedy had recently signed an executive order implementing affirmative action, to bring equal opportunity principles to federal agencies, and NASA was not about to escape scrutiny. So a special assistant to the President sent a memo to his opposite number at the Pentagon requesting information about the possible selection of minority astronauts. It was important, the assistant wrote, that "for symbolic purposes in crossing the frontiers of space, this country have qualified members from minority backgrounds."

The Pentagon's answer was a familiar one: NASA astronaut candidates were required to be graduates of the USAF experimental test pilot school, or nongraduates with combat experience, or extensive flight time, in high-performance fighter jets. No Black candidates from such backgrounds had so far come forward; neither the Department of Defense nor NASA discriminated on the basis of race. The White House was apparently unimpressed; the aide replied with a request for more details and, more pointedly, gave the Pentagon a deadline to come up with the name of a suitable minority candidate: November 1, 1961.

In the summer of the following year, Murrow brought his suggestion for a Black astronaut to the President himself. Kennedy sent a memo about the proposal to the Vice President, in his role as Chairman of the Space Council, asking if "something might be done." Johnson assigned the problem to his aide, former newspaperman George Reedy, who reported back that, while a Black astronaut would undoubtedly be a great asset to the United States, the White House should "dispose of the concept that NASA can just reach out and grab a Negro and make an astronaut candidate out of him."

The public relations risks of attempting to do so were numerous: he might not succeed as an astronaut, making it look as if he had been set up to fail; it was possible he might be killed, leaving NASA open to accusations they were treating Black astronauts as expendable; and—worst of all—the public might discover that he had been selected for reasons other than merit alone, undermining the point of the entire enterprise.

But by the end of August 1962, the White House had found their man: twenty-seven-year-old Captain Edward Joseph Dwight, a B-57 bomber pilot with a degree in aeronautical engineering who had logged more than 1,700 hours of flying time in jet aircraft, was enrolled in the new US Air Force Aerospace Research Pilot School (ARPS) at Edwards Air Force Base. If Dwight could successfully complete the exacting one-year, two-stage training course at Edwards—not just learning to handle the most advanced fighter aircraft in the US arsenal, but progressing to classes in thermodynamics, bioastronautics, and Newtonian mechanics—then the Air Force would deliver him to the gates of NASA, having met all of the criteria necessary to make him a candidate eligible for astronaut selection. Like every member of the new breed of American heroes NASA had anointed so far, Dwight was charismatic, accomplished, and hard-charging; he was also Black.

Ed Dwight was born and raised in a middle-class Catholic family in Kansas City, Kansas, where his parents owned a restaurant; as a boy he earned pocket money delivering newspapers, built a pretend airplane out of orange crates in his backyard, and sketched the aircraft landing at the airport near the house. He dreamed of flying, but believed it forever out of reach, the distant preserve of white men; Kansas City was segregated. But in 1945, Dwight's older sister integrated the local Catholic high school, nine years before *Brown v. Board of Education.*

Dwight was eighteen before he thought he might become a pilot himself, inspired by seeing a photograph of another Kansas man, Lieutenant Dayton "Rags" Ragland, on the front page of the *Kansas City Call.* Ragland, the only African American pilot in his squadron and the first in the Air Force to fly an F-86 Sabre fighter, had been shot down in combat over North Korea and taken prisoner; Dwight's concern over Ragland's fate was surpassed by his astonishment that the government had allowed a Black man to fly a jet airplane.

Just five feet four, and hampered by a stutter, Dwight attempted to enlist in the Air Force; the recruiters rejected him because of his height, his speech impediment—and because of the color of his skin. So he wrote directly to the Pentagon, and was notified that a pilot selection team would soon be visiting his junior college. Dwight was commissioned into the Air Force in 1955, and two years later graduated in aeronautical engineering from Arizona State University. He excelled as a young officer: commended by his superiors for his aggression and skill, he used his free time to log extra hours in the air, took correspondence courses in calculus and electronics, and began work on a master's in nuclear engineering at Berkeley. But Dwight had never considered flying in space—until, in November 1961, he received a letter from the Pentagon, inviting him to enroll in test pilot training at Edwards, the first step on the path to becoming an astronaut candidate. He was elated; news of his entry into the school the following year was trumpeted in the Black press as a sign that an African American would soon join the astronaut corps: "Select Negro for Aerospace School: May Become First of Race in Space," wrote the *Cleveland Call and Post*; in the *Baltimore Afro-American* he was "Tagged as First Tan Astronaut." President Kennedy even called Dwight's parents to personally congratulate them on their son's achievement. Just by being admitted to ARPS, he had apparently become the icon of equal opportunity the White House had wanted: a real-life example of civil rights in action, proving to young African Americans everywhere that with tenacity and education they, too, could join the race to the stars.

Dwight's day-to-day reality was different: on arriving at the school he discovered he was resented by those who felt he had only been admitted because of the President's influence, and believed he had taken the place of better-qualified white pilots. According to Dwight, the ARPS commandant, Chuck Yeager—the first man to break the sound barrier, portrayed in Tom Wolfe's *The Right Stuff* as the archetype of the swaggering experimental aviator—wanted him off the course from the minute he arrived. "We can get him out of here in six months," Yeager told one of Dwight's classmates. "We can break him." Frozen out by his colleagues, contending with the everyday racism of Jim Crow–era

America when he traveled off base, Dwight began the first phase of the training, the eight-month Experimental Test Pilot Course, under mounting pressure.

Yeager would later deny Dwight's account of his behavior as commandant, and criticized Dwight's abilities, describing him as "an average pilot with an average academic background" who couldn't keep up the exceptional standards of his classmates. When it came time to qualify for the second part of the training, the Aerospace Research Pilot Course, explicitly designed to prepare the best candidates for flying in space, Yeager decided that Dwight failed to make the cut, finishing last in his class of twenty-six. But General Curtis LeMay, chief of the Air Force, had already been instructed by Attorney General Robert Kennedy to integrate the test pilot school, and when his office learned that Dwight had not made it through to phase two, Yeager received a telephone call from LeMay himself.

"Bobby Kennedy wants a colored in space," the general said. "Get one into your course."

With additional funding from LeMay, Yeager expanded the size of the aerospace course from eleven students to sixteen, making room to include Dwight. News of his success—announced to the press by a pair of US senators before it was formally released through the Air Force—made Dwight a national celebrity. Captain Dwight appeared on the cover of *Jet* magazine in his Air Force uniform, and on the front of *Sepia* clad in his flight suit beside the headline "US Trains First Negro Spaceman"; the *Daily Defender* went further still: "Negro Astronaut May Be First American to Reach Moon." No matter that Dwight had not yet begun the aerospace training course, or that—even if he completed it—it would merely make him one of dozens of fliers eligible for selection by NASA.

The agency itself, always hungry for good publicity, did little to correct the impression that Dwight had already joined the handful of heroes suiting up to prepare for the Apollo missions; Ed Murrow's US Information Agency sent out photographs depicting Dwight's progress to the newspapers, and used his image to promote the United States' interests abroad, helping secure US missile and tracking installations in Africa off the back of "the future black spaceman." At home, Dwight became as popular in the Black community as John Glenn was with white America, and traveled frequently to Washington, DC, to burnish his connections there. In one year he made more than 170 speeches, at schools and Lions clubs across the country; he had his own private secretary to handle his fan mail: at one point he was receiving 1,500 letters a day, and sent out 5,000 press pictures a month. And in late 1963, Dwight completed the second phase of his course and graduated from ARPS, eighth in his class of sixteen; at last, an African American candidate had unquestionably met all of the qualification criteria for NASA's Astronaut Selection Board.

The agency announced the names of its third group of astronauts at a press conference in Houston on October 18, 1963: eleven men in all, including Gene Cernan and Buzz Aldrin, who would become the second man to walk on the moon. Ed Dwight was not among them. Although his name had been among twenty-six pilots recommended "without qualification" to NASA by the Air Force, he had not made the final cut.

After the new astronauts had filed onstage for presentation to the world's press, one reporter directed a question toward Deke Slayton, who had overseen the selection: "Was there a Negro boy in the last 30 or so that you brought here for consideration?"

"No," Slayton replied, "there was not."

In the days after the announcement, Black newspapers expressed outrage over the choice, but Dwight reacted with equanimity; after all, hundreds of white men—and two women—had also been among the applicants the agency had passed over. "That was entirely up to NASA," he told reporters. "They had their needs and criteria, and I respect their decision."

But the following month, President Kennedy was shot dead in Dallas, and Dwight was soon assigned to a backwater job at Wright-Patterson Air Force Base in Ohio. There, he wrote a fifteen-page report detailing his experience of racism and prejudice in the way he had been treated by the Air Force, which he sent to members of Congress and the White

House, and eventually found its way into the hands of the press. His commanders—particularly Yeager—were furious. What became known as the "Dwight Case" was the subject of a long-running newspaper investigation and prompted questioning by a pair of sympathetic congressmen, but NASA and the Air Force denied any wrongdoing. Finally, in 1966, frustrated and angry, Dwight resigned his commission, took a job at IBM, and eventually became a successful sculptor.

The first African American selected to join the US astronaut program, Major Robert H. Lawrence, was finally chosen in 1967—not by NASA but by the Air Force—as part of the group being trained to fly aboard the Manned Orbiting Laboratory. Six months later, on the afternoon of December 8, Lawrence was flying a training mission with a student pilot in a modified F-104 Starfighter when the aircraft crashed on approach to the runway at Edwards. The landing gear collapsed, the cockpit canopy shattered, and the aircraft scraped along the runway for two hundred feet before bouncing briefly into the air and smashing into the ground a second, and final, time. At the moment Lawrence pulled the handle to eject, the plane had begun to roll over, and the pilot's rocket-powered seat fired him horizontally across the desert floor, likely killing him instantly. His death before he had the chance to fly in space, and the secrecy of the MOL program, helped ensure that Lawrence's achievement went unacknowledged for decades.

CHAPTER SIX

THE FNGs

When it finally came, in a press release issued from the Johnson Space Center in July 1976, the formal announcement that NASA had opened astronaut applications to women and minorities barely made the news. Relegated beneath reports of the impending White House race between President Gerald Ford and Democratic opponent Jimmy Carter, and the exploits of Soviet cosmonauts aboard the Salyut 5 space station, the details were carried dutifully on the inside pages of most newspapers. The stories were brief, but prominently quoted an agency spokesman saying that NASA anticipated no difficulties if any women applicants made the cut. He explained that the Space Shuttle would be equipped to accommodate male and female crew, including a brand-new "waste management system"—a zero-gravity toilet—designed for use by both genders.

The agency had prepared an eight-page pamphlet describing the role of Mission Specialists, which laid out the most relaxed qualification criteria that NASA had ever permitted. The new category of astronaut required no existing flying experience, and only a single degree in physical or biological science, mathematics, or engineering. The stringent physical standards that still applied to pilot astronauts were also loosened for this new category of everyday spacefarer: although they had to be in excellent physical health, applicants would need neither perfect hearing nor vision. Both pilots and Mission Specialists would have to commit to two

years of training in Houston before final selection and—if successful—
would then be expected to sign on to serve as an astronaut with NASA
for a minimum of five years. For all candidates, the pamphlets explained,
the agency was embracing an affirmative action program: women and
minorities were "encouraged to apply."

While the announcement garnered little attention in the media,
Navy, Air Force, and Marine pilots and flight engineers at combat bases
and airfields around the world began studying the NASA rubric carefully; the call for new astronauts had, predictably, electrified the entire
brotherhood of American military aviators.

:::::::::::::::::

In his office at Edwards Air Force Base, thirty-seven-year-old Captain
Dick Scobee found news of the announcement buried on page 22 of the
Los Angeles Times, sandwiched between a story about California Governor Ronald Reagan's potential as a Republican candidate for President
and a half-page advertisement for JCPenney belted radials. He clipped it
from the newspaper and took it home to show his wife.

It was a day June Scobee had dreaded for years, and hoped would
never come. When she and her husband had met almost two decades before at a Baptist hayride in San Antonio, Texas, Dick was fresh from high
school and already in the Air Force. He told her he had long dreamed
of flying—but without the education necessary for pilot training, he had
taken a job on the assembly line near his home at Boeing in Seattle, before enlisting as a mechanic. Fixing the reciprocating propeller engines
of cargo planes on the flight line at nearby Kelly Air Force Base was as
close as he could get to aircraft without being an aviator.

When the couple married in July 1959, Dick was twenty; June just
sixteen. But with his new wife's encouragement, Scobee began taking
night school college courses, and secured a place in an Air Force educational program for enlisted men. In 1965, he graduated from the University of Arizona with a degree in aeronautical engineering, and was
sent for officer training—and flight school at Moody Air Force Base in
southern Georgia. Scobee was a natural, an intuitive pilot who finished

second in his class, giving him his pick of assignments. Now that he could finally realize his dream of flying fast jets, the pilot and his wife sat down in the kitchen; for the first time, they talked about the afterlife, and the possibility of death in the air. June had begun studying for a degree in education, and Dick explained his thinking about the future: because she would soon have a career of her own, if something happened to him he wasn't worried about leaving her alone without the means to support their two young children. June said that if he was concerned about the potential for a fatal accident, then perhaps flying wasn't for him after all. But Scobee wanted his life to have purpose: "Just let me go do what I really want to do," he told her. "I could die out here on this interstate highway in a traffic accident, and my life would have been wasted. Or I could make a difference for my country, and fly."

June finally relented. But instead of choosing to join the swaggering fraternity of high-performance jet fighter pilots, Scobee opted to become a "trash hauler," flying heavy Air Force transport planes around the world. In the cockpits of these lumbering giants, Scobee's odds of meeting an early death seemed much lower—even when, in early 1968, he was ordered to Southeast Asia, on a yearlong tour of duty flying cargo in and out of combat zones in Vietnam. Scobee returned home with a promotion to captain, and was awarded the Distinguished Flying Cross. Yet his old ambitions lingered. Every now and again, he still talked about his dreams of flying faster, and higher.

June was in the spring semester of her first year as a high school science teacher when Dick came home one day to tell her he'd been invited to attend Chuck Yeager's Air Force Test Pilot School in the high desert at Edwards. They talked once again about the risks, and the opportunities that might come with qualifying for experimental flying; they sweetened the prospect of the move for the children—Rich and Kathie, still both under ten—by telling them how close the base was to Disneyland. Two years later, Scobee began work as a test pilot at the NASA Flight Research Center in Edwards, and June took a job teaching English, Spanish, and science at the base middle school. In their spare time, the family embraced life in the wild Mojave landscape: Rich traded his bicycle for a pony, and on weekends Dick and June rode their motorcycles out among the tumbleweeds and creosote bushes, jumping dry creek beds and tearing along trails that wound through the low desert hills. One day, June took her Honda 90 up the steps of the school into the classroom, taking off her helmet and shaking out her hair in front of her astonished students, before announcing that the bike would be the subject of the day's writing assignment.

But Dick's flying—testing new aircraft, from experimental fighters to massive cargo transports—up to, and then beyond their known limits, often terrified his wife, who could see the planes take off from her classroom window. As the pilot of the modified Boeing 747 being developed for use as a replacement for Air Force One, Scobee was conducting an emergency runway test one day in 1974 when the brakes of the plane caught fire, and he scrambled from the cockpit as flames engulfed the tail of the aircraft. Then, in October 1975, Scobee strapped into the seat of a bizarre-looking wingless dart with a flat bottom and a bulbous cockpit that resembled a hypersonic steam iron: the X-24B. The last of the series of experimental rocket planes that had first flown from Edwards almost thirty years earlier, when Yeager himself had broken the sound barrier in the X-1, the aircraft was designed to test the landing characteristics of a spaceplane like the Space Shuttle—already under construction nearby—as it came gliding back to Earth from orbit. Scobee was well trained for his flight, with hours in a simulator and at the controls of other aircraft that could mimic the X-plane's characteristics. Even so, once June saw her husband leave the runway, his plane suspended beneath the wing of a B-52 bomber, she was so anxious she could no longer focus on teaching. Instead, she paced the hallway outside her classroom until the moment she could telephone Dick's office at the base, to make sure he had safely returned.

Her fears were hardly unwarranted: less than five months later, Dick's close friend and fellow X-24B pilot, Lieutenant Colonel Mike Love, was killed when he lost control of his two-seat Phantom and crashed on Rogers Dry Lake bed. Love's memorial service was the first time June ever witnessed planes flown in the missing man formation. Yet, for all her anxiety about her husband's role as a test pilot, June remained assured that at least he would never become an astronaut. At six feet two, Scobee was far too tall to meet the traditional requirements, which had called for men short enough to fit inside the cramped capsules of the Gemini and Apollo spacecraft. But when he returned from the office in the summer of 1976 with the clipping from the *Los Angeles Times* in his briefcase, it didn't take June long to realize that her confidence had been misplaced. Among the constraints that had been relaxed for the new intake of astronauts, the height restrictions had been lifted.

Eventually, it would seem as if almost every officer stationed at the high desert base in 1976 had put their names forward for NASA. Among them was Captain Ellison Onizuka, a thirty-year-old Air Force flight engineer who had graduated from the test pilot school just a year earlier—and lived with his wife, Lorna, and two young daughters in a small, one-story stucco house a few streets away from the Scobees. Onizuka helped devise and monitor research flights, often sitting behind his friend, test pilot Roy Martin, to gather data as they checked that the performance of new aircraft matched their design specifications. Although not qualified to take the controls himself, Onizuka soon logged more than 1,600 hours in the air, in forty-six different types of planes. Diligent and well organized, he showed little fear in the cockpit; he had faith in the pilots to whom he trusted his life on every mission. Sometimes, they allowed him to take the stick in level flight—or to execute "touch-and-go" landings, bringing the aircraft down on the runway at Edwards and lifting off again without ever coming to a halt.

Onizuka's grandparents had emigrated from rural Japan to Hawaii in the nineteenth century and settled on the Kona Coast as independent farmers. He was raised there with the traditional values of the model Japanese child: obedient, quiet, hardworking, and respectful of his family. But his pursuits were those of an all-American boy: he was a zealous baseball and basketball player, an Eagle Scout, and a member of the local 4-H club whose skill at plucking and dressing chickens outmatched all competition. He didn't like to lose.

As a teenager, Onizuka worked summers in his grandfather's coffee fields, and helped out in his parents' general store; but he had fantasized about going to space since he was small. Like others of his generation he had been a boy at the dawn of the space age, and grew up watching the heroes of Mercury and Apollo strive to best the Soviets in the race for orbit and then the moon. But when he mentioned his intention to one day become an astronaut to his scout leader, Mr. Sakata laughed at him. So, even when he left Hawaii to join the ROTC and study aeronautical engineering at the University of Colorado, Onizuka kept his dreams to himself. Instead, for years he had planned patiently for the day when he might submit his own application to NASA. When he finally did so, he didn't even tell his wife.

Six months after their new astronaut recruitment drive had begun, NASA chiefs in Houston and Washington, DC, received initial reports on the first wave of applications. They were alarmed to discover that their plan to remake the astronaut corps in the image of a diverse America had failed among the very groups they had hoped to reach. Of the 870 applications the agency had received to date, only 93 were from women, and just 30—as far as they could ascertain—were from minorities; of these, not a single one was considered outstanding.

In response, Chris Kraft and George Abbey launched a redoubled effort to help distribute the original announcement as widely as possible, and to personally contact women and members of minority groups across the country. Staff from Houston reached out to Black professional organizations, to scientific and technical groups and graduate schools; they sought help from the Association for Women in Science and the American Association of University Women, and publicized the application process through magazines including *Ebony*, *Essence*, *Jet*, and *La Luz*; finding names in the pages of *Who's Who among Black Americans*, *Blacks in Physics*, and *Black Engineers in the United States*, they mailed individual letters to those who seemed qualified to become astronauts. Month after month, members of the Astronaut Selection Board went out on the road to make direct appeals to the public at schools, colleges, social clubs, and even retirement communities; finally, at the beginning of 1977, the agency engaged the actress Nichelle Nichols, who had appeared on *Star Trek* playing communications officer Lieutenant Uhura, to help spread the word.

Nichols toured the country, visiting TV stations and groups of Black engineers alike, to pitch the new astronaut program directly, and appeared in a five-minute recruitment film about the future of NASA alongside astronaut Alan Bean. Clad in a blue shuttle flight suit, Nichols sat at a Mission Control console at the Johnson Space Center and told viewers about the new vehicle that would "put us in the business of space—not merely space exploration . . . built to make regularly scheduled runs into space and back, just like a regularly-scheduled airline."

The actor's script pitched the astronaut program to the broadest imaginable audience, inviting applications to join an agency that now aspired to the kind of diversity previously glimpsed only in science fiction.

"The shuttle will be taking scientists and engineers, men and women of all races, into space, just like the astronaut crew on the starship *Enterprise* . . . the whole family of humankind," Nichols said. "This is *your* NASA."

When Ron McNair found the NASA prospectus in his mailbox at Hughes Research Labs in Malibu one day in early 1977, he wasn't even sure why he'd received it. A twenty-six-year-old staff scientist with a PhD in laser physics less than a year out of MIT, each day McNair basked in the sun-kissed California lifestyle. Every morning he rose early to jog down the beach from his apartment in Marina del Rey to the boardwalk in Venice Beach, and then drove north up the Pacific Coast Highway toward Hughes, a complex of modernist concrete and glass nestled in the canyons high above Malibu, and so futuristic it was often used as a location by Hollywood movie producers.

McNair had already come a long way from Lake City, South Carolina, where he had grown up as the second of three brothers in a rundown wooden house on the poor side of the railroad tracks that divided the Black part of town from the white. As a boy, Ron and his older sibling, Carl, had spent summer days stooped in the fields that surrounded the small town, cropping tobacco leaves for $4 a day, sometimes cooling off in the irrigation pond behind the Pine Bay Dairy. The experience helped convince both boys to stay in school and get an education that would allow them to make a living some other way, despite the many obstacles of segregated schooling in the Jim Crow South of the 1950s—where tattered textbooks were passed down when white schools had finished with them and improvisation was the norm for both students and teachers. Whether at the counter of the town drugstore or in the whites-only section of the cinema, everyone knew their place; Lake City was a Klan town, subject to intermittent spasms of terror. One night in 1956, Walter Scott, the local president of the NAACP, lay with his family on the floor of their living room as shotgun blasts peppered the front door and shattered the picture window over their heads; meanwhile, large crosses were daubed in paint on the walls of the gas station where he worked.

The two older McNair boys were born just ten months apart, so close in age that people often took them for twins—but Ron, gifted with numbers, soon outstripped his older brother in the classroom. Encouraged by his father, an auto-body repairman in town, and his mother—a teacher who, for six years, made a regular two-hundred-mile round trip to South Carolina State University to get her master's in education—Ron was consistently at the top of his class. He was often frustrated that the pace of learning was too slow for him. At the age of nine, McNair inadvertently integrated the local library, after the librarian called the police rather than allow Ron to check out the book he wanted, a privilege reserved for white children. "I'll wait," he replied. When the cops arrived, they refused to intervene, and at last Ron was allowed to take home the book. Growing more aware than his older brother of the disadvantages that they were facing, McNair wrestled with a volcanic temper that only began to subside as he arrived in high school. There, he learned to channel his rage into intensive study, and football.

As a teenager, Ron appeared to have all the trappings of the stereotypical nerd: he was slight and bespectacled, with a shirt pocket constantly filled with pens; while playing basketball over at the McNairs' house one day, a friend looked up to see Ron measuring the shadow of a nearby telephone pole to calculate its height, using his slide rule. But McNair's appearance belied an innate athleticism. He thrilled at the aggression and physical contact on the football field, was named MVP by his coach and eventually captained the Carver High team; but he also excelled at baseball and track. In the meantime, Ron and Carl had both joined the school band, where they shared a single clarinet until the band instructor found Ron a saxophone—worn-out and held together with tape and rubber bands, yet playable nonetheless. The boys from the horn section learned chart hits by James Brown and the Temptations, and at recess took requests in the schoolyard. By the time he graduated from Carver High in the spring of 1967, McNair had become an intensely focused jazz musician whose playing was marked by fluid runs and improvisation on his instrument.

Although he received an offer to attend Howard on a football scholarship, McNair chose to enroll at another historically Black college, North Carolina A&T State University, to study physics. It was not easy; for all his gifts, what he had learned at Carver High did not prepare him to compete with fellow students arriving from better-resourced schools in New York, Chicago, and Washington, DC. "I'm not even in the running," Ron told his brother Carl; after just one week, he was already falling behind. In despair, he decided to abandon his dream of one day garnering a doctorate in physics and switched his major to music. But a counselor reviewed McNair's grades and school transcript, and convinced him to take remedial math and physics classes to catch up, and stick with his ambition. "You're good enough," she told him.

During his first year at college, McNair began studying karate. Martial arts were approaching a cultural high watermark in the United States, and at the start of the semester two hundred students were enrolled in the college karate class. By the end of the semester, there were just five left, including McNair, who eventually became a black belt, and by the time he left A&T he had begun teaching a karate class of his own. He graduated magna cum laude in 1971, and was admitted to the doctoral program in physics at MIT on a four-year Ford Foundation fellowship. His professor at A&T wrote to the review board, "If you can't give the Ford Foundation fellowship to Ron McNair, you can't give it to anyone." The work McNair faced at MIT was the toughest yet: specializing in laser physics, he spent long hours in the basement lab he referred to as "the dungeon," but still found time for karate; eventually he would publish a paper in *Scientific American* on the scientific principles of the sport, illustrated with photographs of the author smashing wooden boards and blocks of concrete with his bare hands. McNair also continued to attend church—a central part of his life since he was a boy in Lake City—and went every Sunday to St. Paul AME Church, close to the MIT campus in Cambridge. There, in the fellowship hall, he organized a karate class that attracted both experienced martial artists and elementary school children who had never tried it before; to his delight, other members of the church began to call him "Minister of Defense."

In Boston, McNair was shocked to encounter racism of a type he had never expected to find in a supposedly enlightened northern city he regarded as "the cradle of liberty": strangers muttered hateful epithets behind his back on the street; on one occasion a gang of white men chased him from practicing karate in the park; his brother Carl's car was stolen, vandalized, and abandoned, left with racial slurs etched into the bodywork. At MIT, McNair was among only a handful of Black students scattered across the campus, and some of his fellow scholars clearly felt that affirmative action had granted him a place that might have gone to a better-qualified white candidate. He fell back on the lesson that both his parents and teachers had instilled in him as a child: if being Black meant that he had to work harder than his white peers to prove himself, then that is what he would do—and without complaint.

McNair was still in his first year in Cambridge when he met Cheryl Moore, a young elementary school teacher from Queens, New York, studying nearby for a graduate degree in education. Introduced to him at a church potluck for singles, Moore had already heard about the karate exploits of the Minister of Defense—but was struck by how gentle he seemed, for someone who could do such violence with his hands, and felt charmed by his scientific view of the world around him; eighteen months later, they were married. The couple honeymooned in Bermuda, where they stayed with friends of their church minister, and toured the island on a motorcycle.

Back in Cambridge, Ron took Cheryl to his lab late one afternoon to show her an experiment he was working on, and to retrieve his notes: a collection of notebooks and computer punch cards charting years of doctoral work, stuffed into an old army duffel bag that he rarely let out of his sight. But as they left the building, the couple stopped to admire the evening view across the Charles River. Ron put the duffel down beside his car, and walked around to where Cheryl stood, gazing at the lights of Boston, glittering in the gathering dusk. By the time McNair returned to the driver's side of the car, it was dark—and the bag had vanished. Perplexed, he returned to the lab in case he'd left it behind, but it wasn't there. The duffel bag, and with it the data from the hundreds of experiments he had conducted as part of his PhD research, had been stolen. The couple sat in the car in silence for a few minutes as the impact of the loss set in: McNair would have to repeat years of laborious and detailed tests; his doctorate, until a few moments ago less than twelve months from completion, was slipping from his grasp. Cheryl waited in trepidation for an outburst of rage and despair that never came. At last, she summoned the courage to ask Ron what he was going to do.

"I don't know," he said. "I'll figure it out."

Refusing to quit or request more time to complete his doctorate, McNair gathered what data he had remaining—from papers left in the lab, or those held by his professor—and systematically began to rerun his experiments. Working around the clock, he repeated three years' worth of research in just three months. In the summer of 1976, Dr. Ronald E. McNair graduated from MIT with a PhD in laser physics; he was twenty-five years old. Later that year, he and Cheryl decided he should take the job at Hughes in California—where, as a staff physicist, he developed lasers for use in isotope separation and for optical communication between satellites in space—in part because they were tired of the harsh winters in Boston. The couple moved into a condo overlooking the ocean in Marina del Rey, and Cheryl began teaching at an elementary school near the Burbank Studios. Ron had been working at the laboratories in Malibu for only a few months, when he came home one day with the NASA prospectus he had found in his office mailbox seeking Mission Specialists for the Space Shuttle program, and asked Cheryl what she thought.

"Well, what do *you* think?" his wife replied. "You've certainly got the credentials. . . . You could do it easily."

"Yeah," he said. "I think I'll fill it out."

McNair completed the initial application, and put it in the mail. Then he telephoned his older brother.

"I'm going to be an astronaut," he said.

"Sure," Carl replied. "And I'm going to be the Pope."

::::::::::::::::::

Judy Resnik, a talented classical pianist who graduated first in her electrical engineering class at Carnegie Tech, filled out her NASA application in the summer of 1977, while lying on her stomach in the sand at Bethany Beach in Delaware, where she shared a vacation house with friends. "You've got to be kidding," one of them said when she saw it. Resnik was twenty-seven years old, and about to earn a doctorate for her research in biomedical engineering at the National Institutes of Health—but was also going through a difficult divorce and searching for a new purpose in her life.

As a teenager, Resnik had been deeply wounded by the disintegration of her parents' own marriage. When she and her younger brother, Charles, were assigned to the custody of her mother, Judy challenged the decision in domestic court, and went to live with her father, creating a rift that never healed. But she was a gifted student. At Firestone High School in Akron, Resnik's teacher in advanced math noticed that her papers were written with astonishing precision, as if they had been printed by a computer. He always checked her work first, to make sure that his own answers were correct; the yearbook photograph of the Firestone Math Club shows fourteen boys, and Judy. Arriving at Carnegie in

1966, she had avoided the campus protests against the war in Vietnam to concentrate on her studies. She skipped classes on Jewish holidays, except for those with scheduled tests, which she attended with some guilt—but tried to avoid being identified by her religion, or her gender. Before making the application to become a Mission Specialist, Resnik had never evinced any particular interest in the space program, but with her academic gifts and experience of applied science—including a job at RCA, where she designed radar control and rocketry circuits—to her friends NASA seemed an obvious fit.

"Daddy, guess what? I've applied to NASA," she told her father over the phone.

"Good," he said. "So you'll become an astronaut."

Resnik prepared for her application with typical thoroughness: she interviewed former Mercury astronaut—and now senator—John Glenn; visited the newly opened Air and Space Museum in Washington, DC, to track down Apollo 11 pilot Michael Collins, who had taken a job there; and devised a new regimen of nutrition and exercise to meet the demands of the physical. At first, she believed she had little chance of making the final cut, but soon learned she was through to the last two hundred of the more than one thousand women who had applied. Meanwhile, she finished her PhD and took a new job at the Xerox corporation in California, and in her spare time she began studying for a pilot's license. Each time Resnik passed another professional milestone, she sent a telegram to NASA to let them know.

In the end, nearly 25,000 men and women requested a copy of the Space Shuttle astronaut application from NASA. When the deadline passed on June 30, 1977, around 8,000 had returned completed documents—the majority of them hoping to become Mission Specialists. Using a points system and a program of background checks and references, George Abbey and the other members of the Astronaut Selection Board gradually narrowed the initial applicants down to 208 finalists, of whom just 21 were women; over three months in the autumn of 1977, they were invited, twenty at a time, to the Johnson Space Center in Houston for a week of evaluation. On November 7, it was Judy Resnik's turn: her instructions were to bring gym clothes, and avoid food, caffeine, and alcohol for twenty-four hours before the date of her physical exam. When her stepsister Sandy Vilseck arrived to drive her to LAX for the flight to Texas, she found Judy was a nervous wreck, afraid that—even after her months of diet and exercise—she would fail the physical.

But the infamously grueling and undignified medical trials that had once been used to prepare previous astronaut groups for the then unknown rigors of spaceflight had long been abandoned by NASA doctors. The Space Shuttle promised a gentle ride into orbit in a shirtsleeve environment; hurtling centrifuge rides and swallowing a rubber hose were no longer necessary. Instead, Resnik took a timed run on a treadmill and submitted to a urine test. Over the rest of the week, she was also given a tour of the facilities at the Space Center, briefings on the shuttle, and attended interviews with two separate psychiatrists; of these, one asked friendly questions, while the other was more inquisitorial, in an attempt to screen out those candidates who might react badly to stress. Another test involved zipping the applicants into a large fabric ball designed to ferry astronauts through the vacuum of space in the event of an orbital emergency: curled tightly in the darkened sphere, none of the candidates was told how long the experience might last. Each handled the claustrophobia in their own way; some fell asleep.

A final hour-long interview, with Abbey and the other members of the Selection Board, was designed, in part, to establish how well each potential astronaut would represent the agency in public. But many felt that there was never a moment in Houston when they were *not* being assessed, so they did their best to behave with rigor and purpose at every turn, even when walking from place to place across the grounds of the Clear Lake campus.

Abbey was looking for individuals who could work well as part of a team, but also generalists who pursued a skill—a musical instrument, or a sport—in addition to their professional discipline. He wanted men and women who were "artists as well as technicians, athletes as well as intellectuals"—a balanced and integrated group of astronauts, not the troublesome collection of competitive mavericks who had first tested the boundaries of space.

But to those he was selecting, Abbey remained as inscrutable as ever. One candidate—George "Pinky" Nelson, a twenty-six-year-old astronomer and formerly a gifted high school athlete once scouted by the Minnesota Twins—had his psychiatric interview unexpectedly cut short, and was taken directly to the mound during a softball match. Told to pitch for the astronauts' team, which Abbey himself coached, Nelson hit two home runs. "I figured that was my real interview," the astronomer said later.

When the day eventually came to notify the members of the first successful new group of astronaut applicants in more than a decade, George Abbey made every one of the calls himself, from his desk in Building 1. Characteristically oblique, once he had the candidates on the line, the Director of Flight Operations often chose to creep up to the subject, bewildering some with small talk about the weather before apparently absentmindedly asking them if they were still at all interested in becoming astronauts. Only one said no. There were thirty-five of them in all: fifteen pilots and twenty Mission Specialists, including six women and three African Americans.

Just before six o'clock in the morning on Monday, January 16, 1978, Judy Resnik was preparing to leave for work at Xerox from her apartment on Redondo Beach Boulevard, when the phone rang. It was Abbey. Afterward, Resnik began calling her friends and family to pass on the news: "I did it," she said. That night, she went out for dinner in LA with two fellow candidates whom she had befriended during the interview process, the emergency room physician Anna Lee Fisher—who had been selected—and Fisher's husband, Bill, who had not. A hundred miles farther north, Ellison Onizuka was already on the bus to his office at Edwards when his wife, Lorna, saw a car pull into the driveway of their house, bringing two officials looking for her husband. The men set off in pursuit of the bus, but lost it. By the time Captain Onizuka finally reached the base, many of his colleagues had already heard from NASA that they hadn't been selected; but the phone in his office was ringing.

With Abbey's call, Onizuka became the first Asian American ever chosen for astronaut training. But his celebrations were muted by the disappointment of his friends, among them Roy Martin, the pilot Onizuka had sat behind on so many test flights. Martin was crushed by his failure to be chosen. "I'm sorry," said Onizuka as the two men shook hands, his eyes brimming with tears. And Lorna was terrified by how dangerous this new assignment might be: flying into space seemed far more hazardous even than test-flying new aircraft. Onizuka tried to reassure her.

"Everything's a risk," he said gently.

The news was officially released to the press at 1:00 p.m. Houston time that day. The six successful female candidates—Judy Resnik; Anna Fisher; astrophysics student Sally Ride; Kathy Sullivan, studying for a doctorate in marine geology and geophysics in Nova Scotia; surgery resident Rhea Seddon; and the biochemist and mother of three Shannon Lucid—provoked a frenzy of media attention and were deluged with often inane questions from reporters. When all thirty-five members of the new class gathered onstage at the Johnson Space Center for a press conference at the end of January, Dick Scobee, wearing a gray checked suit and tie, took a seat in the dead center of the front row, legs spread wide in the posture of masculine authority. But the white men in the group quickly discovered they were of little interest to the assembled media, and many sat around the NASA auditorium patiently waiting for interview requests that never came.

Over the course of a week of induction and orientation, the astronaut candidates were issued security badges, posed for their first official portraits, and drove out to the NASA hangars at Ellington Field, a few miles north of the Space Center. There, they were measured for the flight suits and helmets they would need to begin flying in the agency's T-38 two-seat jets, used by astronauts for both training and travel between Houston, Cape Canaveral, and other NASA facilities around the

country. Many of the candidates also went house hunting, in preparation to move permanently to Clear Lake in time for the formal start of
their training, set for the beginning of July 1978.

NASA tradition dictated that the new class of astronauts choose
a name for themselves; at the suggestion of one of the military pilots,
they settled on the "TFNGs." Officially, the initials stood for Thirty-Five
New Guys. But the acronym was already familiar to those among them
who had served in Vietnam, echoing the one applied to newly arrived
replacement personnel whose life expectancy was regarded as so short
that their individual names weren't worth learning. They were simply
the FNGs. The Fucking New Guys.

As part of their induction, the TFNGs would eventually be divided
into groups and given a tour of Mission Control. There, they were handed
headsets to help them more fully understand the role of the CapCom,
or capsule communicator: those astronauts who remained in Houston
during the course of each mission to relay information over the radio
to their colleagues in space. When one group's demonstration was complete, the veteran astronaut in charge of the tour told them to keep their
headsets on. "Roll the audio," he said to a technician. What each member
of the group heard next, fragmented and hissing with static, were the
final minutes of the console recording from January 27, 1967: the day
Chaffee, Grissom, and White died on the launchpad at Cape Canaveral.

"We've got a fire in the cockpit!"

"Get us out of here!"

"We're burning up!"

⁑⁑⁑⁑⁑⁑⁑⁑⁑

Meanwhile, the vehicle that was supposed to be carrying these new astronauts into space was on the verge of becoming a national embarrassment: overdue, over budget, and still years away from launch. On
his arrival in the White House almost exactly a year before, President
Jimmy Carter had taken office wondering whether it might be best to
shut down the project altogether.

CHAPTER SEVEN

NEVER A STRAIGHT ANSWER

It was a blazing afternoon in the Antelope Valley as the technicians rolled out the first Space Shuttle onto the concrete runway of Rockwell's Plant 42 on September 17, 1976. Although towed by a tractor painted with a Stars and Stripes design and a curling 76, a part of the national celebrations to mark the bicentennial of the founding of the republic, the orbiter was no longer known as *Constitution*. Following a fervent write-in campaign from more than 100,000 *Star Trek* fans, the White House had buckled just the week before to their demands that the world's first reusable spacecraft be christened in honor of their favorite TV show. So the name *Enterprise* now stood out in fresh, black sans serif letters along the vehicle's fuselage. Among the crowd of almost one thousand people waiting to greet the shuttle stood many of the principal cast members of the science fiction drama, dressed in bell-bottom pantsuits and flashy turtlenecks, posing for pictures with NASA Administrator James Fletcher and Florida congressman Don Fuqua. The shuttle shone blinding white in the desert sun, its stubby delta wings and the twin doors of its boxcar-shaped payload bay tessellated with the pattern of the heat-insulating tiles intended to safeguard its return from orbit. An Air Force band played the *Star Trek* theme. "This day," Fletcher announced, "we're about to enter a new era."

In reality, the *Enterprise* was no more capable of flying in space than its imaginary namesake: it had no rocket engines, no life support system, radar, or maneuvering thrusters; its wings were partly fiberglass; and most of the apparently heat-resistant tiles covering its skin were made from molded Styrofoam. A full-scale test vehicle destined to make only atmospheric glide trials, this first orbiter was a sophisticated mock-up. The shuttle vehicle intended to put the United States back into the business of manned spaceflight, *Columbia*, remained an incomplete hulk hidden in a hangar nearby. And although the first voyage of the new spacecraft was still officially scheduled for more than two years away, in the spring of 1979, even this target seemed increasingly out of reach.

When he had first announced the inception of the Space Transportation System from President Nixon's Winter White House in San Clemente at the beginning of 1972, James Fletcher had reckoned on needing six years to get into orbit, with a launch date in 1978. But he soon learned that the financial guarantees of the Nixon administration were no more reliable than the word of the president himself. Inadequate at the outset, the budget promised to build the shuttle was struck by the first of a series of cuts: there wasn't enough money to build the engines, or to fully maintain standards of quality control; the initial dream of a fleet of a dozen orbiters was scaled down to six, then five. The expected launch date slipped back a year. As funding threatened to run dry, the agency was forced first to extend its deadlines to match limited annual budgets, and then to bleed money from its scientific and unmanned programs to keep the shuttle going. At the same time, the development plans for each of the engineering challenges involved—which NASA managers had based on the appealing and financially expedient expectation that their work would proceed without a single failure or setback—had proved to be wildly overoptimistic.

The shuttle's main engines were a problem almost from the beginning. The design called for three in total, powered by a volatile mix of liquid hydrogen and oxygen. The orbiter's reusable design meant that each engine had to fly into space, gimbal and throttle precisely under computer control during ascent, survive the stresses and shocks of launch and reentry, stop and start with 100 percent reliability, burn continuously for eight and a half minutes—and then be ready to do it all again just a few days later. NASA told the contractor, California-based specialists Rocketdyne, that each engine had to be good for at least one hundred missions—but also capable of taming some of the most destructive forces ever harnessed by mechanical engineering. Above all, they had to be small, yet so powerful that their power-to-weight ratio would exceed that of any engine in history; this combination of compact size and astonishing power presented the engine designers with challenges that often seemed insurmountable.

In order to fly in the oxygen-starved environment of the upper reaches of Earth's atmosphere and in the vacuum of space, all rocket engines operate along a similar principle. The internal combustion engines used to drive cars and the first airplanes, or the jet engines that power modern aircraft, burn gasoline or kerosene in the oxygen sucked into the mechanism from air in the ordinary atmosphere. But as aircraft ascend higher, the air becomes thinner and, like the lungs of mountain climbers growing breathless at high altitudes, their engines become starved of oxygen in which to burn fuel; if they fly too high, combustion ceases altogether: the engine cuts out and the aircraft falls out of the sky. To reach space, a rocket engine must not only have fuel, but also bring along its own oxidizer, in order for combustion—and powered flight—to continue.

The shuttle was designed to carry 390,000 gallons of liquid hydrogen fuel and 145,000 gallons of supercooled liquid oxygen, held at temperatures approaching absolute zero inside its giant external tank. The largest of the three main parts of the shuttle—taller than the Statue of Liberty, and weighing almost as much at liftoff as four diesel locomotives put together—the pencil-shaped tank provided the system with its structural spine; attached to the orbiter and the solid rockets by bolts parted by explosive charges on the way to orbit, it contained the hydrogen and oxygen in two pressurized compartments. Separately, the frozen gases were each highly flammable—and if prematurely combined in the presence of the tiniest spark, the resulting explosion would not only destroy the shuttle and everything in it, but level the launchpad and all of its supporting equipment for hundreds of yards in every direction.

Drawing the fuel and oxidizer into each of the main engines was a pair of high-pressure turbopumps designed to force the two liquids together into a combustible mixture burning at 6,000 degrees Fahrenheit—hotter than the boiling point of iron. The engines consumed fuel so voraciously that the pumps were capable of draining the entire contents of a backyard swimming pool in less than half a minute. But size and weight limitations meant that each pump—driving turbines spinning at more than 500 revolutions a *second*, generating pressures of almost 450 atmospheres, enough to shoot a jet of liquid hydrogen thirty-six miles into the air—was no larger than a kitchen trash can. The parameters of other components were no less extreme: the shuttle would eventually have all three engines clustered together in its bulbous tail, and each one would require continuous cooling by its hydrogen fuel to avoid melting under the heat of its operating temperature.

Yet at the Rocketdyne plant in California and on the test stands of NASA's Rocket Propulsion Test Complex in the swamps of southern Mississippi, one after another of the experimental engines caught fire, melted down, or exploded after only seconds of operation. To save money, the agency had opted to conduct "all-up testing," which meant that instead of conducting trials of individual components, they all had to be tested together as part of a complete engine, by firing the entire system—with often catastrophic results. The rocket team destroyed one unit after another, and soon began to run short of engines to test. The process was not cheap; each one cost at least $40 million.

· · · · · · · · · · · · · · · · · ·

The highly sophisticated heat-insulating tiles had proved just as troublesome. Early in the shuttle design process Max Faget and his engineers in Houston had committed to building the airframe of the orbiter—its metal skeleton and skin—from aluminum. This was light, malleable, and cheap, but, unlike the titanium used in top secret spy aircraft and earlier spaceplane designs, it could not withstand the extreme temperatures of hypersonic flight. At just a few hundred degrees Fahrenheit it began to lose structural strength and, if exposed directly to the heat generated by the friction of reentering Earth's atmosphere at more than twenty-five times the speed of sound, an aluminum airframe would melt like butter and catch fire; the orbiter would break up and disintegrate in flight.

The heat protection system the engineers devised to prevent this happening was even more untried and exotic than the engine technology. The areas of the shuttle expected to face the most extreme temperatures—the nose cone and the leading edges of the wings—would be covered by panels of reinforced carbon-carbon; but most of the protection would take the form of a covering of silica-fiber tiles made from purified Minnesota sand, glued to the surface of the spacecraft. Manufactured in a dedicated plant built by Lockheed, the tiles began life as silica mixed with water, pressed into rectangular blocks six inches thick, dried in a microwave oven, and then kiln-baked for two hours at more than 2,000 degrees Fahrenheit, before being cut into cubes. Each cube was composed of more than 90 percent air, and—despite its solid appearance—felt unexpectedly light, like pumice or Styrofoam. Once it was coated with a thin, hard shell of borosilicate glass, this lack of density gave the material extraordinary properties of heat absorption and insulation: Lockheed technicians liked to demonstrate these by placing a sample in a furnace until it was white-hot, removing it with tongs and allowing the surface to cool for a few minutes. At that point, a volunteer could pick the sample up by its edges using their bare hands—even as the inside of the block remained searingly hot; in the same way, the insulation could absorb the high temperatures encountered by the orbiter on reentry, keeping the heat inside the tiles and protecting the aluminum airframe beneath. Yet, although the miracle substance was apparently rugged enough to survive at least one hundred trips into Earth orbit and back, the tiles proved delicate—and attaching them to the skin of the new spacecraft proved even more difficult than expected.

The shuttle designers had initially intended that the cubes be cut into identical rectangles, varying only in thickness, and trimmed to size for installation—but soon discovered that fitting flat tiles to the complex curves and aerodynamic contours of the orbiter's fuselage was impossible. Lockheed was forced instead to embark on an ambitious plan to manufacture some 34,000 tiles, every one of them unique. Each one would have to be specifically shaped to fit a single place on the skin of the shuttle and milled by computer-controlled machines programmed with dimensions provided by the spacecraft designers at Rockwell. Depending on their location and how much heat they were expected to withstand, the resulting pieces might be thin white tiles six inches square, or black bricks as much as eight inches across and three and a half inches thick. But every one had to be measured and precision-fitted by hand, to tolerances measured in thousandths of an inch. "It is a horrendous job," the project manager at Lockheed said after the decision was made. "The manufacturing people were appalled."

Even then, shaping and fixing the tens of thousands of delicate tiles in place proved even more complex than the designers had imagined. If they were placed too close together, the tiles could rub against one another in flight and break up; too far apart, and hot plasma could find its way through to the skin of the orbiter during reentry. The hard outer coating of the insulation was brittle and only sixty-thousandths of an inch thick: an accidental blow from a wrench or a key chain was enough to crack the glass shell; when that happened, the damaged tile had to be removed and replaced with a newly manufactured one.

:::::::::::::::::::

President Jimmy Carter arrived in office in January 1977 facing an energy crisis and a collapsing economy, and determined to rein in federal spending using similar methods to those he'd applied as Governor of Georgia. And although a supporter of space science—he had trained as a nuclear engineer in the US Navy, and had a good grasp of technical concepts—Carter saw no real need to keep sending astronauts into orbit; robotic spacecraft could explore the solar system more safely, and cheaply. He believed that the Space Shuttle was a boondoggle contrived simply to keep NASA alive—and was furious to discover that the project had been in a financial hole since its inception. And now the bills were all coming due at once.

Carter's new NASA chief, Administrator Robert Frosch, was a gifted scientist with a background in Navy research and development and oceanography. But he had no experience in aerospace and—unlike his predecessors—was not bewitched by the heroic quest of human spaceflight. Center Director Chris Kraft and other senior figures at the Johnson Space Center regarded him with suspicion.

A blunt New Yorker born in the Bronx, Frosch had taken the NASA job almost by accident: he and his family spent their summers in a house on Cape Cod next door to Frank Press, Carter's science adviser, who one day casually suggested that Frosch might like to join the government as head of the National Oceanic and Atmospheric Administration, or NOAA. But when Frosch arrived in Washington, DC, in the spring of 1977 as part of the group of new appointees meeting at the White House, Press took him aside and explained that the NOAA job was no longer available; he'd had to give it to Richard Frank, a major contributor to the new President's election campaign. "He's going to be running NOAA; end of story," Press said. "Would you mind NASA?"

When Frosch accepted the job, the first question Press asked him was "Should we cancel the Space Shuttle?"

In 1978, as the shuttle program continued to be dogged by embarrassing delays and cost overruns, Frosch flew down to Houston for a meeting with Chris Kraft and all of the senior shuttle managers. Seated in the large conference room on the ninth floor of Building 1 at the Johnson Space Center, Kraft presented Frosch with a stark choice: either scale back the shuttle into a pure research project, as the X-15 rocket plane had been, and abandon the grand plans for flying into space on an airline schedule, or beg the President for the extra money necessary to finish the revolutionary spacecraft they had imagined.

Eventually Frosch would ask Carter for two separate bailouts, almost half a billion dollars in total, and only the intervention of Defense Secretary Harold Brown dissuaded the President from axing the shuttle altogether. Carter agreed to grant NASA the money, but as a result the shuttle emerged from the budget crisis cemented as a key asset in Pentagon plans for conducting spy missions from orbit. Later, when meeting granite-faced Soviet leader Leonid Brezhnev to discuss the SALT arms reduction treaty in Vienna, Carter would explain that the new US spacecraft would be used to monitor the number of nuclear missiles deployed across the USSR. Nonetheless, the Office of Management and Budget cut down the size of the planned fleet of orbiters once again, to four. In the meantime, Brezhnev, convinced by his generals of the security threat represented by the new US technology, had signed off on a program to develop a Soviet space shuttle—known as *Buran*, or blizzard.

Even as the future of the shuttle program hung in the balance, the thirty-five members of The Fucking New Guys—or, more formally, NASA Astronaut Group 8—were arriving in Houston for the start of their training, due to commence not later than Monday, July 10, 1978. Judy Resnik was the first to begin: in May, she was already taking instructional flights in the back seat of a T-38 out at Ellington Field. Dick Scobee arrived the following month, at the end of a 1,500-mile journey across the country from California on Interstate 10, in a two-car convoy carrying his wife, their two teenage children, a cat, a dog, a motorcycle—and, in a trailer of his own design towed behind the family Pontiac, the half-finished carcass of a two-seat light airplane he and his son had built from scratch on the porch of their house in Edwards. The kids, who had already told their friends that their father was an astronaut, teased Scobee that their arrival looked more like that of *The Beverly Hillbillies*. "Well, I'm not really an astronaut yet," he said: until his two years of training were complete, he would officially remain an astronaut candidate or, in the inelegant jargon of the acronym-fixated agency, an AsCan.

The Scobee family moved into a three-bedroom Texas-style ranch house in Oakbrook—a new subdivision a ten-minute drive from the Space Center built by developers keen to capitalize on the arrival of the latest generation of American heroes to settle in Clear Lake City. But the local realtors had been shocked to discover that the demographics of race and gender were not the only way in which the TFNGs were unlike the astronauts who had preceded them. The star power and public adulation that surrounded the Mercury Seven and the dozen men who walked on the moon had granted them perks unavailable to most test pilots—including brand-new sports cars on dollar-a-year leases, cut-price home loans, and a joint publishing deal on their life stories that provided each one with a generous annual income.

But the men and women of Astronaut Group 8 had to make do with regular civil service salaries, paid on a scale calculated according to their existing earnings and seniority; Anna Fisher, who had been working as an emergency room physician in Los Angeles, and who would arrive for her first day at work at the Johnson Space Center behind the wheel of a freshly purchased two-seat Porsche, chose to accept a 75 percent pay cut as the price of her opportunity to fly in space; the Scobees, after a decade of living in subsidized military housing, were relatively well-off because of Dick's seniority, but took on an enormous mortgage to buy their new home. Driving them around a selection of new houses near the Space Center, one realtor had asked Mike Mullane and Dick Covey, two of the new group joining NASA directly from the Air Force, about their income. They explained to her that, as they had both recently been promoted to major, they each earned around $25,000 a year. The realtor stamped on the brakes. As the car jerked to a halt, she turned to them in astonishment.

"A *welder* makes more than that!" she said.

At the Johnson Space Center, the AsCans were divided into pairs and assigned their own offices, on the top floor of Building 4, adjacent to the main administration building occupied by George Abbey and the other senior NASA managers. With humming fluorescent lighting, linoleum floors, and government-issued wooden furniture, the rooms were bleakly utilitarian, with prefabricated steel walls that could be moved around to change the size of the spaces as necessary; the scent of floor polish, bad coffee, and ancient cigarette smoke lingered in the hallways. With the arrival of the thirty-five new candidates, the Astronaut Office more than doubled in size; they joined twenty-seven veterans of the Apollo and Skylab programs, the most-recently arrived of whom had been admitted to NASA back in 1969.

Among these men were two who had walked on the moon: Alan Bean, and Chief Astronaut John Young—who had made two lunar voyages and spent longer in space than any of his serving colleagues; but there were others who had trained for the Apollo program and seen their dreams snuffed out with the cancellation of the final moon missions. Many of the old guard could be forgiven for regarding the new intake of astronauts, and the competition they would soon represent, with suspicion and resentment. By the time their fresh-faced colleagues joined them in Houston, some—already in their late forties, still walking around the Space Center in plaid jackets and Sansabelt slacks, which had seemed at the apex of fashion during the first Nixon administration—had spent more than twelve years waiting for their first opportunity to fly in space.

:::::::::::::::::

At eight o'clock in the morning on July 10, John Young welcomed the AsCans to Room 3025, the large windowless conference space on the third floor of Building 4, for their first meeting of the Astronaut Office: the administrative ritual that from then on would take place every Monday throughout their time at NASA. Slight and dark-haired, the Chief Astronaut sat at the head of the large table and mumbled a greeting: for all his heroic achievements, first as a Navy test pilot and then in space, in some ways Young, at forty-seven, was the most extreme embodiment of the laconic astronaut archetype. He rarely used two words if one would do, was diffident to a fault, had difficulty maintaining eye contact in conversation, and displayed deep discomfort with almost any kind of public speaking; he was nobody's idea of a natural leader.

But beneath Young's remote demeanor was a rigorous technician with a merciless eye for detail—often revealed in his long and bluntly critical internal memos, already infamous within NASA—and an arid sense of humor. Though born in San Francisco, Young had been raised in Orlando, and spoke with the country twang and down-home idiom of Central Florida; he sometimes smoked a corncob pipe, and often referred to anyone from outside the Astronaut Office as "them boys." When, after his second spaceflight, aboard Gemini X in 1966, the City of Orlando chose to name a new section of state highway John Young Parkway in his honor, the astronaut expressed his feelings with typical economy: "Them boys shouldn't have done that," he said. "I ain't dead yet."

The focus of the morning's meeting was the preparations for the maiden flight of *Columbia*—known as OFT-1, or Orbital Flight Test-1— which was still officially scheduled for June 1979. George Abbey had already chosen Young to command the mission, and the forty-year-old Navy commander Bob Crippen, who had joined NASA from the abandoned Manned Orbiting Laboratory program in 1969, but had yet to fly in space, as his copilot. Crippen, a dark-haired Texan with a vulpine grin who had worked on the shuttle project almost since its inception, was a member of Abbey's inner circle of friends. After a fellow pilot was killed in a plane crash, George's gang had helped the widow to move house, and subsequently named themselves "the Ace Moving Company"; their motto: "We move husbands out, and women anywhere."

Although Young and Crippen had spent months in training for a flight that was ostensibly now less than a year away, neither of them expected NASA to meet the deadline. While the newly arrived astronaut candidates were still—briefly—naive enough to believe the agency's most optimistic forecasts, the veterans in the office had long ago learned that NASA often told the public and Congress one thing, while it quietly made plans to do another. One old hand liked to joke that the letters of the agency's acronym stood for Never A Straight Answer.

The next morning, the thirty-five prospective astronauts' training began in earnest.

::::::::::::::::::
::::::::::::::::::

Standing on the deck of the landing craft in Biscayne Bay, Rhea Seddon glanced down the four hundred feet of towrope connecting her parachute harness to the motorboat idling in the water below, and prepared to run. The fierce Florida sun hammered on the white football-style helmet on her head; beside her stood an instructor from the US Air Force Water Survival School, stripped to the waist, in sunglasses and a peaked cap. As Seddon's parachute flattened against the mesh wall behind her, the instructor rattled through a final briefing: When you see the boat start moving, sprint to the end of the bow and jump; stay on your feet, because if you don't, the nonslip deck will tear up your hands and knees;

the parachute should fill as you go over the edge. As the boat accelerates, you'll ascend; when you see the flag wave, hit the release bar on the towrope; as you drift down, run through your checklist—check canopy, discard face mask, inflate life preservers, deploy life raft.

"But whatever you do, don't release the parachute until you're in the water," the instructor said. "Got it?"

"Yep," Seddon replied.

"Remember how to keep your face out of the water if you get dragged?"

"Yep."

"Any last words for the folks back home?"

"What . . . ?"

"Okay—get going."

The instructor jerked his thumb in the air. The coxswain opened the throttles of the motorboat, and Seddon ran across the deck. The rope snapped taut. A petite five feet two, until recently working as a resident in a Tennessee hospital emergency room, Seddon had suffered from a fear of the water since her teens.

What if the chute doesn't work? she thought; she hoped that she wouldn't be killed.

The ocean yawned beneath her.

Not for the first time, she wondered if she wanted to be an astronaut, after all.

A key part of preparations to fly aboard NASA's fleet of two-seat T-38 jets, the three days of water survival drills were among the first and most grueling parts of the new astronauts' training. Unlike their predecessors, who—regardless of their background or role—were required to qualify as supersonic jet pilots, the new Mission Specialist astronauts were not expected to learn how to fly a T-38 themselves. But they would have to log at least fifteen hours of flying time a month, and their training as astronauts began with lessons on how to navigate and handle communications from the back seat of the aircraft. They made checkout flights to acclimatize them to the g-forces encountered during tight turns and aerobatics, breathing oxygen at altitude, and how to use the ejection seats in an emergency. Now they were being drilled on what to do if they were to bail out over water. Most of the candidates selected from the armed forces had already been through similar exercises during their military careers, so at the end of July just sixteen TFNGs—including all six women—were sent to Homestead Air Force Base in Florida to take the survival course. As well as parasailing into the open ocean weighed down with a full complement of survival gear, they were towed through the water at high speed in a harness to simulate being dragged by a billowing parachute and, in a climactic test, plucked from a drifting personal life raft by helicopter.

NASA's public affairs officers—the agency's press liaisons—had told the sixteen AsCans that the media would be excluded from this part of the training. As it turned out, the exercises were conducted under the gaze of dozens of reporters and cameramen, who filled a small flotilla of boats shadowing the astronauts across the bay, doing their best to capture a telling quote, or a photograph of any action—but especially of the six female trainees. By now, each of the women understood that the novelty of being America's first female astronauts made them a focus of attention, but their patience was already fraying. "I just want to be one of the guys, and not hassled by the press," Rhea Seddon told one reporter; when, as she was being winched aloft by a helicopter, a photographer asked Sally Ride to make a "happy face" for the cameras, she simply yelled, "No!"

Although none of the women discussed it, they all knew they were being watched for signs of weakness. And they were determined to disappoint those who expected them to fail.

Back in Houston, much of the training was a grind: day after day of lectures, on the history of spaceflight, the methods and procedures of the Astronaut Office, spacecraft engineering, flight operations, orbital mechanics, space navigation, and the myriad subsystems of the Space Shuttle itself; as many of these systems were still in development when the classes began, they were often taught by the same engineers responsible for designing and building them. Because the astronauts would be expected to observe Earth from orbit, there were also lectures on geology, volcanology, oceanography, and meteorology from space; they were instructed in astronomy, and in the life and material sciences that would be the subject of experimentation aboard the shuttle. The new intake of astronauts was so large that lecturing them as a single group was impractical, and so Abbey had them divided into two teams, headed by the two senior military officers in the class: Red, led by former Navy test pilot Rick Hauck, and Blue, led by Dick Scobee.

For months, the classes and orientation took place up to six hours a day, five days a week; at the end of each day, the astronauts went home with briefcases filled with printed handouts to study. The volume of information was overwhelming: the candidates' astronomy instructor, a professor from Texas A&M University, told them that he would cover four years of undergraduate and two years of graduate study in just twelve hours of classes. Continuing a practice from the earliest days of the astronaut corps, they never received a single written test—but had to do their best to retain whatever they could from the deluge of knowledge. "Your test," they were told, "will be when you fly."

The candidates' classroom instruction was broken up by frequent T-38 training flights, and by field trips to other NASA facilities and contractors employed by the shuttle program across the country. The thirty-five AsCans toured the pads at Cape Canaveral, where they witnessed the launch of an expendable rocket, and visited the Mousetrap, the Cocoa Beach steakhouse once favored by astronauts and engineers during the glory days of Apollo—now haunted by middle-aged men and women still pining for a glimpse of their 1960s heroes. They flew to the Ames Research Center outside San Francisco, where NASA engineers were conducting aerodynamic testing and wind tunnel analysis to better understand how the shuttle would travel through Earth's atmosphere; and to the Rockwell plant in Downey, California, where the shuttle orbiter had been built. By the time they arrived, *Columbia* had finally been sent for the next stage of its construction, at the Rockwell facility ninety miles away in Palmdale, so the AsCans had to make do with a briefing on its manufacture. At night, the candidates partied together in restaurants and hotel bars: those from the Navy and Air Force were long accustomed to bonding with their fellow officers at often remote bases with booze and pranks, and many of their new civilian colleagues were happy to join in: on the long-distance field trips, it became a tradition to wait until late in the evening, and then one inebriated AsCan would be nominated to telephone George Abbey back in Houston. Abbey, when he picked up the phone in the middle of the night, seemed to be flattered by the attention; his wife, Joyce, was less amused.

The cockpit of the Space Shuttle orbiter had more than 2,200 switches, readouts, controls and circuit-breakers—lining the walls and ceilings of the flight deck, encased in metal cages and often spring-locked to prevent accidental operation—and every astronaut had to learn what each one of them was for. But they also had to become accustomed to the new and potentially deadly environment of space. So—slowly at first, and then with increasing intensity and confidence—the astronauts began work inside the windowless concrete hulk directly behind Building 4, entered through a secure door overlooked from John Young's office window.

Inside the harshly lit, hushed spaces of Building 5, run by a team of six hundred staff and banks of state-of-the-art UNIVAC and IBM computers reading data from punch cards and reel-to-reel tape, were the Shuttle Mission Simulators. Alongside individual devices dedicated to training on each one of the spacecraft's subsystems—computers, electrical, propulsion, life support—was a pair of massive machines designed to reproduce every part of a shuttle mission simultaneously. One, the Motion Base Simulator, was built around a mock-up of the orbiter flight deck, mounted on hydraulically powered legs, which could pitch, roll, yaw, and shudder to mimic the vibrations of launch and ascent; rudimentary computer-generated projections depicted the gantry sliding past the windows as the orbiter cleared the tower, the flare of the explosive charges releasing the solid rocket boosters as they flamed out on the brink of space, or the approach of the runway as the shuttle came in to land. The other, a static mock-up of the entire crew compartment, familiarized the astronaut candidates with what would happen in orbit: in the Fixed-Base Simulator, shuttle crews faced marathon rehearsal flights in which they were linked to the same console operators in Mission Control who would work with them while they were in space.

These simulations, which would run for as long as thirty-six hours, were intended to subject both astronauts and their flight controllers to the worst stresses the simulation supervisors could imagine: compounding computer glitches; shorting electrical systems; leaks; or engine failure. The supervisors, who sat in a nearby control room, triggering malfunctions by sweeping light pens across the screens of their computers, took pride in presenting the astronauts with the most bewildering combinations of failures they could conjure up, and often delighted in catching them out. They aimed to make sure that, by the time each of the astronauts finally left the ground on their first mission, they had rehearsed for almost every conceivable eventuality—except one. Although the simulations often created circumstances in which the shuttle crews' missteps or miscalculations might result in a crash, the supervisors never devised scenarios from which they knew there was no escape, or would inevitably end in what NASA engineers referred to as "loss of vehicle, mission and crew": certain death.

Nearby, in Building 9, there were separate simulators of the new technologies developed for use aboard the shuttle. These included the Remote Manipulator System—the giant robotic crane designed for tasks including satellite deployment and in-orbit repairs, which would be installed in the cargo bay of the orbiter and operated by a specially trained member of the crew from inside the cockpit; and a device resembling an oversize air hockey table, which allowed astronauts frictionless movement—helping them get the feel of the Manned Maneuvering Unit, an experimental jetpack they could fly around in space. To accustom the astronauts to living and working in zero gravity, they also drove out to Ellington Field for trips aboard the "Vomit Comet," a KC-135 refueling tanker plane refitted with a padded floor and walls: as it flew through long parabolic arcs over the Gulf of Mexico, its occupants experienced a few seconds of floating free of gravity, as if cresting an incline on a roller coaster, and often the nausea and heaving that gave the aircraft its name. Inside the drum-shaped hulk of Building 29, there was also a twenty-five-foot-deep pool—the Weightless Environment Training Facility—at the bottom of which lay a full-size mock-up of the shuttle fuselage. Here, the AsCans could don cumbersome pressure suits and submerge underwater to simulate weightlessness, and practice tasks they might have to perform during space walks.

On Friday evenings, the most gregarious of the new astronauts gathered for happy hour drinks, wherever the Building 4 secretaries could find a spot: at Pe-Te's Cajun BBQ House, the restaurant just outside the main gates of Ellington Field, where the walls were soon plastered with space memorabilia; Maribelle's, a Day-Glo pink bar frequented by local shrimp fishermen and infamous for its mud-wrestling competitions and "Naughty Nighties" parties, at which the waitresses served beer wearing lingerie; and, later, at the Outpost Tavern. Entered through a pair of Western saloon-style doors shaped into the silhouettes of bikini-clad women, the Outpost was a dimly lit nicotine-stained dive bar in an old military barracks building that had been moved wholesale from its original location to a potholed lot on the corner of Egret Bay Boulevard and NASA Road 1, a five-minute drive from the Johnson Space Center. George Abbey became a happy hour regular there, and often lingered long into the night among the new recruits he called "the kids." Nursing one long-necked beer or glass of white wine after another, Abbey said little, but soaked up information from those around him until, often, he fell asleep right there at the bar.

At the end of summer 1978, the TFNG class threw a party in Clear Lake for the entire astronaut corps. The candidates laid on entertainment including an elaborate skit that satirized the selection process that had brought them all to Houston: Judy Resnik and Ron McNair, accompanied by one of their white, male colleagues, gathered behind a sheet hung from the ceiling with three holes cut in it. Resnik placed her head through the central hole, while McNair and the second man each poked an arm through those on either side. Clothes pinned to the front of the sheet completed the effect: a single astronaut candidate with a female face, and two male arms, one Black, one white. As other AsCans playing the part of the selection panel posed their questions, the absurd entity waved and nodded its extremities in a ridiculous, ill-coordinated pantomime, until the final question: What makes you qualified to be an astronaut? With both hands wagging in the air for comic effect, Resnik delivered her punch line flawlessly: "I have some rather *unique* qualities."

CHAPTER EIGHT

THE GREAT TILE CAPER

Ordered down to Cape Canaveral at the beginning of April 1979, Larry Kuznetz arrived at the Kennedy Space Center to find the Space Shuttle *Columbia* enmeshed in a web of scaffolding, jigs, and wheeled stepladders. Kuznetz knew little about the Space Shuttle, or what he was expected to do in Florida. Ten years before, the thirty-six-year-old engineer had been part of the back-room team in Mission Control ensuring that the life support systems of Armstrong's and Aldrin's space suits kept them alive during the first moonwalk. Since then, he had completed a NASA-sponsored PhD in bioengineering at UC Berkeley, and recently garnered attention for his work at the University of Houston, bringing his systems engineering expertise to bear on women's sportswear. But his boss in the Crew Systems Division in Clear Lake was mortified by the accompanying publicity—particularly the story in the *Houston Chronicle* headlined "The Bra That Came from NASA."

"We didn't send you to Berkeley to make women's brassieres," he said. He dispatched Kuznetz to the Cape with instructions to report directly to the Orbiter Processing Facility, the pair of massive eleven-story hangars where technicians would prepare *Columbia* and her sister ships to go into orbit and—all being well—refurbish them on their return.

Kuznetz protested that he knew nothing about the shuttle program; but his boss wasn't listening.

"You'll find out!" he said.

After years of delays and desperate scrambling for more funds, the first spacecraft in NASA's fleet of Space Shuttle orbiters had finally rolled out of its hangar at Rockwell's factory in Palmdale at the end of the first week of March 1979. Publicly, Administrator Robert Frosch and other agency chiefs remained confident that the shuttle's maiden voyage would begin in November that year. And the evidence was now there for everyone to see: the first shuttle that would fly in space appeared complete, and was ready to begin making the long journey across the country toward the launchpad in Cape Canaveral. But the unveiling was another public relations feint: when it was towed out onto the tarmac at Palmdale, *Columbia* was still at least six months away from completion and lacked critical pieces of equipment, including engines and its onboard computers. There were still some seven thousand insulating tiles missing from its skin; the contractors had filled the spaces with more polystyrene dummies, held on with glue and double-sided tape.

But NASA had ordered Rockwell to deliver the shuttle to the Kennedy Space Center regardless, and planned to have more than a thousand workers from the contractor flown out to finish the job in Florida. To the agency, it was less important to finish the new spacecraft before it left the factory than it was for the public—and Congress—to see that the program was making progress. But the strategy backfired. After another humbling series of false starts, when the shuttle finally touched down on Merritt Island—perched awkwardly on the back of a repurposed American Airlines 747—the white skin of the orbiter was visibly pitted with a ragged gray patchwork of holes; thousands of the dummy tiles had fallen off en route. And the tile failures weren't confined only to the temporary polystyrene insulation: the real problems were just beginning. Instead of a triumphant milestone, the arrival of *Columbia* at the launch site was the beginning of what Robert Frosch later called the "great tile caper."

Showing his pass to the security guards manning the double doors of the Orbiter Processing Facility, Larry Kuznetz stepped inside and saw immediately that the campaign to finish the Space Shuttle in time for her maiden voyage had begun in a frenzy. Hundreds of NASA staff, and contractor employees from the various divisions of Rockwell International and the Lockheed Corporation, responsible for building the orbiter and manufacturing the troublesome insulating tiles, swarmed the floor. The air was filled with the din of pneumatic hammers and whining electric fans. Yet where the surface of *Columbia* could be glimpsed through the forest of scaffolding, much of the spacecraft's skin remained naked, the aluminum airframe covered by only a layer of green primer and a grid of red guidelines showing the intended site of individual tiles. Kuznetz soon learned that six competing management teams were stepping on one another in the attempt to complete the work that was supposed to have been done before the shuttle had left California; of the more than 31,000 tiles that would need to be applied before launch, less than two-thirds were securely in place. The remainder were "either missing, damaged, or falling off."

Kuznetz received his first briefing from astronaut Bob Overmyer, a veteran test pilot and Marine aviator who had joined NASA as part of the Manned Orbiting Laboratory program, and was now at the Cape to monitor the tile work. He told Kuznetz he would be sent to Rockwell's "tile school," before starting work with a tile team on the floor of the facility, where he would find out from the inside why things were going so badly wrong: a NASA spy, eavesdropping on the contractors.

To keep his role secret, Overmyer suggested that Kuznetz should carry a notebook and covertly record his observations whenever he could, out of sight in the bathroom. "You gotta take a dump now and then, don't you?" the astronaut asked him. A few weeks later, grizzled NASA troubleshooter Kenny Kleinknecht arrived at the Cape, torn away from a comfortable posting with the European Space Agency in Paris, to bring the work under control. Kleinknecht, a six-foot-four cigar smoker with a shock of white hair, was an engineer's engineer. At fifty-nine, he had, among other achievements, been a program manager on Project Mercury, and helped Apollo get back on track after the 1967 launchpad fire. Now he had Kuznetz write a computer program, named Pipeline, to track every stage of the tile process, from initial manufacture to successful application and testing, to find the weak points in the chain—and predict when it might, finally, be finished. In the meantime, Administrator Frosch was forced to admit to Congress that the maiden flight in November was slipping away from him: a launch in the first three months of 1980 was more realistic.

By July 1979, Larry Kuznetz had learned that the problems they faced in the Orbiter Processing Facility were much more serious than snarled supply chains or procedural failures: they lay in the very composition of the tiles themselves. Throughout their early years of research and development, the engineers at Lockheed had focused their testing on the thermodynamic properties of the material intended to protect the skin of the orbiter from the temperature extremes of traveling to space and back. At the time, Tom Moser, head of Structures and Mechanics at the Johnson Space Center, had pushed to have the contractor also test it for structural integrity. But Lockheed refused, citing the unnecessary expense. So Moser conducted his own ad hoc experiments, gluing tile samples to the air brakes of NASA T-38 jets, subjecting them to aerodynamic conditions similar to those he expected them to encounter on the shuttle.

Oh Christ, Moser thought when he saw the results.

That should not have happened.

Each of the samples he'd glued to the T-38s had been torn off the aircraft in flight, disintegrating somewhere in the sky over Texas. Moser knew immediately that there was no way the tiles would survive the journey to space.

....

Back in Clear Lake, on a typically sweltering Friday in August, the TFNGs received a surprise visit in Building 4 from Center Director Chris Kraft. Although they had only just marked the first anniversary of their time at NASA, Kraft told them that their two-year probationary period as Astronaut Candidates would be ending immediately.

"OK, that's it—training is over," John Young announced. "You're all astronauts now."

Columbia itself might still be being built, but there remained so much to do on the myriad systems and procedures necessary to carry the shuttle into orbit that every one of the thirty-five new astronauts was needed on the front line: another year of boozy field trips and geology classes was a luxury NASA could no longer afford. So each of the TFNGs received a silver lapel pin depicting a five-pointed shooting star rising through a halo—signifying their eligibility for spaceflight—and jobs supporting the first voyage of *Columbia* and the subsequent program of orbital test flights.

Each of these test missions would be crewed by the veteran astronauts who had been in Houston when the new class arrived. But developing much of the means to make them possible fell to their rookie counterparts. The tasks ranged from seemingly minor details to those in which failure would be catastrophic; yet all were essential. The Red Team leader, Rick Hauck, was assigned to devise a format for the laminated checklists that Crippen and Young would refer to during their ascent, and provide summarized procedures to follow if something went wrong. Dick Scobee and Ellison Onizuka were part of the team that would fly regularly to Cape Canaveral to assist in setting up *Columbia's* cockpit hardware and preparing the crew for flight, and began calling themselves the "Cape Crusaders." Judy Resnik worked on software development, and trained in operating the robotic Remote Manipulator System—now known simply as "the Arm"; Rhea Seddon and the other medical doctors in the class were sent to rehearse with the crews of the search and rescue helicopters, who would be on standby when the shuttle launched, ready for an aborted mission or crash landing. Others received the most coveted support role of all: the CapCom—or Capsule Communicator—the only person in Mission Control permitted to speak directly to the crew of *Columbia* during each phase of flight, providing Crippen and Young with their primary link to the ground.

In September 1979, NASA sent out a press release announcing a last-minute contingency plan to deal with the tile problem: a group of engineers and astronauts was being assigned to develop an emergency repair kit, which would enable the pilots to go outside while in orbit to patch any holes in the shuttle's heat shield before they returned to Earth. Anna Fisher—still grappling with the transition from emergency room medicine to the problem-solving processes of engineering—was part of a team flown out that fall to aerospace giant Martin Marietta's factory in Denver, Colorado; they were assigned to test a sophisticated caulking gun that a spacewalking astronaut in an Extravehicular Activity suit could use to make repairs. At the plant, Fisher established that it might indeed be possible to perform such a complex task in zero gravity, and even to formulate material that was malleable in the freezing vacuum of space but would harden into a reliable heat shield before reentry; but to do so would take years, not months.

In the meantime, Tom Moser and the engineers of the Structures and Mechanics Division in Houston tried to ensure that the tiles wouldn't come unstuck in the first place. They had known from the outset that the tiles were so delicate that they could never be bonded directly to the surface of the spacecraft: the skin of the orbiter was expected to flex as it flew, and yet, if bent more than one-sixteenth of an inch in any direction, the tiles would shatter. It would be, they said, "like gluing eggs to an anvil." Instead, the aluminum body of *Columbia* had been covered with a soft layer of fireproof nylon felt, to which technicians could stick the tiles. According to their initial computer simulations, the engineers believed that this "strain isolation pad" would protect the tiles from the predicted stresses of flight, and maintain the integrity of the shuttle heat shield all the way from Earth to orbit and back. But now Moser's experiments showed that those stresses were far greater than the models had predicted, and the effect of the pad was the opposite of what was intended: it concentrated the strain in just a few spots on the underside of each tile, until some fractured and came loose.

The remedy was to paint the bottom of the tiles with a suspension of silica that, when baked in an oven, hardened like cement; only then would the bond between the pad and the heat shield hold. But the treatment required technicians to hand-paint each tile with a brush, before it was dried, baked, weighed, checked, waterproofed, and, at last, glued fast to the outside of *Columbia*. From beginning to end, this process might take as long as two weeks for a single tile. And it wasn't until October 1979 that engineers in Houston perfected their method, which they called "densifying." By that time, the tile teams in Cape Canaveral had been gluing untreated tiles to the skin of the orbiter for almost seven months.

From his base outside the Orbiter Processing Facility, in one of a small village of some twenty construction trailers parked near the hangar-like building, Larry Kuznetz tracked the progress of the task through his Pipeline computer program. The work was Sisyphean: tiles that had not been densified now had to be removed and modified one by one, and subjected to quality control tests verifying everything from the size of the gaps between them to the aerodynamic forces required to tear them off. To ensure the glue would hold, technicians used a Rube Goldberg–looking device equipped with articulated legs and vacuum suction cups—that pulled the tile away from the shuttle's skin—and a sensitive microphone to pick up the sound of tearing as the bond approached the point of failure.

The skin of *Columbia* now resembled an enormous three-dimensional jigsaw puzzle, each unique piece of which was inscribed with its own individual seven-digit code number, and assigned a location on one of the many tessellated maps tacked to corkboards nearby. Men and women scurried beneath the orbiter with clipboards, while others continued electrical and mechanical work inside. Many were in their late teens and early twenties, and wore bright orange jackets on the back of which were printed the words PUZZLE PEOPLE; on the front, LAUNCH FEVER—CATCH IT. Each time the press published a negative story about the shuttle, someone posted it on the noticeboard in Larry Kuznetz's trailer, part of a perverse strategy to keep up morale. A workforce that had begun as just a few hundred had now swelled to around three thousand, working three shifts, twenty-four hours a day, seven days a week. But the regimented installation process that the contractors had initially planned had broken down, into an apparently haphazard task so complex and time-consuming that one person might take three weeks to apply just four tiles. As 1979 ground on, the work began going in reverse: in June, Rockwell calculated that they had 10,500 further tiles to fix in place before the job was complete—but even as workers at the Cape glued these in position, the suction-cup tests revealed that thousands of others would have to be replaced. By December 1979, the number of tiles the beleaguered Puzzle People had to install had risen to 13,100.

In Houston and at headquarters in Washington, DC, senior NASA officials regarded the industrial fiascoes of the Great Tile Caper with growing alarm. Every morning at seven o'clock at the Johnson Space Center, Chris Kraft picked up the phone to Florida for a conference call with Kenny Kleinknecht and the people from Rockwell, calling to get an update on the daily tile count. And once a month Kraft and Tom Moser were summoned to Washington, DC, for a meeting with John Yardley, the overall chief of the shuttle operation; he warned them that the stakes could hardly be higher. "Get this son-of-bitch in space," Yardley said. "Get this thing in space, or we're going to lose the program."

∷∷∷∷∷∷∷∷∷

The delays and disappointments of *Columbia*'s stumbling path to orbit couldn't have come at a worse moment for the shuttle, NASA, or the United States. The growing sense of crisis on President Carter's watch had only escalated since the beginning of 1979, when the Islamic revolution in Iran had added a new oil shock and spiking gasoline prices to the already roiling inflation at home and political turmoil abroad. In March, the reactor meltdown in the nuclear plant at Three Mile Island in Pennsylvania added to Americans' feeling that they were at the mercy of forces beyond their control, and their dwindling confidence in their nation's ability to master technology with which it had once led the world.

In July, as the tenth anniversary of the moon landing approached, NASA was buried in weeks of humiliating headlines and public ridicule with the news that the agency's massive, seventy-seven-ton orbiting laboratory, Skylab—which had been mothballed in orbit since the last astronauts left it in 1973—would soon be plunging to Earth in a cascade of flaming metal. Until recently, there had been a plan to use the Space Shuttle in a mission to rescue the rudimentary space station by boosting it into a higher trajectory; but when, after a series of miscalculations, Skylab's orbit began decaying at unexpected speed, while *Columbia* remained in the hands of the Puzzle People at the Cape, the project was quietly shelved. For all their expertise in orbital mechanics, and a chain of radar tracking stations strung around the globe at their command, NASA experts had to admit to the public that they could not say for certain where or when the debris from the doomed station—the largest piece of which was a lead-lined photographic vault weighing two tons, capable of making a crater on Earth five feet wide and a hundred feet deep—might make landfall. They remained concerned it could strike populated areas of the United States; quick-witted entrepreneurs did a brisk trade in "Skylab survival kits" including jokey plastic helmets and bull's-eye targets, and the *San Francisco Examiner* offered a $10,000 bounty for the first person to bring a piece of wreckage to their offices. A few hours before impact, the agency predicted the bulk of the debris would fall harmlessly into the Indian Ocean; when large chunks of NASA's billion-dollar orbital laboratory instead showered down upon the parched outback of Western Australia, President Carter sent a message of apology to the local authorities.

A few days later, while panic-buying motorists gathered in long lines of cars jamming the streets around gas stations across the country, Carter delivered a thirty-minute speech live on television to address the nation's energy problems and what he called a national "crisis of confidence." Frank and impassioned, periodically clenching his fist for emphasis, Carter described a country in the grip of problems deeper than recession and a decade of inflation, more enduring than gas lines and energy shortages—one still crippled by the wounds of Vietnam and Watergate, seized by a moral and spiritual decay, and disillusioned with the institutions of government. "We can see this crisis in the growing doubt about the meaning of our own lives and in the loss of a unity of purpose for our nation," he said. "The erosion of our confidence in the future is threatening to destroy the social and the political fabric of America." The President implored the country to come together to confront the crisis as one, just as they had been united by the noble endeavors of the past. "We are the same Americans who just ten years ago put a man on the moon," he said. "With God's help and for the sake of our nation, it is time for us to join hands. . . . Let us commit ourselves together to a rebirth of the American spirit. Working together with our common faith we cannot fail."

His appeal went unheeded. The following week, Carter fired his cabinet and most of his senior advisers as part of his bid for national redemption, but the effort backfired: the President's polling sagged to new lows, and his policies at home were soon overtaken by events overseas. At the beginning of November, Iranian students overran the walls of the US embassy in Tehran and took fifty-three members of the staff—all of them US citizens—hostage, in what would become the defining catastrophe of Carter's presidency; months of negotiations for their release came to nothing, and in April 1980, news of a bungled rescue attempt was splashed across the front pages of newspapers worldwide: a special operations fiasco marred by equipment failures, bad weather, bad luck, and worse planning left eight US servicemen lying dead in Iran's Great Salt Desert, entombed in the charred wreckage of their aircraft.

Against this backdrop, the failures of the shuttle program were no longer viewed as a heroic national effort to pull off a breathtaking feat of engineering, but became instead emblematic of a once-invincible superpower tangled in its own garments; as the Great Tile Caper limped on, and Administrator Frosch and his staff were forced to publicly move Columbia's date back again and again—first to September 1980, then perhaps November, then probably early the following year—press coverage of NASA turned savage.

In an editorial titled "Aluminum Dumbo," the Washington Post decried the "deeply flawed design" of the shuttle, and expressed doubt that, like Disney's cartoon elephant, Columbia would—or should—ever fly; the column called for the program to be curtailed and postponed for another ten years until it could be realized safely: "It would be courageous to accept the fact that pushing the limits of technology with minimal funding has become too costly and risky to complete as planned." Describing "The Endless Countdown," the *New York Times Magazine* followed the long thread of compromise, cost cutting, and fantastical accounting that had run through the project from the beginning, attributing its survival to its military priorities and, inside NASA, a misguided nostalgia for the triumphs of the past: "the lingering dreams of those in an increasingly retirement-aged management for whom space exploration was mankind's highest destiny."

Most damning of all was the widely syndicated cover story in the *Washington Monthly*, which appeared beneath the headline "Beam Us out of This Deathtrap, Scotty!" In a scornful evisceration spread over eighteen pages of the magazine, writer Gregg Easterbrook cataloged the shuttle program's numerous delays, cost overruns, and expedient assumptions, arguing that the vehicle was misconceived, unprecedentedly dangerous, and, ultimately, almost useless. Finally, Easterbrook pointed out that the shuttle would be the first crewed vehicle in the history of NASA not to make a series of unmanned test flights before being formally "man-rated" as safe to take humans into space.

Courageous as they were, the astronauts who had flown the Mercury, Gemini, and Apollo missions all knew that their vehicles had been safely launched and checked by remote control before they ever strapped into their seats for the final countdown. But John Young and Bob Crippen would have to find out for themselves whether *Columbia*—with its 2.5 million working parts, 230 miles of wiring, its balky high-powered engines, experimental solid rocket boosters, and brittle ceramic tiles— was truly fit for flight on its maiden voyage. It would be the most hazardous mission in NASA's history; Easterbrook's story warned in blunt terms how years of shrinking budgets and expedient decision-making had drastically increased the risk of a catastrophic failure, one in which the lives of astronauts would be sacrificed for a national future in space that was largely illusory. "Here's the plan," he wrote. "Suppose one of the solid rocket boosters fails. The plan is, you die."

By the beginning of June 1980, the troubled President had apparently endured enough embarrassment from the trials of *Columbia*. "We're running out of patience on the Space Shuttle," a White House source told the *Washington Post*. "There's a feeling here that we can't tolerate any more delays in the Shuttle program." But as the summer wore on, the news emerging from NASA at last grew more encouraging: on the giant test stands in Mississippi, two of the three main engines that would help carry the orbiter into space passed their flight readiness trials; in a pyrotechnic simulation with Young and Crippen in the cockpit of the orbiter, the tiles survived a key heat and shock test entirely intact; at the end of July, Administrator Frosch committed to launching *Columbia* no later than March 1981.

Meanwhile, down at the Cape, Larry Kuznetz and the Puzzle People finally began to see real progress. In September, their dogged efforts were rewarded with net gains of more than four hundred tiles attached in a single week—and fewer than five thousand missing from the final total. In October, the tally sheet Kuznetz kept in his trailer showed that they had fewer than a hundred tiles to go. At last, as midnight approached on November 24, the tile teams removed the final jigs and scaffolding, and the high bay doors of the Orbiter Processing Facility ground slowly open. Pulled by a tractor, its black-and-white fuselage brightly lit by spot lamps, *Columbia* rolled out into the shadow of the Vehicle Assembly Building, where it would be hoisted into a vertical position by a giant crane and bolted to the external tank and solid rocket boosters. Two days later, Kenny Kleinknecht gave a triumphant interview to a reporter from *Aviation Week*.

"The vehicle is ready to launch," he said.

::::::::::::::::::

Back at NASA headquarters in Washington, DC, Robert Frosch remained nervous about the flight. There were simply so many things that could go wrong; privately, he believed that the chances *Columbia*'s mission would succeed were less than fifty-fifty. In autumn 1980, Frosch had visited the White House to discuss the first flight. In the Roosevelt

Room, the NASA chief met the President and his senior staff; Carter revealed that he, too, was anxious about the risks of the launch. He asked Frosch if it was possible to convert *Columbia* to automatic control so that the first mission could be flown robotically, without a crew. Frosch told him that although it was technically feasible, the changes would be complex and expensive. Besides, the shuttle was—at last—almost complete. Carter agreed to proceed, but as the meeting broke up, Frosch asked for a moment to speak privately to the President. Carter led Frosch into the Oval Office. Once they were alone, Frosch told him he wanted a guarantee of prompt and direct telephone access to the President throughout *Columbia*'s maiden flight.

"You know, we don't know what's going to happen," Frosch said. "We think we are alright, but it's a risky business. If something goes wrong . . . you better know about it immediately and I want you to know it from me."

PART TWO

The High
Frontier

CHAPTER
NINE

WHEN THOUGHTS
TURN INWARD

April 12, 1981

Up on the third floor of the Operations and Checkout Building at Cape Canaveral, John Young sat in a leather recliner in his burnt-orange-colored pressure suit, a thick umbilical hose trailing away behind him, and gazed down at his boots. As white-coated technicians busied themselves around him in the Suiting Room, Young closed his eyes; his thoughts turned inward. The years of delays and reversals seemed finally at an end. Liftoff was just hours away. Young, the veteran of four spaceflights, the ninth man to walk on the moon, was NASA's most experienced astronaut: he had waited so long for this moment that, now fifty years old, in the cockpit of *Columbia* he would wear a pair of rectangular black-framed spectacles to read the small type on the flight checklists. And apprehension gnawed at him still.

Young tried to focus on the ascent procedures, and the string of emergency protocols devised to allow him to glide the orbiter back to Earth if something went wrong during the crucial first eight minutes of flight. Each was fraught with risk. He thought about the ejection seats, intended to shoot him and pilot Bob Crippen to safety in the event of catastrophe—but which they knew offered only the slimmest opportunity for survival. And with all the problems there had been with the main engines, he believed they would be lucky to live to see orbit. Yet Young knew that he wasn't being paid to worry about all the things that could kill him, and now lay beyond his ability to control. So he tried to keep his fears at bay, concentrating on the routines that he had spent so many years running and rerunning, hundreds, then thousands, of times: every eventuality, in one simulation after another. Besides, there was every chance they would once again be grounded by another equipment fault.

But if everything went as planned, *Columbia* would launch later that morning on its first two-day orbital test flight, just a few weeks shy of the deadline Robert Frosch had forecast the year before—although neither the NASA chief nor the President who had appointed him would be there to see it. Having lost the 1980 general election in a landslide to Ronald Reagan, Jimmy Carter had been chased from office amid a cascade of humiliations, culminating in the release of the Iranian hostages— deliberately held up by the authorities in Tehran until the first day in office of the triumphant new President.

In the months since *Columbia* had at last been mated to its rocket boosters and fuel tank and rolled out to the launchpad at Cape Canaveral, the press and public had come to regard the project with renewed awe, the years of doubt and recrimination about the Aluminum Dumbo suddenly forgotten. In February, *National Geographic* carried an image of the shuttle—stacked on the pad and apparently ready to go—on its cover, heralding a lavish thirty-two-page story examining the epic tale of the spacecraft from every angle, illustrated with dozens of glossy photographs and specially commissioned cutaway drawings depicting everything from the main engines to the parachutes designed to safely return the solid rocket boosters to Earth. "The second space age is about to begin," the magazine announced, before informing its more than 10 million readers about the details of what it described as "our nine-billion-dollar super machine." At the Johnson Space Center in Houston, the staff of the public affairs office were soon overwhelmed, as all ten lines of the newsroom switchboard lit with requests for interviews and information from around the world.

At the end of March, the shuttle program witnessed its first fatalities when, following a launchpad test, a group of technicians climbed into the shuttle's engine compartment unaware that it had been filled with pure nitrogen gas minutes before. Within seconds, three men were overcome by the lack of oxygen, and it was several minutes before they were rescued; two died without ever regaining consciousness, and the third would succumb years later to his injuries. But NASA, now so close to its goal, did not pause. The last of the heat-resistant tiles were still being attached to the skin of the orbiter as the final countdown approached. The first planned launch, on April 9, was scrubbed after a bewildering computer failure but, three days later, just before seven o'clock in the morning Florida time, John Young and Bob Crippen lay strapped into their ejector seats on Pad 39A, their faces lifted skyward. There were less than two minutes to launch. The shuttle had begun running on its own power; the giant oxygen and hydrogen tanks came up to flight pressure. Over the ground-to-air radio link, they heard a Rockwell engineer, three miles away in the Launch Control Center, murmur a fond farewell: "Smooth sailin', baby."

On the flight deck, Crippen turned to Young.

"Hey," he said. "I think we might really do it."

::::::::::::::::::

As the final moments ticked away, and the spacecraft shimmered in the subtropical haze, each of the TFNGs waited anxiously—aware that not just the lives of the two men in the cockpit, but their own future, depended on *Columbia*'s success. Many of the rookie astronauts were scattered across the country to attend to the wider needs of the mission: Judy Resnik had been assigned to provide live commentary from the TV studio NBC had set up at the Cape, and smiled gamely through an early-morning interview in which Tom Brokaw asked her about "romance in space," and whether she was "too cute to be an astronaut," before the first launch attempt was scrubbed; Dick Scobee and Ellison Onizuka had completed their work as "Cape Crusaders" to prepare the shuttle as it sat on the pad, and Onizuka joined George Abbey alongside

Young and Crippen in the dining room of the crew quarters for their traditional preflight breakfast of steak and eggs, coffee and orange juice. Now Onizuka and Scobee stood with their families in the bleachers at the Kennedy Space Center listening to the countdown; overhead, other members of the class, assigned to fly chase plane flights around the Cape, surveyed the scene from 38,000 feet.

At almost precisely 7:00 a.m., the shuttle's main engines lit, followed promptly by the big solid rockets; eight seconds later, *Columbia* roared clear of the tower on a column of smoke and flame. In the sky above, Pinky Nelson gazed from the cockpit of his T-38 as the ungainly troika of orbiter, tank, and solids streaked past him, rolling slowly on its axis as it headed out over the Atlantic, drawing on the Earth's eastward rotation and accelerating toward 15,000 miles an hour as it gathered the speed necessary to break free of gravity. "I'll be damned!" Nelson said to himself. "It worked."

More than eight hundred miles away, inside Mission Control in Houston, Flight Director Neil Hutchinson now took over responsibility for *Columbia* from the Launch Control Center at the Cape, while another TFNG astronaut, former Navy pilot Dan Brandenstein, acted as CapCom, to communicate with the crew. Seated nearby were Chris Kraft and Max Faget, the exuberant spacecraft designer who had first conceived of the shuttle. Around them, the room was as tense and hushed as an operating theater: a soft-lit, windowless space in which two dozen men and women sat behind rows of consoles arranged in four tiers descending toward the front of the room, where the "ten-by-twenty"—a half-inch-thick glass back-projection screen displaying a map of the world ten feet high and twenty feet across—showed the shuttle's flight path as a vivid green line on a black background.

At every console was a pair of computer monitors displaying data from among the hundreds of live streams of information beamed into the building from *Columbia*'s telemetry downlink, and updated every second; at each position was a placard showing the responsibility of the specialists seated there: beyond FLIGHT DIRECTOR and CAPCOM, others included FDO for the Flight Dynamics Officer, the "FIDO," who monitored the trajectory of the spacecraft; BOOSTER, the engineer over-seeing the shuttle's engines and solid rockets; SURGEON, the flight surgeon monitoring the vital signs of the crew; and PAO—the public affairs officer providing live commentary of the flight for the media and the public worldwide. Every controller was focused on the data on the screen in front of them, and the laminated cue cards and ring-bound folders of procedures at their fingertips; after hundreds of hours of simulations, the atmosphere in the room had been engineered to be as unemotional as it could be, deliberately isolated from the thundering immolation and white-knuckle ride of rockets in flight: there were no TV monitors to display distracting images of *Columbia* hurtling through the sky. Hutchinson and Brandenstein had consoles in the center of the second-highest tier, from which they could look down on the rest of the room—but communicated with their colleagues almost exclusively through their headset microphones, each exchange a staccato dialogue of brevity and clarity, begun by everyone first announcing who they were addressing, and then the name of their own position:

"Flight—Booster."

"FIDO—Flight."

"*Columbia*—Houston."

As the shuttle ascended, Brandenstein called out to the crew each of the spacecraft's key milestones on its way to orbit. In the cockpit, Young and Crippen were entirely in the hands of the spacecraft's five IBM computers, chattering silently, making thousands of calculations a second to keep the shuttle on course. While flying on the solid rockets, the shuttle remained aerodynamic only within pitilessly narrow parameters: if *Columbia*'s course deviated even minutely from its calculated flight envelope, it would be torn apart by friction, wind resistance, and gravity. The two test pilots could do little more than keep their eyes on their instruments and acknowledge the machines were performing their tasks as programmed:

"*Columbia*, you are go for roll maneuver."

"Roll program complete."

"*Columbia*, Houston, you're go at throttle up."

"Roger, go at throttle up."

Two minutes into the flight, at an altitude of nearly twenty-seven miles, the two solid rockets, their propellant spent, would soon separate from the spacecraft, ending the most hazardous phase of the ascent.

"*Columbia*, you're go for SRB sep."

There was a crackle over the air-to-ground radio as the bolts blew and sixteen separation motors fired simultaneously; Bob Crippen saw yellow-white flames from the pyrotechnics flash past his cockpit window, and the giant boosters fell away like burned-out Roman candles.

"Roger on the sep, *Columbia*."

At this, Max Faget leapt to his feet: after twelve years of struggle, reversals, and compromise, his design had at last passed a crucial test—and, with the separation of the boosters, Faget no longer had to worry about one major aspect of its test flight that had caused him so much anxiety.

"They're off!" he shouted, sundering the cathedral silence of the room; at the console in front of him, Neil Hutchinson recoiled in alarm.

"Max, for Chrissake, sit down," Chris Kraft said.

Soon afterward, and just four minutes after liftoff, *Columbia* crossed the Karman Line, returning American astronauts to space for the first time in six years. Crippen gazed down as, more than sixty-two nautical miles away, the Earth receded beneath him.

"What a view! What a view!" he said.

In Mission Control, the elation would not last.

:::::::::::::::::::

Unbeknownst to either Crippen, Young, or the hundreds of NASA staff and contractors following the mission from the ground, the problem had begun at the instant the solid rocket boosters had ignited on the pad. For years before the final countdown began, the engineers modeling the effects of the two giant rockets exploding into life on the steel launch platform had known that they could create an acoustic shock violent enough to destroy parts of the gantry or—worse still—rebound upward to damage the shuttle itself. But they had no full-scale test data from which to draw precise conclusions about the scale of the shock, because the boosters—like so many other key parts of the shuttle system— had never been flight tested. Indeed, in the static firing trials that the Thiokol Corporation had conducted out in the Utah desert, the rockets had not even been in the same attitude they would be at launch. To save money, they had only ever been fired horizontally, never vertically.

In the absence of information from the flight hardware, the NASA engineers had conducted miniature tests, using the rocket motor from a Tomahawk cruise missile and one-fifteenth-scale models of the orbiter, tank, and launchpad. Scaling up from the results of these experiments, the engineers had designed a system of water cannon—"rainbirds"—to flood the surface of the launch structure in the seconds before ignition and protect the shuttle from acoustic damage. Only later did the engineers learn that their calculations had been off by a factor of ten: when the giant rockets finally fired for the first time, they belched flame that scorched the grass for a mile in every direction across the Cape; the blast wrenched light fixtures and fire alarm boxes away from the launch gantry, melted or vaporized automatic cameras, and blew a steel handrail through the wall of an elevator cage. And, in an acoustic pulse that lasted just twenty milliseconds, the shock wave from ignition rolled upward across *Columbia*, buckling fuel lines in the Reaction Control System, smashing an aerodynamic control flap vital to aligning the orbiter during hypersonic flight, and shearing away delicate heat-protection tiles from the fuselage, leaving the debris lying in the flame trench of Pad 39A as Crippen and Young thundered away from Earth.

The astronauts were not quite two hours into their mission when they discovered the damage to the heat shield. One of Crippen's first tasks when *Columbia* reached orbit was to open the spacecraft's sixty-foot-long payload bay doors; but as he looked out through the windows of the flight deck to check on their progress, he noticed small black patches on the skin of the orbiter. He called Young over and, as soon as it became possible, the commander described what he could see to Dan

Brandenstein in Mission Control. The damage appeared to be superficial: on either side of *Columbia*'s tail fin, on the surface of the bulging pods that housed the engines of the Orbital Maneuvering System. "We want to tell y'all here we do have a few tiles missing off both of them," Young said, orientating the orbiter's cameras to send live TV pictures of the damaged surfaces to the ground. Examining the tiles through binoculars, the Chief Astronaut seemed unconcerned, and the two test pilots maintained the air of laconic assurance that tradition demanded; what doubts he had, Young kept to himself.

Down in Houston, it was approaching eight o'clock in the morning as Max Faget examined the grainy video images coming down from space. He immediately told everyone that the missing insulation would not be a problem: the location of the damage, well away from the critical areas of the fuselage that faced the fiercest heat on reentry, was not dangerous, and posed no threat to either the orbiter or the astronauts. The public affairs office quickly briefed the media that there was nothing to worry about; "I'm just not concerned about it," Neil Hutchinson told the press at a postlaunch news conference. "It's not going to bother us on the way home."

But when she first glimpsed the TV pictures, Dottie Lee, the veteran engineer who had been working on the shuttle's heat-protection system since being summoned by a mysterious phone call in 1969, began to cry. The truth was that, without further analysis, she couldn't be certain how the missing tiles would affect the complex thermodynamics of the airflow around the Orbital Maneuvering System, and what would happen if the rocket engines inside the pods—needed to navigate *Columbia* home—became damaged during the return to Earth. More important, Lee and her team of engineers in Building 13 at the Johnson Space Center could only assess the problems that were visible. She knew that if the damage was more extensive than that already described by Young and Crippen—particularly if it had reached the flat underside of the orbiter, or the leading edges of the wings, which would face the most extreme heat—there was no means for the astronauts to make repairs in orbit, nor any way of rescuing them: they would either be killed as the shuttle disintegrated during reentry, or marooned in space to die of oxygen starvation.

Lee called Rockwell in California to consult with the men who had helped build the shuttle, and huddled with her colleagues in the Structures and Mechanics Division in Houston to calculate the possible effects of the partial heat shield loss. The work would go on all day and into the night. Meanwhile, over in the windowless hulk of Building 30, a one-star Air Force general seated in the Flight Control Room picked up the red telephone at his console, beginning an effort to use top secret ground-based telescopes in Hawaii, California, and Florida to take photographs of the underside of *Columbia*. As Lee and the other engineers ran computer models and reviewed test data in Clear Lake, one after another, Air Force officers stationed at each of the classified observatories tried to capture images of the orbiter as it passed over their heads at 17,500 miles per hour.

But in their photographs, *Columbia* appeared only as a distant blip.

It wasn't until eleven o'clock on Monday night, after twelve hours of analysis, that Dot Lee joined the materials scientists and structural engineers at a meeting in Building 13; by that time, Young and Crippen were trying to sleep at the end of their first day in space. Together, the technicians in Houston confirmed that—based on the damage they had seen on the orbiter's rocket pods—they agreed with the assessment Lee had gathered from Rockwell in California: "You will not burn up the spacecraft." Spent but relieved, Lee climbed into her car for the short drive to her house; it was midnight.

In the kitchen, she stood in front of the open door of the fridge, searching for something to eat. She found a cold sausage patty, shook some ice cubes from the freezer into a glass, and covered them with Tanqueray. She finished her drink, showered, climbed into bed, and slept for three hours. By 4:30 a.m. she was back at her desk at the Space Center, hollow-eyed and exhausted, for a five o'clock call with NASA headquarters in Washington, DC.

Yet, no one knew if any of the hundreds of brick-like black tiles covering the bottom of *Columbia*, where temperatures were expected to rise to 2,000 degrees Fahrenheit, had been sheared off by the same forces that had damaged the shielding near the tail. At each of the twice-daily press conferences held in the packed Public Affairs Office auditorium of Building 2 at the Space Center, reporters had pressed Gene Kranz, the Deputy Director of Flight Operations, for details of what he knew so far, and what steps he had taken to find out more; on the second day, he brought out Tom Moser to describe the exhaustive analysis done so far— but admitted that efforts to photograph the bottom of the orbiter using Air Force telescopes had failed. "We have no usable photos obtained from the ground stations," Kranz said; but his statements contained a carefully worded obfuscation.

What Kranz couldn't say to the press was that NASA and the National Reconnaissance Office already had long-standing plans to use a KH-11 "Keyhole" spy satellite—one of the Pentagon's most highly classified espionage tools—to refocus its gaze away from Earth and instead shoot images of *Columbia* from orbit. With high-resolution digital cameras capable of identifying objects not much larger than a packet of cigarettes from two hundred miles out in space, the satellite would be able to provide Mission Control with the information it needed about the shuttle's heat shield. But the KH-11's orbit would only cross paths with *Columbia* for a fraction of a second on three occasions before Young and Crippen's scheduled reentry at the end of the two-day mission. In Houston, the engineers hurriedly rewrote the astronauts' flight plan. Sending messages to the teletype machine on the shuttle flight deck—in part to avoid disclosing classified information over the open air-to-ground radio link—they gave Crippen and Young instructions to roll *Columbia* upside down, so the satellite could capture images of the critical areas of the heat shield as it swept past. Finally, late on the second day of the mission, Gene Kranz arrived in the Flight Control Room with a handful of black-and-white photographs depicting in extraordinary clarity the underside of *Columbia*, revealing no sign of serious damage. The two astronauts could safely bring the orbiter home, after all; when asked by members of his flight control team where the pictures had come from, Kranz was characteristically blunt.

"I can't tell you," he said.

::::::::::::::::::
:::::::::::::::::

Two days, five hours, and twenty minutes into *Columbia's* maiden flight, as the orbiter sped east over the Indian Ocean toward Australia, John Young and Bob Crippen made the final preparations to come back to Earth. Once again harnessed tightly into his ejector seat, Young glanced at the green numbers displayed on the TV monitors in front of him, armed the engines of the Orbital Maneuvering System, and reached for the button on the computer keypad marked EXEC, for "execute"; on the screen, a fifteen-second countdown began.

It was not quite eleven o'clock on Tuesday morning in Houston as *Columbia* began its descent. Traveling at almost five miles per second, its nose tilted up at an angle of forty-one degrees just like the balsa wood model Max Faget had demonstrated years before, the orbiter was slowed by the friction of its passage through the thickening air, the energy of velocity sloughed away as heat. Through the cockpit windows, Young and Crippen watched an incandescent shock wave form around the nose of the orbiter, like water breaking at the bow of an ocean liner, gradually enveloping the entire spacecraft in a hot-pink glow. Inside the Flight Control Room, the atmosphere grew tense.

For all the hazards of launch, the path of the shuttle's return to Earth lay through unknown territory; no one had ever attempted to return a winged vehicle from orbit before, and the journey would simultaneously put two of *Columbia's* most sophisticated technologies—the heat-resistant tiles, and the fly-by-wire avionics computers that would control most of the descent—to their first full flight test. While the computer analysis of Dottie Lee's team and the Keyhole satellite photographs assured Max Faget and Chris Kraft that there was no catastrophic damage to the heat shield, there was no guarantee that miscalculations in the original design would not lead to a failure.

Of particular concern was the so-called zipper effect: the hypothetical possibility that if a single one of the tiles came loose during reentry, a finger of hot gas could work its way into the gap and peel away first one piece of insulation, then another—and another—in rapid succession, stripping an entire row of tiles from the shuttle's skin, exposing the aluminum airframe beneath until it succumbed to the heat, wrenching the orbiter apart in flight. And although the NASA engineers had run thousands of wind tunnel simulations to calculate the aerodynamic forces that would bear on the wings of the hurtling spacecraft as it left the vacuum of space and made its transition into an aircraft, until now the work had remained almost entirely theoretical: Kraft wasn't certain *Columbia* would even fly at hypersonic speed.

The shuttle flashed through the sky over the Pacific, heading east toward Hawaii and, beyond, its final destination on the desert landing strip at Edwards Air Force Base. Steeply descending from a height of more than sixty-five nautical miles, now traveling at sixteen times the speed of sound, the orbiter's violent collision with the thickening air caused the molecules in the atmospheric gases around it to dissociate, tearing away electrons and forming a sheath of ionized plasma around *Columbia*—and cutting off all radio communication.

For the most treacherous phase of reentry—as the leading edges of *Columbia*'s wings reached temperatures of 2,750 degrees Fahrenheit and the computers guided the orbiter through a series of dramatic and structurally challenging S-shaped turns designed to shed some of its colossal speed—Houston had no way of contacting Crippen and Young. Inside the dimly lit room in Building 30, Kraft, Faget, and the astronauts manning the CapCom console watched their screens, waiting in taut silence for the end of the radio blackout.

As the sun climbed once more over the Antelope Valley, a quarter of a million people packed into a sprawling lot at the edge of the high desert Air Force base, amid a multicolored patchwork of camper vans, tents, barbecues, bunting, and American flags. Standing on the hoods of cars, holding binoculars, sipping from cold cans of beer in defiance of the early hour, they raised their faces to the sky; across the country, millions more waited in apprehension, watching flickering images from Mission Control on TV screens in bars, stores, and schools. Almost as it had during the first moon landing, everyday life seemed to come to a standstill, as city courts and construction projects alike stopped work and live television pictures were relayed around the world.

In the Flight Control Room, a clock counted down to the planned time for reacquisition of radio contact: three minutes; two minutes; one minute. On all three national TV networks, the commentary from Mission Control was punctuated by the soft hiss of static over the air-to-ground loop. Then, with a few seconds in the countdown still to go, the Flight Dynamics Officer reported two radar contacts at 165,000 feet—traveling at 9,700 feet per second, still 410 nautical miles out from the lake bed landing strip. Moments later, John Young's voice broke through the white noise:

"Hello, Houston. *Columbia* here."

CapCom astronaut Joe Allen, dressed in fine formal wear—a spearpoint-collared white shirt, wide blue tie, and vest—grinned into his headset microphone.

"Hello, *Columbia*. Houston's here. How do you read?"

The shuttle dropped through 151,000 feet and Mach number 8; high-magnification cameras picked out the trembling image of a triangular white smudge advancing across a field of gray. From the cockpit, Bob Crippen saw the coastline ahead of him; the computers had the orbiter flying perfectly down the center of its planned trajectory.

"What a way to come to California!" he said.

In the glassed-in VIP enclosure behind the Flight Director's console, Max Faget and Chris Kraft were on their feet. Kraft rapped on the window and held a piece of copy paper up against the glass so that every one of the controllers in the room could see the message he'd scrawled on it:

WE JUST GOT INFINITELY SMARTER.

Out in the desert, the first spectators glimpsed *Columbia* as a white shape plunging out of the sky like a cinder block with wings, falling nose-down in the attitude of a dive bomber, plummeting from an altitude of three and a half miles to 250 feet in just 91 seconds. Assuming control of the orbiter from the computers, Young brought *Columbia* down for a flawless landing at 10:21 a.m. California time, speeding along Runway 23 at Edwards, on the same shimmering dry lake bed that had witnessed the triumphs and disasters of the X-15, and all the experimental flights before them. In Houston, the Flight Control Room filled with elated cheers; in the Alpine Tavern on Seventh Avenue in Manhattan, the lunchtime drinkers clinked glasses; and a crowd that had gathered in front of the TVs in Gimbels department store in Philadelphia broke into applause.

A bellman in the lobby of the New York Hilton shouted, "We're number one again!"

When at last John Young came jogging down the rolling staircase pushed up against the orbiter's hatch, the awkward and taciturn Chief Astronaut was barely recognizable. Met by George Abbey—wearing a blue one-piece NASA flight suit and a wide smile—who threw a congratulatory arm around his shoulders, Young ran back and forth beneath the belly of his spaceship, punching the air in jubilation like a schoolboy returning from his first roller coaster ride. Standing at the foot of the staircase, Pinky Nelson thought he saw something else in Young's emotional transformation: a man simply astonished to be alive.

::::::::::::::::::

The success of *Columbia*'s first mission was heralded across the world as a new dawn of the Space Age for the United States, a fitting overture for the new President's promised Morning in America, a technological redemption for a nation longing to emerge from the dismal shadow of foreign and domestic failures that had loomed over the country for more than a decade. Recognizing the symbolic weight of the space program and NASA's past achievements, Reagan seized the moment to bind the shuttle project into his message of a resurgent country, recasting outer space as the "high frontier," ripe for conquest by the spiritual descendants of the intrepid pioneers who had gone west a century before.

The press clamored for access to Crippen and Young, who NASA packed off first on a national, then a world, tour; the taciturn test pilots were overwhelmed by the attention: Young, interviewed in his office in Houston, appeared mortified at being made the subject of a BBC documentary, titled *The Ultimate Explorer*; Crippen was recognized in the street even in France, and grew a mustache—but shaved it off when he discovered that it failed to disguise him from fans, to whom he was clearly Bob Crippen, with a mustache.

The same newspapers and magazines that had once attacked the shuttle program for its waste and danger now lauded the heroism of the astronauts and the astonishing sophistication of their spacecraft: "All Americans had the right stuff again," *Newsweek* reported, "and it turned out to be Nomex felt insulation and heat-resistant silica tiles, 31,000 of them fitting together as seamlessly as Arizona and New Mexico." *Time* replayed the shuttle's triumph after years of criticism in epic cinematic terms: "The shuttle had become a kind of technological Rocky, the bum who perseveres to the end, the underdog who finally wins." In the *Washington Post*, the perils of the Aluminum Dumbo were brushed aside, and the paper anticipated a dominant America finding new scientific and business opportunities in low Earth orbit.

Central to these ideas was the way NASA had sold the shuttle as a thrifty and cost-effective workhorse that could make space travel routine and even turn a profit, part of a sleek new agency that would be reshaped for the sharp-toothed, go-go entrepreneurial values of the new decade. The man expected to undertake this task was Reagan's newly appointed NASA Administrator: James M. Beggs, a fifty-five-year-old former executive at General Dynamics, the country's largest defense contractor. A graduate of the US Navy college in Annapolis, an imposing six foot three even without the Stetson hat he favored, Beggs spoke in a booming voice, and liked to quote from Shakespeare and the classics. But he had also served as both an assistant administrator at NASA under President Johnson and as Undersecretary of Transportation for Nixon, where he had publicly advocated the use of space technology in private industry. After his years among the technocrats of corporate

America and his time navigating the hallways of Washington, DC, Beggs seemed the ideal man to bring the bold free-market precepts of Reaganomics to the nation's space agency—even while critics worried that the new President harbored dark plans for its further militarization. "NASA is already beating its plowshares into swords," one congressional aide told a reporter.

Administrator Beggs now pressed for a swift end to test flights for the shuttle, so it could begin carrying commercial payloads into orbit as soon as possible. The original budget forecasts for the Space Transportation System had promised that the spacecraft would launch as often as once a week by the mid-1980s, with ground crews refurbishing each shuttle for fresh flights in days, not months. To meet such a goal required the new NASA chief to transform the agency from the world's most sophisticated research and development laboratory of aerospace engineering into something more like a commercial shipping company—or an airline that just happened to conduct most of its business in the vacuum of space.

With three other orbiters for the program under construction, Beggs now presided over an orbital flight test program condensed into a total of just four launches, and began reconfiguring the complex experimental shuttle operation as a scheduled space trucking service. To do so, NASA officials referred to an organizational chart borrowed from American Airlines. The three remaining research and development flights of *Columbia* were complete by the following summer. On July 4, 1982, *Columbia* touched down on Runway 22 at Edwards Air Force Base, where it was greeted by a crowd of half a million people including President Reagan and the First Lady.

In an adroitly choreographed patriotic spectacle, the astronauts were met on the tarmac by President Reagan, who gave a speech in which he declared the Space Shuttle cleared for operational flight (a designation usually reserved for aircraft that had flown hundreds of test missions—not less than half a dozen), and committed the country to new commercial and military goals in space. Reagan led the crowd in singing "God

Bless America," and Nancy wept as the second orbiter in the shuttle fleet roared overhead on the back of the NASA jumbo jet transporting it to Cape Canaveral for its first launch.

This was *Challenger*, named after a Royal Navy ship that had led a landmark nineteenth-century oceanographic expedition. "The pioneer spirit still flourishes in America," the President said.

::::::::::::::::::

In the meantime, the solid rocket boosters from *Columbia*'s second mission, parachuted into the Atlantic 140 miles off the Cape, had been retrieved from the ocean by a pair of dedicated recovery ships. In a dockside hangar at Cape Canaveral Air Force Station, a team of Thiokol Corporation engineers watched as the rockets were carefully disassembled for an initial postflight inspection. Once the segments of the right-hand booster were separated, the technicians climbed a ladder and probed the opened joints with flashlights. It was then that they noticed a curious new anomaly: one of the synthetic rubber gaskets sealing the sections of the rocket together seemed to have been badly damaged—burned or partly vaporized—at some point during the course of its two-minute flight. Alarmed, the Thiokol inspectors filed a report about the damage to the O-rings with the Marshall Space Flight Center in Alabama. But it wasn't a high-priority problem. Plans for the next flight proceeded regardless.

CHAPTER
TEN

THE FIRST AMERICAN
WOMAN IN SPACE

Early one morning in April 1982, three members of the Thirty-Five
New Guys made their way to George Abbey's office, a large, blue-carpeted
eighth-floor corner suite in Building 1 of the Johnson Space Center.
More than five years since they had been accepted by NASA, the Director
of Flight Operations planned to personally anoint the first members of
the class of rookie astronauts with their flight assignments on the Space
Shuttle. Among them would be the first African American, and the first
American woman, to make the journey to orbit. Summoning them one
by one, Abbey gave the news first to the crew of the seventh Space Shuttle
mission, STS-7, scheduled to fly in April of 1983. But then he pledged
them to secrecy: they could say nothing of their selection before Abbey
made the formal announcement of the assignments, a week later.

Until that moment, despite their individual differences in background and outlook—Vietnam veterans and former antiwar protesters, tightly wound academics, doctrinaire feminists, and unrepentantly
macho test pilots—the TFNGs had shared an easy camaraderie. After
almost four years of training and working together at NASA, and a rolling program of social events often orchestrated by Abbey himself—beer
calls, softball games, and chili cook-offs—the new class of astronauts
had become good colleagues, if not close friends. Those with military

experience extended the social traditions developed at far-flung bases and airfields around the world by gathering at one another's houses for drinks and barbecues. Their children studied together, at the local elementary school named for lost Apollo 1 astronaut Ed White, or at Clear Lake High or Seabrook Intermediate; one July 4, a dozen of the TFNG families packed their kids into station wagons for a weekend near Canyon Lake in the Texas Hill Country, sleeping in stone cabins, waterskiing, tubing, setting off fireworks—and partying late into the night. But competition was never far below the surface.

Many of the astronaut families were neighbors—Dick and June Scobee lived just a few streets away from the Onizukas on Brookpoint Drive, close to the back entrance to the Johnson Space Center and Ellington Field. Dick Scobee's son, Rich, had a paper route with the son of fellow test pilot Fred Gregory, one of the three Black astronauts selected as part of the TFNG group, who lived down the street from the Scobees. When it rained, the two pilots drove the boys along their delivery route so they could stay dry, tossing newspapers from the car window onto tidy lawns and front porches on streets overhung with live oaks, Norway maple and ash trees. A few years after arriving in Houston, Scobee and another friend from the astronaut group, former Navy flier Jim "Ox" van Hoften, had bought a used open-cockpit biplane—so rudimentary that it lacked navigation aids, a radio, or even a reliable fuel gauge—from a farmer out in California and flew it back to Texas. Van Hoften had a private airstrip in his backyard, and the two pilots often hosted parties there at which they'd carry guests up for aerial joyrides, punctuated by loops and barrel rolls. Outfitted with leather helmets and goggles, Dick also gave June lessons in the plane, and their son took to flying with such enthusiasm he began taking acrobatic instruction: the day Rich turned sixteen, in April 1980, he applied for a driver's license and flew solo cross-country to qualify as a pilot; one day, he hoped to join the Air Force himself. Meanwhile, June had begun teaching English at Clear Lake High, and enrolled at Texas A&M University to study for a PhD in education. At lunchtime, June sometimes drove over to the Johnson Space Center to join Dick for hamburgers in the NASA canteen, or go for a run together; often, Captain Scobee and his wife returned from a jog holding hands, like courting teenagers—to the amusement of some of his colleagues from the Astronaut Office.

At first, the handful of Black astronauts often stuck together—finding that race offered a stronger bond than professional background—and socialized with other Black families in the local community: Ron and Cheryl McNair befriended one of the few Black engineers at the Space Center and his family and, when a fourth African American astronaut, Charlie Bolden, joined the program in 1980, Cheryl and Bolden's wife, Jackie, established a local chapter of the Jack and Jill club, the national organization founded to support Black children. Former Air Force pilot Guy Bluford, another African American in the TFNG class, rarely socialized with other astronauts at all, avoiding the happy hours and beer calls to clock off at the end of every day as regularly as if he worked in a bank, to go home to join his family for dinner. Bluford's wife and children showed little interest in his training to go into space, and he scarcely discussed it with them. When he had joined NASA from the Air Force, other Black pilots in the US military were so rare that Bluford knew almost all of them personally; so when he began taking flights around the country with Ron McNair in the back seat of a T-38, he took special pleasure in the expressions on the faces of the ground crews when the canopies lifted to reveal two Black astronauts at the controls of the plane.

And yet—whatever advances may have been made inside the walls of the Johnson Space Center—the world beyond the perimeter fence remained unmistakably part of the Deep South. Charlie Bolden, a US Marine and Navy test pilot, had grown up in Columbia, South Carolina, not far from McNair's childhood home. He had been accustomed to working in a white-dominated environment from the moment he arrived at the Naval Academy; Bolden also knew that Clear Lake had been segregated when the Manned Spacecraft Center had first opened in the early sixties—Black engineers who worked there had been unable to buy homes in the neighborhood, and instead been forced to commute from Houston's historically African American Third Ward. But he still wasn't ready for what he saw when he and Jackie took the drive a few miles up the coast to Pasadena, one day in 1980, to put down the deposit on the electricity hookup for their new house. As they approached the city limits, Bolden caught sight of a billboard looming over the side of the highway and turned to his wife in disbelief.

"You've gotta be *shittin'* me," he said.

Twenty feet across and reaching two stories high, the sign depicted a robed man riding a white horse, his face concealed by a pointed hood. Beneath the figure was the legend:

WELCOME TO PASADENA,
HOME OF THE KU KLUX KLAN

Elsewhere, other members of Astronaut Group 8 continued the hell-raising antics of their predecessors—although no longer threatened by the same fear of scandal that had forced the astronauts of the 1950s and '60s to curb their appetites in the interests of preserving their careers at NASA. By the time the TFNG astronauts arrived in Houston on the cusp of the 1980s, the klieg light of celebrity—and the trembling needle of America's moral compass—had shifted elsewhere. Twenty years before, the handful of heroes anointed by the Mercury and Apollo programs had been the subject of constant media scrutiny, good family men expected to embody the lily-white values of God-fearing patriots everywhere.

But the first astronaut group of the Me Generation was bound by few such conventions. Many were still in their twenties, and single, when they arrived in Houston—and the mixture of men and women injected a new frisson of sexual tension into an Astronaut Office in which previously the only female staff had been secretaries; the unmarried women astronauts—Ride, Resnik, and Rhea Seddon—all dated colleagues inside Building 4, and generated some fear and suspicion among the wives of their married male colleagues. And while John Glenn had once infamously admonished his fellow members of the Mercury Seven to "keep their hands clean and their peckers stowed" to maintain their places in the program, and the image of NASA in the eyes of the public, some of the TFNGs now regarded adultery and divorce as a threat closer to home.

On returning from one field trip on which some of the academic astronauts had been aghast to discover that married colleagues had spent the night entertaining space groupies, Red Team leader Rick Hauck lectured the nonmilitary astronauts on the importance of silence: "Everybody needs to understand their moral standards aren't necessarily shared by others in the group," Hauck told them. "If you see something on one of these trips that offends you, keep it to yourself. It's none of your business. You could damage somebody's marriage."

Yet the singles scene at the Johnson Space Center had also led to two weddings within the new astronaut group. Following a divorce from his first wife—and after dating Sally Ride—former Top Gun pilot Robert "Hoot" Gibson married Rhea Seddon in May 1981; their first child was born in July the following year, and announced in the "Milestones" section of *Time* magazine as the arrival of "the first US astrotot." That same month, Ride quietly married fellow TFNG astronomer Steve Hawley in the backyard of his parents' home in Kansas; the couple had already been living together for months, in a small brick house in Clear Lake decorated with prints of the shuttle and T-38s and, on the wall of the bedroom, a photo of the moon landing.

:::::::::::::::::

By the time George Abbey came to make the first crew assignments within Astronaut Group 8, Judy Resnik had been transformed by her years at NASA. Unflappable, stoic, and schooled by a life excelling in male-dominated environments, Resnik embraced the values of the agency she had made her home. Intensely private, she had come to detest the public relations parts of the job: even more than the other women in the program, she wanted to be seen as just "one of the guys," and her encounters with journalists grew increasingly terse, as she often parried personal questions with single-sentence answers and barely concealed exasperation. She relished the chance to disappear into the sprawling technocratic cult of NASA—where teamwork had become a hallowed rite, and where sloppy displays of feeling were not only deemed counterproductive, but regarded with disdain or horror.

Resnik's skill as an engineer and her perfectionism made her a natural as an astronaut, and she lost herself in the years of training necessary to excel in the complex routines of spaceflight, continuing to focus on operating the Remote Manipulator System that would become central to future shuttle missions. As she stood at the controls of the simulator in Houston, practicing again and again until the robot arm began to seem an extension of her own body, her face betrayed the ferocious concentration of someone determined to succeed at all costs.

Physically, too, Resnik had changed since her arrival in Houston in the early summer of 1978. Although there was no exercise requirement within the program, all the astronauts had to submit to an annual medical to remain qualified for flight—and the intense competitiveness within the corps meant that almost all became avid runners, despite the unforgiving South Texas climate; even at five o'clock on a summer morning, the temperature might already be eighty-five degrees, with an insufferable 95 percent humidity. Resnik began jogging regularly with her friend Sylvia Salinas, one of the secretaries from the Astronaut Office, and told her that she aimed to lose twenty or thirty pounds. Resnik switched to a high-protein diet, and when the women went out to dinner together at a steakhouse in nearby Webster, the astronaut would simply order a steak, with no sides—as rare as possible. "Bring it screaming," she'd tell the waitress.

Four years after sitting for her first official NASA portrait at the Johnson Space Center, Resnik was barely recognizable as the full-faced young woman with the shoulder-length bouffant and a shy smile: lean and self-assured, Resnik now acquired a new nickname—J.R., like the suave villain of the TV soap *Dallas*, then near the peak of its popularity—and took any opportunities she could to vanish into professional anonymity. She regularly took day trips from Houston to Washington, DC, riding in the back seat of a T-38, taking a cab from Andrews Air Force Base to NASA headquarters at 600 Independence Avenue. At lunchtime, Resnik would duck out of the office to meet friends at a nearby restaurant, still wearing her blue flight suit, but pulling the Velcro-backed NASA insignia from her chest before she arrived. Other diners thought she was a mechanic.

Of all the female astronauts in the shuttle program at the start of 1982—a further intake of AsCans in 1980 had included two more women, bringing the total to eight—Resnik and Sally Ride were widely seen as the most likely to become the first American woman in space. Ride, a thirty-year-old physics PhD from Stanford, was also an outstanding athlete: as a nationally ranked junior tennis player, she had chosen to study astrophysics over pursuing a professional sports career. Unlike Resnik and other women in her astronaut class, Ride was an uncompromising feminist who refused to tolerate sexist jokes from her male colleagues with a dismissive eye roll, and campaigned against the slightest perception that men and women might be treated differently in space. Both Resnik and Ride had been chosen by George Abbey to specialize in the development and use of the robot arm, and both were equally determined to be taken seriously as professionals: Resnik designed bright pink bumper stickers, made up in a local print shop, reading A WOMAN'S PLACE IS IN THE COCKPIT; Ride fixed one to the front of her desk in the Astronaut Office.

But for all her earnestness about the program and her apparent brittleness with reporters, among her peers Resnik was known for her sense of humor, admired as "a man's kind of woman" by male colleagues at the Space Center, and regarded as one of Abbey's close confidantes. She could be blunt to the point of rudeness—and curse with the conviction of a Marine drill instructor. While Sally Ride and her new husband liked to spend their evenings at home playing the fantasy-themed video game Zork on their Apple III computer, and would sit down after dinner to run through the shuttle's ascent checklist, Resnik went out dancing in the nightclubs of central Houston, collected art, cooked Julia Child recipes for her friends, and played the baby grand piano she had installed on the ground floor of her two-story townhouse in Clear Lake. On her own desk in the Astronaut Office, alongside stacks of technical manuals and a volume of bathroom-themed cartoons called *The Toilet Book*, she displayed a sign that said A HARD MAN IS GOOD TO FIND.

Officially, the final choice of flight assignments for the seventh, eighth, and ninth missions of the "operational" Space Shuttle—scheduled for 1983—would be the result of discussions between George

Abbey, Chief Astronaut John Young, and the respective heads of department in Houston, and subject to the approval of Center Director Chris Kraft. But everyone in the Astronaut Office knew the decision would be made by the man referred to by his secretary as Don Jorge.

Even so, to help Abbey justify the list of names he presented to Kraft, the Director of Flight Operations cooked up a study comparing the attributes of the chief candidates to become the first American woman and African American in space. In this evaluation, Resnik, Ride, and Anna Fisher—Kraft's favored choice—were the top contenders, with both Ride and Fisher recommended for their "outstanding public presence." But Fisher had only recently been sent for last-minute training on the robot arm, which would be central to the planned missions—while Resnik and Ride already excelled in its operation. Resnik was let down by her hostility to public affairs duties, and Abbey felt that Ride had a technical edge; she had already twice acted as CapCom, the astronaut contact in Mission Control, and was better trained on the shuttle's orbital systems. Abbey's appraisal of the candidates to become the first African American in space was also a closely run competition, between Ron McNair and the softly spoken Air Force fighter pilot Guy Bluford. Abbey's analysis recommended McNair for his engaging public persona. But Bluford won out because of his technical skill and flight experience: a decorated combat veteran, Bluford had flown 144 missions in an F-4 Phantom over Vietnam and Laos before going on to earn a doctorate in aerospace engineering.

On April 19, 1982, Abbey convened a meeting of the Astronaut Office in Building 4. The windowless conference room was packed—not only with the thirty-five TFNGs, but also members of the recently selected Astronaut Group 9. "We've made some new assignments," Abbey announced in his lugubrious drawl. The room fell silent as he read from his list of names:

"The STS-7 crew will be Crippen, Hauck, Fabian, and Ride. STS-8 will have Truly, Brandenstein, Bluford, and Gardner. STS-9 will be Young, Shaw, Garriot, Parker, and two payload specialists. Hopefully we'll get more people assigned soon."

With that, Abbey left the room. Those who had been chosen were ecstatic—but some of those who had not could only plaster hollow smiles across their faces as they pumped their colleagues' hands in congratulation. Fred Gregory wandered despondently from the room, his head and shoulders sagging. Gregory was one of the three astronauts who had just discovered that they would not become the first Black American in space; like him, seven women in the room had also been forced to watch quietly as Abbey consigned their names to the footnotes of history. But the disappointment they felt was almost universal.

"This is bullshit!" Gregory muttered.

Publicly, those astronauts who didn't make the cut gamely maintained that the order of assignments was insignificant: "Firsts are only the means to the end of full equality, not the end itself," Judy Resnik later told a reporter. But others believed the meeting marked an irreversible rupture within the Astronaut Office; never again would every member of the TFNG group unite for a social occasion. Once divided into the assigned and the unassigned, mission selection seeded an undercurrent of jealousy among the new astronauts, just as it had with their predecessors. Almost from the beginning of the space program, there had been more crew than missions for them to fill, and there was little that most astronauts would not have done to win one of the coveted slots in orbit. Competition was intensified by the absence of formal testing in the corps—and by the inscrutability of George Abbey and John Young.

Abbey's refusal to explain his decision-making generated an atmosphere of fear and second-guessing among astronauts constantly uncertain about how their behavior, performance—or attendance at any one of Don Jorge's barbecues, beer calls, or baseball games—might affect their chances of reaching space. Rumor and misinformation now swept through the offices at the Johnson Space Center as anxious, unassigned astronauts sought to win the favor of the sphinx-like Director of Flight Operations. One began keeping a pair of binoculars on his desk to better spot colleagues walking toward Abbey's office; at the end of August 1981, on the day Abbey turned forty-nine, astronaut Jim Bagian talked his way onto the roof of Building 1 by pretending to be a window washer and rappelled down the concrete facade dressed in a Superman costume. When he reached the eighth floor, Bagian knocked on Abbey's office window and serenaded him with a muffled chorus of "Happy Birthday."

The seething competition made Abbey a more polarizing figure than ever. Those he favored—often Navy pilots—received prime assignments within the Astronaut Office and were guided into positions where they might further his plans. These astronauts—including the select group of male acolytes nicknamed the "Bubbas"—grew grateful for his patronage, and respected his wisdom and artfulness. But those who failed to earn Abbey's approval could find themselves not only passed over for flight assignments, but isolated or handled like errant children. On one occasion, Abbey terminated a meeting with an astronaut who had displeased him by simply behaving as if the man was no longer in the room, returning silently to his work and ignoring any further questions. Eventually, the astronaut shuffled from Abbey's office like a whipped dog, his confidence shattered. Those Abbey treated this way came, understandably, to detest him.

::::::::::::::::::

NASA's public announcement of the crews of Space Transportation System flights STS-7 and STS-8 was rewarded with a media frenzy of a kind not seen since the zenith of the moon program more than a dozen years earlier. The news of the selection of Guy Bluford and Sally Ride dominated the headlines—but it was the first American woman in space that most of the press wanted to talk to. The agency received five hundred requests to interview Ride on that first day alone; her gender proved to be bigger news than Bluford's race—and besides, he was regarded as just another pilot, like so many of his predecessors. As training intensified and the launch date grew closer, the attention Ride received only increased, approaching that once afforded Neil Armstrong.

At the same time, with the four test flights over and the shuttle officially operational, NASA pressed forward with its plans to steeply accelerate its flight schedule—now expected to include one mission every month in 1985. With each launch carrying as many as eight astronauts,

Abbey feared that he might begin to run short of crew to fly the new vehicle, and had begun to recruit yet more candidates: the agency would soon announce that a new intake would begin arriving at the Johnson Space Center every year, "depending upon mission requirements and the rate of attrition in the existing astronaut corps."

Yet Abbey—who prided himself on closely following the path of all the astronauts he had selected and who always accompanied each crew toward the launchpad at the beginning of every mission, and met them when they landed—had long harbored reservations about the safety of the shuttle. Both he and John Young were especially concerned about the lack of a crew escape system. They had argued vehemently to have the ejection seats installed on *Columbia* before the four test flights. But these could only ever save two astronauts seated on the flight deck—the commander and the pilot—and, with any larger crew, in the event of disaster could leave as many as six people behind to die. Once the orbiter was declared operational, they were deactivated.

But even as the shuttle fleet expanded, senior NASA officials continued to argue that—for technical and moral reasons—the spacecraft should be redesigned to include an escape system like those of the Mercury and Apollo programs. One after another, these efforts foundered in the shallows of cost or bureaucratic inertia. Eventually, a serious initiative to study the issue at the agency's flight research arm in Langley, Virginia, was delayed, and then shelved—not on account of any engineering obstacles, but because of how such an effort would play with the public. NASA managers apparently feared that, if the American people learned that it now required an emergency escape system, they might realize that the Space Shuttle was more dangerous than they had been led to believe. The astronauts themselves, however, had few illusions about the perils of their new spacecraft.

The reliability of the shuttle's main engines concerned many of them from the start. Both they and the senior engineers in Houston believed that if there was to be a catastrophe, the complex high-performance engines, and especially the furiously spinning high-pressure turbopumps inside them, would be the cause. Although the astronauts had worked with NASA technicians to devise a series of abort plans intended to return the shuttle to Earth if one of the three engines failed during the ascent to space, these were viable only within a narrow and hazardous envelope of circumstances. Perhaps most important, not one of the plans could begin during the first two minutes of flight, while the shuttle's solid rocket boosters were still burning.

But once the boosters had burned out and fallen safely away from the spacecraft, the mission commander could turn a rotating selector switch on the console in front of him to initiate one of three preprogrammed abort modes, instructing the shuttle's onboard computers to take an abrupt course change.

In the first two of these scenarios, the astronauts were trained to drop the massive external fuel tank and then turn back for either a glide landing at Cape Canaveral or at a handful of sites scattered around the globe where there were airstrips long enough to accommodate the orbiter's high-speed touchdown; for the projected military launches over the Pacific, the only available emergency runway was on Easter Island. John Young believed that one plan was so perilous that for the crew to survive would require "ten consecutive miracles followed by an act of God."

If, for some reason, no landing strip was available, the procedures called for a water landing; it was recognized by both astronauts and engineers alike that this option—known as a "contingency abort"—would result in the immediate disintegration of the shuttle. The astronauts joked among themselves that the cockpit cue cards laying out the procedures for this eventuality were there not to help save them and their spacecraft, but to give them something to do during their final moments awaiting death.

Finally, if a single engine failed when the shuttle was already at too high an altitude and moving too fast to return safely to Earth, there was a further option—Abort to Orbit, or ATO—in which the crew would continue into space, trying to reach a lower orbit than planned, and then end the mission early with a premature reentry. But if at any point two or more of the main engines failed, or one simply caught fire or blew up during the ascent, the entire orbiter was almost certainly doomed.

The solid rockets were another story.

CHAPTER ELEVEN

THE SQUEEZE

Whiplashed by waves of funding cuts, tormented by the caprices of Congress, public opinion, and five different Presidents, by the time the first regular missions of the Space Shuttle began, NASA was no longer recognizable as the same organization that had put men on the moon more than a dozen years before. The young engineers who had embarked on the journey to space at the dawn of the sixties were approaching retirement, or looking for better pay elsewhere: Max Faget left his position in Houston to start his own company after the second shuttle flight in 1981; Johnson Space Center Director Chris Kraft would announce his retirement early the following year. From his aerie on the eighth floor of Building 1, George Abbey watched the agency's bureaucracy grow slowly more tangled and sclerotic. Down at the Cape, the seasoned zealots who had worked through the Apollo years found that a new breed of technicians and middle managers were no longer eager to toil around the clock for another countdown, but preferred to go home and spend time with their families; they might as well have been working for the post office. At the top, too, there was acknowledgment that the days were gone when the reflected glory of the moon program had burnished the prestige of even the lowliest manufacturer making parts for NASA—and those who failed the country in its time of need would be held to account. Instead, the agency's new administrator believed that the agency and its contractors were partners, and not even occasional adversaries.

No part of NASA had been more profoundly transformed than Wernher von Braun's former empire of rocket propulsion at the Marshall Space Flight Center in Alabama. Von Braun had resigned his position as director just as the first of the post-Apollo cutbacks began, spirited away to NASA headquarters to advocate for his dreams of moon bases, Martian conquest, and planetary exploration. In his absence, the powerful research and development operation he left behind in Huntsville began to unravel, struck by layoffs, management reorganization, and even threats of total closure.

By the time von Braun's handpicked successor, Dr. William Lucas, finally took over as director four years later, the semiautonomous powerhouse of rocket engine technology in Huntsville had been hollowed out. The team of German engineers who had provided the backbone of the United States' expertise in space travel had been purged or demoted, and the workshops and laboratories in which they had been able to build and test their own prototypes shuttered; new hires were hard to find, the average age of the workforce gradually increased, and morale plummeted; embattled and insecure, the managers at Marshall found themselves in an increasingly desperate competition for resources, and the rivalry that had always existed with NASA centers elsewhere in the country—especially in Houston—intensified. Only an entrepreneurial campaign orchestrated by Lucas to diversify the center's expertise into new business, from materials science to coal mining, saved von Braun's legacy from extinction.

Bill Lucas had come to Huntsville directly from graduate school, and early on received his first practical lessons in the high stakes of rocket engineering. At the first launch he ever saw, Lucas watched as a Redstone rocket rose slowly from the pad—and traveled less than thirty feet before falling back to Earth, obliterated in a fireball of blazing kerosene. Over the next twenty-two years, he learned that there were a million things that could go wrong with a rocket engine—and discovered that at any moment, a single lousy twenty-five-cent part could cost you the whole ball game; it could go wrong, and there was nothing you could do about it. But that was the price of progress. Now, as Center Director, Lucas adopted the same management approach he had learned from his mentor, drawing on a working culture developed while building missiles for Hitler: hierarchical and conservative, with a reputation for secrecy and a reluctance to share information with colleagues beyond the fence in Huntsville.

To the German engineers who had come to America with von Braun, this had felt like teamwork; to others, it looked like iron discipline, rigid and unforgiving, and an environment in which no individual felt free to take action without explicit approval from above. Dr. Lucas had excellent credentials as an engineer, but von Braun had commanded devotion with charisma and humility; Lucas had little of either. He insisted instead on a rigid application of rules of his own devising, took a shoot-the-messenger approach to bad news, and came to prize personal loyalty above all else. Many engineers at Marshall feared him; some regarded him as little short of tyrannical.

Wernher von Braun would last less than two years in Washington, DC, the failure of his lobbying for interplanetary travel giving way to the consolations of a thrifty and reusable space truck. Dispirited at the speed with which the American public had soured on space exploration, his vision of mankind's destiny among the stars shattered, he retired from NASA for a place on the board at Fairchild Industries. Five years later, von Braun was dead from colon cancer, at sixty-five. In the meantime, it had fallen to Bill Lucas to oversee the development of the new main engines for the Space Shuttle—and the solid rockets to boost it into orbit.

:::::::::::::::::

Although Max Faget had all along regarded the use of solid-fueled rockets to launch the shuttle as a dangerous compromise, his rivals at Marshall determined that the risks could be minimized by keeping their design as simple as possible. The men in Huntsville were justly proud of their years of pioneering work in liquid-fueled engines like the mighty F-1, which had powered the Saturn V to the moon, but had almost no experience in building solid rockets. So—in contrast to their work on the experimental technologies of the Space Shuttle Main Engines—the Marshall team had been at pains to keep the design of the solid rocket boosters safely within the boundaries of previous experience. George Hardy, the veteran engineer in charge of the project at Marshall, said that his objective was simple: "Avoid inventing anything new."

Yet, for all its apparent simplicity, the form the rockets took had involved another series of questionable trade-offs in the name of expedience. When fully assembled, each of the boosters stood nearly fifteen stories high and weighed 590 tons—the largest solid rockets ever built; before each shuttle flight, a pair of them had to be transported on eight heavy-duty flatbed railcars across the country, from the manufacturing facilities of the contractor, the Thiokol Corporation, in northwestern Utah, to Cape Canaveral: a twelve-day journey crossing eight states, covering more than two thousand miles.

However, experience showed that when laid horizontally such giant rockets could become deformed under their own weight, making them unusable. One solution was for NASA to pay for a dedicated manufacturing facility to be built near the Kennedy Space Center and fabricate the rockets on-site in Florida. Although this same approach had been used to smooth production difficulties during the Apollo program, such spending was now an unthinkable extravagance. NASA opted instead to have Thiokol make the rockets in several segments, which could be shipped cross-country and put together by the company's ground crews inside the giant Vehicle Assembly Building at the Cape.

The solid rocket contractor came to the Space Shuttle program with decades of expertise in defense and aerospace work. The Thiokol Chemical Corporation had first arrived in Utah from New Jersey in the late 1950s, and begun building a manufacturing and test facility on thousands of acres of land north of the Great Salt Lake, an arid wilderness of abandoned railroad towns, sagebrush, rice grass, and snakeweed, baked beneath a relentless sun in summer and scoured by snow and wind in the winter. Although without water or electricity when the company arrived, the remote site was a perfect place to make and test volatile rocketry: separated from populated areas by miles of barren hinterland—to ensure the safety of the local population in the event of an accident—but close enough to a railhead to provide convenient transportation. Construction on the plant began within sight of Promontory Summit, where the two ends of the nation's first transcontinental rail link had been joined with a golden spike in the second half of the nineteenth century.

An aggressive player in the military-industrial complex that flourished during the hottest years of the Cold War, Thiokol expanded quickly—manufacturing motors for the United States' first mass-produced intercontinental ballistic missile, the Minuteman. Beginning with a cluster of open-plan buildings so rudimentary that at first power cables dangled from the ceilings in the engineers' offices, the Thiokol complex eventually sprawled through twenty thousand acres of the Wasatch Mountain foothills, employing more than three thousand people, encompassing its own test range, internal road network, and an airstrip for the company planes.

The plant was more than twenty miles away from the nearest town, and workers and management alike carpooled there each morning from distant homes in Ogden, Tremonton, or Brigham City. An early Mormon settlement named for Brigham Young, the church leader who had first guided the religion's pioneers west to the Utah Territory, the city mushroomed into a company town with new schools, roads, and subdivisions, its fortunes waxing and waning with the defense and aerospace contracts upon which Thiokol's new Wasatch Division relied to stay in business. As the company churned out rocket motors for nuclear missiles in the Kennedy years, the prairie boiled with activity day and night. But by the early seventies the arms race with the Soviet Union had abated, Pentagon orders dried up, and the layoffs began; the contract to build the boosters for the shuttle—worth the better part of a billion dollars—was a godsend for Thiokol and the local community. In 1982, when the corporation merged with the Morton Norwich Company—known in households across the country for their blue packages of table salt—to create Morton Thiokol, Inc., the new conglomerate's Aerospace Division would become one of its most profitable operations.

Yet, despite its rising national profile, Thiokol's Wasatch operation did not run with the same pitiless rigor as its corporate competitors. With its homogeneous staff and decades-old routines, the Promontory plant could seem more like a mom-and-pop business than one of the country's leading aerospace contractors: it was not uncommon for two generations of the same family to work at the plant at the same time, as sons and daughters followed their fathers into the company; quality assurance and engineering changes were often pursued informally, in one-on-one phone calls between company engineers; and paper trails could be fragmentary or nonexistent.

In keeping with the imperative to invent nothing new, the design and manufacture of the Space Shuttle solid rocket boosters drew heavily on previous experience with similar strap-on rockets developed in the 1960s for use with Titan III missiles—built for the US Air Force for military satellite launches. Although much smaller and less powerful than those intended to take the shuttle into orbit—and never intended to be reused, in the way the shuttle motors were—the Titan boosters were simple, and among the most reliable solid rockets ever built. So the Thiokol designers simply scaled up the Titan design, opting to add new features only when strictly necessary.

The main body of the rocket was formed from eleven drum-shaped casing segments, twelve feet across and thirteen feet high, forged from giant blocks of solid-steel alloy by a specialist contractor in Wisconsin. On delivery to Thiokol in Utah, these pieces were partially assembled into four larger segments. The tightly sealed connections between these smaller sections were known by the engineers as "factory" joints; the three between the remaining segments, which would be put together at the Cape, they called the "field" joints.

Inside Thiokol's Promontory plant, workers lined the casing segments with a layer of rubber insulation before filling them with a pyrotechnic slurry of aluminum powder and ammonium perchlorate mixed, like explosive cake batter, in gigantic bowls equipped with remote-controlled blending paddles. Combined with a polymer binder and a curing agent, the viscous liquid was poured around a Teflon-coated steel mold that, when removed, left each filled segment with a hollow core: this space would form the combustion chamber of the completed rocket. The shape and diameter of the chamber was molded to determine the power and duration of the booster's flight in advance—a complex star shape created a greater surface area, leading to more thrust; a smooth bore, with less surface to burn, led to correspondingly less thrust. Once ignited, the aluminum compound would burn fiercely and completely, from the inside out, belching flames and hot gas out through the rocket's exhaust nozzle at temperatures of up to 6,500 degrees Fahrenheit. When spent, the rocket's burning propellant left behind only the empty steel casing and the charred remnants of the insulation designed to protect its metal skin from being consumed by the heat.

The molding of each booster's core was precisely calculated by Thiokol engineers to ensure that the flaming propellant would burn most intensely at the beginning of ascent—when the shuttle required the greatest lift—and slowly subside as it drew closer to space. And the exhaust nozzles of the two rockets, laminated with ablative carbon fiber intended to burn slowly away during its journey to orbit, could be gimballed by the shuttle's onboard computers to help steer the spacecraft as it rose. Otherwise, once lit, there was no means of controlling the boosters in flight.

The astronauts' lack of authority over the rockets was reflected aboard the orbiter where, among the hundreds of dials and meters visible on the flight deck, not one provided any data about their performance. The onboard computers relayed pressure readings from inside the boosters' combustion chambers down to the consoles in Mission

Control, providing a real-time forecast of when their burn would be complete; but the flight controllers knew that any attempt to jettison them before the fuel was consumed—a little more than two minutes after launch—would result in the prompt immolation of the spacecraft. With no control and no means of escape, this made the first 122 seconds of flight the most dangerous part of any mission aboard the shuttle.

:::::::::::::::::

Baked in massive casting pits for four days at 135 degrees Fahrenheit until it set into the rubbery consistency of a pencil eraser, the propellant mix for the boosters was so volatile that the first workers at the Wasatch plant wore safety shoes or stuck adhesive patches on the heels of their boots, as a single spark from a nail on the concrete factory floor could cause a catastrophic explosion; in an accident, flash fire could propagate so quickly inside the buildings that some crews' emergency evacuation routes were not through doors, but enclosed chutes down which they could escape in seconds. Once filled with propellant, each finished segment was plastered with a large cherry-red warning label reading LOADED before being prepared for shipment to the Cape, where a separate Thiokol team would assemble them into finished rockets.

The four fueled motor segments, each of which weighed as much as 300,000 pounds, had been designed to fit together in a similar way to those of the Titan missile boosters, using a simple tang-and-clevis joint: a tongue machined along the bottom edge of each upper segment—the tang—fitted inside a groove—the clevis—in the top of the one below it. Once the parts were slotted together, the joint was fixed in place with high-strength nickel-alloy pins driven through 177 holes drilled around the circumference of both tang and clevis. The integrity of this seal—especially vulnerable at the moment of ignition, when the pressure inside the rocket rose to around one thousand pounds per square inch in less than a second—was crucial to the safety of the shuttle boosters. If any of the high-pressure gas burning inside the rocket leaked through the joint, it could quickly eat through the half-inch steel casing of the rocket like a monstrous blowtorch, cutting it apart in flight.

In the absence of an escape system to rescue the astronauts during the boosters' two-minute burn, NASA managers had to reduce the risk of such a calamity as far as possible; the rockets would simply have to work perfectly every time. So the Thiokol engineers modified the design taken from the Titan III joint, with the intention of making it safer. In the Air Force booster, the surfaces of the tang and clevis were covered with a layer of insulation that fitted tightly together when the rocket was assembled, and sealed with a flameproof putty. In addition, there was also a single synthetic rubber gasket encircling the joint, similar to those used by plumbers to seal bathroom or kitchen faucets: an O-ring.

Designed to fit into a narrow groove machined around the edge of the tang, the diameter of the O-ring was slightly larger than the gap between the two halves of the joint, so that—when the tang and the clevis slid together—the rubber would be snugly compressed between the metal surfaces of the casing, ensuring an air-tight seal. This compression—the engineers called it the "squeeze," and measured it obsessively to tolerances of four thousandths of an inch, or the thickness of a sheet of copy paper—was vital to the safe functioning of the joint. If the diameter of the O-ring was too small, or the space between the metal surfaces of the joint too large, then the squeeze would not be tight enough to hold back the sudden rush of hot, high-pressure gas surging through the rocket at ignition, and the joint would spring a leak.

To upgrade the design to make it safe enough to carry humans— "man-rated," in NASA jargon—the Thiokol engineers added a layer of redundancy. They made the clevis a little longer, and added a second O-ring, settled in its own groove behind the first. Now if the primary O-ring leaked for any reason, the secondary ring, directly behind it, would make sure the joint remained sealed. And, in pursuit of a yet wider margin of safety, the engineers' modifications went still further. Because the seal between the layers of insulation in the new joint was not as gastight as in the Titan III, it would be packed with larger amounts of flameproof asbestos-filled putty.

By the time the design was complete, both the Thiokol engineers and those supervising them at Marshall still believed they were employing a tried-and-tested joint for their new rocket: they had avoided any potentially dangerous innovation. But this was a convenient delusion. Not only were the Titan and the new solid rocket booster quite different in scale and performance, but in creating a man-rated, fail-safe joint they had also modified it so extensively that what they produced was, in effect, a quite new and experimental design.

:::::::::::::::::

The first full-scale test-firing of the Space Shuttle solid rocket booster took place at Thiokol's Promontory test range one hot afternoon in July 1977. The 150-foot-long white rocket lay on its side amid the sagebrush, its nose anchored against a massive concrete block embedded in the hillside, its casing hung with temperature, pressure, and load sensors carrying data to a bunker half a mile away. In the valley below, dozens of parked cars straddled the sides of the state highway, where plant staff and curious locals from Brigham City and beyond crowded along the perimeter of the range to watch. As the countdown reached zero, they saw the flash of ignition, and four hundred feet of flame bloomed from the throat of the rocket. The blacktop trembled, and a few seconds later a heavy roar rolled over the crowd; a voice carried down to them by loudspeakers counted off the duration of the test: 30 seconds . . . 60 seconds . . . 90 seconds. Black smoke and clouds of dust rose a thousand feet into the sky as the shackled rocket blasted impotently at the empty mountainside, the heat transforming the surface of the sand into quartz. At two minutes and four seconds, its fuel exhausted, the yellow-white flame vanished, and blooms of white vapor enveloped the rocket as an automatic quenching system flushed carbon dioxide gas through the casing.

Demonstration Motor One was a stunning success, a testament to the hard-won experience and ingenuity of everyone at the Thiokol plant. To mark the occasion, jubilant managers gave everyone at the complex a commemorative brass belt buckle, cast with engraved images of the Space Shuttle, and of the prone rocket motor firing beneath the jagged silhouette of the Wasatch Mountains.

But when Thiokol engineers conducted further trials on the rockets two months later, they discovered a problem: the crucial field joints between the four propellant sections of the motor did not perform quite as they had expected. Their design studies had told them that, in theory, the sudden pressure increase inside the rocket at ignition would close the two halves of the joint together, compressing the rubber O-rings, increasing the squeeze and securely tightening the entire assembly against the force of expanding gas. In practice, the opposite was true.

In a series of pressure tests intended to verify the strength of the rocket's steel casing, the bewildered engineers discovered that, at the moment of ignition, the walls of the casing ballooned outward like the cheeks of a trumpet player hitting his highest notes—and the two halves of the tang-and-clevis joint sprang apart. This opened the space around the O-rings, compromising the squeeze.

It was true that, after a few milliseconds, the wave of pressure from inside the rocket rolled the first of the two O-rings out of its mounting groove and forced it into the gap, sealing it successfully. But, for the crucial moments before that, the joint could leak. Furthermore, even if it did eventually seal, it was not how the joint was supposed to work: the O-ring had been designed to remain in its groove during ignition. Still, it seemed to do the job, so the Thiokol engineers decided to accept it as it was. They gave the expansion process a benign-sounding name—"joint rotation"—and suggested that the phenomenon was due to the nature of the tests, which had gone beyond the design limits the rocket was expected to reach in flight, increasing internal pressure repeatedly until the booster casing was destroyed. They decided that any other anomalies they had discovered could be corrected by making sure that the technicians assembling the rockets at Cape Canaveral before a shuttle launch did so with extreme care—and by using small steel shims, hammered gently into the joint with leather-covered mallets, to tighten the squeeze.

But down in Huntsville, the technical specialists in the labs of the Marshall Space Flight Center were not convinced. In a pair of memos written in 1978 and 1979, they said that the joint had been badly designed, and needed to be modified. They called for the problem to be addressed as soon as possible—failure to do so would be unacceptable—and wrote that the issue required urgent attention to "prevent hot gas leaks and resulting catastrophic failure."

The Marshall laboratory engineers went so far as to visit the Parker Seal Company in Kentucky, manufacturer of the O-rings, whose managers were astonished that their seals had performed as well as they had so far, and explained that they were not being used in the way they had intended.

But the memos from Marshall were swallowed in the furious paper chase of NASA's growing bureaucracy: one received no response at all; neither ever reached Thiokol; and the engineers in Utah never learned of the disconcerting visit to the Parker Seal Company. When they continued to object, the Marshall engineers were overruled by their managers in Huntsville, who trusted the Thiokol experts more than their own laboratory staff, who—after all—knew a great deal about the exotic and complex world of liquid-fueled spaceflight, but had next to no experience of solid rocket motors.

:::::::::::::::::::

Meanwhile, static firing tests of further Demonstration Motors on the mountainside at the Promontory range continued to produce more reassuring data about the booster joints: although the Thiokol engineers noted that the tang-and-clevis joints did sometimes gape open during the first fractions of a second after ignition, one after another, the giant rockets burned for the required two minutes without leaking. For all the troubling data about joint rotation, and the occasional dissenting voice, the majority of the test results looked good.

As part of the shuttle system's final flight qualification before the maiden launch of *Columbia* in 1981, managers from both the Marshall Space Flight Center and Thiokol presented their findings about the solid rocket boosters' performance to a NASA certification committee. The members of the committee expressed concern about leaking O-rings in the booster joints, and worried that the leaks could grow worse, to the point where a hot jet of escaping gas might cut all the way through the rocket casing to reach the giant external tank and its load of pressurized oxygen and hydrogen fuel. The committee recommended a thorough program of further tests, including full-scale static firings of the rocket at a range of temperatures covering the expected seasonal extremes of weather at Cape Canaveral, to confirm that the joints would always seal safely at ignition.

But Marshall and Thiokol managers assured the members of the committee that the existing data they had seen was not reliable—and besides, a series of fixes, including thicker O-rings and rigorous assembly procedures, were already under way; these changes would surely take care of the potential leaks. Apparently satisfied, the committee decided that none of the rigorous tests they had suggested was necessary after all. In October 1980, they certified the solid rocket boosters for flight. On the Critical Items List—the roster of every one of the thousands of shuttle components crucial for flight, which dictated the degree of attention each received from the engineers before launch—the joints were designated as Criticality 1R, or fail-safe.

At around the same time, a separate NASA department had commissioned an overall risk assessment of Space Shuttle safety. Completed in 1982 by the J. H. Wiggins Company of Redondo Beach, California, the study focused on the possibility of an orbiter being lost during launch. The Wiggins analysts concluded that the solid rocket boosters posed the highest risk in the entire shuttle system: based on records of past solid motor performance, they found that individual rockets had experienced a catastrophic failure once every thirty-four to fifty-nine launches. Thus, a new vehicle riding on a pair of solid boosters could be expected to undergo a failure resulting in the loss of crew and spacecraft once every eighteen to thirty missions.

Yet NASA officials had long ago learned that they could disregard such quantitative number crunching when it proved inconvenient: during the Apollo program, the agency had contracted a branch of General Electric to produce a probabilistic-risk assessment of its plan to land men on the moon and return them safely to Earth. When the GE analysts reported that they estimated the chances of doing so were less than 5 percent, Administrator Tom Paine had the study quietly buried, for fear of damaging the program: if it got wind of the statistics, the American public would never support such a self-evidently perilous endeavor.

Now the Space Shuttle Range Safety Ad Hoc Committee had no qualms about dismissing J. H. Wiggins's conclusions. Arguing that they were based on outdated information, the NASA officials persuaded the analysts to adopt a far more optimistic view of risks to the shuttle system: they suggested that they would expect to see a failure once in every one thousand launches. In fact, when taking into account the technological improvements made to the boosters by Thiokol—including the modified joint, with its backup O-ring—the committee said that the odds could, perhaps, be improved by a whole order of magnitude. Eventually, they coaxed the analysts to conclude that a catastrophic failure would occur only once in *ten* thousand launches.

If these estimates proved accurate, it would allow NASA to fly a new shuttle mission once a week for a hundred years before witnessing a disaster during launch.

Still, when the Thiokol technicians at the Cape began stacking the rocket segments together inside the great cathedral spaces of the Vehicle Assembly Building, they did so with extreme care. The O-rings themselves were especially delicate. Although in principle no different to the gaskets used to seat a kitchen faucet or to seal a carburetor in an automobile engine, the similarities ended there. Milled from lengths of Viton, a synthetic rubber compound, the finished rings were only a quarter of an inch thick, but thirty-seven feet in circumference, so large that the material was shipped in twelve-foot sections from Kentucky to Utah. There, a small family company in Salt Lake City employed a team of half a dozen women—part of a local polygamous sect, who wore the bonnets and long skirts of nineteenth-century Western dress—possessed of the extraordinary patience necessary to splice the pieces successfully together. After curing in a cluster of commercial ovens, the completed rings were greased and sealed into wax-sealed paper bags for delivery to Cape Canaveral.

In the Vehicle Assembly Building, the booster-stacking process became an elaborate and fastidious ritual carried out over the course of more than five days by a team of fourteen technicians from Thiokol, usually at night, when the building was quiet, to reduce collateral damage in the event of an explosion. The mechanical tolerances between each segment of the rocket were so tight that the smallest contaminant caught between the O-ring and the metal faces of the joint—a fleck of lint, a single human hair—could cause a leak. Wearing surgical gloves, hats, and masks, the team carefully removed each ring from its packing, wiped it down with alcohol, inspected it for nicks, and checked with a NASA quality control supervisor before re-greasing it and stretching it very gently to fit over a waiting rocket segment. After that, they prepared to pack the joint with asbestos-filled putty, which they kept in a freezer, coiled inside drum-shaped cans like movie film, to prevent it growing sticky in the Florida heat.

::::::::::::::::::

The meticulous sacraments of solid rocket motor assembly had been shaped, in part, by the discoveries made by the engineers at Marshall and Thiokol after the second shuttle flight in 1981. Their analysis of the charred O-ring found in one of the booster joints from the mission revealed that the damage had probably been caused by gaps in the asbestos putty packed around the seals, created accidentally when the segments had been stacked at Cape Canaveral. Air trapped in the joint had formed bubbles in the putty, which became compressed as the two pieces of the casing were assembled. Under greater and greater pressure as technicians pushed the tang and the clevis together, the bubbles eventually burst, leaving behind a small hole in the putty, no bigger than the width of a pencil. Thiokol engineers had spotted these "blow-holes"—which resembled tiny volcanic craters on the inside surface of the joint—when preparing demonstration motors for testing on the mountainside in Utah. But the engineers there had assumed that the phenomenon was caused by the way the rockets were assembled at the Promontory range—horizontally, instead of vertically, as they would be at Cape Canaveral. So before each test, a senior member of the program team at Thiokol crawled inside the combustion chamber of each test rocket and tamped the blow-holes closed with a broom handle. The engineers neither confirmed their assumptions with experiments, nor reported the fault as part of the testing regime. They never expected to see it in flight.

But when the rockets of that second shuttle mission ignited, a blowhole in one of the booster joints had allowed hot gas to leak out of the combustion chamber and—focused into a searing jet burning at 6,500 degrees Fahrenheit—scorch the surface of the primary O-ring seal. The leak lasted just a few hundredths of a second, and stopped as soon as the pressure in the space around the ring equalized with that inside the chamber of the rocket. But even that was enough time for the gas jet to vaporize part of the Viton gasket, which had never been designed for direct exposure to such high temperatures.

The sterile language the technicians used to describe this phenomenon—they called it "erosion"—made the process sound slow and gradual, like the passage of water wearing smooth the pebbles on a riverbed. This was misleading: the erosion of the Viton seals was violent and sudden, a flash of destruction complete in less time than it takes to blink. And if the problem continued, it would transform those moments immediately after each solid rocket motor ignited into a potentially deadly race between the O-rings and the hot gas. As long as the seals could roll into position before they were too badly burned, and the gap in the joint became too large, the shuttle would fly safely. But if the hot gas had time to reach and vaporize enough of just one synthetic rubber ring before it could squeeze tightly into the gap opening in the joint, the ring could become too severely damaged to provide a seal. Then, disaster might be inescapable.

Back in the labs in Huntsville, the engineers conducted tests on samples of the O-rings. Their findings seemed reassuring. They established that the seals could endure three times more erosion damage than they had seen on the second shuttle flight before they failed. And the redundancy designed into the joint meant that, even if the first seal did leak, the shuttle would still be safe: the secondary O-ring would always be there as a backup. As far as NASA's solid rockets chief George Hardy was concerned, the system had worked exactly as intended. "You don't build in redundancy and never expect to use the back-up," he said. "If you never use your back-up, you're wasting money."

In the wake of the engineers' analysis, the blame for the blowholes settled on the consistency of the asbestos putty, which had apparently become unworkable in the heat and humidity of the Vehicle Assembly Building. They reformulated the putty, and instructed the stacking crews at the Cape to put it in a freezer until they were ready to use it. With these new procedures in place, postflight inspections after subsequent missions revealed no further charring of the O-rings. The Thiokol engineers believed they had solved the problem.

In November 1982, with the shuttle about to begin the first of the "operational" flights of what excitable futurists had taken to calling America's National Spaceline, George Hardy was promoted into a more senior engineering role. His replacement would face different priorities: no longer ironing out the glitches of an experimental spacecraft in development, but instead making the rocket program run so smoothly that NASA could soon launch twenty-four shuttle flights a year.

The new manager of the solid rocket booster project at the Marshall Space Flight Center seemed just the man to make that happen: Lawrence B. Mulloy, forty-eight, had been in Huntsville for twenty-two years, working first on the Apollo project and then, with the advent of the Space Shuttle, as chief engineer for the shuttle's external tank. A big and gregarious Cajun from Shreveport, Louisiana, Mulloy had the ursine presence of a linebacker, and the self-confidence to match. With a private pilot's license and a degree in aeronautical engineering from LSU, he was technically able, and still excited by spaceflight; as a former officer in the 11th Airborne, he understood the chain of command, and his decades at NASA made him a practiced hand in the puissance of agency bureaucracy. Larry was good-humored and collegial—he liked to spend weekends on a houseboat he shared with friends, shucking oysters or barbecuing on the banks of the Tennessee River. But he could also be overbearing—and seemed untroubled by reflection or doubt.

Despite his decades at Marshall, the new job would mark Mulloy's first involvement with the solid rocket booster project. So before he started, he received a series of briefings: some from George Hardy and the other engineers who worked on the solids in Huntsville; for others, he flew out to Utah, where the managers and engineers at Thiokol talked him through the history of the rockets. Mulloy learned that the boosters did not behave quite as designed—he heard about the joint rotation, and the charred O-rings—but he also understood that the problems had been investigated and analyzed, and the rockets cleared for flight. Just a few days after he started work, the first of the operational missions of the Space Shuttle lifted off from Cape Canaveral, with the fifth voyage of *Columbia*, on November 11, 1982.

When the boosters from the launch of STS-5 were recovered from the Atlantic, the postflight inspection revealed no anomalies at all. Astronauts and senior NASA managers alike came to regard the solid rockets as simple but reliable—little more than massive steel tubes filled with fuel; in an otherwise complex and unforgiving system, it was the one element no one had to worry about too much.

CHAPTER TWELVE

THE BLACK CAT

At 1:30 p.m. on April 4, 1983, after months of delays caused by a faulty main engine, the Space Shuttle *Challenger* lifted off from Pad 39A at the Kennedy Space Center in Florida, on its maiden flight—becoming the second orbiter in the growing fleet to reach space. The small team of veteran astronauts aboard launched a satellite from the payload bay, and conducted the first-ever space walk from the shuttle—and the first by an American in nine years—before landing in California five days later. "What the shuttle really is," NASA's program chief told the press the next day, "is a means of making dreams come true." He predicted that one day soon, seats aboard *Challenger* and her sister ships would be available to take paying passengers into orbit. "The future is in space," he said.

From then on, the pace of shuttle missions increased, passing a further series of landmarks in American space exploration: the next flight, just two months later, made Sally Ride the first female astronaut, and an international celebrity; at the end of August, Guy Bluford became the first African American in orbit. Afterward, Bluford embarked on a national tour of schools and colleges, was honored by the NAACP alongside the first Black Miss America, Vanessa Williams—and photographed partying in New York at Studio 54.

Leaving Cape Canaveral at a little after 2:30 a.m. Eastern time, Bluford's mission marked the first night launch and landing of the shuttle, and the first use of a pair of redesigned high-performance solid rocket boosters, on what NASA chiefs initially boasted was "the cleanest mission yet." But when the spent casings of these boosters arrived back in Morton Thiokol's plant for refurbishment, the engineers in Utah made a frightening discovery—an entirely new problem unrelated to their previous worries about O-rings or joint rotation. Their inspection revealed that the laminated lining of one of the two rockets' nozzles, designed to be eroded slowly away during flight to protect it from the plume of superhot gases escaping from the end of the booster, had been eaten almost completely through by flaming exhaust.

They calculated that if the rocket had burned for just eight or nine seconds longer, the spacecraft would have been destroyed, killing all five astronauts on board: the most catastrophic accident in NASA history. Frantic Thiokol engineers worked around the clock for weeks to understand the problem and refit the boosters. Not only did they have to make them safe to fly, they also had to prepare the data to conclusively prove that they were—under the withering cross-examination they knew they would face during the next Flight Readiness Review.

::::::::::::::::::

The bureaucratic climax of the long preparation for every Space Shuttle mission, the Flight Readiness Review process began when the shuttle was already sitting stacked on the pad at Cape Canaveral, some six weeks before each launch—and was a key part of calculating what NASA engineers called "acceptable risk."

Regardless of what the politicians might say about operating the shuttle as if it were a space-going airliner, the engineers who worked on the program understood that they had built an experimental vehicle working within the narrowest margins of safety in an unforgiving environment; flying into orbit had always required striking an often knife-edge balance between the hazards inherent in exploring the unknown, and the morally acceptable levels of diligence and safety expected of a US government agency.

And the shuttle was so complex that those at the top of the NASA pyramid like Administrator James Beggs or Marshall Chief Bill Lucas recognized that risk could never be eliminated from the program; it could be calculated and minimized but, by the time the orbiter left the launchpad on each new mission, some residual potential for failure or catastrophe would always remain. It was left to the agency engineers and their contractors to determine how much risk they could tolerate.

To do so, NASA had formalized a process by which engineers assessed each component of the shuttle to calculate how likely it was to go wrong, and whether the risk of it doing so was acceptable. If the engineers discovered an outstanding fault in any element of the shuttle, it could not fly until the problem—or "anomaly," in the bloodless technical jargon they favored—had been understood and corrected, or the project engineers had conducted enough testing to satisfy NASA managers that it wouldn't threaten the lives of the crew. Only once that happened could the component be designated an acceptable risk and cleared for flight.

The idea was that, even considering the accumulated risks of all of the millions of parts and systems in the shuttle system together, it remained a safe-enough vehicle in which to fly men and women into space. The process began years before the first shuttle flight in 1981, and continued as the program went on. It was necessarily complicated and time-consuming: a tabulation of the acceptable risks in the shuttle compiled before that first launch alone filled six volumes.

The Flight Readiness Review was the forum in which outstanding anomalies were considered before each and every new shuttle mission. In a long series of increasingly inquisitorial board meetings, the project engineers were asked to prove to their managers that the equipment they had built matched the design criteria approved by NASA; to report any existing or potential deviations or problems; to explain how they had been addressed; and, ultimately, demonstrate that it was safe for the purpose for which it had been devised. The review was strictly hierarchical, beginning with internal conferences at the headquarters of the contractors responsible for each of the shuttle subsystems—the main engines, the external tank, the solid rockets, the orbiter itself—and inching upward through the four tiers of NASA's bureaucracy. As the review ascended and the launch date drew inexorably closer, the engineers resolved more and more of the existing anomalies, thinning out the number of problems that would eventually be discussed with senior NASA leadership. At last, two weeks before the launch date, the process culminated in the ultimate Flight Readiness Review, a meeting of representatives of every branch of the agency's sprawling network of technicians and experts, in a large conference room at the Kennedy Space Center.

Here, Administrator Beggs himself would often hear his engineers' final assessment of acceptable risk before one of his deputies placed his signature on a document giving his approval for launch: the Certification of Flight Readiness. Beggs had become infamous at the Cape for his merciless eye for detail, and closely tracked almost every failure and anomaly in the shuttle. Between launches, he hounded his staff with lists of those problems that weren't being resolved to his liking: "The brakes . . . the auxiliary power unit . . . the computer. You're not getting it fixed fast enough," he told his immediate deputy for technical matters, Mike Weeks. Little seemed to escape Beggs's attention, but those beneath him—especially the Center Directors like Lucas—resented his scrutiny. They were angered by his micromanagement; his constant questions; his attendance at the reviews and launches; and the administrator's intrusion into their affairs. But Beggs wanted to send a message about the shuttle program: "Be very, very careful about anomalies."

::::::::::::::::::

As the final countdown approached, further, smaller meetings were convened to resolve remaining issues two days before launch—"L minus 2"—and one day before launch—"L minus 1." By that time, there was usually little for the contractors to do but give their verbal agreement that they were ready to go. But before anything could approach that point, hundreds of engineers involved in the program at every level—at the contractors across the country and within NASA's offices in Washington, DC; Houston; Huntsville; and at Cape Canaveral—had to agree on what constituted acceptable risk for the mission. The Flight Readiness

Review meetings were designed to hammer out this consensus upon the anvil of the scientific method, using data and engineering logic to exclude disagreement, emotion, and faulty analysis from the launch decision chain. The meetings were intended to be adversarial, and the questioning could be harsh: the contractors had to submit packages of charts in advance, showing their risk assessments and their recommendations for launch, before delivering an oral presentation at the meeting.

The NASA managers—many of whom were also highly qualified engineers—came prepared to probe these findings, with the aim of exposing any problems that might prove fatal. With such high stakes, the cross-examination was often ruthless, and continued right up to the final link in the chain. One senior engineer concluded his presentation in a meeting weeks before one mission at Cape Canaveral—explaining why the experimental robotic system designed to land the shuttle under computer control should not be employed—only to have James Beggs issue his own curt judgment: "I think your arguments aren't worth shit," the Administrator said.

Within this exacting framework, the contractors and the government staff to whom they answered existed in uneasy symbiosis. The engineers and managers at Morton Thiokol in Utah working on the solid rocket booster project did so under the direction of the NASA team at the Marshall Space Flight Center, and the company relied heavily on Huntsville for its continued prosperity. Yet the reputations of both Marshall and Thiokol were bound to the success or failure of the shuttle program. The two teams of engineers worked closely together, sometimes in the same offices: some employees from Marshall were based permanently in Utah, while Thiokol specialists flew frequently to Alabama. But they also maintained parallel testing facilities that produced dueling results, and often disagreed about what they meant.

Corresponding tensions simmered within Marshall itself, where the laboratory engineers of the center's Science and Engineering Directorate were often at odds with the project managers of the solid rocket booster program. While the managers' decision-making was based in part on technical considerations and on flight safety, they also had to consider the financial cost of their choices, and their impact on the shuttle launch schedule; by comparison, the lab engineers were in a caste by themselves—purists untroubled by cost or budget concerns, whose loyalties lay in the most conservative safety margins, the highest tolerances and principles of multiple redundancy. Nor could the men of the Science and Engineering Directorate, who had been bureaucratically insulated from management interference since Wernher von Braun's time, be easily silenced.

Instead, they regarded it as their responsibility to seek out faults in the booster program, and pressed to have them corrected. It was these engineering watchdogs who had first questioned the design of the solid rocket joints in the late seventies, writing critical memos and visiting the Parker Seal Company before eventually having their objections overruled. With their potential to ask awkward questions that could require costly equipment redesign and launch delays, the engineers from the Science and Engineering Directorate were detested by the managers at Thiokol, who regarded them as muckraking troublemakers. Behind their backs, some of the Utah engineers took to calling them the "bad news guys."

By the end of 1983, as one Flight Readiness Review had followed another and still Thiokol failed to provide a satisfactory explanation of why the nozzles of its rockets had begun to burn through in flight, the engineers' overseers in Huntsville were becoming impatient. But worse was to come.

On the morning of March 2, 1984, a Morton Thiokol company jet touched down in Huntsville, carrying Allan McDonald, a forty-six-year-old project engineer in the company's Wasatch Division, and two of the most senior executives from its aerospace arm. The three men were on their way to a meeting with Bill Lucas at the Marshall Space Flight Center, about plans to repurpose the shuttle booster design for a new

Air Force spaceflight project. They had barely stepped off the aircraft when they saw a figure sprinting toward them across the tarmac, shouting as he approached. It was the manager of Thiokol's field office at Marshall, bringing shocking news from the plant in Utah: a major fire . . . explosions in the shuttle rocket casting pits . . . early reports of several deaths.

Sitting around the impressive wooden desk in Lucas's office, discussion of any future projects was cast aside as the executives struggled to explain what had happened, and how it might affect the shuttle production schedule. Over the phone to Utah, they established that—although most men had miraculously escaped with only minor injuries, and no one had been killed after all—the damage to Thiokol's facilities was extensive; it was clear they needed to see it in person.

The three men raced back to the airport for the cross-country flight to the plant. The sun was setting as the plane came in over Promontory Point, and the pilot dropped into a series of low-level passes so that Mc-Donald and his colleagues could peer from the windows to survey the destruction below; a company car waited by the runway to take them directly to the scene. At what remained of the casting facility, blackened wreckage lay everywhere across the snow-covered ground; the blast had scattered bent and broken steel beams around the remains of the pits, where workers had been pouring propellant into new solid rocket motor segments for the shuttle; some segments had been partially melted by the intense heat; the two-hundred-ton overhead crane assembly had been torn from its mountings by the explosion, and the building surrounding it had been obliterated.

Almost a quarter of a million pounds of propellant had caught fire and been thrown from the steel casing of one rocket motor segment, exploding across the site, starting secondary blazes and igniting a second segment curing nearby; both segments—containing nearly three hundred tons of propellant—had been destroyed. It would take months to rebuild; the effect on the production schedule almost certainly calamitous. Later, listening in on the meeting convening an investigation into the accident and determining what to do next, McDonald felt quiet relief that he had no direct responsibility for Thiokol's work on the shuttle program.

By the time he sat down in the Senior Vice President's conference room that evening, Al McDonald had worked at the plant for almost twenty-five years: his entire professional career. A third-generation descendant of Scottish-Irish immigrants, he had taken the job at Thiokol over offers from Standard Oil and Esso in New York to stay closer to his childhood home in Montana, and soon after arriving in Utah married Linda, an Italian American from Detroit who worked for the company over in Plant 78.

Coolheaded and capable, McDonald was an expert in the art of technical presentations—but also an iconoclast, not shy about his accomplishments and unafraid to speak his mind. Until a few months earlier, he had been Manager of Project Engineering for the Wasatch Division, heading up a team of some seventy specialists at work on strategic and tactical missile programs, orbital satellite motors, and research and development. He enjoyed his work, and had recently been elected as chairman of the committee dedicated to solid rocket technology at the American Institute of Aeronautics and Astronautics, the country's most prestigious trade body for rocket engineers.

But Thiokol's work on the shuttle rocket motors, managed as an autonomous operation at the plant, was something with which McDonald had never had any involvement. Then, in the summer of 1983, he was given temporary responsibility for handling a looming crisis that threatened all of the company's financial future: after receiving complaints from Hercules, Inc., a Thiokol competitor based near Salt Lake City making explosives and solid rocket motors, NASA had begun considering opening up the shuttle booster contract to competing bids from other manufacturers. This idea had sent shudders of alarm through Thiokol's senior management, who had just completed long-term profit projections based on the assumption that they would be the sole supplier of the shuttle's solid rockets for the next twenty years, expecting to ship twenty-four pairs of boosters to NASA every year from 1988 onward. If the agency now opened up the contract to competing tenders, Thiokol would stand to lose hundreds of millions of dollars.

McDonald immediately set about learning everything he could about the shuttle program and the solid rocket boosters, and began assembling what he believed was a persuasive case showing why it would be a mistake—both financially and politically—for NASA to open the contract to a competitor. Then, in August, came the almost-catastrophic failure of the booster nozzles on the nighttime *Challenger* launch. Suddenly, NASA's suggestion that it might need to take its business elsewhere didn't seem so unreasonable after all.

When he arrived in Huntsville aboard the corporate jet for his meeting with the head of the Marshall Space Flight Center early in March the following year, McDonald knew that Thiokol's multimillion-dollar contract with NASA remained in jeopardy; the fire and explosions back at the plant were the last thing he needed.

Less than a month later, soon after the Thiokol investigation team had delivered its report into the accident, one of the company's top executives called McDonald into a meeting. "Al," he said, "starting next Monday, you are being reassigned to be the Director of the Space Shuttle Solid Rocket Motor Project in project management, reporting to Joe Kilminster, Vice President of Space Booster Programs."

This was not what McDonald wanted to hear, and he said so: his assignment to work on the solid rocket booster contract had been only temporary, and he had been looking forward to returning to his old job on missiles and satellites. Now, he explained, it seemed that he was being offered the chance to step into management of a program for which a large part of the manufacturing facilities had recently burned to the ground; the sole customer was looking for an alternative supplier; and its principal product had demonstrated a potentially catastrophic fault the cause of which the company's engineers still did not fully understand. Under the circumstances, McDonald suggested he might pass.

"Al, this is a good opportunity for you to get some program management experience," his boss said; McDonald knew he had no choice.

It was the first time in his life he had felt depressed after receiving a promotion.

::::::::::::::::
::::::::::::::::

Back in Houston, the shuttle program continued, and gathered tempo. After the ninth successful shuttle mission, the simple numerical designations of the early flights were replaced by a Byzantine system of code numbers and letters. Officially, the change was designed to avoid bureaucratic confusion when missions were postponed and reshuffled. But many NASA veterans believed it was a result of Administrator James Beggs's superstition about the number thirteen, a fear that had become widely ingrained at the agency after the near catastrophe of Apollo 13—infamously launched at 13.13 Houston time and scheduled to reach lunar orbit on April 13. So Beggs issued instructions for an entirely new numbering system. "I don't care what you do," he said, "but we're not going to fly an STS-13."

Using the new method, what would have been the thirteenth shuttle mission was now renamed STS-41-C: it would be Dick Scobee's first flight, taking the copilot's seat in the cockpit of *Challenger* beside Commander Bob Crippen. Both Crippen and Scobee thought the new flight designation system ridiculous—and Dick had always told his wife that thirteen was his lucky number. So the astronauts submitted to NASA their design for an official patch to stitch on their flight suits, embroidered with the 41-C code, depicting the shuttle in orbit beside its main payload, a package of long-term space experiments locked in a container the size of a school bus. But Scobee also sketched out his own maverick design, which he had made up and distributed to the crew: this showed a crazed-looking black cat, its back arched, a bright green number *13* emblazoned on its flank, with a shuttle orbiter swooping between its legs.

In the end, James Beggs's superstitions were apparently misplaced. The launch of STS-41-C looked flawless. From the spot reserved for crew families on the roof of the Launch Control Center at Cape Canaveral, June Scobee and her two children watched, exhilarated, as *Challenger* tore its path to orbit through a clear blue sky. With a landing planned six days later on the shuttle landing strip on Merritt Island, Dick had told her to enjoy herself while he was away in space. "You and the kids go have fun," he said. "Because *I'll* be having fun." June, Rich, and Kathie spent the next week together at the beach, and visiting Walt Disney World; meanwhile, Dick Scobee's first shuttle mission went almost entirely as planned, until the last minute.

After bad weather forced the crew to divert the landing from Florida to California, *Challenger* touched down safely just after 5:30 a.m., on Friday the 13th. The five men flew back from Edwards Air Force Base aboard a NASA jet to Houston; their wives were waiting for them on the tarmac. When June at last saw Dick coming down the aircraft steps, she ran out ahead of the other women and leapt into his arms.

CHAPTER THIRTEEN

THE HUMAN SATELLITE

One afternoon in August 1983, Kurt Heisig was at his workbench at his house in Rio Del Mar, a picturesque village on the coast of Northern California, when the phone rang: it was a clerk from the music store Heisig owned down in Monterey, specializing in woodwind and brass instruments. Heisig's clientele included scientists and venture capitalists from Silicon Valley; he built and repaired instruments himself, but also played in a band and offered saxophone lessons. The store was often so busy that he found it easier to work from home.

But now the clerk reported that they had a customer asking peculiar and detailed questions that only Heisig could answer. Visiting from Texas, the customer was looking for a very specific type of soprano saxophone—small, and curved, not straight. But he didn't want to buy it; he wanted to rent it instead. This was a strange request, and Heisig wasn't sure he could help; yet the caller seemed earnest, smart, and sincere. There was a silence on the line as they both struggled to find a compromise. At last, the customer made a confession:

"Well, you see, Kurt, I'm an astronaut," he said. "I want to take the instrument into space—but you can't tell anyone this, I haven't even told NASA."

Ron McNair had received his first mission assignment to fly on the shuttle seven months earlier, when he was selected for the crew of

STS-41-B, the fourth launch of *Challenger*, scheduled for early 1984. The mission would include two satellite launches, and the initial test flights of the Manned Maneuvering Unit—the bulky jetpack propelled by puffs of nitrogen gas that would allow two astronauts to fly away from the shuttle once it reached orbit and make the first untethered space walks in history.

McNair was assigned to conduct a series of seventeen different experiments, operate the robot arm, take photographs, and film the mission, using a fish-eye-lens Cinema 360 movie camera designed to produce imagery that could be projected in a planetarium. His only previous experience of photography was taking snapshots at birthdays and other family gatherings with a point-and-shoot camera—nothing like the complex film camera or the large-format Hasselblad he would be using in space. So he began reading everything he could find on the subject, and practiced with the equipment at home, using it to shoot pictures of his son, Reggie, not yet two, attempting to imitate his father's karate moves.

But, in spite of all the demands on his time since he had first joined the school band at Carver High as a teenager—from teaching martial arts to the frantic effort to repeat the hundreds of experiments necessary to complete his doctoral research on time—McNair had never relinquished his devotion to jazz and, even after he reached NASA, practiced the saxophone whenever he could. Now he had seen an opportunity to become the first person to play the instrument in space, and wanted to figure out the practicalities even before he sought permission to do so from the agency. On the telephone, McNair explained to an astonished Kurt Heisig that he didn't want to jeopardize his chances of NASA approval with premature publicity about the idea, so he asked him to keep it a secret. The Yanagisawa soprano sax he wanted was expensive—and a significant investment on a government salary—but it fitted the tight stowage requirements aboard the shuttle perfectly. Heisig agreed to let the astronaut rent it, just as he had asked.

Over the next few months, in a series of long-distance phone conversations, the musician coached McNair through the potential challenges of playing the instrument in orbit, drawing on the astronaut's expertise in physics and Heisig's own study of acoustics to address the many unknowns of playing music in this strange new environment: How would spit propagate in zero gravity, or the cabin pressure affect the way air moved over the reed?

As the launch date approached, McNair called in for sax lessons over the phone with Heisig whenever he had a gap in his training schedule, and the teacher became accustomed to receiving calls at all hours, in the store and at home, sometimes in the middle of the night.

::::::::::::::::::
::::::::::::::::::

The first shuttle mission of 1984, the launch of *Challenger* on the cool, windless morning of Friday, February 3, was apparently perfect—and eagerly trailed by a national press anticipating the first flights of what they called the "Buck Rogers jetpacks," a spectacle that reporters hoped would approach the theatrics of watching men walk on the moon more than a decade before. But, once in orbit, the astronauts experienced a trickle of humiliating failures: first, McNair oversaw the launch of the Westar 6 communications satellite, which spun like a top out of the payload bay as planned, but then apparently vanished. Mission Control sheepishly admitted that they had simply lost it somewhere in space. "We can't find it," they told the astronauts over the radio link. "It's not where it's supposed to be."

Then an experiment designed to test the shuttle's facility for orbital rendezvous, using a Mylar balloon inflated with gas, also ended in farce when the balloon launched, but promptly exploded; meanwhile, the shuttle toilet—which had always been troublesome—stopped working altogether. After waiting forty-eight hours to assure themselves that Palapa B-2, the second satellite due for release on the mission, would not suffer a similar fate to Westar 6, Mission Control gave the crew instructions to proceed with launching it. Seconds later, they lost contact with that, too. The total bill for the two mislaid satellites was at least $180 million.

By the time two men—Colonel Bob Stewart and Captain Bruce McCandless, one of the veteran astronauts who had signed on to NASA during the Apollo program, and waited eighteen years for this moment—stepped into the air lock to test the Manned Maneuvering Unit on the fifth day of the mission, NASA was desperate for good publicity.

They were not disappointed. At 8:25 on Tuesday morning, Houston time, McCandless fired the thrusters of his jetpack, rose slowly from *Challenger*'s cargo bay, and flew clear of the spacecraft. He ran through his flight checklist, touching the joysticks with his fingertips to verify that the pack was working as it should: "Pitch down, pitch up, roll left, roll right . . ." he began, enunciating each word clearly into his headset microphone. Behind him, the pack trembled and shuddered like a nervous pony, as its onboard computers automatically corrected his attitude with tiny whispers of gas from its dozen nitrogen jets. Despite all his years of training, and the bitter cold inside the suit, his palms prickled with sweat; his heart quickened. "It may have been a small step for Neil," he said, "but it's a heck of a big leap for me."

Moving backward at no more than a foot a second, to preserve valuable fuel, McCandless watched as the gulf between him and *Challenger* steadily expanded. His space suit became so cold his teeth began chattering; he switched off the suit's cooling unit, and continued sailing out into space. He looked for stars, but saw only an enveloping darkness. The astronaut held a crude range finder—an aluminum bar etched with marks against which to measure his diminishing view of the shuttle's cargo bay—to estimate the distance from the orbiter, and make sure he didn't stray too far. Inside the cabin, Ron McNair stood at the controls of the shuttle's robot arm, ready to snatch McCandless to safety if necessary, and kept a laser tracker and *Challenger*'s TV cameras trained on him, transmitting live pictures to Houston and television stations across the planet.

Still gazing back toward the shuttle, McCandless at last reached his destination, and brought the jetpack's progress to a halt: some 320 feet out in space, 170 miles above the South Atlantic—a human satellite in orbit, traveling at twenty-three times the speed of sound. Yet the astronaut felt no sense of movement until he looked down and saw the planet rotating beneath his feet, a pin-sharp relief map unspooling at four miles per second: "Looks like Florida. It *is* Florida!" he said, his composure ebbing momentarily as he spotted Cape Canaveral below. "It really is beautiful."

Looking out from the cockpit, pilot Hoot Gibson centered the distant figure in the viewfinder of his Hasselblad—but then pulled the camera away from his eye, briefly startled by the power and clarity of the image he had framed. Gibson saw McCandless suspended alone in the abyssal blackness—angled slightly from the vertical, his white suit seeming to glow beneath the unfiltered sunlight of outer space, the luminous blue band of Earth's atmosphere curving away beneath him. Gibson checked the settings on the camera, and then checked them again; he pointed the lens through the triple-paned cabin window and tilted it to level the horizon. He squeezed the shutter.

Meanwhile, Captain McCandless continued to test the abilities of his experimental flying machine: he sailed back toward *Challenger* and then drew away once more; he dipped and rose, and turned somersaults. Like the other astronauts before him who had spoken rapturously of walking in space—Ed White had been so reluctant to return to his Gemini capsule that he described doing so as "the saddest moment of my life"—McCandless had hoped, if only for a few moments, to experience the noiseless solitude of being alone in the heavens; but the relentless chatter of the three audio feeds in his headset made it impossible.

The experimental plan had called, too, for the astronaut to orient himself to turn away from the shuttle when he reached the limit of his journey, and face out into the void. McCandless, a grizzled and shaven-headed forty-six-year-old third-generation Navy officer whose father and grandfather had both been awarded the Medal of Honor, was well acquainted with the limits of fear; as a Naval aviator, he had often landed his Phantom on the pitching deck of a carrier at night, and believed that he would never do anything more dangerous. Yet, despite his intentions, not once in the entire spacewalk did he turn his back on *Challenger*, his sole means of returning home.

Almost six hours after their experiment had begun, McCandless and Stewart clambered back inside the spacecraft, sealed the door of the air lock behind them, and removed their helmets. Down in Mission Control, the two astronauts' wives fell into each other's arms and wept.

::::::::::::::::::

In spite of its myriad technical failures, Ron McNair's first shuttle mission proved a public relations bonanza for NASA. The photograph that Hoot Gibson shot of Bruce McCandless was, in the words of space historian Andrew Chaikin, "one of the great mind-blowers of the 20th century." Appearing on the front pages of newspapers and magazines for months afterward, Gibson's picture of the lone astronaut hanging in orbit captured the imagination of a new generation of Americans, became emblematic of new frontiers of courage and technical ingenuity, and soon joined those photographs taken during the Apollo missions among the most iconic images of manned spaceflight.

And although unreported during the course of the mission, by the time he returned to Earth, McNair had realized his musical dream to become the first man to play the saxophone in space. On their second day in orbit, while the *Challenger* crew were forced to kill time waiting for clearance to launch their second satellite, McNair had retrieved his instrument from stowage and serenaded them with three songs he had practiced for the occasion: "America the Beautiful," "What the World Needs Now Is Love," and "Reach Out and Touch (Somebody's Hand)."

But the musical interlude took place during one of the mission's regular communication blackouts, so no one on the ground heard it. And although another member of the crew recorded his performance on video, before the weeklong mission was over McNair had inadvertently taped over the footage—erasing the moment forever; all the evidence that remained was a few photographs, later released by NASA. In one, the astronaut was captured floating on his back in *Challenger's* middeck, the golden saxophone raised to his lips, cheeks flared, the fingers of his right hand poised over the polished keys, his eyes fixed in concentration.

When McNair discovered that the record of his long-prepared performance had been destroyed, he handled the disappointment with the same composure with which he'd faced the theft of his research at MIT. He fell silent, and set to work on a plan to fix it.

A week or so after *Challenger* touched down on Runway 15 at the Kennedy Space Center, Kurt Heisig received a letter, scribbled on a single piece of paper torn from a yellow legal pad, written by his most ambitious student. In it, McNair explained his intention to take the saxophone with him on his next mission—and broadcast his performance live from orbit. "We must remember," McNair wrote, "since you are now up to your chin in this, that the project is *NOT YET COMPLETED*. Having a picture of a sax in space and having the world hear one being played from space are two entirely different events with as different impacts. I have a green light from everyone to play live next time."

In the months that followed, the astronaut and his family basked in the acclaim accorded the second African American in space. Having already featured on the cover of *Jet* magazine, he was now awarded the Order of the Palmetto—South Carolina's highest honor—by Governor Dick Riley, and dispatched on a national publicity tour by NASA. But for McNair, the unquestionable climax of his return to Earth was a triumphant visit to his hometown in the second week of April 1983. Dressed in his sky-blue flight suit, and seated in the back of an open-topped car with his wife and son, the astronaut led a parade down Highway 52 in Lake City; the same high school band he had once played in marched behind the car, and Black and white citizens alike lined the streets to watch them pass, holding balloons and tossing streamers. In a speech marking his arrival, the mayor had declared that April 12 would be designated Dr. Ronald McNair Day, and that the stretch of the highway running through the center of town would be named Ronald E. McNair Boulevard in his honor. "This native Lake Citian has done this town proud," he said.

The astronaut's septuagenarian grandmother, Mable Montgomery, a small, white-haired woman whose father had been born into slavery, shook her head in disbelief. "Who would have believed that such a day would come?" she asked Ron's brother Carl. "All these people shouting and cheering for one of our own."

Over the course of the three-day visit, McNair toured the local schools. The astronaut was at his most relaxed in the classroom, a natural teacher whose enthusiasm for learning was infectious. At Lake City High, he spoke to the children about the value of education; it was, he told them, the stepladder to success: the more you learn, the higher you can climb. "If you are happy cropping tobacco and picking cotton, you can drop out of school right now. Anybody here want to spend the rest of your life in the fields?"

Not one of the students raised a hand; the astronaut feigned surprise.

"No?" he asked. "Picking cotton is great fun, especially after your hands start bleeding."

Later, as he awaited his next mission assignment, McNair found ways to relive the thrill of his first flight. Visiting Carl at his house in Atlanta, he brought with him a handful of videotapes of the 41-B mission; Carl enjoyed watching the footage of life in space, and the astronauts' antics in microgravity.

But the part that Ron watched repeatedly was the film of *Challenger*'s liftoff. With the TV connected to Carl's home hi-fi, he turned the volume up to maximum, and cranked up the bass, recalling the adrenaline rush he'd felt as the shuttle's main engines had lit. The floorboards of his older brother's living room trembled with the shocks of ignition as the SRBs rumbled into life on videotape, and Ron lost himself in the enveloping roar of the rocket engines as *Challenger* rolled away from the tower once more. Like a child with his favorite cartoon, Ron watched the launch sequence again and again, so many times that eventually he blew out the cones of Carl's new speakers. But nothing could recapture the transporting rush of the real thing.

One day in 1984, Ron telephoned Carl to talk to him about the future: he told his brother that he didn't intend to stay at NASA forever. He wanted to try something new, to pursue a career in teaching—or politics, perhaps. But before he quit, McNair still needed to take a second ride into space.

"One more flight," he said.

The notice pinned to the bulletin board in the lab didn't look like much: a flyer printed on regular copy paper and illustrated with a simple sketch of the shuttle orbiter in flight, it might have been soliciting donations for a coat drive, or players for the company softball team. But when he saw it, Greg Jarvis could hardly believe what he was reading:

WANTED—HUGHES EMPLOYEES
TO RIDE SPACE SHUTTLE

The announcement explained that a selection board from the Hughes Aircraft Company was now accepting applications from among the staff of its Space and Communications Group for two "payload specialists." A new category of crew conceived by NASA to fly aboard the shuttle, payload specialists would be scientists or engineers with expert knowledge of a piece of equipment or set of experiments the orbiter carried as cargo, and would not require years of astronaut training at the agency. Instead— since they could not be expected to know how to operate the shuttle or any of its systems—these specialists would be well-qualified passengers, given a few months' flight training before launching into space.

Hughes planned to recruit two of its employees to go into orbit aboard the shuttle as payload specialists alongside two of the company's newest satellites, on missions scheduled for February and July 1985. To qualify, candidates would need to pass a NASA medical, and enlist for three months of full-time training in Houston; the successful applicant would require maturity, good judgment, and the "ability to make sound decisions and act under stress." The notice concluded in bland corporate style: *If you're interested in taking advantage of this opportunity, pick up an application from Ralph Rhoads, Bldg S66. . . .*

Greg Jarvis, thirty-nine, had been an engineer at the Space Communications Group for eleven years, and loved his job: as a project manager building satellites for both US military and commercial customers, he worked long hours, staying late and sometimes rising in the middle of the night to return to the big Hughes plant in El Segundo to deal with urgent problems. He regarded the satellites he worked on as his babies, and often brought his wife, Marcia, to the plant to show off his latest project. The couple wandered through the workshop in white lab coats, and Marcia marveled at the Leasat satellites Greg was working on— massive glinting cylinders fourteen feet high and weighing more than one and a half tons, sheathed in solar panels—amazed that her husband could create something so complex that would start its work so far beyond reach, 180 miles out in space.

Greg and Marcia had met when they were in their early twenties, as students at the University of New York at Buffalo, in the autumn of 1965. They were both working to pay their way through college, and their paths often crossed in one of the libraries where they took jobs for extra money. Greg was a striver: academically, he had to study hard to make his grades, and more than once had to repeat courses he failed. Slowly, his friendship with Marcia grew into something more, and in 1967, while out at a restaurant having dinner with his parents, Greg broke off from the meal to call her and propose over the phone.

They married in the summer of 1968, right after Marcia's graduation, and moved into an apartment on Beacon Street in Boston while Greg completed a year of graduate study at Northeastern University and worked at the defense and electronics manufacturer Raytheon in his spare time. At college Greg had been a member of the ROTC, and part of a program in which the Air Force paid some of his expenses— in exchange for which he had to do four years of military service after he graduated. Although the war in Vietnam was then approaching its height, his training as an engineer meant that he was assigned to an office job in California, with the Air Force Space Division in El Segundo.

They left Boston in the summer of 1969, setting out across the country in a brand-new, no-frills Dodge Dart. They drove at night to stay cool: the car lacked air-conditioning, and only had a radio because Greg installed one himself; they arrived in Los Angeles on the afternoon of July 20—in time to listen in, spellbound, as Armstrong and Aldrin's lunar lander touched down in the Sea of Tranquility.

Jarvis's work at the Space Division was dull and bureaucratic—evaluating proposals submitted by contractors for defense satellite projects—and he disliked the military discipline; upbraided by a senior officer for wearing scuffed shoes with his tan Air Force uniform, he replied, "You want performance or you want style? You don't get both." When his four years were finally up, Jarvis worried that he wasn't well qualified for the jobs he applied for, but he received offers from every one.

In 1973, he began work at Hughes; Marcia took a job as a dental assistant, and the couple moved into a mustard-colored stucco house six blocks from the ocean in Hermosa Beach. Marcia had never done anything athletic before they arrived in California, but together they now took up skiing, hiking, sailing, and snorkeling. They joined a bicycle club—the Los Angeles Wheelmen—and bought a tandem bike, taking long trips all over California. Tandems were not for everyone, and could be a litmus test of a marriage; Marcia and Greg knew several couples who invested in the two-seat bikes, but later sold them, rather than continue riding and get divorced. Yet Marcia was happy to let Greg take the lead on their journeys together, and in the summer of 1980 they rode one thousand miles across Canada, from Kamloops to Revelstoke to Calgary, sleeping in motels and swimming in ice-cold mountain lakes.

Intense and competitive, Greg had an apparently inexhaustible enthusiasm for new experiences, signing up for one community college course after another, and sticking tenaciously even to those pastimes for which he had little natural aptitude: he kept up a course of lessons on the classical guitar, although his playing remained terrible. The couple decided against having children—choosing, instead, the freedom to travel and indulge their hobbies; but one day on the beach the couple came across a litter of five puppies being given away by their owner, and Greg picked one out to take home: a black mutt they named Syrah, after the wine.

From the very start, Greg was captivated by the Space Shuttle. He and Marcia watched every launch, rising in the small hours of the morning to catch the early countdowns on TV. Many of his colleagues at Hughes enjoyed their work with satellites because of the proximity it gave them to the space program; with their lab coats and slide rules, they knew they weren't cut out to be astronauts, and developing the sophisticated cargo that NASA would carry into space was the closest they imagined they would ever get to orbit. So when in April 1984 the corporation offered the chance of a ride aboard the shuttle to two lucky employees, there was a deluge of applications: in total, six hundred men and women from across the company submitted their names. Yet when Jarvis returned home to his wife with news of the competition, he seemed to make little of it.

"Are you going to apply?" Marcia asked.

"I'm thinking about it," he said.

And although some of his colleagues thought his chances of being chosen were excellent—his former supervisor described him as "a sentimental favorite"—still he hung back. A few weeks passed, and Marcia asked if he'd been by Building S66 to collect an application form. "I keep meaning to pick one up, and I forget," he said. She began to suspect that he was simply afraid of disappointment. It was not until April 30, 1984—the final deadline—that Jarvis at last submitted his details. The following month, Greg learned that he had made it through to a final group of ten, and was called in for an interview; then, one day in early June, Marcia had an odd premonition that Greg was going to receive a piece of important news. She was at her desk in the dentist's office when he called. He didn't say much, because neither of them liked making personal calls at work.

"I got it," Greg said calmly.

Marcia stifled a shriek of excitement, and stepped out into the alley beside the office to whoop at the top of her lungs. She wished that she could go, too.

* * *

It was still dark in the early morning of Monday, June 25, 1984, when the crew of Space Shuttle mission STS-41-D climbed aboard the Astrovan for the drive to the launchpad, and rode the elevator to the 195-foot level of the tower at Cape Canaveral's Pad 39A. Judy Resnik, who suffered from a fear of heights, felt a sudden rush of anxiety as she stepped out onto the steel gantry. Looming above her, bathed in the crossed beams of powerful xenon floodlights, was the Space Shuttle *Discovery*, the third and most sophisticated of the orbiters, fueled and ready for her maiden journey into orbit. The mission commander, silver-haired Henry "Hank" Hartsfield, led the way to the White Room, the boxlike clean space clamped to the outside of the orbiter; one by one the six members of the crew were summoned by ground support staff, until at last it was Resnik's turn. Inside the White Room, she crouched in the opening of the crew hatch and, on her hands and knees, crawled inside.

Just a few days earlier, Resnik had sat on the steps of her townhouse in Houston with her friend Sylvia Salinas, who she had invited over for dinner to talk before the mission. Resnik told her that—because she wasn't married and had no immediate dependents—NASA was pressing her to make a will, and she wanted help deciding who would receive her belongings if something went wrong: there was the house; her car; and a life insurance policy that, as a civilian astronaut, obliged the agency to pay out in the event of her death.

"I don't have a lot," she said. "But who do I name?"

With the document in her hand, Resnik talked over the decision with Salinas. Eventually, they resolved that Judy would leave everything to her niece and nephew, her brother Charles's two children. But Salinas tried to reassure her that the will wouldn't be needed any time soon.

"Nothing's going to happen," she said.

"It might, it might."

"It might. . . . Are you ready for that?"

"Oh yeah," Resnik replied. "I know what's involved."

Inside the cabin of the orbiter, everything shone. Unlike those in the scuffed and grubby simulators in which they'd all been training for sixteen months back in Houston, the brand-new keys of *Discovery*'s computers showed no wear; there were no gaps in the instrument panels where redundant controls from previous missions had been replaced. It even smelled new.

Of the half dozen crew on board *Discovery* that morning, only fifty-year-old Hartsfield—who had been with NASA since the late sixties and piloted the final test flight of *Columbia* in 1982—had been into space. As the ground support staff who formed the five-man "closeout crew" helped them into their seats and cinched their harnesses tight, Resnik and the others tried to mask their fear and anticipation with banter and small talk. But as the final countdown began, the jokes and asides dried up. Each of the rookie astronauts wanted nothing more than to fulfill their ambition to reach orbit for the first time, but all were aware of the risks: the external fuel tank was little different from a giant bomb, packed with more than half a million gallons of propellant. As the minutes passed and the clock drew closer to the moment of ignition, the mood in the cabin tautened until—with barely more than thirty minutes to go—a voice from the Launch Control Center three miles away flagged a problem. One of the orbiter's backup computer systems had hung up. The countdown entered a temporary hold while the technicians searched for a solution—but to no avail.

"*Discovery*," the Launch Director announced over the radio, "we're going to have to pull you out and try again tomorrow."

It was 8:35 a.m.

The crew felt crushed; wrung out from the hours of tension, and physically spent. For their families watching from the Launch Control Center, it was no better. Accompanied by the astronauts designated as family escorts—on hand to help with accommodations and travel before launch, for the duration of the mission and, if the worst should happen, to gently usher them into widowhood—the wives and children of the married crew were taken off to lunch, exhausted. Yet, the following morning, the crew and their loved ones had to endure the entire process once more: the early-morning knock at the door, the breakfast of steak and eggs, the walk out to the Astrovan under the glare of the TV lights, the long elevator ride up the gantry.

On the second attempt, the countdown proceeded smoothly through the same moments when the launch controllers had called a halt the day before; in the cockpit, Hank Hartsfield fell into a doze. At T-minus five minutes, the crew felt a slight vibration as, far beneath them, the orbiter's three Auxiliary Power Units came on, generating pressure for its hydraulic system; at two minutes, they closed their helmet visors; Hartsfield and the pilot shook hands. "Good luck, everybody," he said. "This is it. Do it like we've trained. Eyes on the instruments."

T-minus 31 seconds: "Go for auto-sequence start."

Discovery took over control of the countdown program from the launch center computers.

T-minus 10 seconds: "Go for main engine start."

Six high-pressure turbopumps began to whir, forcing one thousand pounds of propellant into the combustion chambers of the three main engines every second.

At T-minus 6, the cabin began to rattle violently as the first engine lit; it was finally happening.

5 . . . 4 . . .

Then the master alarm sounded. The vibration abruptly ceased and the cockpit fell silent: with just three seconds to go before launch, the engines had shut down. Hartsfield reached out to cancel the alarm. Two engine shutdown lights glowed red, but the third was unlit; the pilot stabbed at the button to close it down, but the status lamp remained dark. Could it still be running? More frightening still was the possibility that the solid rocket boosters might ignite. If that happened, then the entire shuttle—rockets, external tank, orbiter—would be torn apart in seconds.

Launch control announced a fire on the pad; water from the Firex gantry fire suppression system sluiced down the windows of *Discovery*. Hartsfield told the crew to prepare for an emergency evacuation, using the slidewire baskets installed on the launch gantry. Down in the middeck, Resnik unstrapped herself, crawled over to the door through which they would make their escape, and peered through the porthole. The access arm had swung back into position; she could see no evidence of fire.

"Henry, do you want me to open the hatch?"

The commander told her to wait; he had no idea what was going on outside, or how dangerous the situation might be. Besides, no one had

ever tested the slidewire—intended to deliver the astronauts to a concrete bunker four hundred yards away in thirty seconds—with a human
being on board.

"We're not going anywhere," he said.

On the roof of the Launch Control Center, watching through the
early-morning haze, hearing only the boom of the shuttle's main engine
start, followed by sudden silence, some members of the crew families
broke down, convinced that they had witnessed a catastrophic explosion; Resnik's mother bent her forehead to rest on her hands, as if to
pray. It was more than forty minutes before her daughter, soaked by the
pad's sprinkler system, bedraggled and terrified, was helped from the
orbiter by members of the closeout crew.

Two days later, NASA organized a press conference at the Cape with
Hartsfield, who delivered a textbook performance of astronaut sangfroid. Discovery's engines had shut down due to a faulty valve, and the
fire had been caused by hydrogen propellant escaping from the rocket
nozzles; as the gas burned with a transparent flame, it had been invisible to the crew—but could nonetheless have killed them if Resnik had
opened the orbiter's hatch. Yet Hartsfield insisted that the crew had remained calm throughout the entire ordeal. "It was an orderly shutdown,"
he told reporters. "There was no reason for concern." Resnik gave her
own assessment in characteristically bloodless language: "I was disappointed," she said. "But I was relieved that the safety systems do work. It
was unfortunate we had to check them out. But it built confidence in the
whole system." Much later, Mission Specialist Mike Mullane, who had
sat directly behind Hartsfield on the flight deck during the abort, would
insist that every member of the crew had, in truth, been petrified—but,
modeling their behavior on the stereotypes set down by the first astronauts, knew better than to say so in public.

It would be another two months—after one of the main engines
had been replaced, and the crew had strapped in for two further
scrubbed launches—before Discovery lifted off from Cape Canaveral
on its maiden flight. On August 30, 1984, Judy Resnik at last became the
second American woman in space. Once in orbit, Resnik concentrated

on conducting experiments on a one-hundred-foot-long solar array designed to generate power for future satellite projects. And, during one live broadcast to Earth, she held up to the camera a message she had scrawled in marker on a clipboard: HI DAD.

By then, the cascade of launch delays had pushed NASA's ambitious calendar of shuttle missions further behind schedule, and the agency announced that it was cutting the number of flights for the year from ten to seven.

CHAPTER
FOURTEEN

ACCEPTABLE RISK

Less than two weeks after *Discovery* landed safely at Edwards Air Force Base at the end of its maiden flight, a freight train rumbled up the rail spur toward a Morton Thiokol facility in Clearfield, Utah, carrying the burned-out cases of the two solid rocket boosters that had launched the shuttle into space. Before preparing them for refurbishment, a team of company engineers inside Thiokol's Building H-7 conducted a final disassembly of the rocket casings and inspected them for unusual damage.

Soon after, the phone in Al McDonald's office began to ring.

Since reluctantly agreeing to lead the company's shuttle operation back in March, McDonald had been in his new job for barely six months; but he was already overwhelmed. Drowning in an alphabet soup of unfamiliar NASA acronyms, swamped by constant telephone calls with engineers and managers from the Marshall Space Flight Center in Alabama, half his workweek was disappearing in unpaid overtime—with no end in sight.

Almost as soon as he had started, McDonald had become the main presenter for Thiokol at the punishing Flight Readiness Reviews scheduled before each shuttle mission. An increasing number of these conferences were now held over the phone, "telecons" conducted between meeting rooms wired for sound at the Thiokol plant and Marshall; a single one of the group calls could easily last four or five hours, as each of the engineers made their technical presentations and then faced cross-examination over the long-distance lines. And in the background loomed the continuing threat that NASA might soon offer Thiokol's valuable sole-source contract on the solid rocket boosters out to a competitor. McDonald carpooled to the plant with his colleagues each morning, but at the end of the day often remained stuck in his office, struggling to finish his work, keeping his friends waiting in the plant parking lot as darkness fell. Even then, he stuffed his briefcase full of documents to review when he got home; his wife and children barely saw him.

McDonald's new tasks included heading the internal Thiokol panel that appraised any new technical anomalies discovered in the solid rockets—and placing his signature on the document that accompanied each piece of hardware that the company delivered to NASA, testifying to its flightworthiness. He found this part of the job a source of corrosive anxiety: often, the gathered Thiokol engineers' opinions were not unanimous, and McDonald had to make the final decision; those anomalies he accepted gnawed at him long afterward, as he wondered whether a fault that he deemed an acceptable risk might one day lead to the loss of a shuttle and its crew.

But McDonald found those anomalies he deemed *unacceptable* equally stressful: in doing so, he sometimes had to convince both his own management at Thiokol and those at NASA to scrap pieces of hardware they had previously considered completely flightworthy, yet might now have to be replaced, often at a cost of millions of dollars. At the end of the flight readiness process, McDonald also had to take his turn to travel to Cape Canaveral for the final countdown, where he would give the company's final "go" or "no-go" recommendation for launch. The scale of these responsibilities staggered him. His hair seemed to grow grayer every day.

Even so—despite having spent his entire career at Thiokol, despite his years of expertise in solid rockets, and his seniority as an engineer— it wasn't until after taking up his new position that McDonald learned of the problem in the seals of the shuttle boosters. He was startled to discover that such a fault existed at all—let alone one so serious that it had gone apparently unresolved for years, going all the way back to the second shuttle launch in 1981.

:::::::::::::::::

In the months since President Reagan had stood on the desert runway at Edwards and declared the Space Shuttle operational, no one at NASA headquarters had heard much about the O-rings in the solid rockets. Administrator Beggs, with his unsparing eye for detail, was focused on the problems he found elsewhere. If he thought about the seals at all, even Beggs's deputy for technical matters—the most senior engineer in the entire agency—considered the rubber seals as little different from those in the trusty Titan III missile, which had served the Air Force so well for almost two decades.

And at Thiokol, there was a long list of other problems—the near-calamity of the disintegrating rocket nozzles, the simmering concern over the NASA contract, and the pressure to manufacture new sets of boosters to meet the agency's planned launch schedule—for the engineers to worry about. Besides, the tweaks they had made to the manufacture and assembly of the rockets after discovering the strange charring in the seals following the second shuttle flight—the careful stacking procedures, the steel shims, the putty kept in the freezer—seemed to have worked. Three years later, the shuttle had flown a further seven missions, and the Thiokol technicians' postflight inspections at the Cape and in Clearfield had revealed no further damage to any O-rings. In spite of all of the other setbacks they now faced with the rockets, the gas leaks in the booster joints seemed to be exactly what the engineers had first said they were: an isolated and harmless anomaly, with causes they well understood, which they had learned from and addressed; the system of nimble empiricism that had served NASA so well since its inception was working just as it should. The issue had never even been discussed at a single Flight Readiness Review, and no word of any trouble with the O-rings had ever reached the Astronaut Office.

So when the Thiokol inspection team in Florida had begun to examine the two boosters that had taken Ron McNair and his crew into space in February 1984, it seemed unlikely there would be anything new to report. As he walked around the boosters with his notebook, the engineer in charge logged the same kind of minor damage often caused when the burned-out rockets parachuted onto the surface of the ocean at around sixty miles an hour: battered external insulation; nozzle actuators torn loose from both left and right motors; and missing patches of paint.

But once technicians had prized the segments of the motor apart, the engineer noticed unmistakable signs of burning across an inch-long section of the primary Viton ring inside the forward field joint of one rocket—and more in the seal of the nozzle joint of the other. This was alarming news: it meant that the O-ring erosion discovered in 1981 was no longer an isolated exception, but a recurring issue. And this time it had happened in two places on the same flight.

Even so, the Thiokol engineers soon arrived at a likely explanation, and were gradually convinced that they knew what had happened. Immediately before the February flight, they had doubled the pressure of the safety tests they conducted at the Cape for leaks in the booster joints, part of a routine of prelaunch checks made on the fully assembled rockets before the shuttle was stacked and rolled out to the pad. The increased pressure of the new leak check had almost certainly blown a hole through the asbestos putty packed around the joints—opening a path for a tongue of hot gas to reach the rubber gasket when the rocket ignited.

And, although this new kind of blowhole constituted yet another unexpected failure of the joint design, the Thiokol engineers had several reasons to believe that the O-ring erosion it caused remained harmless. First, the charring they had seen in February still wasn't as bad as the damage on the second shuttle mission—and both those launches had made it safely into space without further mishap. Second, this new erosion still lay well within the margins of safety they had already established for the rings, margins that they had confirmed with fresh experiments. In the laboratories at Thiokol, the engineers deliberately gouged large chunks of rubber from sample seals and then ran them through pressure tests; they found that even the mutilated O-rings could contain more than three times the force they would be subjected to at launch.

Finally—and most importantly—the evidence from the recovered boosters suggested that the leaks in the seals had remained "self-limiting." The engineers knew that the superhot jet of blazing propellant escaping from inside the rocket's combustion chamber had burned the O-rings for only a fraction of a second: just as long as it took the high-pressure gas to enter the joint, race around the thirty-seven-foot circumference of the rocket casing, and fill up the narrow annular space around it. As soon as the hot gas came full circle, the pressure inside the joint equalized with that in the chamber of the rocket and the O-ring sealed tight against the metal surfaces of the joint. Then the leaking stopped. Only if the gas could somehow completely burn through or bypass the first O-ring and then *also* breach the backup seal, giving the gases from inside the rocket a direct route into the atmosphere, could an unlimited leak begin; then the problem would become catastrophic. But the possibility of that happening seemed vanishingly remote.

When they filed their analysis with the program managers at the Marshall Space Flight Center, the Thiokol engineers were satisfied that they had continued to learn more about how the joints worked as they went along, and understood the nature of the problem. Now they knew not only why the erosion occurred, but that they could expect to see it in the future; the leak check that had caused the blowholes was an essential safety procedure, and the charring to the O-rings—while not a part of the design, and certainly nothing to be proud of—caused a negligible danger to the spacecraft and its crew. So the leak checks, and shuttle flights, could continue. "Remedial action—none required," the Thiokol engineers wrote in March 1984. "Possibility exists for some O-ring erosion on future flights. . . . This is not a constraint to future launches."

So when erosion had once again occurred during Dick Scobee's first launch in April, it was neither a surprise nor a cause for alarm in Utah; instead, it confirmed the engineers' working hypothesis. Even so, NASA Deputy Administrator Hans Mark instructed Larry Mulloy—now settling in to his second year as the manager of the booster project at Marshall—to deliver a full review of the joint problems to headquarters. Mulloy passed the message on to the managers at Thiokol, who promised to conduct a program of further research and testing. In the meantime, Mulloy had also moved to have the solid rocket joint reclassified on NASA's Critical Items List; from 1982 onward, headquarters officials no longer regarded the field joint as hardware with built-in redundancy, but as a "Criticality 1" item: a potential single-point failure, the destruction of which would result in "loss of vehicle, mission and crew." However, this was yet another technical issue about which the rival teams at Marshall and Thiokol could not agree, and—regardless of the official change—in both Alabama and Utah many engineers and senior managers continued to regard the joint as fail-safe.

::::::::::::::::::

Five months later, the technicians at the Promontory plant were still working on the tests ordered by headquarters when Allan McDonald heard from the engineers disassembling the rockets recovered after Judy Resnik's first flight, aboard *Discovery*. This time, taking apart the nozzle joints of the boosters, they had found the erosion they expected, but also a disturbing new phenomenon: traces of soot in the space between the first ring and the secondary O-ring beyond it. This suggested that hot gas from inside the rocket had not only blown through a hole in the heatproof putty to scorch the gasket, but then the primary O-ring itself had—for a few moments at least—failed to seal. Although the secondary ring in the joint held, and showed no sign of any heat damage, it was clear the engineers' confidence had been misplaced: Whatever was going on in the booster joints was getting worse. And they didn't understand why.

With the next scheduled shuttle flight less than a month away, McDonald scrambled to assemble a team to analyze what had gone wrong, and prepare a report for the impending Flight Readiness Review. He had just one week to determine whether Thiokol could recommend NASA proceed with the launch.

At a meeting on September 12, McDonald presented the team's

results to his NASA managers in Huntsville, and reconfirmed the engineering rationale from earlier in the year: the damage wasn't so bad; it was still within their safety margins; it remained "self limiting." And while the hot gas getting past the primary ring—a new phenomenon they called "blow-by"—was unprecedented, the escape had lasted for less than a second before the seal had done its job and the joint had fully pressurized; there hadn't been time for the gas jet to cause any damage to the second O-ring. As far as they were concerned, the joint design remained fail-safe. The Thiokol engineers' opinions were unanimous: McDonald pronounced the boosters "go" for flight. In the meantime, they planned more tests, and more minor fixes. The analysis passed into the decision chain of the Flight Readiness Review, and a week later Larry Mulloy used the data for his own presentation to his chiefs within NASA.

The package of information Mulloy handed out to those at the meeting included one chart that itemized five engineering anomalies discovered since the previous mission, of which the joint failure was the last. In the column dedicated to the official resolution of the issue, the document classified the seal as an ACCEPTABLE RISK; in the next, headed TO BE DISCUSSED, there was one word: NO.

And yet: beneath these crisp certainties, obscured amid the blizzard of charts, data-filled binders, and Viewgraph slides, the rocket engineers failed to realize that they had reached a critical inflection point. Over the course of the years they had been developing and flying the solid rocket motors, the men at Thiokol and Marshall had slowly expanded the parameters of what they regarded as acceptable risk in the joints. Incrementally, they had begun to accept as normal problems that deviated dangerously from the original design standards set for the boosters—and the seals that constrained the seething power of their volatile propellant in flight.

Erosion of the O-rings by hot gas should have presented a grave warning of calamity for the engineers. But—proceeding step-by-step up the bureaucratic ziggurat of the Flight Readiness Reviews, reassured by the data that, mission by mission, had widened the envelope of what their hardware could apparently withstand without catastrophe—it did not.

Instead, for the first time since the shuttle program began, in meetings with his managers, Larry Mulloy now began to discuss the degree of destruction permissible in the O-rings—while continuing to vouch for their safety, even after they were known to have failed in their primary purpose, of sealing the joint after ignition.

Back in Utah, these issues continued to worry the beleaguered and overworked Al McDonald. But he was reassured by the rigorous process, the tests, and the findings of his own heat-transfer experts and structures specialists. Above all, McDonald well understood that riding rockets was a dangerous business; and he believed that after each mission, the men and women flying aboard the shuttle reviewed exactly the same kind of postflight data that he and his engineers worked so hard to produce—and, as a result, astronauts like McNair, Scobee, and Resnik understood exactly where the relative risks posed by his solid rockets lay amid the constellation of hazards they faced on their route to orbit.

He was mistaken.

In the middle of September 1984, just as McDonald and Mulloy were presenting their latest risk assessments of the boosters, Greg Jarvis and the other three astronaut candidates from Hughes flew to Florida to visit the launch facilities at the Kennedy Space Center in Cape Canaveral. Jarvis had been chosen by Hughes as the first of their payload specialists to fly into space; the second, John Konrad, a thirty-four-year-old project manager from the systems engineering laboratory, was a long-standing space nut who had grown up watching *Star Trek* on TV. Jarvis and Konrad were accompanied by their respective "alternates," Bill Butterworth and Stephen Cunningham, who would train beside them—and take their places if a last-minute accident or illness forced them to withdraw from their assigned mission. The four men did not know one another well, but were united by their glee and astonishment at their selection for the payload specialist program.

Posing for publicity photographs in front of the gigantic Hughes Leasat 3 satellite that Jarvis had helped build, the four men could hardly have presented a better advertisement for the democratization of space travel, sundering some of the last rock-ribbed astronaut stereotypes. At six feet five, John Konrad—too tall to stand upright in the shuttle—towered over Jarvis, who was just five feet nine, and draped his arm casually on the shorter man's shoulder like one-half of a little-and-large comedy act; Konrad was nearsighted, Jarvis was color-blind, and—despite his athletic pastimes—slightly overweight; both were losing their hair. Beside them, Butterworth and Cunningham smiled out from behind heavy window-pane spectacles, shirts and ties covered by nylon lab coats: they resembled garrulous radiologists, or a pair of affable loss adjusters. The Hughes public relations people did their best to school the four men on how to conduct themselves in press interviews: don't touch yourself, or put your hands in your pockets—do this, do that. But the four men were giddy at the prospect of being selected to take the trip of a lifetime, and reveled in their unlikely elevation into the company of the heroes of spaceflight. Jarvis grinned in astonishment from almost every photograph in which he appeared. "You think of astronauts as being cool," said one reporter after interviewing them. "These guys were not cool. They were so excited."

At the Cape, a team of three NASA engineers took the Hughes specialists out on a tour of one of the orbiters aboard which Jarvis expected to fly early the following year: *Challenger*, already sitting on the launch-pad ready for its next mission, scheduled for November. They rode the elevator to the top of the gantry and began stopping at intervals all the way to the ground, examining the full length of the spacecraft as they went.

Holy shit, thought Konrad.

"The only thing small is us," Jarvis said.

They were overwhelmed by the monstrous scale of the vehicle towering overhead. Only now did the concrete reality of their undertaking dawn upon each of them: they knew that if anything went wrong while they were on board this thing, it was over.

That night, the four men gathered in the Cocoa Beach Holiday Inn and sat up into the early hours discussing the risks involved. They recognized the hazards of spaceflight, and understood that the shuttle had the potential to kill them. But none of them considered it any riskier than flying aboard an airplane. As engineers, they had confidence in the hardware—and in the end, the rewards of being a part of the space program more than justified the risk.

Jarvis told the others that he wasn't particularly concerned about his own safety if something went wrong. But he did worry about what would happen to Marcia if he didn't come home. Calm and businesslike, they all agreed that Butterworth and Cunningham, as the alternates, would take care of those that Jarvis or Konrad left behind.

:::::::::::::::::::

Less than two months later, soon after dawn on November 8, Greg and Marcia sat side by side in the bleachers at Cape Canaveral to witness a shuttle launch in person for the first time. They felt the ground shake beneath them and held their breath as *Discovery*, now embarking on its second mission, rose toward a pillowy bank of low cloud, tinted gold by the rising sun; awestruck, Greg kept his eyes on the shuttle as it slowly dwindled into a shimmering speck and, at last, disappeared high above him into the cobalt sky. He found it almost impossible to believe that one day soon it would be him up there, riding a rocket into space.

As mission STS-51-A began, the mood in the sterile concrete buildings at the Johnson Space Center in Houston was tense: it was, one official told the press, "the most challenging flight we've flown." The five astronauts aboard *Discovery* were about to undertake an audacious effort at orbital salvage, a complex task fraught with risk, and one unlike anything yet attempted in the history of space exploration.

What had originally been planned as a routine shuttle flight to deliver a simple commercial payload to orbit had been hurriedly reorganized by NASA managers around a plan to locate and recover the two communications satellites that had been lost in space during Ron McNair's mission earlier in the year. Overseen by Tom Moser, the engineer who had played such a key role in resolving the Great Tile Caper, the plan called for the *Discovery* crew to locate the two satellites in orbit and then draw alongside them in the shuttle, a hazardous exercise of close-up celestial navigation known as "proximity operations"; two of the astronauts would then step out into space and, propelled by the Manned Maneuvering Unit and using a specially constructed device they christened the Stinger, capture the satellites before wrangling them back into the cargo bay of *Discovery*. The crew would then fly the salvaged hardware back to Earth, where it could be refurbished before being returned to space to complete its work as originally intended.

Moser, who had only a few months to design and build the Stinger in time for the astronauts to test and learn how to use it in space, was confident that the idea would work. He believed it was an excellent demonstration of the capabilities of the agency's space trucking vehicle: a practical and highly visible return on the taxpayers' investment in the shuttle program. Yet Chris Kraft, who—despite his resignation as Center Director—still made occasional visits to his former fiefdom, advised him to abandon the effort, on the grounds that it was too dangerous. "It's not worth the risk," Kraft told Moser one day as they examined the Stinger—a five-foot steel spear surrounded by a circular cage, like a monstrous shuttlecock—in the workshops of the Center's Technical Services Center. It was one thing to brainstorm a desperate solution to rescue men from certain death, as the NASA engineers had with Apollo 13; quite another to choose to improvise one that gambled on the recovery of two pieces of heavily insured electronics—no matter how much good publicity its success might bring.

The astronauts, too, had their doubts. As launch day approached, the crew of *Discovery* were surprised to read a *Florida Today* newspaper report in which they were quoted by an unnamed NASA spokesperson saying that the likelihood of recovering both satellites was "very high." A few days later, mission commander Rick Hauck took the opportunity of a crew meeting with the agency's newly appointed head of public affairs to set the record straight: he explained that no member of the crew had said such a thing; nor did any one of them believe it. "I will personally tell you that my assessment is, if we successfully capture one satellite, it will be remarkable," Hauck said. "And if we get both satellites, it will be a fucking miracle."

Not only were proximity operations innately perilous—flying two spacecraft toward one another while both were traveling around the Earth at 17,400 miles per hour made even the slightest collision

Maxime Faget, Director of Engineering and Development at the Johnson Space Center in 1969, holding his balsa model of a prototype shuttle orbiter. *NASA*

Faget's original shuttle design envisioned the astronauts aboard the orbiter carried to the edge of space by a crewed booster vehicle, which would return to Earth and land on a runway like an airliner. In this artist's impression from 1970, the orbiter is peeling away in the background. *NASA*

Dorothy "Dottie" Lee, photographed in 1976. As a Senior Aerothermodynamics Engineer, Lee helped develop the heat shield designed to protect the shuttle as it reentered Earth's atmosphere. *NASA*

The shuttle's heat shield—composed mostly of delicate ceramic-coated tiles—presented so many unforeseen technological challenges that it delayed the launch of the spacecraft for years. *NASA*

To save money and time, Faget's manned booster vehicle was eventually replaced by a pair of strap-on solid-fueled rockets, designed and constructed by the Thiokol Chemical Corporation. Here, a rocket is prepared for a test at the company firing range in Utah in 1979. *NASA*

The first three Black astronauts, photographed wearing the modified pressure suits similar to those worn in space shuttle flight tests, in 1979. *Left to right*: Ron McNair, Guion Bluford, and Fred Gregory. *NASA*

The first six women astronaut candidates gathered in the Public Affairs Briefing Room at the Johnson Space Center to meet the media on January 31, 1979. *Left to right*: Rhea Seddon, Anna Fisher, Judy Resnik, Shannon Lucid, Sally Ride, and Kathy Sullivan. *NASA*

Director of Flight Operations George Abbey, "the Thomas Cromwell of the Johnson Space Center" (*second from left*), the day before the first successful test flight of the Space Shuttle *Columbia*, in April 1981. With Abbey are (*left to right*) *Columbia* pilot Bob Crippen, Chief of Aircraft Operations Division Joseph Algranti, and astronaut Joe Engle. *NASA*

Columbia's successful touchdown at Edwards Air Force Base in California on April 14, 1981. A crowd of 250,000 people stood in the desert to watch the triumphant return to Earth of the world's first reusable spacecraft. *NASA*

James M. Beggs, appointed NASA chief by President Reagan shortly after the first shuttle flight in 1981. *NASA*

Bill Lucas, Director of NASA's Marshall Space Flight Center in Huntsville, Alabama, oversaw the contractors' design and construction of the Space Shuttle's main engines and the solid rocket boosters. *NASA*

Chris Kraft (*middle*), Director of the Johnson Space Center in Houston, celebrates the completion of the second shuttle mission in November 1981. Beside him are Deputy Director of Flight Operations Gene Kranz (*left*) and Head of Structural Design Tom Moser (*right*). *NASA*

Public Affairs Officer and Chief Mission Control Commentator Steve Nesbitt (*foreground*) hosting a prelaunch press conference in Houston with the crew of mission STS-41-D before the maiden launch of the third shuttle orbiter, *Discovery*, in 1984. Behind Nesbitt (*left to right*) are astronauts Charlie Walker, Judy Resnik, Steve Hawley, Mike Mullane, Mike Coats, and Commander Hank Hartsfield. NASA

Booster Systems Engineer Jenny Howard was one of the few female flight controllers at NASA when the shuttle program began. She narrowly saved the shuttle from disaster during the launch of *Challenger* in July 1985. NASA

The second orbiter in the fleet, *Challenger*, on the launchpad at Cape Canaveral on January 26, 1983, during preparations for its maiden flight three months later. *NASA*

"One of the great mind-blowers of the 20th century." Bruce McCandless II, flying free in space using the nitrogen-powered jetpack known as the Manned Maneuvering Unit on February 7, 1984. *NASA*

potentially lethal—but the unhurried way in which astronauts appeared to float around in space using the Manned Maneuvering Unit belied the unforgiving realities of celestial mechanics. As long as their relative speed and orbital trajectories remained closely matched, the shuttle and the untethered astronaut flying his jetpack were seemingly able to move around each other as elegantly as dancers on a ballroom floor; but with the smallest error in choreography, their plans could come unraveled at bewildering velocity: if they strayed too far apart, or if one of the nitrogen jets on the MMU failed or became stuck in the "on" position, their flightpaths would abruptly diverge and the stricken astronaut would spiral away into deep space at hundreds—or thousands—of miles an hour.

Hauck calculated that if that happened, he would have only fifteen seconds to disengage from his rendezvous with the satellite and take the shuttle after his crewmate before he or she disappeared forever into the void, where they would asphyxiate as soon as their personal life-support system gave out.

On the fifth day of the mission, ABC-TV broke into its regular breakfast programming with live pictures from space. "This is the world's first capture of a runaway satellite," space correspondent Lynn Sherr announced to her viewers as astronaut Joe Allen, strapped into the bulky jetpack and bracing his torso behind the extended lance of the Stinger, set out from the cargo bay of the shuttle in pursuit of his quarry. Rick Hauck and Mission Specialist Anna Fisher had spotted the Hughes Palapa satellite from more than one hundred miles out, and the commander gradually closed the gap to less than forty feet; now the colossal black cylinder, nine feet long with a mass of more than half a ton, spun lazily against the luminous blue backdrop of Earth—its surface pebbled with delicate patterns of white cloud. Allen swore under his breath as he approached the satellite from behind, and found himself temporarily blinded by the sun; but the rest of the maneuver proved easier than he had expected. He soon had the Stinger locked into the nozzle of Palapa's rocket motor. "Stop the clock. I've got it tied!" he said.

Yet just as the engineers and the satellite's insurers were celebrating down in Houston—"It's almost money in the bank," muttered the representative from Lloyd's of London, part of the insurance syndicate that had already paid out $180 million on the lost hardware—the crew discovered that Palapa had not been built to exactly match the blueprints they had studied on the ground. The equipment that Hughes engineers had given them to bring it home wouldn't fit; stuck out in space with the satellite halfway back to the shuttle, the astronauts realized they would have to resort to further improvisation. Eventually, Allen held the satellite over his head, steadying it as best he could in both hands, for an hour and a half—the time it took him to complete one entire orbit of the Earth—while fellow astronaut Dale Gardner struggled to find a way to secure Palapa inside the cargo bay. Finally, after an exhausting five-and-a-half-hour space walk, it was over. When word of their success reached the headquarters of Lloyd's in England, the underwriters ordered the ringing of the Lutine Bell, traditionally struck twice to mark the return of an overdue ship. Forty-eight hours later, the two men recovered the second satellite. Afterward, President Reagan called from the White House to congratulate the crew: "I just want to tell you how proud we are of what you've achieved on this mission. You've demonstrated that by putting man in space aboard America's Space Shuttle, we can work in space in ways we never imagined were possible," he told them.

The *Discovery* crew's triumphant celestial salvage operation put the Space Shuttle back on the front pages of the papers for the first time in months, accompanied by yet more astonishing photographs of the astronauts in space; the *New York Times* described Allen, mounted in his nitrogen-propelled jetpack, Stinger at the ready, as "a modern-day Sir Galahad, a knight in shining Nomex, Teflon and aluminized Mylar." From the VIP room at Mission Control, Tom Moser saw a more recent historical parallel, and had the Space Center graphics department run off a souvenir print, overlaying the picture of Allen on a nineteenth-century etching of Nantucket whalers chasing their quarry across the ocean from the prow of a whaleboat—"See, that's a guy with a harpoon going after a creature in a harsh environment," he said.

On the day Hauck and his crew touched down on the shuttle landing strip at Cape Canaveral, returning satellites from orbit for the first time in history, Chris Kraft sat down at home to write a letter to the commander.

In it, Kraft expressed his admiration for the same mission that, a few months earlier, he had tried to halt. "Your performance ranks with the greatest I have witnessed in all of the space missions that NASA has conducted and I am indeed proud to have been one of those who was associated with your selection as an astronaut," Kraft wrote. "Please accept the highest congratulations I can offer—you are truly a great American."

Yet in their postflight press conference in Houston, the five members of the crew embodied all the traditions of aw-shucks astronaut modesty, making light of their achievement as they ad-libbed their narration of an eighteen-minute film of what Hauck introduced as "a very fun mission." One of the highlights of the movie was a slapstick clip in which Dale Gardner created the illusion that he had stowed the elfin Allen in one of Discovery's middeck lockers, unfolding the diminutive astronaut from storage like a piece of collapsible equipment suddenly needed at work.

After receiving the letter from Kraft, Hauck invited him to lunch with his crew at the Hilton hotel overlooking the water in Clear Lake, where the former Center Director reiterated his congratulations on their training and improvisation, in a mission that had been filled with risk. But Kraft also chided the astronauts for suggesting that it had been easy. "Because it was difficult," he told them. "I know it and you know it, and your making it look too easy is not helpful to the space program in the long run." Now, Kraft implied, the public would expect perfection from the agency every time.

But Hauck thought little of the warning Kraft delivered, and the conversation moved on. It was only much later that he realized that the STS-51-A mission represented a kind of zenith of NASA's image as an institution of almost infinite cleverness and ingenuity, one that could routinely extend the limits of human achievement. Like many of his colleagues, the commander had come to think of himself and the agency around him as almost infallible. And if Kraft was concerned that NASA was in danger of exhausting its supply of heroic feats and grand spectacles, he need not have worried. At the agency's headquarters in Washington, DC, senior officials had long been at work on a new and ambitious plan with which to dazzle even a jaded audience.

CHAPTER FIFTEEN

THE TEACHERNAUTS

On the last Wednesday of July 1985, Johnny Carson swept aside the multicolored curtains behind the stage in Burbank and strode out to begin his opening monologue. Silver-haired and dapper in a blue blazer and tie, he was in his twenty-third year of presenting *The Tonight Show*; his ad-libs were swift, his material as wry and effortless as ever. "There's nothing earth-shattering in the papers today," he said, before skipping through a handful of topical riffs: an invasion of killer bees in California, Michael Jackson and his single sequined glove, the half centenary of Porky Pig, and the success of Sylvester Stallone's *Rambo: First Blood Part II.* There was even a Rambo action figure, he said, equipped with a bow and arrow and a machine gun, for kids. "Interesting doll," Carson deadpanned. "You wind it up, and it breaks into Ken and Barbie's room looking for communists."

With the commercial break out of the way, Carson announced the evening's lineup. "My first guest tonight, I'm sure you have read about or heard about—she has about a three-page article in this week's *People* magazine. She's a high school teacher from Concord, New Hampshire," he said to scattered applause and cheers, "and she was recently selected by NASA from a field of over eleven thousand applicants to be the first—as they call it—average citizen in space."

Backstage, Christa McAuliffe patted her hair and straightened the skirt of her sensible white suit, and a few minutes later stepped out to kiss Carson on the cheek. At thirty-six, McAuliffe had been teaching for fifteen years, and projected poise and confidence: she joked, and parried the host's sardonic asides, and explained the outlines of what NASA called its Space Flight Participant Program; she regaled him with anecdotes about how she had been selected to make the experience of spaceflight more comprehensible to the public; she described her experience of overnight fame, and how it helped her continue to reach the students she had left behind; she told him what she expected of her mission: "It's going to be six days, and every ninety minutes I'm going to be orbiting the Earth at seventeen thousand miles an hour," she said with a giggle. "I really hope to get on the treadmill around California—and jog across the United States."

Carson was clearly charmed.

"It sounds absolutely fascinating, and you seem to be handling it really well and I think they made a very good choice," he told her. "Because I think you can communicate this to most of us who really can't understand all of it. You're excited to go, huh?"

"Oh, I really am," McAuliffe said. "I can't wait."

::::::::::::::::::

Ordinary Americans had begun petitioning NASA to carry them into orbit from the moment Alan Shepard first left the atmosphere in 1961, and for the next twenty years the agency received thousands of letters every year from men and women anxious to become Everyman astronauts. But the Space Shuttle, with its explicit promise of making space travel routine, brought public requests to new heights. Proposals poured in from artists, real estate developers, poets, and policemen; from those who hoped the shuttle would take them closer to God, and women who wanted to deliver the first baby in space. Rockwell International, the contractor who built the orbiter, had even produced a blueprint for a version of the spacecraft that could carry seventy-four passengers.

By the time the initial test flights of the shuttle were taking place in 1981 and early 1982, more than ten thousand pieces of mail from would-be space travelers were arriving at NASA facilities every year.

Behind the big green doors of his office on the seventh floor of the agency's headquarters on Independence Avenue in Washington, DC, Administrator James Beggs hosted visits from organizations pressing to send the first citizen astronauts into space: for a while, Beggs considered choosing a photographer from *National Geographic* magazine, and was especially taken with the idea of selecting a suitable Boy Scout.

Yet Beggs soon concluded that a formal procedure would be necessary to handle the rising torrent of competing requests. In July 1982, just as the fourth and final test flight was completed and President Reagan declared the shuttle "operational," NASA appointed a six-man Citizens in Space Task Force to examine the issue. The group, which included shuttle pilot Dick Truly and blockbuster novelist James Michener, set out to decide whether there was a valid argument for flying private citizens on the shuttle, and—if so—how they should be selected. To help it decide, the group spent a year meeting and touring NASA facilities around the United States. But it also solicited advice from more than two hundred leading figures in aerospace, business, and the arts, of whom more than half took the time to respond; their collected opinions covered a spectrum from the aridly unimaginative to the outright fanciful.

Former agency Administrator Robert Frosch suggested that members of the general public should be given the chance to become astronauts through buying tickets in a national lottery, while at the same time NASA could invite artists and performers to join the program in a separate category, through which traveling in space would inspire or inform "a creative act." Writer Norman Mailer said that the selection process should address the public perception of the agency as a "private and intolerably starchy club." Princeton physicist Freeman Dyson suggested that what the agency really needed was someone who could undermine the solemnity of space exploration by focusing on the absurdity of humanity's place in the universe; Woody Allen, he wrote, would be ideal: "A good clown is what you need."

Meanwhile, former Apollo astronaut Bill Anders, who had spent years lobbying for NASA on Capitol Hill, recommended choosing a candidate from among defense and aerospace executives, or members of Congress. Many other respondents opposed the idea of a lottery, not least because the random selection process it promised would open the agency to the risk of introducing "an unemployed, undereducated, ne'er-do-well" into the hallowed company of astronauts. One scientist at Johns Hopkins instead proposed inviting only the wealthiest members of society to become gentlemen space travelers, suggesting that NASA help create a network of exclusive "Shuttle Clubs"; modeled on the New York Yacht Club, these would solicit funds from the richest cliques in America to underwrite their own trips into orbit, "much," he suggested, "as they succeeded in providing sailing ships to compete in the America's Cup."

The final report of the Citizens in Space Task Force reached James Beggs's desk in July 1983, granting cautious approval to the idea of sending passengers into orbit aboard the shuttle. "Flight of private citizens," they wrote, "is both feasible and desirable." But they warned that there were clear risks: the group feared, for instance, that the public may see the project as "a self-serving public relations gimmick that trivializes the space program." Every member of the task force also understood the dangers inherent in spaceflight: they knew that the odds of losing astronauts in an accident were significant. At one point they discussed how the agency's image might suffer if a beloved figure like Walter Cronkite— the then sixty-six-year-old anchorman and dean of TV news, who had already put his name forward to fly—was killed aboard the shuttle. The group suggested that NASA start slowly.

By early the next year, the agency's newly named Space Flight Participant Program was already under way. Staff had identified a total of thirteen impending shuttle flights that could carry a passenger, with scheduled launch dates stretching from August 1984 to October 1986. And while the press luxuriated in rumors that the first citizen in space would be a celebrity—bespectacled country star and spaceflight fanboy John Denver was a favorite—an internal committee concluded that the best candidates would be either teachers or journalists. They recommended to James Beggs that a teacher should go first, and the NASA chief agreed. "The biggest receptive audience we have in this country are the kids," he said. "Kids love space. A teacher could give you an introduction to those kids that no one else could."

In July 1984, Beggs sent the idea to the White House for approval, and President Reagan—grasping the popular appeal of the initiative in an election year—made it his own. In August, Reagan announced the program following a visit to a resource-starved junior high school in Washington, DC. "Today, I'm directing NASA to begin a search in all of our elementary and secondary schools and to choose as the first citizen passenger in the history of our space program one of America's finest—a teacher," he said to polite applause. "When that shuttle lifts off, all of America will be reminded of the crucial role that teachers and education play in the life of our nation. I can't think of a better lesson for our children and our country."

Across town, in an auditorium at NASA headquarters, James Beggs watched on a giant video screen as the President concluded his remarks—and then launched a press conference dedicated to the details of the "Teacher in Space" program. Applications would open in November, he explained, and be due by the end of the year. NASA clerks would weed out those hopefuls who didn't meet the basic qualifications, and forward the rest to the Council of Chief State School Officers, usually responsible for the Teacher of the Year awards, and who NASA had contracted to assist with sorting the mountain of applicants. The council's committees in every US state or territory would each choose two candidates, contributing to a total of more than one hundred teachers, who would travel on to Washington, DC, for interviews with a national panel. The panel would choose ten finalists to be sent to Houston for a week of appraisals and testing before the winner—and his or her backup—was announced later in 1985. After their flight, the successful applicant would be required to spend a year meeting the public to discuss their experience. "We don't expect them to keep it to themselves," Beggs said. "They'd better be aware they're going to be a hot property for us."

That same day, Christa McAuliffe was driving through Concord with her husband, Steve, when they heard Reagan announce the program on the car radio. Christa said nothing, but felt a quiver of anticipation. She turned to her husband with a smile. "It's a don't-miss," he said. "Go for it."

The next morning, she picked up the *Concord Monitor* from the porch of the couple's brown-shingled Victorian on Pine Ridge Drive and glanced at the headline on the front page: "Reagan Wants Teacher in Space." Beside it was a photograph of Judy Resnik, climbing from the cockpit of her T-38 in Houston as she prepared for her mission aboard *Discovery*. McAuliffe had included Sally Ride as part of the course she taught her high school class on women's history; now Resnik would be America's second woman in space, with a teacher to follow; it seemed hard to believe. *There may be hope yet*, McAuliffe thought. But with the new school year a little more than a week away, there was a lot to do to prepare for it; losing herself in work, she quickly forgot all about the Teacher in Space program.

Meanwhile, citing his lack of spending on education, the nation's largest teaching union dismissed the President's announcement as an electioneering gimmick. "We don't need to send one teacher into space," the union president said. "We need to send all teachers into their classrooms fully equipped and ready to help students learn. Sending a teacher into outer space won't solve the problems in schools on earth." On Capitol Hill, former astronaut Senator John Glenn opposed the initiative as dangerous and frivolous.

Even so, many teachers across the country were galvanized by the prospect: at least 45,000 of them requested application forms, and some 11,500 applied. The entry requirements were so rudimentary that it seemed almost anyone could qualify—candidates needed to be US citizens, and to have five years' teaching experience; but they were not subject to stringent health requirements, or even an age limit. "We're not looking for Superman," a NASA public affairs officer said. "We're looking for the person who can do the best job of describing their experience on the shuttle to the most people on earth."

It wasn't until November that Christa McAuliffe thought again about the competition. The weekend before Thanksgiving 1984, she was attending a teachers' conference at the Hilton hotel in Washington, DC, when she came across a table covered with copies of the initial application for the Teacher in Space program. When no one was looking, she filled a folder with a stack of them and, on her return to Concord High, handed them out among her friends—and then completed one herself. A week later, a blue-and-silver folder with an image of the shuttle on the cover arrived in the mail. The formal Teacher in Space application package was hefty and daunting: it included a twenty-five-page booklet, with eight separate essay questions to answer, as well as references and background information; it could take as long as 150 hours of work to complete. McAuliffe saw no point in kidding herself; the packet looked ten times worse than a term paper: she knew that if there was one thing she was going to put off until the last minute, this was it. And the final deadline for completed applications was still months away. She stuffed the package into a drawer.

Sharon Christa Corrigan was born in Boston in 1948, the first of five children in a family of Kennedy Democrats. In her early years, while her father was studying at Boston College, she slept in the single bedroom of a cold water flat in a Dorchester housing project, while her parents slept on the couch. In junior high, she sat with other students in the cafeteria to watch Alan Shepard's Redstone rocket blast off from the Cape on television—but was more focused on piano lessons, and spending time with her family, than the developing space program. She met Steve McAuliffe at their small Catholic high school in Framingham, Massachusetts; he proposed soon after she turned sixteen—but they agreed not to get married until after college. Zealous but headstrong, Christa may not have been at the top of her class, but made the National Honor Society with a combination of hard work and infectious enthusiasm. In her junior year, she won a place in the annual Girl Scout Roundup competition, beating other applicants from all over the world. Enrolling at Framingham State College in 1966 to major in American history and secondary education, she maintained a long-distance relationship with Steve while he studied at Virginia Military Institute, and entered what she called her "radical period": she wore her hair in a bleached Afro, went to see Jefferson Airplane play in Boston, protested against the war in Vietnam, and spoke out in support of rights for women and Black and Native Americans. But she also captained the debate team, made the dean's list three times, and rarely missed a class.

She and Steve were driving through a rainstorm in Pennsylvania the night Neil Armstrong set foot on the moon, and they both cheered as they heard his voice crackle over the car radio, reporting his leap to the lunar surface. Afterward, when Pan Am announced the opening of reservations for the first commercial flight to the moon, she encouraged her fiancé to join her in signing up. The couple married in August 1970, eight weeks after graduation, and moved to Washington, DC.

McAuliffe began work as a substitute teacher, and then spent seven years teaching American history, English, and civics to the largely Black students of a junior high school in a lower-middle-class Maryland suburb, while Steve studied for the bar in DC. Christa earned a master's in education supervision and administration, became a strong and outspoken advocate of change at the troubled school, and led a campaign for better union representation of teachers. Her husband encouraged her to become a lawyer, too—but she believed she could make a bigger difference in society in the classroom.

Their first child, Scott, was born in 1976; their second, Caroline, followed three years later. By that time the family had moved to Concord, the state capital of New Hampshire. Christa described it as "a Norman Rockwell kind of place"—a placid family town on the banks of the Merrimack River, surrounded by fields of corn and strawberries, where she eventually took a job at the local high school, a rambling redbrick building three blocks from home.

At Concord High, McAuliffe taught American history, law, and economics, establishing herself as a well-liked and inspirational teacher who took risks in the classroom. With two young children at home, she often arrived late, her hair still wet from the shower—but she kept her students engaged with role-playing and frequent field trips, and combined her feminist principles with an interest in the role of ordinary people in history in a course she titled "The American Woman." If colleagues had any criticism of her, it was that she was overcommitted: she mentored students outside school hours, as well as leading a Girl Scout troop, volunteering at a family planning clinic, appearing in community theater, and raising funds for the hospital and the YMCA. And while McAuliffe was still working through her multiple to-do lists in the fall of 1984—she kept one copy on the fridge at home, another taped to the dashboard of her Volkswagen bus, and a third in her pocket—across the country other teachers had already begun vying for NASA's ticket into space. Before applications were even open, a group of more than forty of them attended an event at Space Camp in Huntsville hosted by agency officials, where they donned astronaut flight suits and sat aboard a shuttle mock-up. Elsewhere, others approached the competition as a political campaign, taking to the streets and handing out buttons, or encouraging their students to petition the public on their behalf; in the meantime, thousands of students across the country wrote letters to NASA explaining why their teacher should be the one chosen to fly aboard the orbiter.

McAuliffe might have forgotten altogether about the contest without the prompting of her husband. Steve, by now a trial lawyer in private practice, who had always dreamed of flying, but had been stopped by poor eyesight, began reminding her soon after Christmas 1984 of the approaching deadline. Yet still Christa stalled until, ten days before the application was due, she finally sat down with the package and the list of eight essay questions. Staying up late in the evenings after the rest of the family was in bed, McAuliffe wrote and rewrote her responses to "Why do you want to be the first U.S. private citizen in space?" and "Describe your teaching philosophy"; she signed a contract waiving her rights to government benefits including life insurance; and she agonized over the project she would choose to conduct in space. At the last possible minute, on February 1, 1985, she dropped off the completed package at the post office in Concord. She was one of seventy-nine teachers to apply from New Hampshire, but when the members of the state selection board met her a few months later, they knew right away that McAuliffe would be on their list of finalists. "She had that girl-next-door quality," said one. "A kind of wholesome American look."

Meanwhile, in Texas, the training of those passengers NASA had already promised a ride aboard the shuttle had begun.

::::::::::::::::::

Greg and Marcia Jarvis had landed at Hobby Airport in Houston on a Muse Air plane from California one crisp afternoon in the first week of January 1985. With their dog, Syrah, languishing in a pet carrier, they took a cab to the Peachtree Apartments, a long-stay complex ten minutes from the back gates of the Johnson Space Center, where the couple planned to live for the next three months. Marcia had taken a sabbatical from work while Greg trained for his spaceflight. Jarvis's backup, Bill Butterworth, and his wife, Jenny, moved into an apartment a few doors down.

Jarvis was scheduled to fly on the shuttle in March, on a mission set to carry into orbit one of the big Hughes satellites he had helped build out at the plant in Los Angeles, and as the weeks went by his preparation for the launch intensified. First, he and Butterworth took a trip aboard the Vomit Comet. The two men flew forty parabolas in total: Butterworth, who had occasional vertigo, was sick on all but six; Jarvis didn't suffer at all. The following week, the engineers spent a morning in firefighting practice with the shuttle crew they'd been assigned to, learning to extinguish gasoline fires, wrangling the powerful hoses as part of a team. Back home at the Peachtree Apartments, Marcia and Jenny Butterworth watched the black smoke rising on the horizon before going out for an afternoon of riding lessons with Bob Crippen's wife, Ginny, at the stables she ran nearby. Two days after that, it was Greg's turn for ejection seat training.

Over the months Jarvis and Butterworth spent together in Houston, the two engineers grew close to their shuttle crew, under commander Dan Brandenstein—a former Navy test pilot, and the only one among them who had flown in space before. Amid the acrimonious competition for flight assignments, the payload specialist program had long been regarded with suspicion by some astronauts. Although embraced by George Abbey, the initiative had been fought almost from the beginning by Chris Kraft, who believed that their jobs would be better done by fully trained members of the astronaut corps. Yet—while it soon became clear that Jarvis would have little to do with placing the Hughes satellite in orbit—Brandenstein and his crew tried hard to make both Greg and Bill feel a valuable part of the team, treating them as aerospace professionals with a key role in the mission. After hours, the astronauts and their families welcomed the two engineers and their wives to cookouts and dinners, with parties for birthdays and Super Bowl Sunday.

By late February, the final preparations for the mission were under way: in Building 5 at the Space Center, Jarvis joined the crew in the shuttle mock-up for a thirty-four-hour flight simulation, involving a full-strength team in Mission Control, that took them through everything from liftoff to landing in real time; arriving home at eight-thirty one Wednesday evening, he slept for as long as he could, before rising to join the simulation again at 4:00 a.m. Jarvis posed with the crew for a series of official NASA portraits, and at the end of the month received his mission patch—featuring his name among those of his fellow astronauts, embroidered beneath an image of a golden eagle and the shuttle soaring through space.

But the ambitious shuttle flight schedule continued to be hobbled by delays and worrying technical problems. At the beginning of the year, the fifteenth shuttle launch—and the first classified military flight in NASA's history—of *Discovery*, had been pushed back due to record-breaking cold weather that swept down the Eastern Seaboard. The public was kept in the dark about much of the mission: NASA refused even to release precise launch or landing times in advance, and ground-to-air communication with the crew would be encrypted throughout the flight. But agency engineers did admit to fears that three consecutive nights of freezing conditions had caused icing on the launchpad—threatening to damage *Discovery's* heat-shield tiles—forcing them to postpone the planned liftoff.

The second mission of the year, aboard *Challenger*, had been initially slated for February—but the launch was delayed again and again,
at first as technicians at Cape Canaveral worked to replace four thousand of the orbiter's tiles, which had come loose on its previous voyage;
the problems made news in part because the crew included Republican
Senator Jake Garn, who aimed to become the first politician in space.
Congressmen and senators alike had lobbied NASA since the inaugural
shuttle launch for an invitation to fly into orbit, but Garn had been both
persistent and hard to ignore: as chairman of the Senate committee that
controlled funding for NASA, he could make life awkward for Administrator Beggs if he didn't get what he wanted.

Garn maintained that it was a constitutional necessity for him to
fly aboard the shuttle as an "observer"—a payload specialist whose expertise was to understand where taxpayers' money was going. This ra-
tionalization fooled no one. The press derided Beggs's formal invitation
to Garn to become a politician-astronaut as "the ultimate junket"; but
when, at last, the *Challenger* mission was finally canceled, it meant that
the Senator was first in line for the next available seat into orbit.

On March 1, 1985, with less than three weeks to go before he expected to go into space, Greg Jarvis learned that his mission with Dan
Brandenstein's crew had also been postponed; the payload specialist
positions for forthcoming flights would have to be reshuffled. After an
agonizingly long weekend of uncertainty—leavened only by a trip to
the movies in Houston to see the new Eddie Murphy film *Beverly Hills
Cop*—Jarvis was at home sick with flu when the phone finally rang. It
was Bill Butterworth with the news that Greg was out: Senator Garn
would be taking his place as a payload specialist on the next shuttle
flight; Jarvis would just have to wait for a new assignment, perhaps at the
end of the summer. The decision was announced to the public at noon
the following day, and—although NASA insisted it had come under no
political pressure—in Houston the evening news carried calls for Garn
to be investigated by the Senate Ethics Committee. No one from the
agency bothered to officially inform Jarvis he had been cut from the
mission. Nonetheless, he booked a function room in their apartment

complex for a party to bid goodbye to Brandenstein and his crew, who he and Marcia had come to consider good friends. While Greg made the arrangements for their farewell dinner, Marcia called the movers to take them back to California. On the day they left for the airport, rain sluiced from the skies.

::::::::::::::::::

In her room at the Hyatt Regency hotel in Washington, DC, Christa McAuliffe unpacked her clothes for the week, including the eight new outfits her parents had bought her for the occasion, and then sat down to survey her competition. NASA public affairs officials had given each of the Teacher in Space semifinalists a set of capsule biographies of everyone involved, describing their background and achievements. Now, the more McAuliffe read of the 112 other teachers with whom she would be competing in what amounted to an extended talent contest, the further her spirits fell. She reached for the phone and called her husband at home in Concord.

"Steve, these people are doctors and authors and Fulbright scholars and teachers of the year and a woman who climbed the Himalayas and . . ." She trailed off. "I'm out of my league," she said, her voice flat. "I haven't got a chance."

"Hang in there," Steve said. "You're doing fine. You wouldn't be there if you didn't have merit. Just relax and have a good time."

The Teacher in Space events had begun earlier that June evening with a banquet in the grand ballroom of the L'Enfant Plaza—a Watergate-era hotel and shopping mall a few blocks from NASA headquarters—and a welcome speech delivered by James Beggs. The candidates, chosen from all over the United States—as well as territories in the South Pacific and the Arctic Circle—had been invited to the nation's capital to attend lectures from astronauts and space agency staff, workshops and parties, and a special screening of the shuttle documentary *The Dream Is Alive*.

At the climax of it all, they would each face interviews with judges from a panel of fourteen men and six women, which included a former professional basketball player, the inventor of the artificial heart, astronauts Deke Slayton and Gene Cernan, and the actress Pam Dawber—who had starred opposite Robin Williams in the screwball science fiction TV show *Mork & Mindy*. The ten finalists chosen by the panel would be sent to Houston for a week of more intensive tests and interviews, before James Beggs himself selected the winner—who would then go into orbit aboard the shuttle *Challenger* early the following year.

The Teacher in Space mission would be the twenty-fifth flight of the shuttle, designated STS-51-L and scheduled for launch from Cape Canaveral in January 1986, with an experienced crew already selected by George Abbey. Commander Dick Scobee and Mission Specialists Judy Resnik, Ron McNair, and Ellison Onizuka would all be making their second flights into space. Only the pilot, Mike Smith—a soft-spoken Navy flier and former test pilot who had arrived at NASA with the second group of new astronaut recruits in 1980—had yet to cross the Karman Line. In the ballroom of L'Enfant Plaza, Smith followed James Beggs up to the podium to explain the details of the 51-L mission. He discussed the other members of the crew, and told the audience how much he was looking forward to joining the select fraternity of those who had seen the Earth from orbit. If space flight was even half as remarkable as his fellow astronauts had described it, he said, both he and the winning teacher were in for the adventure of a lifetime.

Over the next few days, the 113 candidates attended a series of NASA lectures intended to reveal a less idealized picture of the commitment the world's first teachernaut would be expected to make. The winner of the competition would report to Houston in early September 1985 for three months of flight training, and would be required to publicize their experience in the program for at least a year after their return from orbit. Frank Johnson, NASA's director of public affairs, explained that the winning candidate would become one of the most recognizable faces of the agency, a national spokesperson helping to boost public enthusiasm for its work—and justify the continuation of its $7.6 billion budget. There would be more interviews and autographs than they would want; their lives would no longer be their own as they were plunged into the fishbowl of celebrity; the media would come at them from all sides.

"Nobody's going to be out to destroy the space program," he said, "but, you know, when you get fingers into the pie, sometimes things can get messed up."

The teachers received a briefing on how to deal with the press—and the director of the agency's Public Service Division, Chester Lee, was frank about the difficulties they were having keeping the shuttle on its launch schedule, and how the delays were undermining the agency's commercial satellite delivery business; but he emphasized that there would never be pressure to fly at the cost of safety. He explained that there were repeated Flight Readiness Reviews leading up to every launch, and each system on the shuttle had multiple redundancies, designed to keep the spacecraft flying even after multiple failures; those that could not be equipped with fail-safe systems were built to take twice as much punishment as might be expected on a normal flight. He used the solid rocket boosters as an illustration: they had double the insulation necessary to prevent hot gases burning through the casing, because the engineers were especially concerned about the two minutes during which the shuttle relied on them to carry it into orbit. "If one of those rockets goes, why, it's pretty bad," he said.

Still, an air of unreality prevailed. At a press conference later that same day, the candidates were paraded before the TV cameras as dozens of reporters peppered them with questions, and the teachers raised their hands to answer. Amid the thicket of waving arms, no one called upon Christa McAuliffe. But Sophia Clifford, a chemistry teacher from Alabama, told the journalists that she thought she might win the contest because she'd recently received a fortune cookie at a Chinese restaurant in Birmingham. Clifford explained that, cracking it open, she had found an auspicious prophecy on the slip of paper inside:

You've been promised a ride on a starship by the galactic wizard.

CHAPTER SIXTEEN

ABORT TO ORBIT

By the time he stepped up to address the banquet audience in Washington, DC—offering a quotation from ancient Greek philosopher Epictetus by way of advice to the roomful of aspirant Teachers in Space— James Beggs was in more desperate need of support for NASA than ever.

Publicly, President Reagan remained an enthusiastic backer of the space program, and had recently given his blessing to Beggs's plans for building a space station in Earth orbit. But Reagan's second-term White House was now packed with officials who actively opposed manned spaceflight, and a few who detested Beggs himself. Although a shrewd political operator, since assuming control of the agency the NASA chief had alienated key West Wing insiders—those he considered "right-wing nuts"—with perceived slights that included inviting conservative bête noire Jane Fonda to the Cape to watch Sally Ride become the country's first woman in space, and going behind their backs to win Reagan's backing for the space station.

These men—including the President's Science Advisor, who derided the idea of a permanent orbital base as "a motel in the sky for astronauts"—were reactionary ideologues with little interest in NASA's peaceful objectives of exploration for all mankind. Instead, they wanted to annex space as a new theater of Cold War confrontation with the Soviet Union, to be occupied by the nuclear-powered death rays and antimissile beam weapons of the Strategic Defense Initiative—the "Star Wars" program. Meanwhile, many members of Congress continued to regard much of NASA's work as dangerous, wasteful, and expensive.

And the shuttle program itself had reached a crossroads: its future, and therefore the future of all US manned spaceflight, had now become dependent on proving that the orbiter really was the cost-effective space truck NASA had promised it would become. That meant delivering commercial satellites into orbit, and achieving the audacious launch schedule it had predicted back in 1972, flying as many as forty-eight missions a year.

But when NASA officials in Washington, DC, looked over those early projections in the spring of 1985, they could only close their eyes and sigh in disbelief: the true launch rate was a fraction of that. And the poor reliability and repeated delays of the shuttle had both impatient commercial customers and other government agencies looking elsewhere.

The upstart European Space Agency had already begun launching satellites at bargain prices aboard their unmanned, single-use Ariane rockets from a remote site in the jungles of French Guiana; both the Air Force and the National Oceanic and Atmospheric Administration had made arrangements to go back to the old way of doing things, launching their satellites on the same expendable military rockets the shuttle had been supposed to replace. In the meantime, the voodoo accounting that had underwritten the shuttle project from the beginning had finally been exposed for what it was: the true costs of the program were orders of magnitude higher than the Panglossian forecasts of Beggs's predecessors. On some launches, the price of a payload flown aboard the orbiter was not the $270 a pound NASA had once promised, but as much as $5,200: almost twenty times more. "Nowhere was the fabric of the Emperor's New Clothes thinner than on these missions," *Reader's Digest* space columnist Malcolm McConnell would observe.

The only way to bring down costs was to ramp up the pace of launches. Now Beggs had the ground crews at Cape Canaveral working around the clock in an attempt to rotate *Columbia*, *Challenger*, and *Discovery* more frequently into space, and prepare a fourth orbiter, *Atlantis*, for its maiden launch in the fall of 1985. He announced his intention to fly missions at least once—sometimes twice—each month in the coming year. But the reality was that the most complex machine in the world could not be flown on round trips into orbit like a jumbo jet, especially if the entire operation was being run on a shoestring. NASA estimated that, collectively, the crews at the Cape had to put in the equivalent of three years of work on the ground for every minute each orbiter spent in space.

Instead of the returning spacecraft being subjected to a few routine checks, cleaned up, and cleared for another flight, as the designers had once imagined, now every part of the orbiter had to be examined carefully before each mission, and often disassembled into its component parts. This intricate screening revealed unforeseen problems: the brakes and steering of the landing gear weren't holding up to repeated use; the tires were at risk of explosion; the hand-built components of the main engines had proved even less robust than expected; and the tiles were an enduring worry. When *Discovery* returned from a mission in April that year, the astronauts discovered that part of the heat shield on the spacecraft's left wing had come loose on liftoff, and during reentry hot gas had melted through a control flap; the resulting hole in the aluminum surface was a foot wide. Meanwhile, cost cutting meant that the program was chronically short of spare parts, so the engineers at the Kennedy Space Center were often forced to cannibalize components from one orbiter that had just returned from space to install on another waiting to leave on the next mission.

Inside the hangars of the Orbiter Processing Facility, work went on day and night, as three successive shifts of engineers and technicians labored around the clock to keep the shuttle fleet on schedule. As the flight rate increased, staff once used to the systematic focus of readying a single, unique, spacecraft for orbit were now assigned duties on two or three at once. In the interests of implementing airline-style efficiency, Beggs had placed control of the complex flow of tasks required to prepare the orbiters for launch in the hands of a single contractor, California-based aerospace giant Lockheed. Under the new regime, the number of quality control inspectors and safety checks in the process fell, while overtime hours increased.

Among the fourteen thousand contractors and NASA employees at the Kennedy Space Center, many senior managers and engineers were soon working twelve-hour days, seven days a week—often for months at a time, without a day off. Numbed by overwork, driven by the relentless drumbeat of the launch schedule, the staff did their best to keep up with the laborious procedures and paper trail the agency required for every minor modification and repair performed on the orbiter's hundreds of systems and subsystems. But mistakes were inevitable.

In the meantime, the three shuttles kept launching—and soon the long-promised goal of flying two missions a month no longer seemed quite so impossible; the fruits of the effort on the ground were realized in April 1985 when *Discovery* and *Challenger* took off from Cape Canaveral within seventeen days of each other. But in the all-important sphere of public relations, the work was counterproductive. Although it was still far behind schedule, in just four years NASA had all-too-effectively fulfilled its promise of making space travel seem quotidian; and just as had happened with the moon landings, the grand spectacle of shuttle launches now seemed to be taking place so frequently that they were rarely carried live by the major networks.

In the same week the competing teachers gathered in Washington, DC, the agency successfully returned the one-hundredth American astronaut from space, aboard *Discovery*: a mission chiefly notable for carrying a member of the Saudi royal family into orbit. And with each successive launch, there were fewer and fewer journalists assigned to the Cape; many of those who remained were on a "death watch"—ensuring that they would be there to provide on-the-spot coverage in the event of a catastrophe.

:::::::::::::::::::

Christa McAuliffe had prepared with characteristic care for her interview with the judges in the L'Enfant Plaza, researching their backgrounds and rehearsing her answers to the questions she imagined they would ask. Although the interview itself was only fifteen minutes long, the panel also reviewed the application forms every candidate had so laboriously completed earlier in the year, and a conversation on video they had each filmed in advance. When the judges met to decide on the ten finalists who would be sent to Houston for physical and psychological evaluation, their deliberations went on for hours—but the one candidate they agreed on almost immediately was McAuliffe; they were all impressed with her spirit, her eloquence, and her ideas about communicating the experience of spaceflight. Of the six women and four men—aged between thirty-three and forty-five—they chose, McAuliffe did not have the most impressive résumé, or the most elaborate idea of what she would do if she reached space. Among the nine others was a woman who had climbed the Andes, crossed the Atlantic in a thirty-one-foot sailboat, and was planning an expedition to Antarctica; a former Air Force fighter pilot from Greensboro, Vermont; another candidate who had graduated Phi Beta Kappa from Stanford; and a charismatic local activist who had written prize-winning plays, gone to jail twice as part of a campaign to improve conditions for local teachers, and recently published an essay in *Newsweek* defending the Space Flight Participant Program against its critics. Between them, the projects they had suggested carrying with them into orbit included the study of space sickness and experiments surveying the gravitational field of the moon. After all her deliberations, McAuliffe had simply proposed that she would keep a journal of her experiences, like the pioneer women of the Old West.

Yet NASA, already overstocked with seasoned adventurers, multidisciplinary geniuses and scientists, was looking for something else. What the agency required from the first teachernaut it chose to ride the shuttle was a charismatic cheerleader for manned spaceflight, an ordinary person with a gift for communication who could carry the broad promise of the High Frontier into living rooms and classrooms across the United States. And when she and the other candidates arrived in Houston early in July 1985, McAuliffe's performance stood out.

The week of medical and physiological tests, and public appearances before the press, helped dispel any romantic notions the candidates may still have harbored about what lay ahead—there were lessons on decompression sickness and spatial disorientation, blood work, X-rays, and musculoskeletal analysis; a trip on the Vomit Comet and fifteen long minutes spent zipped into the Personal Rescue Enclosure. The press, with a newfound enthusiasm for the space program energized by the novelty of the search for a teachernaut and its high-stakes talent contest dynamics, turned out in force. After the ten candidates dined in the Johnson Space Center cafeteria with Bob Crippen and Judy Resnik, autograph hunters swarmed after the teachers, yet the two astronauts walked away almost unnoticed.

But McAuliffe seemed at ease with the attention. She went through one newspaper interview after another with "a nervous giggle and the gee-whiz bounce of a camp counsellor," handling questions on everything from the safety of spaceflight to teachers' salaries with an engaging openness. In one training session, the ten teachers were sealed in a hypobaric chamber to experience the symptoms of high-altitude oxygen starvation. One member of the group became so belligerent under the influence of hypoxia that the trainers had to overpower him, forcing him back into his breathing apparatus; McAuliffe sat through it all with patient serenity. When the same psychologist who had examined the TFNG astronaut candidates years earlier subjected each of the would-be teachernauts to a two-hour interview and an hour-long written test, he rated McAuliffe the most well-adjusted of them all. "I know this doesn't sound very scientific," he said, "but I think she's neat."

On the last of their seven days in Houston, the finalists were escorted to Building 30, to witness the latest launch of the spaceship that would soon take one of them into orbit. Gathered alongside Bob Crippen in a row of theater seats at the back of the Flight Control Room, the ten teachers listened as a voice from the Launch Control Center in Cape Canaveral described the final countdown for *Challenger*'s eighth mission. McAuliffe leaned forward in anticipation, resting her face in her hands, as the shuttle's three main engines lit—but then, just three seconds before the solid rocket boosters reached ignition, abruptly shut down again. As the spigots of the pad fire suppression system spewed water across the hot engine bells and steam billowed into the air, the teachers sat transfixed. Crippen began to explain what was happening inside the orbiter, and the NASA official running the Teacher in Space program once more brought up the risks of spaceflight; would any of them like to withdraw?

They would not.

:::::::::::::::::::

The name of "the first private citizen passenger in the history of space-flight" was due to be revealed by James Beggs and Vice President George Bush in an event attended by all ten teachers at the White House at 1:00 p.m. on July 19, 1985. By that time, the story was such hot national news that reporters tried everything they could—phoning the candidates in their hotel rooms in the small hours of the morning, prowling the corridors of NASA headquarters, appealing to contacts in the Senate—to find out the name of the winner before it was announced.

The teachers themselves hid from the press in an unoccupied office at NASA, where the chairwoman of the selection committee finally told Christa McAuliffe: "You're the one." Drawn together by the relentless pressure of the previous two weeks, McAuliffe's nine colleagues gathered in a crescent around her and, one by one, held her in an embrace. Kept incommunicado, the ten teachers were then driven to the White House in a pair of gray government station wagons.

Only when Beggs leaned into the microphone in the Roosevelt Room, facing the clattering shutters and flashes of a phalanx of photographers, did McAuliffe begin to grasp what was about to happen. Her previous life as a small-town schoolteacher was over; she would soon be more famous than Gene Cernan, the last man on the moon. When the Vice President announced her name—and that of her backup, thirty-three-year-old Idaho elementary school teacher Barbara Morgan—she was overwhelmed.

"It's not often that a teacher is lost for words. I know my students wouldn't think so," she said. "I've made nine wonderful friends over the last two weeks. When that shuttle goes, there might be one body—" At this, her voice broke; her eyes brimmed with tears. She stopped and steadied herself, pressing a finger to her lips.

"But there's going to be ten souls that I'm taking with me. Thank you."

Back in Concord, inside the shingled Victorian on Pine Ridge Drive, McAuliffe's mother-in-law answered the first of hundreds of telephone calls from across the country; outside, reporters waited for the opportunity to talk to the world's most famous teacher, while two helicopters from TV stations in Boston buzzed the house from the air. Christa's five-year-old daughter, Caroline, ran around the house in a bathing suit, greeting visitors with a bag of popcorn. She told them she was looking forward to going to Cape Canaveral to watch her mother's launch; she pointed through the kitchen window into the sky beyond.

"They go wa-ay up in space," she said.

Her brother Scott, eight, already wanted the whole thing to be over.

"Nana," he asked his grandmother again and again, "when is the phone going to stop ringing?"

Two weeks after the ten teachers had witnessed the hair-raising last-minute launchpad abort, on July 29, 1985, *Challenger* blasted off from the Cape bearing a crew of seven and a package of some of the most sophisticated experiments ever carried into space. Although some of the scientists aboard had spent seven years preparing for the mission, it had attracted more attention for what NASA called the Carbonated Beverage Dispenser Evaluation. In the first orbital battle in the so-called Cola Wars between Coke and Pepsi, the astronauts had instructions to try out competing can designs the two corporations had devised to make it possible to drink the first carbonated soda in space.

The Coca-Cola Company, reeling from the disastrous launch of its reformulated New Coke earlier in the year, had initially persuaded NASA to take only its drink on the mission, until Pepsi intervened by directly lobbying the White House to add its to the flight manifest. As the launch approached, senior officials told the crew to take photographs of the respective containers they tested—and to make sure that both brand logos were visible.

For the first three and a half minutes after launch, *Challenger*'s ascent went exactly as planned: the engines lit, the shuttle rolled away from the tower, accelerating downrange, out over the Atlantic, high into the blue sky over the Cape. But down on the "booster" console in Mission Control, twenty-nine-year-old Jenny Howard was watching closely as the columns of gray figures on the three screens flickered, rose, and fell. And, three minutes and thirty-one seconds into the flight, the fluctuating readout from one temperature sensor in the central main engine was replaced by the word FAIL; a second followed two minutes later: the first signs of a problem that would soon threaten the lives of everyone aboard the spacecraft.

Howard had been in Houston for five years, a graduate of the aeronautical and astronautical engineering course at Purdue University, and one of only a handful of women on the consoles in Building 30. She had been fascinated by the space program since she was a teenager in Indianapolis, collecting newspaper clippings and mail-order NASA newsletters about every launch, which she stuffed into a suitcase in her bedroom. Above all, she wanted to be an astronaut; by the time she arrived at Purdue to study rocket propulsion in 1973, Howard still had no idea that girls weren't even expected to become engineers. She was astonished when, during one parental visiting day, a professor took her mother and father aside to tell them that he knew Jenny had only enrolled there to find a husband—and that by doing so she was depriving a serious male student of a valuable place.

When she began her internship in the wind tunnel department at NASA's Langley Research Center in Virginia, at first the men there refused to speak to her: "We had a girl once, and she didn't work out," one explained. But when she arrived in Houston in 1980, she was welcomed into the close-knit team at Mission Control, training among the veteran flight controllers and rocket engineers who had put men on the moon. Assigned to the booster console, Howard became responsible for all of the Space Shuttle's propulsion during ascent: the three big main engines, the solid rocket boosters, and the numerous thrusters and maneuvering jets known collectively as the Main Propulsion System.

The role of the Booster Systems Engineer was one of the most stressful in the Flight Control Room. It took just ten minutes for the shuttle to travel from the launchpad into orbit—but command of every ounce of power that the spacecraft relied on to get there ran through Howard's console; for that brief time the lives of the astronauts rested in her hands. And Booster decisions had to be made in seconds; there would be no time for discussion: she had to understand and prepare for every eventuality long before the mission clock began to run.

In years of simulations and months of on-the-job training, in trips to the manufacturing plants of Rocketdyne and Thiokol in California and Utah, and meetings at the Marshall Space Flight Center in Alabama, she learned everything she could about the shuttle's main engines and solid rockets: gathering data about their idiosyncrasies, their software, design issues, and manufacturing problems. Among the anomalies that bothered Howard most were persistent problems with the temperature sensors in the high-pressure fuel turbopumps of the main engines and repeated reports of failures in the joints of the solid rocket boosters. But she and her colleagues talked often about the dangers of the Space Shuttle Main Engines, and all shared the conviction that if a catastrophe ever occurred, the engines would be the culprit. Computer resources in Houston were scarce and strangled by bureaucracy, so Howard spent more than $2,000 of her own money on one of the first Apple computers and ran technical analyses at home, writing reports in the evenings and on weekends, and tracing performance trends among the key engine components that might lead to a failure.

By the end of July 1985, when *Challenger* took flight with its cargo of hard science and experimental soda, Jenny Howard was a veteran of fourteen shuttle launches; she felt the quirks of the main engines with the intuition of an old friend. Before the boosters lit, she was always nervous. Yet as soon as *Challenger* cleared the tower, the rocket engineer was overtaken by a calm born of years of experience: her mind fell into the patterns of her training, programmed as reliably as any computer processor by years of simulations, the stages of ascent called out in a familiar cadence: *roll program; go at throttle up; SRB separation*... And even the failure of the two temperature sensors caused her little alarm. Howard had seen it happen half a dozen times before; the damned things were just unreliable: a redesign was in the works. The exchanges over the Mission Control communications loop were crisp, but relaxed.

"Booster—Flight," the Flight Director said. "How're the engines looking?"

"Lookin' fine," Howard said.

But at five minutes and forty-five seconds into the flight, with the shuttle traveling at three thousand miles per hour, she heard the commander on the loop. In a detached, emotionless voice he said, "We show a center engine failure."

"We copy. Stand by."

There was a brief pause, and a crackle of static. Howard suspected there had been nothing wrong with the engine itself—but the shuttle's onboard computers had shut it down because faulty sensors had failed. Still ascending, and at an altitude of seventy miles, flying on two engines: she knew it was just possible for the shuttle to make it into space.

"*Challenger*—Houston," said the CapCom. "Abort ATO. Abort ATO."

On the flight deck of the shuttle, the commander reached for the abort switch, twisted it to the setting for ABORT TO ORBIT, and pressed the button to engage the emergency flight program.

But just two minutes later, Howard saw the same sensors, this time in the right engine, exceeding their temperature limits. To prevent a fire or explosion, the shuttle's avionics computers were now programmed to shut this engine down, too. Flying on one main engine alone, *Challenger* could never make it into orbit, and would be at risk of tumbling out of control, or falling back under its own weight to crash into the ocean.

Inside the shuttle cockpit, flight engineer Story Musgrave opened his checklists, preparing to try for an emergency landing on the other side of the Atlantic, at the designated military airstrip in Zaragoza, Spain. His fingers snapped rapidly through the pages, rehearsing the procedures he would have to recite to the pilot if a second engine went out.

As he worked, he sensed the eyes of the Mission Specialist seated nearby, fifty-eight-year-old Karl Henize, boring into him. Henize was an astronomer, an astronaut-scientist who had personally discovered two thousand stars; he hadn't trained for this, and had little idea what was going on. But it looked bad. Through the shuddering noise filling the cabin, he finally spoke:

"Where are we going, Story?"

Musgrave didn't look up.

"Spain, Karl."

But down in Mission Control, Jenny Howard believed this was just another sensor malfunction: the rest of the data on her screens looked good. She didn't want to lose one more of her perfectly good engines, and she didn't want *Challenger's* crew to attempt a return to Earth that could end in the ocean—and almost certain death.

Howard could recommend that the shuttle commander override the computers—"inhibit limits," in NASA jargon—and keep the engine running for long enough to reach orbit. But if she was wrong and the turbopump really was overheating, it would soon catch fire and explode—destroying the orbiter and killing everyone aboard.

She had only seconds to make a decision. But when she spoke, her voice betrayed no urgency; she might have been discussing weekend plans over an afternoon coffee.

"Flight—Booster," she said into her headset mic. "We lost another sensor. I'm tempted to inhibit limits."

The Flight Director barely hesitated; it was already too late for the shuttle to return to Earth. "Inhibit limits," he said.

In the shuttle cockpit, the commander flicked the switch to override the onboard computers, allowing the two remaining engines to burn on for nearly a minute longer than planned—and, at last, *Challenger* limped safely into orbit. Down in Houston, it was only then that Howard began to think about what had happened, and asked her colleagues in the Booster section whether she had made the right call. The rocket engineer would eventually receive an award from the National Space Club for her quick thinking, and her hometown newspaper ran a front-page interview with her in which Howard was characteristically self-effacing: "I earned my pay, I guess," she said.

But few beyond the perimeter of the Johnson Space Center understood how close the shuttle had come to calamity—still less how it had been saved by the split-second decision-making of one person.

Two days later, 1,300 miles away in Burbank, California, Christa

McAuliffe was sitting onstage with Johnny Carson, her hands folded neatly in her lap. After the first few minutes of genial banter—discussing her students, the idea of the Space Flight Participant Program, and the elimination process that led to her selection—the host turned to more serious matters. Carson, a keen amateur astronomer, had begun at *The Tonight Show* in the same year that John Glenn became the first American to orbit the Earth, watched the liftoff of Apollo 11 from the bleachers at Cape Canaveral, and interviewed the crew of Apollo 13 from behind the same desk at which he sat now; he understood the risks of space flight only too well.

"When is this scheduled for?" he asked. "Do they have a target date for you to, as they say, blast off?"

"The target date is January 22," McAuliffe said. "But with the missions being bumped up a little bit with the problems that they've had, I would assume that it's probably the beginning of February now."

"Are you in any way . . . frightened of something like that? . . . Because just the other day it was kind of frightening—I mean, one of the engines went out."

"Yes," McAuliffe said carefully, and began to formulate just the kind of oblique answer with which astronauts had dismissed such inquiries since the earliest days of manned spaceflight. "I really haven't thought of it in those terms, because I see the shuttle program as a very safe program. But I think the disappointment—"

Carson cut her off with a smirk and a knowing glance at the camera.

"Who was it once said—Deke Slayton, I think. . . . I may be giving credit to the wrong astronaut. But they said, How do you feel when you're up there in that capsule? And he said, 'It's a strange feeling to realize that every part on this capsule was made by the lowest bidder.'"

It was an old line, but a good one.

Christa laughed, and the studio audience laughed, too.

CHAPTER SEVENTEEN

THE MYSTERY OF HANGAR AF

January 28, 1985

It wasn't until 1:30 a.m. that Lieutenant Colonel Onizuka at last returned home from his first journey into space. The street in front of his family's small brick house in Clear Lake was cold and deserted; the grass in the front yard glistened in the moonlight, just as it had when he had left Houston for the Cape two weeks before.

But Ellison Onizuka was no longer the same. Inside, he took a place on the living room couch between his two teenage daughters, and his wife, Lorna, sat opposite. Yet no one said a word; a deep hush filled the room. Lorna understood: it was the culmination of the dreams her husband had harbored since he was a Boy Scout in Hawaii, and a struggle that had overshadowed all of their lives for almost ten years. Now Onizuka was at peace, but unable to articulate the magnitude of what he had witnessed in the vast emptiness that lay beyond the Earth's atmosphere. While his eyes filled with the rapture of space, he felt unable to describe the experience, even to his family—and not just because he lacked the words; his mission, as part of the five-man crew of the shuttle *Discovery*, had also been the first entirely military flight in NASA history, its payload classified, a "black" operation so secret that almost all of Mission Control's conversation with the crew was scrambled, and even the launch time had been kept from the press and the public until nine minutes before the engines lit. Onizuka remained forbidden from discussing the details of his spaceflight with anyone lacking the appropriate Pentagon clearance.

As dawn approached, it was Lorna who finally broke the silence.

"What are your goals now?" she asked.

"I haven't really thought about it," he said quietly. "I'm just grateful for what I have."

By the time the sun rose, Janelle and Darien had both fallen asleep in their father's arms, but the astronaut remained awake, electrified by the memories of orbit: of the sixteen sunrises and sunsets in every day, of his glimpses of the Hawaiian archipelago, a handful of green jewels scattered on the deep blue backdrop of the Pacific. At last, Lorna saw her husband's eyes close. She rose to turn off the lights. The fear she had once felt for him had left her; there would be no need for further anxiety in their lives: he had become what he wanted to be.

Two days later, Morton Thiokol engineer Roger Boisjoly strode into Hangar AF in Cape Canaveral with a notebook in his hand, ready to complete his inspection of the solid rockets that had helped deliver Onizuka into orbit. The two boosters had been retrieved from the ocean using the same elaborate process scheduled after every shuttle launch. After burning out, and reaching the end of their flight trajectory at a zenith of thirty-eight nautical miles, the rockets had tumbled back toward Earth—in a descent that took as long as five minutes, slowed by parachutes, before finally splashing down into the Atlantic some 150 miles southeast of the launchpad. Tracked by Air Force radar operators, the boosters were equipped with radio beacons that transmitted details of their position to the crews of the two purpose-built recovery ships—the *Liberty Star* and the *Freedom Star*. Teams of divers from the vessels tethered the rockets and filled them with compressed air so that they could be towed back to Hangar AF beside the quay at Cape Canaveral Air Force Station, on the bank of the Banana River.

There, NASA workers had lifted them by crane into custom-built railroad cars and begun the days-long task of dismantling them for Boisjoly's postflight inspection: they rinsed them clean of seawater, sluiced away any lingering toxic propellants, and delicately removed the explosive charges installed as part of the rockets' self-destruct system. Only then could Boisjoly begin his examination, walking around the two boosters with a Thiokol staff photographer, recording any damage sustained during their journey to the threshold of space. Once his work was complete, the burned-out segments of the rockets would be shipped back to Utah, where they would be refurbished, refilled with propellant, and prepared for another mission.

A big, quiet man who shaved his head almost completely bald, Boisjoly enjoyed his assignments at the Cape. The work at Hangar AF often stretched over seven or eight days, including hours spent waiting around while the technicians overcame snags in the disassembly process, and Boisjoly liked to spend his free time wandering along the ocean, before joining his colleagues for the all-you-can-eat buffet at Guppy's in Cocoa Beach; the team often returned to Utah feeling a few pounds heavier than when they had left. The trips usually began the same way: back at the Promontory plant, where it was a company ritual to watch shuttle launches on the big projection-screen TV in the second-floor conference suite they called the Management Information Center—the MIC room. To save money and time—to make certain they weren't stranded in Florida sitting out repeated launch delays—the members of the postflight inspection team waited in the MIC room, plane tickets in hand, and watched the first two minutes of the shuttle ascent, until the moment they saw the solid rocket boosters separate. Then, as their colleagues stood to applaud another Morton Thiokol success, they left for the airport.

This trip, however, had been different from the start: Boisjoly flew down early from Utah to brief technicians at the Space Center on the correct ways to handle and assemble the solid rocket segments as they prepared them for a mission, and arrived at the Cape several days before the scheduled launch of *Discovery*. And when he checked in at the Holiday Inn in Cocoa Beach on January 21, he was unprepared for the biting chill in the air. The next day, he drove to Kmart to buy a sweater, as temperatures plunged to record-breaking lows: the coldest weather in Florida history, sinking overnight to fourteen degrees below freezing.

Back in his hotel room, Boisjoly heard local TV news announcers explain that the cold had devastated the crops of oranges and grapefruit in the commercial groves around Merritt Island—killing thousands of trees across Florida outright. The state Secretary of Agriculture described the freak weather as "the freeze of the century," and the Governor declared a state of emergency; across the Eastern Seaboard, at least 126 people died as temperatures plunged; in Washington, DC, it sank to minus seven, and the outdoor ceremony to inaugurate seventy-three-year-old President Reagan for his second term in office was canceled, for fear the members of the marching bands scheduled to parade down Pennsylvania Avenue would suffer frostbite in their fingers.

At Cape Canaveral, the water in the plumbing of the Fixed Service Structure on Pad 39A froze solid, fracturing and bursting pipework, and NASA staff worried that ice on the gantry and *Discovery*'s external tank might break off during the planned launch and cause potentially deadly damage to the shuttle's heat-resistant tiles. They were forced to delay the liftoff for twenty-four hours to wait for it to melt, and so that they could make repairs.

By the time Boisjoly took a prime spot outside the Vehicle Assembly Building to watch *Discovery* thunder into the sky on the afternoon of January 24, the air temperature had risen into the low sixties; he was in shirtsleeves. But it was still some of the coldest weather in which the shuttle had ever launched.

Although an experienced engineer and an active member of the Church of Jesus Christ of Latter-day Saints, in other ways Boisjoly was quite unlike his colleagues at Morton Thiokol. Born and raised in Lowell, Massachusetts, he had arrived at the Promontory rocket plant in the summer of 1980 at the age of forty-two, marking him out as a relative newcomer in an engineering department dominated by lifelong company men. He had taken the job after a two-decade career in aerospace spanning more than a dozen different companies, working on everything from intercontinental ballistic missile research to the environmental control system of the Apollo lunar lander. Previous employers had not always found Boisjoly an easy man to have around. Although he thought of himself as an introvert, he was also intemperate and emotional, and from the outset of his professional life had pursued rigid principles of honesty and rectitude in his job, learning to call out engineering errors and miscalculations, even when they might prove inconvenient or expensive.

In the mid-1970s, he had spent two years at Rockwell International's Space Division in California, working on the crew compartment of the Space Shuttle orbiter; conditions in the Downey plant were difficult— long hours without paid overtime in what Boisjoly and his colleagues liked to call "the armpit of the industry." He became central to efforts to organize the workforce—soliciting support from the Teamsters Union— before eventually moving on in 1976. But before he left, Boisjoly had one encounter that would profoundly shape his attitude to work and his personal responsibility as an engineer.

One Saturday in 1974, a usually genial colleague from the design engineering department approached Boisjoly's desk and began haranguing him with such vehemence that Boisjoly feared he was about to start a fistfight. It seemed so out of character that, afterward, Boisjoly went to his supervisor for an explanation.

"Haven't you heard?" the manager asked.

He then told to Boisjoly that, before coming to Rockwell, his colleague had worked on the development of the McDonnell Douglas DC-10, a new triple-engine airliner that had entered commercial service just two years earlier. The engineer, part of the team of subcontractors responsible for the aircraft's cargo bay doors, had become concerned about flaws in the design of the door-latching mechanism, which he suspected could cause them to open in flight. He pushed for a redesign, and—when that failed—fought to stop the plane's certification by the

Federal Aviation Administration. But he couldn't get anyone in manage-ment or at the FAA to listen to him, and the DC-10 entered service in 1971. Over the next two years, McDonnell Douglas executives ignored a series of warnings and close calls with the DC-10 cargo doors, and instituted a desultory fix; but the aircraft kept flying.

On the afternoon of March 3, 1974, Turkish Airlines Flight 981 left Orly airport in Paris with almost every seat filled with passengers, many of them anxious to arrive in London before a scheduled strike shut down Heathrow. The ground staff in France were in a hurry to get the plane off the ground, and the baggage handler responsible for closing the cargo doors did not speak either of the languages in which the bilin-gual instruction decals for the latches were printed. The mechanism had already become damaged in use, but when the handler forced the door shut he believed it had locked securely.

Nine minutes after takeoff, Flight 981 had reached an altitude of 11,500 feet when the left rear cargo door of the aircraft burst open, caus-ing an explosive decompression of the fuselage. Almost the entire floor of the passenger cabin collapsed, instantly severing all of the hydraulic cables leading to the plane's tail and number two engine. Six passengers, still strapped into their seats, were sucked out of a ragged hole in the fuselage into the skies above France. Seventy-seven seconds later, the crippled plane plowed into a wooded valley in Bosquet de Dammartin at almost five hundred miles an hour, where it disintegrated in a fireball of blazing aviation fuel. There were no survivors. In total, 336 passen-gers and crew lost their lives: it was the deadliest accident in the history of commercial aviation.

In the aftermath of the disaster, blame fell on the aerospace execu-tives who had concealed what they knew about the dangers of the cargo doors, scrimped on a redesign, and gambled with the lives of their pas-sengers. But the engineer at Rockwell who had worked on the door was emotionally shattered; he felt personally responsible for the deaths of more than three hundred men, women, and children. He blamed him-self for not doing enough to force his managers to approve a redesign of the faulty mechanism, or to have the aircraft grounded. Although prescribed medication to treat his depression, after the crash of Flight 981, he was no longer the same and remained unstable, tormented by guilt. When Roger Boisjoly finally left Rockwell two years later, he was determined that no one would ever saddle him with such an unbearable burden.

::::::::::::::::::

On his arrival in Utah in July 1980, Boisjoly was assigned to work as a structural design engineer on the Space Shuttle solid rocket boosters, and was ready to settle down. Anticipating retirement, he aimed to see out the remaining twenty years of his career at the Promontory plant: with two daughters in high school, he and his wife, Roberta, were looking forward to a life of stability after a decade in which he had bounced from one position to another in the California aerospace industry. The family bought a house in Willard—a tranquil rural community of orchards and fields bordering the northeastern corner of the Great Salt Lake—which Boisjoly began renovating and landscaping himself; he had been there a little more than a year when, after leading local opposition to a gravel pit being dug nearby, he was elected town mayor.

At the same time, Boisjoly had learned that his zeal for improvement was not always welcome in his new job. He soon began to fear that coming to Thiokol had been a terrible mistake. Despite being a prime contractor on the world's most sophisticated spacecraft, to Boisjoly the company's solid rocket division seemed stuck in the past, operating late-fifties equipment, with antiquated ideas to match: he found that the computer technology used for modeling the behavior of the boosters was hopelessly primitive, and his initial suggestions for modifications that would make the rockets safer and more cost-effective were met with disinterest by his colleagues. Boisjoly repeatedly took his ideas to Bob Lund, his boss in the Thiokol engineering department—an affable and capable engineer, a passionate golfer, and a senior figure in the Church. But after initial enthusiasm and scheduling more meetings to discuss the proposals, Lund always found a way to stick with the status quo. Boisjoly brought one suggestion after another about the rockets to

Lund, always with the same result; at last, after the fifth or six attempt, he lost patience. Furious, he told Lund he would no longer be wasting his time with any more new ideas: "Take a good look at this face," he said, "because it's the last time you'll see it in your office."

Soon afterward, Thiokol's top management picked Bob Lund for promotion to Vice President of Engineering for Wasatch Operations— chief engineer for the entire plant, and one of the most senior positions in the company. Boisjoly concluded that, although he seemed open and approachable, Lund too often proved obedient to the wishes of upper management, his convictions malleable when he was pushed around—a weakness the new executive unconsciously acknowledged in a self-deprecating phrase he sometimes used when he changed his mind: "Wishy-washy Lund," he would mutter, as if in abasement for reversing himself.

Down in his office on the first floor of Building A-2, Boisjoly had little patience for his own management tasks, preferring to lose himself in the day-to-day problem-solving of engineering work. He bridled at the seemingly pointless bureaucracy his job involved, and soon took to reading only the opening lines of the memos that flooded his desk each week before balling them up and filing them in the trash can. He considered resigning from Thiokol, and even interviewed for a similar position nearby. But Boisjoly feared that he was growing too old to make another new start, and his job hunt went nowhere. In the meantime, he had become popular and well respected by other colleagues in the engineering department, and a frequent member of the group sent to Huntsville to attend Flight Readiness Reviews at the Marshall Space Flight Center; in 1982, he received a company award for his contributions to the program, and a small US flag, flown into space aboard STS-2, to commemorate his achievement.

Boisjoly had been witness to the O-ring failures of the solid rockets since the very start. As part of the postflight inspection team assigned to the second Space Shuttle mission, he had examined the boosters in which the first isolated instance of charring in the seals came to light. Later, he learned from talking to the engineers at Marshall that the booster joint had never worked as designed, but he refused to believe the data they gave him about how it fitted together; instead, he made his own calculations.

When the erosion to the seals emerged as a recurring problem, Boisjoly was already an old hand at the postflight inspections and—although initially concerned—held fast to the engineering logic with which he and his colleagues analyzed the fault: the degree of damage always seemed within acceptable limits; they had calculated a good safety margin; and there was always a second seal to provide redundancy—a backup if the first one failed. While they worked on finding a long-term solution, he was confident to continue flying. Besides, it had long been clear to him that there was no way NASA was going to ground the Space Shuttle while they waited for a team of engineers in Utah to fix an O-ring problem.

By the time he stepped out of the Florida sunshine into the cool shadows of Hangar AF one Wednesday morning at the end of January 1985, the forty-six-year-old engineer had been studying the seals for years, and was regarded by staff at both the Cape and the Promontory plant as an indispensable scientific troubleshooter. He knew the O-rings better than anyone.

Inside the hangar, Boisjoly stood and watched as the technicians separated the battered segments of the left booster: first they prized apart the aft field joint, then the center; everything looked normal. But when he climbed onto the hydraulic scissor lift and began to inspect the forward joint, the engineer glimpsed something that filled him with horror; he felt his heart hammer in his throat. By the beam of his flashlight, he could see that the honey-colored grease that had been packed between the primary and secondary O-rings before launch was now coal black—colored by traces of scorched putty, vaporized rubber, and burned propellant—evidence of an unprecedented failure in the seals, and a grotesque escalation in the degree and complexity of damage inside the rockets' most critical joints. The blackened path stretched around almost a third of the circumference of the booster, revealing that an O-ring in a field joint had, for the first time, lost its race with the hot gas escaping from inside the rocket combustion chamber—and failed to seal, allowing a tongue of flame to surge through the gap between the booster's steel casing and the edge of the ring. The condition of the joints in the second rocket was even worse: in the center segment, Boisjoly found a black trail of sooted grease extending for more than seven feet. And in both joints, he could see that the hot gas blowing by the O-ring had incinerated some of the primary seal as it passed, causing some of the worst damage he had ever seen—and, in one rocket, had even reached the secondary seal beyond. With leaks on such a scale, he was astonished that *Discovery* hadn't been blown to pieces on the launchpad.

And he thought immediately of the cold.

Boisjoly didn't wait to finish his inspection, but went to the nearest phone to report what he had seen to project management in Utah. They called him back with instructions to fly to Alabama and brief the NASA engineers at the Marshall Space Flight Center on his findings. Using cotton swabs to gather samples of the scorched grease from inside the joints, he stripped the damaged O-rings from around the motor segments, sealed them in a set of polyethylene evidence bags, and headed for the airport.

The next day, Boisjoly presented his preliminary data to a packed conference room in Huntsville: an audience of thirty or forty engineers and senior NASA managers on the solid rocket program listened as he described the burn damage, the "blow-by" erosion, and the flat sheen of brief heat distress he found on the secondary seal in one joint. But he was hesitant to share his conclusions, because—if he was right—he knew it would endanger NASA's hectic schedule of year-round shuttle launches, and could even jeopardize the entire future of the program. At last, Boisjoly sucked in a deep breath and told them what he thought: that the rockets had almost been crippled by the freezing weather at the Cape.

"Guys," he said, "you probably don't want to hear this, but I'm going to tell you anyway. It is my technical opinion that the precipitating cause of this event was temperature."

There was silence; then, from somewhere deep in the room, Boisjoly heard a single voice respond.

"You're right, Rog," an unseen engineer said. "We don't want to hear that."

:::::::::::::::::::

Back in Utah, Boisjoly and a team of specialists began an urgent search to find a scientific explanation for what had gone wrong, before the next Flight Readiness Review. Once again, they were under pressure to provide answers in time for Allan McDonald, the director of Boisjoly's project office at Thiokol, to make his formal recommendations to NASA for an impending shuttle launch—scheduled for just a few weeks away. The engineers issued instructions to compile and scrutinize in fine detail the records of the solid rockets that had flown on all previous shuttle flights. They examined the quality control reports filed when the O-rings and drums of fireproof putty had arrived at the plant from the manufacturers, and the assembly logs kept by the Thiokol workers at the Kennedy Space Center who had put the segments of the rockets together before each mission. They compared different lots of the putty, and samples of the rubber used to make the O-rings, to make sure there were no chemical or physical inconsistencies that might have caused a failure; they even interviewed technicians in the Thiokol labs and on the production line who had worked on the joint components months earlier, to see if they could recall any manufacturing anomalies that the standard tests might have missed. But the bewildered engineers found nothing: the seals on the solid rockets that had failed so acutely on *Discovery* seemed no different from those on missions where the boosters had returned to Hangar AF without any evidence of leaks at all.

That left the one aspect of the January launch which they felt sure was unprecedented. Although the air temperature at Pad 39A was 62 degrees Fahrenheit at the moment *Discovery*'s engines fired, the Thiokol engineers now calculated that the temperature inside the solid rockets had been far lower. The massive bulk of dense propellant filling the boosters, chilled down to below freezing for three days before the final countdown began, had retained the cold even as the weather had warmed, chilling the rubber O-rings bound deep within the rockets' steel casings. When the motors lit, the temperature of the seals was still only 53 degrees Fahrenheit. McDonald, Boisjoly, and many of their colleagues in the structural engineering department at Thiokol now believed that the record low temperatures at the Cape had caused the synthetic rubber of the O-rings to shrink and harden in the cold, reducing the squeeze. And when the rocket motors ignited, the lack of elasticity in the rings had made a critical difference to the speed with which they'd responded to the abrupt increase in pressure in the joint—delaying their ability to spring into the gap between the tang and clevis of the booster, allowing jets of hot gas to shoot past them until, a few moments later, they finally sealed.

But Boisjoly lacked the data to prove his hypothesis. Although it stood to reason that the weather had made the O-rings less elastic—that was a matter of elementary physics—he could not prove that the cold had *caused* the leaks, because no one had ever tested the seals' effectiveness across a range of temperatures. In the strict regime of NASA's Flight Readiness Reviews—and especially the rigidly Teutonic traditions of the Marshall Space Flight Center—this absence of causal evidence was unscientific, tantamount to a hunch.

In the past, Boisjoly had been reprimanded for saying, "I think" and "I feel" in presentations at Marshall—and when he came to deliver his conclusions at the first formal Flight Readiness Review regarding the January launch, his reception was no different. Larry Mulloy objected to the link the Thiokol engineer made between the leaking seals and the weather, and told Boisjoly and Allan McDonald to make sure they tempered their assertions before they reached the next level of the review process. So the two men agreed to say only that the cold "enhanced the probability" of erosion in the joints—and anyway became convinced that they needn't worry about the problem recurring as it had in January. The circumstances surrounding Ellison Onizuka's launch aboard *Discovery* had been exceptional, and the arctic cold a once-in-a-century meteorological fluke. When they suggested to Mulloy that he place a restriction on future shuttle missions, to forbid launching if the air temperature was below what they'd seen in the O-rings after that record low in January—fifty-three degrees Fahrenheit—the Marshall engineer angrily refused. "Come on—this is a hundred-year event, for Christ's sake!" he said. "Do you think that this is going to happen again before we fix the joint? *No!*"

So while the Thiokol engineers now at last ordered up a series of tests at the Promontory plant to see how the O-rings might function at low temperatures, the experiments were hardly a priority. In the meantime, they waved the troublesome joints through for use on the next shuttle launch, the first of the two back-to-back flights scheduled for April 1985: CONDITION IS NOT DESIRABLE, they wrote in their risk analysis report, BUT IS ACCEPTABLE.

Carried up through the long review process in a series of presentations by Mulloy himself, gradually trimmed and condensed as the rules required, by the time the Thiokol recommendations reached NASA headquarters on February 21, 1985, they contained no mention of low temperatures at all.

::::::::::::::::::

Yet while the twinned April launches of *Discovery* and *Challenger* eventually took off successfully from Cape Canaveral a little more than two weeks apart, back in Utah the experiments requested by Boisjoly produced disquieting results. In the laboratory at the Wasatch plant, technicians placed samples of the O-rings in an Instron compression testing machine, which squeezed the sections of synthetic rubber between two steel plates before releasing the pressure, allowing a gap of a few thousandths of an inch to open around the edges of the gasket; they then monitored how long it took for the rubber to fill the space at different temperatures. At 100 degrees Fahrenheit, the seal made contact with both metal surfaces almost instantaneously, but at 75 degrees—room temperature—it took a full two and a half seconds; and as the mercury fell further, the numbers grew still more unsettling: at 50 degrees, the technicians waited in vain for ten minutes for the rubber to return to its original shape; eventually, they abandoned the test.

It was clear that Roger Boisjoly's suspicions had been correct: the performance of the O-rings—and the squeeze—was profoundly affected by temperature. In cold weather, the rings would become so inflexible that they might fail to seal against the metal surfaces of the casing—and not just in the first moments after the rockets lit, but at any point during their two-minute flight. But when Boisjoly reported the findings to his manager in the engineering department at Thiokol, he told him to keep the data to himself; it would be too damaging to the company if anyone at NASA learned what they had found. While the engineers at the plant continued with a program of material evaluations to address the existing problems with the O-rings and the sealing putty, the results of the resiliency tests remained a secret—known only to Boisjoly, his colleague Arnie Thompson, and the handful of technicians who had conducted the experiment.

Meanwhile, as the weather warmed down in Florida, the freak conditions of January seemed less and less relevant to the Thiokol engineers' evolving calculus of risk. And, sitting in his office overlooking the administrative parking lot from the main floor of Building A-2, Al McDonald's attention was now elsewhere. Just as it was in the massive hangars at Cape Canaveral, work at the Utah plant was humming at maximum capacity in an effort to keep up with the hastening tempo of shuttle missions, and McDonald's responsibilities remained overwhelming. The US Air Force was now close to completing its own dedicated shuttle launch facility in California for its planned roster of dedicated military flights, and McDonald had to oversee the production of a new, lightweight booster rocket design for the expanding Pentagon shuttle program—while also continuing to participate in the Flight Readiness Reviews for launches at the Cape. And in June 1985, there was another catastrophic propellant fire at the Wasatch plant, burning an entire building to the ground, causing $3 million worth of damage and bringing a team of NASA investigators out to Utah to determine what had gone wrong.

In the meantime, the Thiokol engineers' schedule of postflight inspections had begun to lag behind the relentless pace of the Flight Readiness Reviews. Once the boosters from the second shuttle flight in April had been recovered at sea, towed to the Cape, shipped back to Utah, and disassembled for final analysis, the beginning of July was approaching, and the next mission on the crowded flight manifest had already left the ground. When the Thiokol team finally opened up the nozzle joints of the spent boosters for inspection, they were—once again—startled by what they found. Inside the joints from flight STS-51-B was by far the worst seal damage they had ever seen. In both rockets, the O-rings sealing the aft segments of the rockets to their nozzles had been burned by hot gas: and this time the technicians found charring not only to the primary seal, but also in the backup; a portion of the secondary ring itself had been eaten away by the gas plume. And, unlike previous launches—in which the primary O-rings had failed at first, but sealed after a few fractions of a second—the damaged gasket in the left-hand booster seemed not to have worked at all. In three separate places, it had burned through completely; in one area, 80 percent of the O-ring had disappeared, vaporized by searing gas. The Thiokol engineering analysis revealed that the seal had failed on the launchpad, and then leaked continuously for 122 seconds of flight: the quarter-inch gap between the primary ring and its backup was thick with black soot.

This was clear evidence that the lives of the shuttle crew had relied on the backup seal alone for the entire duration of the rockets' journey. The flight had been a perilously narrow escape: only much later would the astronauts learn that they had been as little as three-tenths of a second away from an explosion in the solid rocket booster that would have torn apart the orbiter and killed everyone on board.

For Roger Boisjoly, the news destroyed what remaining confidence he had that the booster joints were fail-safe. He knew that the crew of flight STS-51-B had been doubly lucky, because the leak had occurred in a nozzle joint—which was bolted together on the factory floor in Utah before shipping to the Cape, and designed in such a way that hot gas leaking past the primary O-ring had to go around a corner in the casing before it reached the secondary ring. But the seals in the field joints put together at the Cape were much more vulnerable to hot gas damage. If a similar leak had happened there, the secondary O-ring would have evaporated just as the primary had done. *My God*, he thought, *if that ever happens in a field joint, we've bought the farm.*

The new damage report caused panic in Thiokol management: the next scheduled shuttle launch was, once again, less than three weeks away. To avoid delaying the liftoff, Al McDonald immediately assembled an emergency "anomaly team" to investigate the problem. The NASA managers in Huntsville were so concerned that they instructed Thiokol to fly to Alabama within the week and present them with a full review of the history of O-ring failures since the beginning of the program. They now had to decide whether to ground the entire shuttle fleet while the joint design was fixed.

In Washington, DC, Mike Weeks, the Deputy Associate Administrator for the Office of Space Flight, Technical—the engineer closest to the top of the agency's pyramid of bureaucracy—was alarmed to so belatedly discover that they had nearly breached the backup seal on a shuttle booster. In mid-July, Weeks flew to Utah and questioned the Thiokol staff about what they knew of the problems with the O-rings. He didn't like what he heard: the rocket engineers still didn't seem to him to have much information about what had gone wrong. Weeks told Al McDonald to come to NASA headquarters the following month, prepared to make a comprehensive presentation on the solid rockets, the joint failures—and Thiokol's efforts to correct them.

In the meantime, the members of the anomaly team at the plant had used computer modeling and laboratory testing to conclude that they understood how the latest O-ring damage had occurred. Shortly before the July 4 holiday weekend, the team flew down to Alabama to deliver a presentation to the program managers at the Marshall Space Flight Center. Along with their analysis, for the first time they disclosed the results of the temperature testing on the O-rings that had been conducted in the spring, but kept secret ever since.

In spite of the disturbing experimental data demonstrating how the O-rings responded to low temperatures, the Thiokol engineers' computer projections showed that, even in the worst-case scenario of secondary seal erosion, the backup O-ring would never be damaged so badly that it would fail completely, so the joint remained safe. NASA managers instructed Thiokol to begin studying a redesign of the joint, but left the presentation satisfied that the problem didn't warrant grounding the shuttle. And, because the contractors were already at work on a long-term solution, it would no longer be necessary to discuss O-ring erosion at Flight Readiness Reviews for future missions. The Thiokol assessments of acceptable risk once again passed up the agency's long chain of bureaucracy, and the launch schedule ground inexorably forward. A few days later, another shuttle crew received clearance to fly.

Back in Utah, the work on longer-term fixes for the solid rocket failures began. On July 19, just as Vice President Bush was announcing the winner of the Teacher in Space competition at the White House, Roger Boisjoly and a team of engineers at the Thiokol plant formed an unofficial task force to find a permanent solution to the O-ring issues. Convinced of the need to move quickly, over the course of the next seventy-two hours, Boisjoly sketched out more than thirty different potential new designs for the joint, which he had printed and distributed to his colleagues. Yet no one in management seemed interested; two weeks later, he had still heard nothing in response. He grew angry, and frustrated, then frightened. But he was not alone.

::::::::::::::::::

At the end of a long meeting at Marshall one sultry evening in late July, Al McDonald asked Larry Wear, Manager of the Solid Rocket Office in Huntsville, to join him for dinner; McDonald wanted to wrap up their outstanding business before flying back to Utah early the next morning. The two men talked at length and, at one point, Wear admitted his fear that the shuttle would suffer a disaster on their watch. "It's going to happen sooner or later," he said. "I think it's inevitable." McDonald agreed, and confessed how uncomfortable he felt about signing off on the flightworthiness of the solid rocket motors before every mission.

But Wear and McDonald remained less worried by a potential failure in the boosters than by a fault in the complex and overtaxed machinery of the shuttle's three main engines. Wear, who had come to the solid rocket office at Marshall after spending years working in the engine program, had seen enough catastrophic fires and test stand explosions during development to fear the worst. "NASA is just not prepared to deal with a shuttle failure, and neither am I, no matter what the cause," he said.

Both men recognized the responsibility they would bear—and how traumatic it would be—if something did go wrong, and only hoped to do their best to head off a disaster before either of them reached retirement. But McDonald confided that, although he was frightened by what he understood about the engineering limitations of the shuttle program, he was kept awake at night by thoughts of what he might not.

"It's what I don't know that scares me the most," he said.

Just a few days later, the two rocket engineers' worst suspicions seemed to be confirmed when *Challenger* scraped into orbit after its premature engine shutdown, saved from calamity by the quick thinking of Jenny Howard in Mission Control.

At the Thiokol plant, Roger Boisjoly, still waiting for an official response to his long list of potential solutions to the O-ring problem, was growing desperate: he couldn't get anyone in management to listen to his concerns; the informal five-man task force had so far accomplished nothing; and he was worried that their effort would simply be allowed to languish until it was too late. Apprehension of a disaster caused by the seals—and his potential responsibility for it—tormented him daily, making it hard to concentrate; he had not yet forgotten the victims of Flight 981. At last, Boisjoly decided to put his fears on the record. On the final Wednesday of July 1985, he sat down and wrote an internal memo addressed to Bob Lund, and copied to other senior engineering managers at the plant.

"This letter is written to insure that management is fully aware of the seriousness of the current O-ring erosion problem," Boisjoly began, and went on to explain why it was a mistake to think that NASA could safely continue flying astronauts aboard the shuttle while they reevaluated the joint design. Given the damage he had seen on the second flight back in April, he said he could not rule out the total failure of a joint in the solid rocket motors on a future mission—in all probability, within the first second after their ignition on the pad at Cape Canaveral. "The result would be a catastrophe of the highest order—loss of human life," Boisjoly wrote. Although he still stopped short of urging an immediate halt to flights, he called for a team of engineers to be dedicated full-time to the O-ring problem and given the authority to carry out the work as quickly as possible. He closed with a stark forecast: "It is my honest and very real fear that if we do not take immediate action . . . we stand in jeopardy of losing a flight along with all the launch pad facilities."

By the time a copy of the memo landed on his desk, Al McDonald had already begun preparing the review he was due to give at NASA headquarters on August 19; now he decided to alter its scope and focus, to make sure it included an emphasis on Boisjoly's fears about the weakness of the Viton seals in the field joints of the boosters. But before leaving for Washington, DC, McDonald called Larry Mulloy at the Marshall Space Flight Center to notify him about the presentation. Mulloy was furious that headquarters hadn't consulted him about the meeting, and told McDonald that he couldn't deliver the briefing unless its content was approved with him first. The Thiokol engineer faxed a copy down to Marshall and, four days before he was due in Washington, Mulloy scheduled a teleconference. He had, he said, a few amendments to make.

On the call, Mulloy told McDonald that he wanted two principal changes to the briefing: the first was to add a clause indicating that Thiokol recommended that it would continue to be safe to fly the shuttle as long as certain inspections and checks were performed on the boosters before each mission. The second seemed to McDonald to be barely significant; it once again concerned his observation about potential failures of the seals in cold weather. "Data obtained on resiliency of the O-rings," he had written, "indicate that lower temperatures aggravated this problem."

Mulloy instructed McDonald to remove this statement from the presentation in its entirety.

CHAPTER
EIGHTEEN

THE BACK GATE

Wreathed in the scent of cut grass, and sweating from the heat, Alison Smith was mowing the lawn down at the Cannons' place when she saw her father arrive, riding down the street on his yellow Schwinn ten speed. The landscaping business had been his idea: a way to teach the kids some important lessons about the value of labor, of doing things the right way; of serving others. Working weekends and after school, Alison and her older brother, Scott, had more than two dozen yards in Timber Cove, where they would drive the ride-on mower from house to house, taking turns using the weed-eater in the awkward ditches around each one. Sometimes, their father even helped out: Mr. Cannon kept a photograph of him, posing in sunglasses and blue Adidas running shorts, halfway through cutting the grass in the backyard of his fieldstone house on the corner of Harborcrest Drive; Cannon liked to tell people, "I have an astronaut who mows my lawn."

At forty, Mike Smith had the lean athleticism and blue-eyed sangfroid of the early astronauts, but little of their ego. By the time he was accepted by NASA in 1980, he had already spent more than a decade in the Navy—as a flight instructor, test pilot, in combat over the skies of Vietnam and, eventually, an officer being groomed to command his own squadron at Naval Air Station Oceana in Virginia. Growing up on a poultry farm in North Carolina, all he had ever wanted to do was fly. As a high school junior, Smith had worked Saturday mornings in the town's building-supply store, and in the afternoons spent the money he made on flying lessons at Beaufort–Morehead City Airport; he earned his pilot's license the same day he turned sixteen, before he was even allowed to drive a car. He played quarterback for the Beaufort High Seadogs, but was also a tinkerer and perfectionist, a keen carpenter who once used an Erector Set to automate his room so that he could control the lights, raise the blind, and open and close the door without leaving his bed. Sharp and methodical, he graduated third in his class and, in 1963, with a recommendation through a family friend, Smith won a place at the US Naval Academy in Annapolis. He was a quick study, with an intuitive grasp of engineering and aerodynamics—and his flying experience gave him a head start on his classmates. In his junior year, he received a postcard from a girl he had met in Atlantic Beach years before, now living in Alexandria and working as a stewardess for Eastern Airlines; on their first date, they went to Howard Johnson's for ice cream. Jane Jarrell and Smith made a striking, all-American couple: the petite blonde with the dazzling smile and the tall, dark-haired midshipman; when he proposed, he made no secret about his plans for the future: "I'd like to be a Navy pilot, and after that I'd like to be either a Blue Angel or an astronaut," he said. "Are you going to have a problem with that?" Jarrell didn't blink. They married ten days after he graduated from Annapolis, in June 1967.

While Mike embarked on an accelerated class in aeronautical engineering at the Naval Postgraduate School in Monterey, California, Jane joined the world of the Navy wife; a life of excitement and dislocation, and close friendships bound tighter by the danger that stalked their husbands—and the fear that they all shared, but never discussed. The couple's son, Scott, was born in January 1969, and Mike immediately wrote a will; he received his aviator wings four months later, and was soon practicing carrier landings in preparation for a tour of Vietnam aboard the USS *Kitty Hawk*. In daylight, setting down an A-6 Intruder on the deck of the ship was like landing "on a tabletop on a lake"; at night, it was among the most dangerous tasks a pilot could ever face. Smothered by darkness, ocean and sky melted into an ink-black void: it was like flying with your head in a sack; the lights of the carrier were a trembling dot on the horizon, and judging distances all but impossible. Pilots kept their eyes on their instruments and concentrated on the guidance coming in over the radio: only when they were less than a mile from the ship would they hear the instruction from the controller: "You're three-quarters of a mile, call the ball," and know that they could raise their gaze from the dials. For Smith, the next few moments required every scintilla of his concentration; a fraction of a second of doubt or hesitation would be enough to send the aircraft skittering across the carrier deck, or over-shoot, or fall short, dooming both him and his bombardier to death in the ocean below.

The couple's second child, Alison, was just six months old when Mike shipped out from San Diego as part of Attack Squadron 52. Jane didn't dwell on what might happen, and never asked her husband to sugarcoat the truth. By then Mike had begun to regard risk as simply another variable to be monitored and managed in flight, like airspeed or remaining fuel. He placed his trust in the Navy "plane captains"—the key ground staff in charge of maintaining individual aircraft—to keep the equipment airworthy, and relied on his own skill at the controls to make sure he returned alive from each mission he flew over North Viet-nam. But before he deployed he made meticulous preparations in case he didn't come back, leaving a package with his brother containing a list of insurance documents, telephone numbers, and the names of the pallbearers for his funeral. He returned from Southeast Asia decorated for his heroism under fire, but it would be years before he told Jane about his medals. By the time he qualified as a test pilot at the US Navy's Patuxent River school in 1974, she was surrounded by a tight-knit group of tough, resilient women who believed that death was something that happened to other people. But still she had learned to dread the day she might arrive home to find the Navy chaplain's black car waiting for her in the driveway.

By 1979, Mike Smith had risen to the rank of Lieutenant Com-mander, serving as the maintenance and safety chief for a squadron of carrier aircraft deployed to the Mediterranean with the Sixth Fleet; in a series of superlative annual fitness reports, his superiors recognized him as the best aviator in the squadron, a "dynamic, aggressive and extremely productive" officer who should be fast-tracked to his own command. But the overseas cruises were already taking him away from the family for months at a time, and a promotion would only mean longer absences; back at home in Virginia Beach, he sat down with Alison and Scott and told them he had decided to leave the Navy. "OK, kids," he said, "I'm either going to apply to be a pilot at FedEx and move to Memphis, or I'll apply to become an astronaut and we'll move to Houston."

Smith joined NASA as part of the second class of shuttle astronauts, selected in the summer of 1980, and the family—now including a younger daughter, Erin, born in 1977—settled into a lakefront house in Timber Cove. The neighborhood was a close community with customs and relationships established in the early years of the space program, when it had first been populated by astronauts and NASA staff: four of the original Mercury Seven had lived there, and the community swimming pool was built in the shape of the capsule that carried them into space. When Jim Lovell passed over the dark side of the moon aboard Apollo 8 on Christmas Eve 1968, his neighbors lit lanterns in their yards made from candles shaded in paper bags, beginning an annual tradition that would continue in Timber Cove for decades. When Lovell narrowly escaped death aboard Apollo 13 a little more than a year later, many of those same neighbors waited for his return at the entrance of the subdivision; holding torches and flashlights, they met the car that brought him home from Ellington Field and walked silently beside him all the way back to his house on Lazywood Lane.

When they arrived in Houston, the Smith family found themselves among familiar faces: Mike already knew all but two of the eighteen other astronauts selected beside him as part of the new class, either from the Navy or from test pilot school; among them was his friend Bob Springer, who had been stationed alongside him in Virginia Beach and moved with his family into the Lovells' old house in Timber Cove. For their children, the new neighborhood was a suburban paradise, with quiet, tree-lined drives overhung with Spanish moss they could roam by bike: the capsule-shaped swimming pool was just a few blocks down the street from their new house; and their backyard sloped gently into the green waters of Taylor Lake. The main gate of the Johnson Space Center was only a fifteen-minute drive away, and when Smith came home from astronaut training he would often strap on a tool belt to work on a carpentry project in the garage. Thrifty and practical, he made picture frames and furniture for the children, and within a few years of moving in, he and Scott built a boathouse for the family runabout, the *Tuf Enuf*; Alison and her brother had learned to water ski by the time she was in fifth grade. Smith's fellow astronauts often dropped by to take the boat out on the lake: Anna Fisher liked to ski during her lunch hour, driving back to the world of simulators and checklists with her hair still wet. As a farm boy and a Navy aviator, at home or at work Smith remained fascinated by meteorology, alert to the signs and portents of the weather, and watched it constantly. When the eye of Hurricane Alicia swept over downtown Houston in August 1983, Smith stood in his backyard as the cyclone howled around him, the wind whipped and tore at his black plastic poncho, and the storm surge drove the waters of the lake higher and higher. The house he had chosen sat at the top of a rise; when other homes in Timber Cove flooded, the Smith place stayed dry.

Although he wasn't the type for the happy hours, late-night beer calls, or Naughty Nightie Parties at Maribelle's or the Outpost Tavern, Smith quickly became one of George Abbey's favorites. Within a year of arriving in Houston, the Director of Flight Operations had made Smith his new technical assistant—his "Bubba"—transferred to an office on the eighth floor of Building 1 and responsible for fielding memos and telephone calls and, frequently, delivering bad news on behalf of his boss. Abbey may have seen some of himself in Smith: a fellow Annapolis graduate, the astronaut gave little away, but was a sponge for information, and believed fiercely in the mission of manned spaceflight; over the next few years, the two men developed a deep mutual trust and respect. One Christmas, Abbey made Smith the gift of a large silver tankard, engraved with the astronaut's name; beneath, in italics, was a line from a recent speech made by President Reagan, inspired by the final scene of James Michener's

Korean War novel, *The Bridges at Toko-Ri.* Upon the flight deck of an aircraft carrier in the Sea of Japan, a US admiral watches as Navy planes catapult into the darkened sky for another combat sortie over Southeast Asia. Realizing that many of the pilots embarking on their mission know that they may not return, he asks himself: *Where did we get such men?*

But Smith's closeness with Abbey did nothing to move him up the ladder of astronaut flight assignments, or save him from the many caprices of the launch schedule: instead, he waited patiently as, one after another, other pilots from the 1980 astronaut group took their places in the right-hand cockpit seat of the orbiter and blasted into orbit. In June 1984, NASA publicly announced that both Smith and Bob Springer would be part of the crew aboard *Discovery*, launching from Cape Canaveral in November the following year, but the mission was canceled before they had even begun training together. At the beginning of 1985, Smith began preparing to join the same secret Pentagon mission that would take Ellison Onizuka into space for the first time—as a potential replacement for another astronaut undergoing tests for a serious medical problem. But when the flight surgeons gave the first pilot a last-minute all-clear, Smith stood down again. So when Abbey first told him at the end of January 1985 that he would be part of the five-strong crew of Space Transportation System mission 51-L—along with Scobee, Resnik, Onizuka, and McNair—there was no celebration in the house on Shorewood Drive. By that time Jane had lost track of the number of missions her husband had been attached to, and this one seemed no more likely to lead to orbit than any of the previous assignments. Training for the flight—long hours in the array of shuttle simulators over in Building 5 at the Johnson Space Center—didn't start for four more months.

It wasn't until the middle of the summer that, while his older daughter was busy mowing the Cannons' lawn in Timber Cove, Smith heard the news that a sixth name had been added to the *Challenger* crew roster. Cycling over to Harborcrest Drive in the drowsy afternoon heat, he told Alison that the Vice President had just announced the winner of the Teacher in Space competition; he would be joined on his first flight into orbit by Christa McAuliffe, a middle school teacher from Massachusetts.

"What do you think about that, Dad?" she asked.

"All the attention will be on her," he said. "So I'll be able to do my job."

:::::::::::::::::::

On Monday, August 19, 1985, Allan McDonald delivered his presentation about the history of the solid rocket booster seals to a team of NASA's most senior managers in Washington, DC. It was the first time McDonald had ever given a briefing at the agency's headquarters, and he took great care over the materials he had prepared for the meeting. There were more than a dozen officials in the room, led by NASA Deputy Associate Administrator Mike Weeks, who had first raised the alarm in Washington about the seals and summoned McDonald to give the briefing; there, too, were McDonald's corporate bosses from Thiokol, and Larry Mulloy, the manager of the Solid Rocket Program at the Marshall Space Flight Center. As he handed around packets of multicolored Viewgraph slides, the engineer was intent on making sure his concerns about the rocket joints were escalated to the highest level.

The meeting was supposed to have been chaired by the head of the entire shuttle program, Associate Administrator for Manned Spaceflight Jesse Moore—who reported directly to NASA chief James Beggs. But the plan had changed after the close call during the *Challenger* launch a few weeks earlier; since then, Moore's attention had been focused on investigating the frightening malfunctions in the Space Shuttle Main Engines that had forced the mission to abort to orbit, apparently a far more pressing flight safety matter than the solid rockets. At the same time as McDonald's briefing was due to start, Moore was scheduled to report to Beggs about the engine investigation. So it fell to Weeks—Moore's immediate deputy—to attend instead and report back.

Weeks was no DC bureaucrat, but a technical-detail man—an almost thirty-year veteran of the space program who had worked with NASA since the dawn of America's attempts to reach orbit, and an engineer who had sat on a console at the Cape for all but two of the shuttle launches since 1981. The briefing lasted a full three hours, and included more than fifty charts. It was the most candid report on the solid rockets Thiokol had ever presented to NASA, and Weeks and his team of propulsion specialists paid close attention to McDonald's agenda: a history of seal erosion, followed by a list of major concerns and recommendations.

The Thiokol engineer talked through the crucial difference between the performance of the troublesome field joints and the "factory" joints connecting the nozzle to each booster rocket. McDonald explained that the primary seal in the nozzle joint was more prone to erosion than that in the field joint, but—even so—the secondary O-ring in the nozzle would always seal. But the field joints between the propellant segments—put together in the Vehicle Assembly Building at the Cape—were far less predictable. Experience showed that if the primary seal in one of the field joints failed for any reason, then the secondary O-ring probably wouldn't hold, either. Indeed, McDonald reminded his audience that—although there had been disagreement about the issue—for more than two years the field joint seals had been officially classified by NASA as Criticality 1 components, which had no redundancy. If they failed, the malfunction would prove catastrophic: the "loss of vehicle, mission and crew." McDonald recommended that, as soon as possible, a way be found to prevent the field joint flexing open at ignition, and that they step up the pace of work to stop the seal erosion.

Still, overall, the numbers seemed so far to be on their side: of more than one hundred primary O-rings in the booster field joints, they had seen erosion in only five; of forty-seven primary rings in nozzle joints, just twelve had been damaged. In keeping with the analysis done at Thiokol—and Larry Mulloy's last-minute amendments to the presentation—McDonald's eventual conclusion was a familiar one. He told the room that it remained safe to continue flying while they worked to fix the problem; despite the engineers' numerous caveats, they agreed that the seals remained an acceptable risk.

* * *

McDonald flew home from the meeting that night believing his briefing had gone well; one member of the headquarters propulsion team had even stopped him afterward and asked to keep his Viewgraph slides as a model for how future presentations should be done. But Mike Weeks felt there was something missing. At the end of the day, he went to Jesse Moore's office to brief his boss on the meeting; as the two men were leaving for the evening, Weeks admitted his disquiet.

"I'm still not satisfied," he said. "I've got to talk to somebody."

He told Moore there was only one man left in the agency he believed would give him a straight answer about the solid rockets: George Hardy, at the Marshall Space Flight Center. Not only had Hardy overseen the design and manufacture of the boosters, and run the rocket program during the first shuttle launches, Weeks also knew him as a superb engineer, calm and mild-mannered, with the clinical objectivity necessary to cut through the noise and agency politics and tell the truth.

The next day, Weeks phoned Hardy in Huntsville.

"George," he said, "how bad is this problem?"

But Hardy explained that he wasn't worried about the O-rings, and Weeks felt reassured by everything the veteran engineer told him; he also knew that Hardy's team had already approved the manufacture of a modified casing designed to limit the flexing of the joint at ignition, and that testing of a new seal design was scheduled for late January of the following year. By the time Weeks hung up at the end of the call he was, at last, satisfied that everything that could be done to fix the defects in the rockets was already being done.

Seven days later, another shuttle mission departed from Cape Canaveral upon a pillar of fire and smoke, carrying five astronauts into orbit on the twentieth successful voyage of America's Space Transportation System.

Back in Utah, Roger Boisjoly's desperate "loss of human life" warning letter had finally goaded the management of Morton Thiokol's Wasatch Division into more urgent action. On August 20, Bob Lund circulated a memo giving the ad hoc Seal Team Task Force official status, appointing Boisjoly and four other engineers to work full-time on solving the problems of the rocket joints, emphasizing the crucial nature of their work and calling for the support of everyone at the Wasatch plant. The team began meeting every day at 8:00 a.m., formulating a plan to develop and test new design solutions coordinated with their colleagues at NASA. Boisjoly approached the task with renewed intensity and started to keep a daily logbook, tracking meticulously the progress of the group in a spiral-bound ledger. And, at first, the mercurial engineer was optimistic. At the start of September, he and the team flew to Huntsville for a two-day conference at the Marshall Space Flight Center. "MSFC and our management *finally* realize how serious the O-ring problem is," he wrote. "Hooray for the management."

But Boisjoly was soon disappointed. In spite of Lund's memo, the members of the task force found it almost impossible to get anything done. The rocket plant wasn't set up for research and development—and almost everyone around them was working feverishly to manufacture the number of booster rockets necessary to meet NASA's breakneck new flight schedule; no one had the time or resources to spare for their effort. The team was hamstrung by bureaucracy, and couldn't buy the new pieces of equipment they needed to conduct tests without going through the bidding process required of any government contractor. In October, Boisjoly's colleague Bob Ebeling complained that they were generating more paperwork than test results.

Ebeling, who had been assigned to the task force from his full-time role in the final assembly of the solid rocket motors at the plant, was a hydraulics engineer who had come to Thiokol from the San Diego water department, and so far stayed for almost twenty-five years—long enough to have learned out on the test range that there was no such thing as a small leak in a solid rocket; if a seal malfunctioned, you knew you'd soon be wandering through the desert collecting pieces of your motor in a bucket. But, like Boisjoly, Ebeling was seen by colleagues as overly emotional, hotheaded—even a little crazy; in one early-morning meeting he became so enraged by the lack of priority afforded the task force that he suggested they should each be given a scarlet blazer and a pair of Western-style six-guns to wear, to set them apart from other engineers on the shop floor—and shoot dead anyone who got in their way.

At the beginning of October, Ebeling laid out his concerns about the roadblocks in their work in a two-page report to Al McDonald. The memo opened in block capitals: "HELP!" Ebeling wrote. "The seal Task Force is constantly being delayed by every possible means," he explained. "We wish we could get action. . . . This is a red flag." But his message made little difference. The team sat by as another three shuttle missions blasted off from Cape Canaveral. With each successive shuttle launch, they watched the countdown feed from Florida with anxiety knotting inside them, convinced that it was only a matter of time before they witnessed a tragedy.

As the weather grew colder, Roger Boisjoly became so consumed by the fear of flying that he thought he would come unglued. During his lunch hour, he stopped eating at his desk and instead began taking long walks around the facility to calm himself. Eventually, Boisjoly could hardly bear to go to work at all. He started to call in sick, just to stay away from the plant.

By late autumn, Bob Ebeling, too, was approaching his breaking point. One day, he emerged from his office down the hall from the main conference room, arms flailing above his head, insisting that the company should halt rocket production immediately—before astronauts were killed.

"We need to shut the back gate," he shouted, "and stop sending this shit out of here!"

His colleagues barely raised their heads from their desks.

"Oh," they said wearily, "that's Bob."

CHAPTER
NINETEEN

THE ULTIMATE FIELD TRIP

Christa McAuliffe arrived in Houston to begin her astronaut training on September 8, 1985, just a few days after dropping her daughter, Caroline, off for her first day at kindergarten in Concord. Before
leaving, she had bought her husband, Steve, a microwave and stocked
up on cornflakes, to help him start his planned twelve months as a single parent. In Clear Lake, McAuliffe and her backup, Barbara Morgan,
checked into a motel, and the next morning drove to the Johnson Space
Center to collect their NASA flight crew passes. The months since her
selection had done nothing to dim the celebrity of the Teacher in Space,
and as she signed the paperwork for her security badges McAuliffe was
mobbed by press and TV cameras. Later that day in Building 4, in the
large office allocated to the astronauts of mission 51-L, the two teachers
met the other members of the *Challenger* crew for the first time. They
lined up awkwardly together in front of a row of desks while a NASA
photographer took their picture. Dick Scobee posed holding a small red
apple—for the teacher—with McAuliffe. Judy Resnik leaned over between clicks of the camera shutter: "That's the apple with the worm in
it," she said, and smiled.

Like many of the veteran astronauts who still publicly opposed it as
a hazardous publicity stunt, Resnik had little time for the Space Flight
Participant Program. Not only did amateur astronauts pose a potential

danger to the professionals in orbit—getting in the way, panicking if things went wrong—but they were also taking hard-won opportunities away from Mission Specialists like her: every seat on the shuttle assigned to a teacher or a politician was one snatched from a scientist or engineer who had spent years training, and waiting, for their chance to fly in space. McAuliffe's own astronaut training was scheduled to take just a few months, including 114 hours of instruction packed into the time remaining before *Challenger*'s January launch: half of that would be spent reading NASA manuals, which provided instructions on how to enter and exit the orbiter, and how to conduct herself in zero gravity and use the space toilet. Much of the rest would be spent in the shuttle mock-ups in Building 5, where McAuliffe began to choreograph the two twenty-minute lessons she was scheduled to teach live from orbit: one, titled "The Ultimate Field Trip," would conclude with a five-minute Q&A with her class back in Concord. She would also conduct a series of six science demonstrations, to be recorded on video and distributed by NASA to an audience of 18.5 million US schoolchildren.

But over the weeks McAuliffe spent in Houston, Resnik's feelings about the teacher began to change. Watching the way her new crewmate navigated the exotic demands of the shuttle training program, as well as handling the burden of media attention and the expectations of being an articulate spokesperson for millions of American teachers—all while isolated from her friends and family—the astronaut's sympathy for McAuliffe grew. And when the social studies teacher discovered that— for all her intelligence—there were some scientific concepts underlying the lessons she planned to teach from space that remained elusive, Resnik stepped in to help. Meeting McAuliffe for coffee, and between training sessions, the engineer explained celestial mechanics, the finer points of magnetism, and the combustion process of the shuttle's main engines. "It's not as hard as they make it sound," Resnik told her.

Greg Jarvis had spent much of the summer back at work in the Hughes Space and Communications plant in Los Angeles, waiting for a new mission assignment; when they passed in the hallway, his colleague John Konrad would ask, "Have you heard anything?" But no word came. At last, in September, Jarvis learned that another Hughes satellite would be part of the cargo for the last shuttle mission of the year, scheduled to launch on December 20 aboard a refitted *Columbia*—and he had been selected to escort the company's hardware into orbit. Jarvis began commuting between California and Texas to train several days a week with a new crew, led by Commander Hoot Gibson, the charismatic Navy aviator who had piloted Ron McNair's first mission into space; the other five men assigned to the flight teased him that by now he was already the best-trained payload specialist at NASA. At the end of the month, the Hughes engineer joined Gibson and the rest of the crew to pose in sky-blue flight suits for their official portrait. But just a few days later, George Abbey summoned the crew to his office in Building 1. The Director of Flight Operations told Jarvis that he would—once again—be bumped from his seat aboard the shuttle by a politician.

This time, it was Representative Bill Nelson, the boyish Florida Democrat whose district included Cape Canaveral—and who had recently maneuvered his way into the chairmanship of a House subcommittee on space science. Nelson had long made it plain to Administrator James Beggs that he wanted a ticket to orbit, and his position gave him a crucial grip on NASA funding. Beggs, seated precariously at the head of a chronically cash-strapped agency facing congressional suspicion, public apathy, and growing hostility from inside the White House, could hardly say no.

At the beginning of October, NASA Associate Administrator for Spaceflight Jesse Moore personally called Nelson to tell him he had been assigned to fly the five-day mission in December with Hoot Gibson's crew, as a "Congressional observer." Anxious to avoid being seen as merely a glad-handing joyrider, the politician signed up for a series of make-work orbital tasks, including photographing drought-wracked Ethiopia—made the subject of global attention by the summer's Live Aid concerts—and conducting a crystal-growth experiment as part of a University of Alabama medical school cancer study. Behind his back, the astronauts mocked these endeavors as Nelson's personal bid to "end the famine in Ethiopia" and "find the cure for cancer."

With Jarvis's seat on *Columbia* now taken by Representative Nelson, George Abbey shuffled his shuttle trip back once more, to the first launch scheduled in the new year. On November 11, Greg Jarvis was officially assigned to fly aboard *Challenger* in January 1986, as the seventh and final member of the crew of mission STS-51-L.

::::::::::::::::::

As the launch date drew closer, training now intensified—to the tempo of a schedule devised to ensure that every astronaut had drilled exhaustively in their mission responsibilities by the time the countdown clocks reached zero. Dick Scobee and Mike Smith flew hundreds of mock ascents and landings in the Shuttle Training Aircraft, the Gulfstream executive jet that NASA had modified to mimic the unforgiving aerodynamics of the orbiter, and in the simulators of Building 5; there, Ellison Onizuka and Judy Resnik sat behind them in the positions they would take for launch—checklists and thick binders of flight information open on their knees, ready to call out prompts and procedures at key moments on the ride to orbit.

Onizuka and Ron McNair both trained to take space walks in an emergency, maneuvering the cumbersome Extravehicular Activity suits in the shuttle mock-up on the bottom of the WETF pool. Each of the Mission Specialists prepared, too, for the experiments they would conduct in orbit: McNair and Resnik rehearsed using the shuttle's robotic arm to launch a small satellite designed to gather data from Halley's Comet, which that winter would be making its closest visit to Earth since 1910. Onizuka practiced the use of a camera designed to take photographs of the comet as it approached the sun, while Greg Jarvis would be conducting his own experiments in fluid mechanics, intended to help develop more stable fuel tanks for satellites carried aboard the shuttle.

But by far the most important cargo aboard *Challenger* would be the second of the agency's sophisticated Tracking and Data Relay Satellites— the largest of its kind ever built. Together with similar hardware launched aboard an earlier shuttle mission, the communications satellite was intended to provide the astronauts in orbit with almost continuous radio contact with Mission Control, removing their reliance on NASA's far-flung network of ground stations—and the intermittent blackouts previous crews experienced as they circled the globe.

At the same time, McNair was collaborating in an ambitious new plan to play his saxophone in space—this time with the help of French electronic musician Jean-Michel Jarre. As part of celebrations to mark the 150th anniversary of the state of Texas—and the twenty-five years since the Johnson Space Center had come to Houston—Jarre had been invited to stage an outdoor concert in the city in early 1986, and aimed to give NASA a central role in the performance. Jarre's idea was to write a piece of music that he would play during the show with an instrumental part contributed live, over an audio downlink from space, by an astronaut orbiting the Earth in the shuttle. After learning that McNair was an able jazz musician, Jarre composed a piece for him to perform on saxophone during the impending *Challenger* mission. Inspired by the idea that, in the silence of Earth orbit, a spacewalking astronaut would hear nothing but their own pulse in their ears, Jarre arranged the song around a rhythm provided by a thirty-second tape of McNair's own heartbeat—recorded during physiological tests at the Space Center and turned into a continuous loop in the musician's Paris studio.

After the two men met in Houston, Jarre sent the score for the piece he titled "Rendez-Vous VI" to Ron to learn before the launch. Each week, the two men rehearsed the song together—either in person in McNair's garage in Clear Lake or, more frequently, over the phone. In its almost improvisational form and its wandering melody, Jarre had sought to capture the spirit of McNair's own style on the saxophone, as much as the sensation of being adrift in space—and did so with such success that, after he played it for the first time, the astronaut felt sure he had heard it before. Over the months that followed, McNair made tapes that he sent to Paris so Jarre could fine-tune the piece to better suit the astronaut's style. They also made contingency plans in case the audacious live linkup proved impossible: in that case, they would record McNair's performance and simply play the tape during Jarre's Houston show. Even as the pressure of training mounted over the closing months of 1985, McNair remained committed to realizing the plan to perform the composition in orbit.

"You have no idea how important it is to me to do this," he said.

And yet, whatever the scientific or artistic potential of Space Shuttle mission 51-L, the public increasingly regarded it simply as "the Teacher Flight." Since her debut on *The Tonight Show*, Christa McAuliffe had become a nationally recognized celebrity: featured by all three national TV networks, and interviewed by magazines from *Time* and *Newsweek* to *Ladies' Home Journal*; a photographer from the *New York Times Magazine* began shadowing her during training, and *People* started work on a Teacher in Space cover story to coincide with launch day. Deluged with interview requests, NASA assigned McAuliffe a dedicated publicist, but cut press access to two hours a week, and turned away inquiries from as far afield as Japan and India. Fan mail gathered in drifts behind the door of her home in Concord and arrived by the sackful in Houston; although NASA eventually prepared a form response letter and paid for an auto-sign pen, McAuliffe insisted on signing replies in person. In October, she was invited to dinner at the White House, attending a state banquet for the Prime Minister of Singapore. Other guests included Sylvester Stallone, Raquel Welch, and Michael J. Fox—but it was McAuliffe who was seated next to the President.

Back in New Hampshire, the chairman of the local Democratic Party began to groom McAuliffe for office, and while, in public, she was careful to disavow any interest in party politics, privately she admitted that one day she'd like to be able to influence education policy from the top. In Hollywood, the producers of the late-night talk shows had already begun competing over who would be first to have the Teacher in Space appear to describe her experience as soon as she returned from orbit.

Each morning that autumn, McAuliffe and Barbara Morgan left the temporary homes they had found at the Peachtree Apartments for the short drive to the Space Center and the office they shared, in a building adjacent to the astronauts' in Building 4. Practice for the planned lessons and experiments in orbit moved into zero-g rehearsals aboard the Vomit

Comet, and Dick Scobee took McAuliffe aloft in a T-38, to accustom her to the pitch, yaw, and g-forces she would experience during the shuttle's ascent and descent. Hurtling through the sky over the Gulf of Mexico in the back seat behind Scobee on one trip, she was surprised when he suddenly told her she had control of the aircraft.

"What do I do?" she asked.

"Take the stick."

"Then what?"

"Anything you want."

Mrs. McAuliffe fearlessly swung the plane into a barrel roll.

By mid-November, all seven members of the *Challenger* crew had begun training together, conducting escape drills from an orbiter mock-up to practice for an emergency landing, and running through the procedures for the journey to orbit and reentry in the simulator. When McAuliffe joined him on the flight deck, Scobee told the teacher simply to make sure she didn't touch anything. He was only half joking; the commander had repeatedly warned the two teachers of the real risks of flying aboard the shuttle, explaining how complex the technology was and how many things could go wrong; but, he said, the parts of the system most likely to let them down were people. In their free time, June Scobee often invited McAuliffe and Morgan over for dinner at their house on Brookpoint Drive, and the entire crew met for meals at Frenchie's restaurant, where their pictures hung on the wall alongside those of other astronauts, and the proprietor called Scobee "*il comandante*," and Resnik "*mi bella mora*."

That same month, McAuliffe's family came to visit her in Clear Lake, and during a tour of the Space Center, she took Steve, Scott, and Caroline to visit the mission commander in his office in Building 4. At first, Caroline was nervous about meeting him, and outside the door clung to her father's leg, but soon discovered there was nothing to fear. Now forty-six years old, Scobee could still turn on a steely military bearing when he needed to, but had mellowed during his years in Houston. Earlier that year, his daughter Kathie had given birth to a son, Justin, making him a grandfather for the first time—an occasion marked by Scobee's colleagues in the Astronaut Office, who gave him a new name patch for his flight suit that read GRANDPA. Scobee had already decided that the *Challenger* mission would be his last in space; on his return, he hoped to take a desk job, in charge of the recently completed facility dedicated to military shuttle launches at Vandenberg in California.

The walls of the commander's cluttered office were hung with photographs, comic signs, and momentos of a long career in aviation. Among them was a framed copy of "High Flight," the poem written by Royal Canadian Air Force pilot John Gillespie Magee, Jr. a few months before his death in a midair collision over England in 1941. First made famous during World War II, Magee's sonnet would become a favorite of US military pilots and, later, of astronauts: Michael Collins had carried a copy on his first journey into space, aboard Gemini X, fourteen lines of verse typed onto a small file card by his wife, Pat:

Oh! I have slipped the surly bonds of Earth
And danced the skies on laughter-silvered wings;
Sunward I've climbed, and joined the tumbling mirth
Of sun-split clouds,—and done a hundred things
You have not dreamed of—wheeled and soared and swung
High in the sunlit silence. Hov'ring there,
I've chased the shouting wind along, and flung
My eager craft through footless halls of air. . . .

Up, up the long, delirious, burning blue
I've topped the wind-swept heights with easy grace
Where never lark nor ever eagle flew—
And, while with silent lifting mind I've trod
The high untrespassed sanctity of space,
Put out my hand, and touched the face of God.

On his return to Earth, Collins had marveled at the transcendent impressions of flight Magee captured from the cockpit of a Spitfire.

"What could he have said after one orbit?" he wrote; he wept over the young pilot's death.

When it came time to choose her own small selection of personal items to pack aboard *Challenger*, McAuliffe included the watch her grandmother had given her; Scott's favorite stuffed animal—a plush frog named Fleegle; a cross and chain for Caroline; Steve's class ring from the Virginia Military Institute; a selection of pins and medallions for friends and family; and her own copy of "High Flight."

· · · · · · · · · · · · · · · ·

Behind the green doors of his office on the seventh floor of NASA headquarters in Washington, DC, Administrator James Beggs knew he was in trouble. The first hint that something was amiss had come at the beginning of September, when a pair of FBI agents had shown up at 600 Independence Avenue with a letter for him. They handed over the document, signed by a federal attorney in California, which explained that Beggs was the subject of a grand jury investigation into a defense contract he had overseen while still at General Dynamics, four years earlier. The case seemed bizarre. It was overseen by the same US Attorney whose office had infamously mishandled the case against maverick auto executive John DeLorean, who—despite being caught on videotape selling cocaine to the FBI—had eventually walked free.

Flabbergasted, Beggs immediately called the legal counsel at General Dynamics, who told him that the suit was one of several government anticorruption cases that had been rumbling on for years; it didn't seem that important. Even so, the attorney advised caution: "I would recommend you get a lawyer."

But by that time, President Reagan's clique of far-right apparatchiks had concluded that Beggs—a lifelong Republican, but an old-fashioned conservative, not a damn-the-torpedos anticommunist ideologue—was not the kind of man they wanted at the head of the nation's space agency. Hoping to militarize NASA and cannibalize the agency's budget for the Star Wars program, they were plotting to impose one of their own on Beggs, in the vacant position of his immediate deputy. Beggs believed that their choice, William R. Graham—a humorless, gimlet-eyed technocrat with a background in defense consulting and nuclear weapons development who had never supervised an operation that approached NASA in size or complexity—was hopelessly underqualified for the job; colleagues warned him that Graham was a "kook." Beggs agreed to interview him only to keep the White House quiet while they found a better alternative. But, before he knew it, the President's signature was on Graham's formal nomination.

Too late, Beggs realized that he had been outmaneuvered.

Graham arrived to take up his post as the second-in-command of NASA in the final week of November 1985. In the meantime, Beggs received written notification that he had become a central target of the grand jury investigation. When his lawyer flew out to California to meet the government prosecutor, he was offered a plea deal if his client pleaded guilty and turned on his codefendants. "Tell him to go to hell," Beggs retorted. As soon as he returned from the Thanksgiving holiday, on December 2, James Beggs and a trio of General Dynamics executives were indicted on felony fraud charges, which alleged that they had conspired to bilk the US Army out of millions of dollars of payments in an antiaircraft system contract. The White House clearly expected Beggs to resign, but eventually agreed to let him to take a leave of absence while he organized a defense. Exactly one week after starting his new job as his deputy, William Graham stepped into Beggs's place as Acting Administrator of NASA.

Beggs now had no doubt that this was the culmination of a deliberately orchestrated scheme to remove him from office and replace him with an extreme Reaganite who could hand the entire agency to the Pentagon. He warned Graham that the job might be tougher than he imagined: it was not an easy institution to run, and in the coming year he would have to get more than a dozen shuttle missions off the ground; he encouraged him to call if he needed advice. And publicly, Beggs offered his successor his support. But privately, he appealed to his friends on Capitol Hill to stop Graham's position becoming permanent: "You can't let him be Administrator. He is a disaster. He could do terrible damage to the program," he said. For his part, Graham had little intention of sharing his newfound power with Beggs. "I'm in full charge," he told a reporter on the day of his appointment, "and I intend to run this agency as though I am."

And yet, although Graham moved into the Administrator's office, Beggs did not leave the building, but took a desk down the hall—while senior members of the management down at Cape Canaveral and beyond continued to make clandestine reports to him about what his successor was doing. The open hostility between the two men only added to simmering strife at the top of NASA. More disruption would soon follow—including the departure of the agency's experienced Head of Public Affairs, and of the Director of the Johnson Space Center in Houston. Morale at the agency plummeted to a new low. In the meantime, Graham's first launch as Administrator was still scheduled to take place later that month: Commander Hoot Gibson's *Columbia* mission, carrying Congressman Bill Nelson on his quest to cure cancer and bring hope to the starving of Africa.

::::::::::::::::::

By the beginning of December, training for the *Challenger* crew was approaching a crescendo, with the most intensive period of drills and simulations scheduled for the final month before launch. On December 12, they assembled with their ground support team in Houston to meet the press for two final days of interviews. The media junket began with a formal press conference, in which Dick Scobee introduced each member of the crew and their responsibilities on the flight, followed by a second day in which they faced reporters one by one.

By now Christa McAuliffe was a seasoned veteran with the press, still filled with enthusiasm and folksy humor, but changed by her experience of celebrity. When asked about the journal she was to have been keeping since her selection, she admitted that she had tussled with NASA over her privacy: she told a reporter from *USA Today* that she had never agreed to make the entire diary public. Instead, she would release only some portions of it after the mission: "the golly, gee-whiz

stuff," her publicist said. What she didn't mention was that no such journal yet existed. She had recorded her thoughts only on scraps of paper
and in letters to friends, and planned to use a Dictaphone to record her
impressions while she was in space.

There was more controversy when it was Ron McNair's turn before
the press, and he revealed that NASA had refused him permission to
take his saxophone into space: his collaboration with Jean-Michel Jarre
was off; he refused to explain exactly why. "Well, let's just say there's
some objection," he said. "Someone in the chain of command objects to
it this time."

"Too frivolous for a space mission?" asked the science correspondent from the *Dallas Morning News*.

"No, it's not frivolous at all. As a matter of fact, it's everything but
frivolous. It's something meaningful."

McNair expressed his support for the idea of flying civilian passengers aboard the shuttle, and one reporter asked about how he would
calm any launch day nerves the Teacher in Space might have. "You never
hear astronauts taking about being afraid," another added.

"Well, I guess you don't hear about it because—" McNair began, and
then stopped himself. "Well, it's not a frightening thing. It's more like
fun, a joyride. It's not frightening."

Ellison Onizuka told the reporters that he hoped for some technical
glitch with the orbiter while they were in space so that he would have the
opportunity to don a pressure suit and perform a space walk to fix it. He
wanted to have the same experience he had enjoyed on his first flight,
only better: "I saw things I never dreamed I'd see. I saw some of the most
beautiful sunrises and sunsets you can imagine. I'll remember those pictures forever," he said. "God, I will remember them." Greg Jarvis was
so enthusiastic about the fluid mechanics experiment he was going to
conduct in orbit that he overran the allotted interview time, and never
managed to complete his explanation.

Judy Resnik—who by now had confided to friends that talking to the
media made her feel "like a potted plant"—took a seat looking pale and
brittle. She rolled her eyes in exasperation at one reporter's question, and

stared at the floor between sips of Diet Coke. After almost eight years in Houston, Resnik knew she would soon face important decisions about the future: although she had often said that she would stay with NASA as long as it would have her, she would have to choose whether to keep flying after her second mission was complete, move into management, or leave the agency altogether. But her personal life had also become complicated: she was dating fellow astronaut Frank Culbertson, who was recently divorced, but had three young children, and seemed more serious about the relationship than Resnik. At Christmas she would accompany Frank and his daughters to a performance of *The Nutcracker* at the Houston Ballet, but the day proved hard work. Accompanying the couple on a double date, Resnik's friend Sylvia wondered whether Judy was ready to be a stepmother. The astronaut's ambivalence would be clarified at a party on New Year's Eve, when she confessed to June Scobee that she was in love with someone else—someone rumored around the Space Center to be another astronaut; a hero of the shuttle program, and a married man.

Sitting in the junket, and asked once more about the experience of being a woman astronaut, Resnik gave her traditional answer: "I'm just a person doing my job." She couldn't wait to get away.

Dick Scobee and Mike Smith arrived together, and joked good-naturedly about their expectations for the flight. By the time *Challenger* left the ground, Smith would have waited five years for his first mission, but had been promised another before the year was out. Asked about who had vetoed McNair's musical performance from space, Commander Scobee cheerfully admitted it had been his call: "Let's put it this way," he said. "I decided Ron could bring his sax if Judy could bring her piano."

Ten days later, *Challenger*—mated to the giant orange external fuel tank and its towering solid rocket boosters, and mounted on the Mobile Launcher Platform—rolled out of the Vehicle Assembly Building at Cape Kennedy. It took eight hours for the massive caterpillar-tracked crawler-transporter to move the orbiter the four and a quarter miles to Pad 39B, where it was positioned at the launch tower to await its moment of final departure.

PART THREE

The Face of God

CHAPTER
TWENTY

FRIDAY, JANUARY 10, 1986

7:00 a.m.

Strapped into the left-hand seat on the flight deck of *Columbia*, Hoot Gibson lay on his back and stared up into the bruised dawn sky above Pad 39A. Fierce wind and heavy rain lashed the orbiter, and water pooled across the cockpit windshield; lightning flickered through the clouds, and the sound of thunder rolled across the Cape. Beside him, pilot Charlie Bolden heard the electricity from the storm crackle in his headset. Each member of the seven-man crew remained harnessed in place as the weather swooped and howled around them; dark shapes caught by the wind flashed over the windows, and the cramped cabin flared with explosions of blue-white light. The voices from the Launch Control Center reassured the crew that the shuttle was safe—protected by the lightning rod built into the tower above them. But they all knew that previous spacecraft had been crippled by electrical storms during launch. And no one needed reminding of what would happen if the shuttle's external tank, with its cargo of more than half a million gallons of liquid hydrogen and oxygen, received a direct hit. With each fresh thunderclap, the astronauts' jaws tightened in apprehension.

Over in Firing Room 1, the controllers recognized that the biblical weather made a liftoff that morning impossible—and Launch Director Gene Thomas had wanted to keep the astronauts in their quarters, rather than risking their lives on the pad. But he had also been told that Acting Administrator Bill Graham had come down to watch the launch, and Thomas's superiors—the men from Level I and Level II in Washington, DC, and Houston—had made their feelings explicit: *The boss is here to see it happen. They want you to go ahead and take a shot at it.* Always the good soldier, the Launch Director reluctantly complied. As the Astrovan carrying the crew had passed the Launch Control Center, he had looked out the window to watch them pass; gray sheets of rain obscured his view of the launchpad beyond. Thomas felt like a fool.

It was the *Columbia* crew's fourth trip out to the pad in less than a month and—by the time the Launch Director at last canceled the effort, after four tense hours waiting in vain for the word to go—it marked the seventh occasion on which the takeoff of shuttle mission STS-61-C had been scrubbed; a humiliating new record for the US space program. To signal their defeat, as they clambered from the hatch two of the astronauts waved white handkerchiefs in the air, pantomiming their wish to surrender.

::::::::::::::::::::

Despite the turbulence and scandal at headquarters, and the nagging technical problems of the shuttle, 1986 was projected to be a landmark year for NASA—one in which the Kennedy Space Center would be busier than it had been in its entire twenty-five-year history. The agency had now scheduled fifteen shuttle flights over the coming twelve months, including missions that would launch nine communications satellites and two major scientific probes of the solar system, as well as three secret military payloads and several space manufacturing and Earth-monitoring experiments. Three missions in the first months of the year—including the one aboard *Challenger*—would carry equipment to observe Halley's Comet, with the third expected to beam live color TV images of the phenomenon back to Earth, three days ahead of a competing Soviet unmanned probe scheduled for a similar rendezvous at the beginning of March. In October, the shuttle would carry the $1.2 billion Hubble Space Telescope into orbit, allowing scientists to gaze fifty times farther into space than was possible using terrestrial instruments, and enabling them to look back in time to the birth of the universe.

And the Teacher in Space mission on *Challenger* was only the first of a series of publicity-grabbing trips planned in NASA's Space Flight Participant Program: the next, also scheduled for October, would carry a journalist. The agency had not yet announced the final selection of candidates, but almost every one of the reporters on the space beat at the Cape—and many who were not—had applied. Walter Cronkite, sixty-nine years old, but still considered a favorite to be the first journalist in orbit, had thrown everything he could into his campaign for a seat on the shuttle; granting an exclusive interview to *Life* magazine, the venerable anchorman was photographed parading the streets of Martha's Vineyard wearing a NASA jumpsuit and helmet, "waving to all the kids like a costumed Donald Duck at Disney World."

But the agency's audacious plans were barely under way before the system began to buckle under the strain. The flight of the newly refitted *Columbia*—the subject of an eighteen-month-long, $42 million overhaul intended to bring the orbiter up to the same standards as its three sister ships and return it to space for the first time in almost two years—had originally been scheduled for December 1985. But the launch was first postponed for twenty-four hours after the ground crew at the Cape fell behind in the countdown checkout process; then it was aborted again, fourteen seconds before liftoff, after onboard computers reported a failure in the steering system of the solid rocket boosters: this, the first time in the history of the program that the solid rockets had been the cause of a delay, caused consternation among the engineers from the Marshall Space Flight Center. Director Lucas was furious, and Larry Mulloy, as the project manager for the boosters, shouldered the blame. After that, Acting Administrator Graham had postponed the launch until early January, to give the exhausted ground crews a rest over Christmas.

Even so, as the men and women at Kennedy prepared *Columbia* for a third launch attempt on the first Monday of the new year, some of them had worked forty straight twelve-hour days without a break. The technicians had begun routinely skipping hundreds of maintenance requirements, and some began to fear for their jobs if they reported accidental damage. Something seemed certain to go wrong. On January 6, *Columbia* was five minutes from launch when a console operator in the Launch Control Center misinterpreted a computer signal and flicked a switch that drained eighteen thousand pounds of liquid oxygen from the shuttle's external fuel tank. The countdown continued regardless, until the siphoning propellant triggered a low-temperature alarm. Barely thirty seconds before the three main engines lit, the shuttle computers halted the launch. If *Columbia* had left the ground without the lost fuel, it would have been unable to reach orbit, necessitating one of the aborts described by John Young as requiring "two successive miracles followed by an act of God": the crew would have been unlikely to survive. By the time the terminal countdown had started, the team in the Launch Control Center had been on duty for eleven hours on their third successive twelve-hour shift in order to prepare for the mission. NASA public affairs officials nonetheless told the press the launch had been postponed due to a "mechanical failure."

Forty-eight hours later, after another launch had been called off because of bad weather over the emergency landing strips in Africa and Spain, the ground crew were working through the night to prepare *Columbia* for one more try, when they discovered a jammed valve in one of the shuttle's main engines, postponing the launch once more. This proved to be a lucky escape: in the process of emptying the shuttle's fuel tanks, the engineers found that the valve was not jammed by an electrical fault but because a five-and-a-half-inch-long pencil-shaped temperature probe had snapped off from inside the fueling equipment and been sucked through a liquid oxygen line into the orbiter, where it had lodged in the complex plumbing of *Columbia*'s number two main engine. If not for the poor visibility in Dakar and rain over Seville, the launch might have taken place with the valve stuck open, and the engine could have operated normally through almost its entire ascent; but then, eight minutes and thirty seconds into flight, it would have exploded at shutdown, blowing the orbiter apart on the threshold of space.

While the Kennedy ground crews struggled to get *Columbia* into orbit, a mile to the north *Challenger* stood waiting on Pad 39B, rising blindingly white above the palmetto and banana palms of Merritt Island by day, brilliantly lit by the beams of the xenon searchlights at night—and apparently ready to carry her own crew into orbit. It was the first time two shuttles had ever sat on their launchpads at the same time. From a distance, this spectacle seemed at last to make concrete NASA's long-held promise of "routine access to space" in a single, photogenic image; the reality was more precarious.

The facilities at the Cape were already stretched to the breaking point by the overlapping launch preparations: the dress rehearsal for the 51-L launch, with Dick Scobee and the crew strapped into their seats on the launchpad for a mock countdown, was delayed twice when *Columbia* failed to take off; there weren't even enough beds in the crew quarters to accommodate the fourteen men and women on both missions simultaneously, forcing some of them to fly repeatedly back and forth to Houston in mid-January. The domino effect of *Columbia*'s delays also forced back *Challenger*'s launch date, adding mounting pressure to get both crews off the ground as soon as possible—not only to keep to the crowded 1986 launch schedule, but also because the scientific missions planned for later in the year were subject to the inflexible timetable of celestial mechanics. The path of Halley's Comet through the solar system was 7.6 billion miles long—carrying it out beyond Neptune and back—and the window to observe it from Earth's orbit was narrow; missing it would mean waiting for the next opportunity, sometime in 2061. Yet the lack of spare parts for the fleet of four Space Shuttles meant that *Challenger* could only truly be made ready to launch once *Columbia* had returned from its five-day mission—and key components had been cannibalized from the older spacecraft for installation on its sister ship.

When shuttle mission STS-61-C finally left the ground on the eighth attempt, at 6:55 a.m. on January 12, its trailing exhaust plumes glowing red and orange in the dawn, pandemonium broke out among the technicians and console operators of the Firing Room: the Cape Canaveral Launch Control Center hadn't seen such jubilation since the shuttle's maiden flight back in 1981. As *Columbia* roared into space at last, the ground crews at the Cape awaited the all-clear to return to Pad 39B and get back to work. *Challenger* was now due to launch in just twelve days, and the manager in charge of the operation knew time was short. "It's going to be very tight," he told the press.

::::::::::::::::::

Late in the morning of Thursday, January 23, Greg Jarvis, his wife Marcia, and Christa McAuliffe clambered into a rental car outside the Peachtree Apartments, drove out past the Putt-Putt miniature golf course, up Galveston Road, and north toward Ellington Field; just a few minutes later, they bumped over the railroad tracks and through the gates into the airfield parking lot. Greg stopped the car in the visitor's section and climbed out, wearing a pair of well-polished black penny loafers with a NASA flight suit that barely reached his ankles. Like him, McAuliffe, wearing sky-blue Nike sneakers instead of military-issue flight boots with her fireproof coverall, the sleeves too long and folded back at the cuffs, could have been mistaken for someone on her way to a costume party.

"Hey, nice shoes!" a passing airman shouted.

Reporters crowded around the car, but were interested in talking only to the Teacher in Space. As McAuliffe stopped to answer questions, Greg and Marcia made their way across the tarmac to the twin-engined Gulfstream, NASA-1, waiting to fly them down to Cape Canaveral. They were joined on the plane by McAuliffe's backup, Barbara Morgan, and the wives of the rest of the 51-L crew, casually dressed in sweaters and slacks and carrying bouquets of yellow roses brought for the occasion by June Scobee. The mood on board was lighthearted as they reached the climax of the great adventure that had drawn them together; the women snapped photographs of each other clowning in the aisle, and McAuliffe used the short flight to write letters to her students and her family. Seated alone at the back of the cabin, George Abbey kept to himself.

It was late afternoon when the plane alighted on the runway of the Shuttle Landing Facility, a fifteen-thousand-foot-long concrete airstrip surrounded by alligator-infested wetland in the heart of Merritt Island. Waiting for them in the languid heat was another scrum of reporters, gathered to witness the ritual of Astronaut Arrival, a set piece of NASA pageantry almost as old as the manned space program itself, in which the departing crew would be greeted by their wives on the apron at the Cape. Corralled behind a yellow rope, the photographers raised their cameras toward the whine of jet engines as three T-38s touched down and taxied in formation across the tarmac, their canopies open to reveal the five remaining members of the *Challenger* crew. At the controls of the lead plane was Dick Scobee, with Ellison Onizuka seated behind him; then came Mike Smith and Judy Resnik; in the third, piloted by Chief Astronaut John Young, sat Ron McNair. As McNair climbed from the cockpit to join the others, Young hung back, meticulously filling out his logbook.

Bathed in amber light from the sinking sun, their shadows lengthening across the runway, the astronauts greeted McAuliffe and Jarvis, and together the seven *Challenger* crew members approached the assembled press. Scobee stepped up to a microphone to introduce his team, hands on his hips: "As usual it's a real pleasure to be here at the Cape," he said, "to come down here and participate in something that the Cape does better than anybody in the world—and that is launching space vehicles . . ."

One by one each astronaut took their turn to speak, emphasizing their boilerplate optimism about the coming mission, admiration for their crewmates, and how happy they were to be returning to the Kennedy Space Center. Among the watching crowd, one of the local photographers grinned sardonically at the 51-L lineup; he was old enough to recall the World War II films in which the ethnicity of the troops onscreen seemed deliberately composed to reflect the melting-pot ideal of twentieth-century America.

"Looks like a war-movie propaganda platoon," he said. "Got one of everything."

By now each of the astronauts was well aware that the press corps was waiting for a single member of the crew in particular: "I'd like to introduce you to perhaps the person you came to see," said McNair as he invited McAuliffe up to the mic. Nearly eight hundred journalists had requested accreditation for the Teacher in Space launch—almost twice as many as the KSC public affairs team had come to expect—their number swollen by TV and newspaper correspondents from New England keen to capture the triumph of a local hero.

McAuliffe was as effervescent as ever. "Well, I am so excited to be here," she said. "I don't think any teacher has *ever* been more ready for two lessons. . . . I've been preparing these since September, and I just hope you all tune in on Day Four now to watch the teacher teaching from space."

:::::::::::::::::

As the photo opportunity unfolded on Merritt Island, in the motels and condos of Orlando, Titusville, and Cocoa Beach, the friends and extended families of the 51-L crew were already busy with their plans to celebrate the mission. Between them, the seven members of the crew had asked a total of more than three thousand people to witness the launch, each guest receiving an austere NASA invitation decorated with a black-and-white action shot of *Challenger* blasting off from the pad on a previous mission. Ellison Onizuka, who had been able to share little about his first flight with his family because it was shrouded in such secrecy, this time had brought a delegation of more than sixty people over from Hawaii for the launch—and organized a blowout landing party to mark his crew's return to Houston at the end of the month; he had already arranged shipments of kalua pig, lomi salmon, and barbecue meat from his hometown on the Big Island for the occasion. Although Ron McNair's grandmother had proved too infirm to make the trip from South Carolina to see his second launch, his brother, Carl, and his pregnant wife, Mary, had flown in from Atlanta; Cheryl's parents were also on hand to help her look after the children: Joy, just eighteen months old, and Reggie, not quite four.

Many of the guests had arrived at the Cape expecting the launch to take place on Saturday, and had just learned of yet another postponement.

FRIDAY, JANUARY 10, 1986 :: 297

This was in part because the ground crews out at Pad 39B were still laboring around the orbiter to install the pieces of equipment stripped from *Columbia* and rushed back from California in the back seat of a T-38 jet—but also, once more, because of the weather. Dust storms had again descended over the shuttle's emergency landing strip in West Africa, while Air Force meteorologists also predicted that a cold front would bring rain across the Space Coast during the weekend, pushing the launch window back to the morning of January 26: Super Bowl Sunday.

Mike Smith had warned his friends to expect delays in a computer-printed letter he had sent out weeks in advance, including a guide to hotels and restaurants near the Cape. "Dear Folks," he wrote, "if you really want to see the launch, plan to be prepared to spend up to two extra days in case of bad weather or shuttle problems (I'm betting we launch on time)." He also explained that he would spend the week preceding launch day in quarantine, permitted only to see a handful of officials and close family members once they had passed a medical screening for colds and infections, and been issued with a NASA Personal Contact Badge.

Christa McAuliffe's relatives had prepared for a long stay. Her parents, Ed and Grace Corrigan, had booked an apartment in the Cape Winds condos in Cocoa Beach, where they could host events in the run-up to the launch, as well as a block of rooms at the nearby Econo Lodge motel for their friends and extended family. Steven McAuliffe and the children—along with Scott's teacher, thirteen chaperones, and seventeen classmates from his third-grade class at Kimball elementary school in Concord, flown in for free by United Airlines—were also in town for a week. While Steven and Caroline, now six, planned to stay in a villa at Disney World, Scott and his friends checked in to the Econo Lodge. Apparently unfazed by the press, the third-graders posed for pictures by the pool and told reporters of their plans to visit Disney and the shuttle exhibits at the Kennedy Space Center. "They realize the privileged situation they're in, being in Scott's class," their teacher told a reporter. But the boy himself missed his mother, and was more weary than ever of the attention; aboard the plane on the first leg of his journey toward the Cape, one writer had asked Scott what he was most eager to do in Florida.

"I'm looking forward to Sea World and the launch," he said.

"Not one more than the other?"

"No, both are about even."

"But your mother is taking off in the Space Shuttle. Isn't that more interesting?"

"Well, I'm looking forward to the launch, sure, but I've never seen a whale. I'd really like to see the killer whale."

In addition to the astronauts' friends and relatives, the *Challenger* launch was set to attract a longer list of VIPs than recent missions: the public affairs staff expected more than fifteen thousand special guests at the Cape on Sunday morning. Representatives Jake Garn, Bill Nelson, and Don Fuqua, chair of the House Committee on Science and Technology, which had oversight of NASA, were all planning to attend, as well as New Hampshire Governor John Sununu, the presidents of the country's two largest teaching unions, a delegation of scientists from the People's Republic of China, and the Moroccan ambassador to the United States. And while Acting Administrator Graham had apparently purged the guest list of Democrats and other troublesome liberals to avoid any repetition of the Jane Fonda fiasco, the White House would also be sending its most senior representative to a manned launch since Nixon had watched Apollo 12 leave for the moon in 1969: on his way to a diplomatic engagement in Honduras, Vice President George Bush had scheduled a stopover in Florida to see *Challenger* take off on Sunday.

Elsewhere, the carnival atmosphere surrounding the Teacher in Space mission was reaching its peak; now embarking on their own journey as national ambassadors for NASA's education initiative, the 112 finalists who Christa McAuliffe had bested in the competition for the seat aboard *Challenger* had also arrived in the Cape to cheer her on. While McAuliffe remained in quarantine, Barbara Morgan hosted the first episode of *Mission Watch*, a live TV show explaining the details of the shuttle and its unfolding voyage to elementary school children, which would be broadcast daily on PBS once the crew were in orbit. The trinket stores of Cocoa Beach were stacked with Teacher in Space memorabilia: T-shirts and pajamas, patches and posters, coffee mugs and

shot glasses, penknives, pencils, tiepins, statues, stickers, and spoons depicting McAuliffe and the shuttle. Her father bought dozens of buttons
featuring his daughter's smiling face, and proudly pinned the largest one
he could find on his lapel. At the Kennedy Space Center, tourists stood
three deep at the counter of the Gift Gantry to stock up on Christa keep-
sakes. One, a school librarian from Texas, told a reporter that she had
timed her vacation to coincide with the launch; McAuliffe, she said, was
"the closest thing we have to a hero these days."

On Friday afternoon, as the forty-three-hour countdown displayed on
the giant digital clock outside the Launch Control Center continued, the
Challenger crew were driven by jeep out to a remote clapboard cottage
nestled in the dunes beyond the launchpad. Known simply as the "Beach
House" by the astronauts and staff at Kennedy, the weather-beaten red
building was one of the last surviving vacation homes built on the Cape
in the years before the area had been swallowed by eminent domain and
closed to the public. It remained in use as a venue for NASA management meetings and for entertaining visiting dignitaries, despite its rustic
decor: the mismatched Naugahyde furniture, the wood paneling, and
the bottles washed up on the beach and collected on the mantelpiece
over the years. Since the early days of the program, the Beach House
had also become the traditional prelaunch destination for shuttle crews,
a place for them to meet their families, and the flight team one last time
before liftoff, isolated from the relentless gaze of the media: a "kind of
group therapy session," a NASA spokesman explained.

Steve McAuliffe, who had become increasingly anxious about the
launch and was having trouble sleeping, was surprised by how relaxed
the crew seemed: the months of simulations were over; now they were
ready to fly. In a T-shirt and sunglasses, Dick Scobee ate barbecued
ribs and potato salad on the deck with June and their son, Rich; inside,
Ron and Cheryl McNair sat side by side in a pair of armchairs, holding hands. Ed and Grace Corrigan had arrived early, and were walking
by the ocean with Judy Resnik's father, Marvin, when they saw a slight

figure in a purple sweat suit emerge from the house and sprint down the sand toward them; it was Judy, who didn't stop running until she had hurled herself into Marvin's arms, just as if she was still his little girl.

But not every member of the crew was so at ease. After all the uncertainty, the repeated delays and last-minute reassignments from one crew to another, and almost a year of waiting, it was only now that Greg Jarvis realized that it was finally going to happen—he was going to ride a rocket into orbit. He knew all too well from his satellite work with Hughes how fraught with danger spaceflight could be, and understood the risk he would be taking. The reality of what lay ahead now settled heavily on him, and Marcia knew he was nervous. As they walked along the beach together, Greg reassured her that he wanted to fly. "Don't worry about me. Everything's going to be fine," he said. "But if something happens, just be happy. Remember everything that we've done together."

The sky was bright and clear. White-capped Atlantic breakers pounded on the sand. A strong, steady wind was blowing from the north.

::::::::::::::::::

At around 10:30 a.m. on Saturday, a red NASA van carrying a contingent of rocket engineers from the Marshall Space Flight Center sped up Florida Route 3 from Patrick Air Force Base; they were in a hurry. Led by Center Director Bill Lucas, the team was due at Kennedy for the morning's L-minus-one-day launch review for the *Challenger* mission, the meeting at which the agency's top management would decide whether to give their final approval for liftoff. But the Marshall group's flight from the Redstone Arsenal airstrip in Huntsville had been diverted to the air base by heavy rain over the runway at the Shuttle Landing Facility, threatening to make them late—and they all knew how Dr. Lucas felt about punctuality.

That same bad weather now overshadowed the timing of the *Challenger* launch, still scheduled for early the next day. Not long before Lucas and his men walked into Room 2201 in the headquarters building on the Cape at eleven o'clock, just as the review was about to start, NASA had issued a press bulletin describing an approaching cold front, "threatening to deliver rain showers and thunderstorms to the area Sunday morning." If the forecast was accurate, it would be enough to scrub the launch again. But there remained enough uncertainty about the movement of the front that the NASA official from headquarters running the review meeting—Assistant Administrator for Spaceflight Jesse Moore—resolved to wait until later in the day to make the call. In the meantime, preparations for the flight continued as normal: inside the Operations and Checkout Building, the crew studied the flight plan, underwent their final medical checks, and sat for briefings about the condition of the shuttle and its payload. To accommodate the morning liftoff, their bedtime would be at 5:00 p.m.

In Room 2201, the engineers from Marshall and the Johnson Space Center in Houston faced one another across an open square of tables and began to run through any problems that remained in the *Challenger* hardware with less than twenty-four hours left in the countdown. The meeting seemed routine; it was over quickly: all the issues that had been classified as still "open" at the end of the final Flight Readiness Review the week before had now formally been resolved. The Marshall engineers reported no concerns about the flight worthiness of their parts of the shuttle system—neither the external tank, the main engines, nor the solid rocket boosters. It was approaching 12:30 p.m. when Jesse Moore collected the official certificates of flight readiness from each of the delegations attending the meeting; he then took a verbal poll of all the project managers in the room for the final clearance to launch; each of them recorded a "go."

At 3:00 p.m. on Saturday, Moore went out to meet the press and told them that he expected to launch the next day—although the weather could still pose a problem, he wouldn't be able to make a final call until later that night, after consulting the meteorologists and Acting Administrator Graham. To prepare for a morning liftoff, the filling of the shuttle's external tank with fuel was due to begin sometime after midnight: Moore would have to make his decision before that happened.

The crew, now through their last medical checks, were held in strict quarantine on the third floor of the Operations and Checkout Building and permitted no further visits except from their spouses. Steve McAuliffe stopped by to see his wife that afternoon and then drove up to the Holiday Inn in Orlando, where he joined June Scobee, Marcia Jarvis, and Ed and Grace Corrigan to host the traditional prelaunch party for astronauts' friends and family. As the sun began to set, McAuliffe, dressed for the party in a jacket and tie, stood beside the ornamental fountain outside the hotel to give his final interviews to TV reporters before liftoff; he told them that he was excited, and nervous, but reassured that NASA was proceeding cautiously.

"They don't delay unless it's not perfect," he said. "And that's fine with me. I'd be much happier if it went up when everybody thought it was perfect, than they go up on a chance basis."

Inside, a few hundred people gathered in the hotel ballroom for cocktails and hors d'oeuvres, stopping at the tables displaying pictures taken of Greg Jarvis and McAuliffe over the years, and writing messages in a visitors' book; down at the Patrick Air Force Base officers' club, Jane Smith threw a party with champagne and canapés for more than four hundred guests, reuniting extended family and old friends from across the country, a grand celebration including a show of Mike's military citations, and pictures of him as a Navy midshipman and aviator. Then, as darkness gathered across the swamps and lagoons of Merritt Island, the astronauts' immediate relatives boarded a NASA bus for the scheduled highlight of the evening: a trip out to Pad 39B for a night viewing of *Challenger*. As the bus rounded the corner toward the launch complex, everyone on board fell silent. The massive spaceship loomed above them from the subtropical blackness, its tessellated flanks glowing white in the glare of the floodlights. Lining up with the others at the pad perimeter fence, Grace Corrigan reached for her husband's hand. She felt a pang of apprehension; would her daughter really be strapped into that?

At 9:00 p.m., Jesse Moore gathered a small group of top-level NASA officials in a conference room on the third floor of the Launch Control Center. With the exhaustive proceedings of the Flight Readiness Review complete, and less than eleven hours left to run on the countdown clock, it now fell to this handful of men—the Mission Management Team—to handle any last-minute decisions leading up to the *Challenger* launch. Sitting around a T-shaped polished teak table, Moore's group included the head of the shuttle program from Houston, Arnie Aldrich; Bob Sieck, who, as the Director of Shuttle Operations at the Cape, had overall responsibility for getting the spacecraft to the pad; and Launch Director Gene Thomas. They were also joined by Acting Administrator Graham, who had, that afternoon, flown down from Washington, DC, aboard the agency's Gulfstream to observe his managers at work, and watch the next morning's launch—if it happened.

But when Lieutenant Scott Funk, shuttle meteorologist at the Cape Canaveral Air Force Station, was patched in over the teleconference line, his briefing made it clear that this was unlikely. Funk explained that the cold front was still surging across Florida from the northwest, with a leading edge expected to arrive over the Cape at around five in the morning; behind it there would be low cloud, poor visibility, fog, rain— and even lightning. By launch time on Sunday, the center of the front would be only fifty miles up the coast in Daytona Beach, the weather at Kennedy deteriorating fast. The chances of launching under those conditions were less than fifty-fifty.

But Funk's forecast for the following day, Monday, was excellent: scattered cloud and good visibility would provide ample time for a morning launch window before gusty winds rose later in the day. Beyond that, things looked bad: he said that a mass of frigid air behind the cold front had swept down from the Canadian arctic, and his forecast for Monday night and early Tuesday was for extreme cold. The temperature in Tampa would fall to 14 degrees Fahrenheit, 21 in Orlando; at the Cape, he predicted it would be slightly warmer, but still in the mid-twenties. This was colder even than the record-breaking weather that had bedeviled Ellison Onizuka's mission aboard *Discovery* almost exactly a year earlier. What had seemed then to be a once-in-a-century fluke of meteorology seemed almost certain to be surpassed just twelve months later: the night of January 27, 1986, would once again see the coldest weather in Florida history.

Jesse Moore consulted with his project managers to determine what other constraints there might be to a Monday launch. But the only concern was the weather. With Graham's approval, the senior managers agreed to reschedule *Challenger's* liftoff for 9:37 a.m. on Monday.

At 10:00 p.m., Moore made his announcement to the media.

"We're not going to launch this thing and take any kind of risk because we have that schedule pressure," he said. "We'll sit on the ground until we all believe it's safe to launch."

Out at the pad, the Rotating Service Structure was rolled back into place, to protect the orbiter from the anticipated bad weather; the giant countdown clock was stopped. By then, the crew of 51-L were supposed to have been asleep for hours—but NASA publicist Linda Long called McAuliffe in the crew quarters and gave her the news.

"It's a scrub," she said.

"Make sure you tell my parents, so they don't get up at 4 a.m.," McAuliffe said.

She was obviously disappointed.

"I'm sorry. I know how you feel," Long told her.

"No you don't. I'd rather be floating around up there."

On Sunday morning, Launch Director Gene Thomas rose early and drove the fifteen miles from his house on Merritt Island toward the Kennedy Space Center to report for work; there were no days off once the countdown started. As soon as he arrived, his assistant told him that Dick Young, from the Kennedy Public Affairs Office, had called; the press had been asking for Thomas's comments on the day's weather. At around 8:20 a.m., Thomas, his assistant, and Young climbed the stairs up through the Launch Control Center building toward the fourth floor, down the hallway, and through a cluster of office cubicles, until at last they emerged onto the roof. Standing at the rooftop railing from which the families of the *Challenger* crew soon expected to watch the astronauts depart for orbit, the three men gazed out toward Pad 39B. They had a perfectly clear view across the more than five miles of scrub and lagoons toward the ocean, where the shuttle sat on the pad, surrounded by the gray steel cocoon of the Rotating Service Structure; the sky yawned above them, a fathomless blue. A few hundred yards away to the southeast, the indigo waters of the Turning Basin—used by the barges delivering the shuttle's external tank to the Vehicle Assembly Building— were as smooth as a mirror. The temperature was in the mid-forties; there wasn't a breath of wind.

Young, a genial Virginian who had once been an environmental reporter on the *Orlando Sentinel*, turned to his colleagues. "Gene," he said, "do you have any words of wisdom to pass along to the media on this weather?"

"Well, Dick, what can I say? It would have been a great day to launch. But I guess we missed a good opportunity."

Mike Smith had woken that morning just before seven. Looking out through the windows of the crew quarters recreation room, the astronaut's face was washed in bright sunshine; it was an ominous sign. Smith knew that the cold front—perhaps briefly stalled, but soon to move inexorably south—would simply arrive over the Cape later than expected: now the delayed bad weather could interfere with the scheduled liftoff on Monday. But there was no going back: the new launch schedule was set; the laborious, hours-long process of fueling the shuttle wasn't even due to begin until later that night.

The scrub was another black eye for the space agency; the headlines the next day seemed sure to be humiliating. Many of the guests who had been waiting to witness a weekend launch began leaving the Cape and heading home: Governor Sununu made his apologies and returned to New Hampshire; Christa McAuliffe's younger sister Betsy and her brother Stephen left; Vice President Bush would now fly straight in to Central America without stopping in Florida. In Washington, DC, Congressman Don Fuqua was furious that an almost-perfect launch window—with the chance that the Vice President could have been in attendance—had been squandered.

"Who the hell called that shot off?" he asked Bill Graham when he returned to the city on Sunday.

"I did," Graham said. "The weather people swore in blood it was going to be bad."

That afternoon, Smith and Dick Scobee drove out to the Merritt Island airstrip and flew a series of simulated landing approaches in the Shuttle Training Aircraft. As they took off, they could see thin wisps of cirrus cloud high in the sky to the northwest. As the day went on, these thickened and descended over the Cape in a gray overcast, forcing the two pilots to cancel the T-38 flights they had planned for later. By the time they arrived back at the Operations and Checkout Building, a light drizzle was spattering on the windshield of their car.

After watching the twentieth Super Bowl on the big color TV in the dormitory lounge, the astronauts ate dinner with their spouses; no one seemed very hungry. At the end of the meal, Ron McNair pulled a magnum of champagne from the shopping bag Cheryl had brought over from the motel, an image of the shuttle etched on the bottle. He passed around a pen so that everyone could sign their names on the glass—his contribution to the festivities planned for the landing party they would be having in Houston the following week. It was still early when the crew said good night to their families, and neither Greg Jarvis nor Christa McAuliffe could sleep. Together, they went out into the parking lot, where they found a pair of bicycles and set off for a ride around the island. The crew of a local TV van passed them as they pedaled along Route 3, and captured a shot of McAuliffe cycling toward the Vehicle Assembly Building, smiling and waving for the camera. "Don't come real close," she shouted. "We're in quarantine!"

Back in his room, Mike Smith remained concerned about the weather, and called his old friend Bill Maready up in North Carolina.

Maready, an attorney, was also a keen pilot, who had flown his own family down to Florida on Friday in his twin-engine Beechcraft Baron to watch the launch, but returned home when Sunday's attempt was scrubbed. Before leaving, he had told Smith that he could give him a firsthand report on the cold front when he got home, because he'd be flying through it on the way up the coast. Smith asked for details on everything he'd seen: wind speeds, precipitation, cloud formations— but was particularly worried about the extreme cold carried behind the front. He told Maready that the weather at the Cape had been perfect for a launch all morning; he remained as frustrated as he had been when he awoke. "You know, Bill," he said, "you've got people down here making decisions who've never even flown an airplane before."

Over in the Launch Control Center, the lower level of Firing Room 3 was already crowded with engineers and technicians. Backs to the launchpad, at their seats behind the three semicircular rows of mint-green metal consoles, they adjusted their headsets. The low susurration of chatter subsided, and the room fell silent. On the wall above them, three separate digital clocks displayed Universal Time, Local Time, and the Shuttle Count in bright red numbers.

In their ears, the members of the Kennedy launch team heard the calm voice of the test conductor speaking.

"Short count: four . . . three . . . two . . . one. Mark. T-minus eleven hours and counting."

On each of their console screens, the amber blocks of letters and digits began to flicker. Three miles away on Pad 39B, 535,000 gallons of liquid hydrogen and oxygen would soon start flowing into *Challenger's* external fuel tank.

The final countdown of Space Shuttle mission STS-51-L had begun.

CHAPTER TWENTY-ONE

MONDAY, JANUARY 27, 1986

8:00 a.m.

From the big floor-to-ceiling windows of Launch Director Gene Thomas's office on the fourth floor of the Launch Control Center, Alison Smith looked out over the parking lot toward Pad 39B, where her father was already cinched tightly into his seat on the *Challenger* flight deck. The rooms around her were crowded with the families of the 51-L crew awaiting the liftoff, including her mother; her elder brother, Scott; and her younger sister, Erin, who played on the floor with Christa McAuliffe's daughter, Caroline. Three-year-old Reggie McNair sat at the conference table, wearing a miniature sky-blue NASA flight suit specially made to match his father's, finding a space for his coloring book between the stacks of paper plates and boxes of breakfast pastries. Above their heads hung a banner made from computer printer paper: WELCOME 51-L FAMILIES FROM THE KSC LAUNCH TEAM. In the corner, a color TV set silently relayed images of the pad closeout crew, making the final preparations for takeoff inside the White Room on the 195-foot level of the launch gantry.

Outside, the weather was crisp and clear; the numbers on the countdown clock were flickering silently down from three hours, and the concluding prelaunch checks were proceeding smoothly. Arriving just before dawn aboard the Astrovan, the *Challenger* crew had ridden the elevator up the gantry, donned their flight helmets and escape harnesses, and one by one crawled into the orbiter, before the pad crew closed the shuttle's side hatch behind them. Two technicians in white overalls stepped forward with the tool used to lock the hatch in place. There was no rush; they were running well ahead of schedule.

Behind them, Johnny Corlew, NASA's Quality Assurance Inspector, consulted his checklist. His next task was to record the signal from a pair of electronic switches inside the door mechanism, confirming that it had latched. But when Corlew turned to take his readings from the instrument panel on the wall of the White Room, the ohmmeter revealed a problem: the hatch had not locked. While one of the other technicians radioed in the fault to the Launch Control Center, Corlew used a lighted mirror to peer inside the lock mechanism. He could see that the latches were exactly where they should be. *Must be a switch malfunction,* he thought; *just needs a little adjustment.* The time was just after 8:30 a.m.; a rising wind rattled the thin aluminum walls of the White Room.

It was another hour before Corlew and the closeout crew received approval to disregard the malfunctioning switches—but their problems were just beginning. When the technicians at last began to remove the temporary handle they used to close the hatch, they found one of the bolts holding it in position spun in place. They tried again and again to work it free, but it was no good; the handle was stuck. They radioed news of the fresh setback to the engineers in Firing Room 3, and decided to use a drill to remove the trapped bolt—but there was no drill available on the pad.

Back in Gene Thomas's office, Alison Smith grew tired, and bored with waiting; she had stayed up late the night before catching up with friends, and then woken at 5:30 a.m. for the drive to the Space Center. Now she and her brother settled themselves on the couch in an adjoining room and fell asleep. It took almost another hour before the drill the White Room crew needed could be found and driven out to Pad 39B—but then they discovered the battery in it was almost dead; they sent for more batteries, and those proved useless, too. They ordered up another drill. That wouldn't work, either. Finally, Johnny Corlew asked for clearance to use a hacksaw to cut off the troublesome bolt. They waited twenty minutes while the engineers in the Firing Room conferred with experts from Rockwell, who had built the orbiter. Soon *Challenger's* two-hour launch window would close; around them, the wind was still strengthening. Over the cockpit radio link, Dick Scobee was already asking them about the changing weather: it was still clear enough for launch, but the conditions were growing more marginal with every second. Corlew turned to the technician still waiting for approval to saw off the bolt.

"Look," he said, "just go ahead and cut it."

At last, more than two hours behind schedule, the closeout crew secured the shuttle's hatch for departure, and prepared for the access arm to swing back. But it was too late. By noon, crosswinds around the Cape had risen to twenty knots, flattening the saw grass around the shuttle's landing strip, exceeding the limits for a safe launch. At 12:36 p.m., after the crew had spent almost five hours growing stiff and numb in their seats, they heard the voice of Gene Thomas in their helmet earpieces as the word went out across the NASA communications network: it was a scrub. As Dick Scobee clambered out of the hatch and shrugged off his emergency escape harness, Johnny Corlew began to apologize. But Scobee gave him a pat on the back.

"Hey, forget it, Johnny. That's the way things go sometimes. We'll just try it again tomorrow."

:::::::::::::::::::
:::::::::::::::::::

Yet over in the Launch Control Center, the engineers waiting at their consoles weren't sure about firing on Tuesday, either. The weather forecast predicted record cold, with temperatures at the Cape expected to fall to 23 degrees Fahrenheit overnight—9 degrees below freezing—with the possibility of further high winds. As the families of the crew drifted from Gene Thomas's office in disappointment, he and the other weary NASA senior managers convened once more in the conference room down the hall to consider their options. They talked at length about the forecast temperatures, and Arnie Aldrich from the Johnson Space Center—number two in the overall launch decision chain at the Cape—gave the go-ahead to implement the cold-weather plan they had formulated after pipes on the launch gantry had burst the previous winter. This time, technicians would drain some portions of the fire extinguishing system and leave the faucets of the emergency showers and eyewash baths slightly open, allowing water to keep trickling through the plumbing to protect it against icing up. But, other than the hard freeze overnight, to Gene Thomas the weather for the following day seemed set to be almost perfect for a rocket firing: crystal clear skies, with unlimited visibility; as the meeting broke up, the Launch Director felt good about the prospect of getting *Challenger* aloft the next morning. Aldrich agreed to reschedule the launch for 9:38 a.m. on Tuesday.

Back at the Marshall operations center in Huntsville, Solid Rocket Program Senior Manager Larry Wear remained troubled by the mention of low temperatures. He felt sure that the Thiokol engineers had mentioned something about the potential effect of cold weather on the boosters after a shuttle launch back in January of the previous year. The details eluded him, but it seemed it could be important. He turned to Boyd Brinton, Thiokol's man on the ground in Alabama, who was seated beside him: What exactly had his engineers said about the effect of low temperatures on the boosters? Should they be worried?

Brinton admitted he couldn't recall what they had discussed twelve months ago. Wear told him to refresh his memory. He felt certain the subject would come up again before the end of the day.

::::::::::::::::::

It was not yet noon in Arnie Thompson's office at the Morton Thiokol plant in Utah when the phone rang. Boyd Brinton came quickly to the point. He told Thompson that there could be a problem with the scheduled *Challenger* launch: a cold front coming in, with temperatures below freezing. Perhaps as low as 18 degrees.

"Are you concerned?" Brinton asked.

"You bet," Thompson said. "Yes, I am."

A supervisor in Thiokol's Structures Division who served on the Seal Team Task Force with Roger Boisjoly, Thompson thought immediately of the O-rings. He put down the phone and sent out the word to each of the senior engineers on the solid rocket project at the plant. They'd need to pool their data on previous cold-weather shuttle launches.

Until that moment, Boisjoly himself had known little about the 51-L launch except that it was due to take place that week. Since returning from his Christmas vacation, he had been entirely focused on seeking a solution for the O-ring failures; just that morning he had received yet another proposal from a seal manufacturer with drawings of a potential modification for the joint that might eliminate the leaks. The new year had so far been consumed by meetings, phone calls, memos, and analyses he documented meticulously in his logbook; his minutes, action items, and to-do lists filled page after page with neatly ordered block capitals as he gathered samples of replacement seal materials from manufacturers across the country and formulated test plans. But as January wore on, his observations had increasingly flashed with frustration. He logged delays in analysis, botched contracts, and missing data; he continued to despair at the lack of urgency given the problem by almost everyone he dealt with at the Wasatch plant. By the third week in January his spirits had sunk to a new low: "I give up trying to get something done in this place," he wrote. Still, he resolved to stay at it, and do the best he could.

Now Boisjoly sat in Bob Ebeling's cramped office on the first floor of Building A-2 along with the other members of the O-ring task force, listening as Ebeling described the phone call from Boyd Brinton. When he heard that the weather at the Cape would be far colder even than the freeze he had witnessed the previous January, Boisjoly was astonished. How was this possible? A *second* one-hundred-year weather event in Florida—barely twelve months after the last one? With the image of what he had seen in Hangar AF seared in his memory, Boisjoly was adamant: he told his colleagues that no one in their right mind would fire the shuttle boosters in temperatures lower than they had witnessed the year before; the launch must be stopped. Over the next hour, Boisjoly and the other seven engineers in the room went over the evidence of seal failure on previous flights, the temperatures at launch, and the results of the tests on the O-rings from the year before—which had shown the seals would be practically useless when the weather was cold. As the meeting went on, other technicians from the department drifted in to join the conversation, while Brinton listened in over the speakerphone from Alabama. But they all soon agreed with Boisjoly, and resolved to take their concerns down the hall to Bob Lund, now the head of engineering for all of Thiokol's rocket programs.

::::::::::::::::::

It was after 4:00 p.m. at the Kennedy Space Center when Al McDonald first heard of the growing alarm in Utah. Taking his turn as Thiokol's senior representative at the launch site, McDonald had risen at four that morning and sat at a console in the Firing Room throughout the failed launch attempt. As he left the building, he struggled through gusts of wind so fierce that he was repeatedly fought to a standstill in the parking lot. The consecutive launch delays had made it impossible to find a motel room, so McDonald took a thirty-minute drive from the Space Center to Titusville, where he was staying with Thiokol's resident local manager and his wife. They were making plans for dinner when Bob Ebeling called, and told him that the local weather report had predicted the arrival of an extreme cold front in Central Florida. The two men shared their worries about the O-rings, and Ebeling assured McDonald that the engineers at the plant had been working on the problem since lunchtime. He explained that, in order to generate an accurate analysis of the effect of the cold on the solid rocket joints, they would need a precise hour-by-hour forecast of temperatures at the launchpad; he thought the numbers he'd heard reported were so low they were hard to believe. McDonald promised to call back with the data. But when he did, the news he gave Ebeling was still not encouraging: NASA meteorologists at the Cape now said that the weather would freeze before midnight, reaching a low of 22 degrees Fahrenheit by 6:00 a.m. They expected the temperature to rise to just 26 degrees at the opening of the launch window: less than half what it had been when the crew of *Discovery* narrowly escaped tragedy the year before. He gave Ebeling instructions to have the engineering team in Utah establish predictions for the actual temperatures of the O-rings inside the rockets when the window opened—and calculate what effect it would have on the field joints at ignition.

McDonald told Ebeling that he wanted Thiokol's engineers to come up with a recommendation for the minimum temperature at which it would be safe to launch, and present their conclusions during a teleconference with NASA. And, knowing how much pressure the company's program managers were under to maintain their schedule, he insisted that the final recommendation come directly from the division engineering chief, Bob Lund. "This has to be an engineering decision, *not* a program management decision," he said.

It now fell to McDonald to alert his NASA bosses that Thiokol had a serious problem. First, he tried to reach Larry Mulloy, who was staying at the Holiday Inn on Merritt Island for the launch. But there was no reply from his room. So McDonald looked up the number for Cecil Houston, Marshall's resident manager at the Cape, and found him at home, where he was just sitting down to dinner with his wife. McDonald explained that the engineers back in Utah were worried about the forecasted cold, and the O-rings sealing the field joints in the solid rockets. "We know they've been looking at it all afternoon, and they've got this concern," McDonald said; they weren't certain whether they should launch.

As resident manager, it was not Houston's job to try to solve problems himself—an electrical engineer, he wasn't qualified for that—but to act as the contact for senior managers from Huntsville, who could make the decisions. If faced with a routine issue that he had seen on previous missions, he could sometimes clear engineers to keep working on a solution until he could approve it with his boss later; but this issue was certainly not routine.

McDonald didn't say so explicitly, and Houston wasn't sure exactly what to think—but right now it sounded to him as if they were going to have to call off the countdown.

"Well, Al," he said, "I'll have to get my people together and let them talk to you."

Pushing his dinner plate aside, Houston tried phoning his immediate supervisor in his room at the Holiday Inn. But he had no luck reaching either him or Larry Mulloy. Dialing the special long-distance direct number to Marshall, Houston called the deputy head of the Science and Engineering Directorate, George Hardy—an old colleague who he had known for years. But Hardy had just left the office for the day. Increasingly anxious to tell someone in senior management what was going on, Houston tried his supervisor's room at the motel once again; by the time he finally picked up, it was around six in the evening. The two men agreed to hook up a six-way teleconference—connecting Allan McDonald in Titusville with four different Marshall managers at their homes, offices, and motel rooms in Alabama and Florida, and with Bob Lund at the Promontory plant in Utah. When the call began fifteen minutes later, the Thiokol engineers began to explain their concerns about the booster rocket seals, and suggested that the launch might have to be delayed, at least until noon on Tuesday, for the O-rings to warm up. But the long-distance telephone lines hissed and wavered; some voices were all but inaudible: there was no way the group could have an ordinary conversation, let alone a complex technical debate. Besides, the Thiokol engineers still had not yet formulated a coherent presentation.

Together, the far-flung rocket engineers agreed to relocate to nearby offices wired for conference calls and restart the meeting at 8:15 p.m. Florida time, or 6:15 p.m. in Utah. This gave the Thiokol team just two hours to formulate their arguments and gather the evidence to support a potentially momentous and unprecedented step: the first time in the history of the program that the rocket contractor had ever called to stop a launch.

Over in the crew quarters suite on Merritt Island, the *Challenger* astronauts were again eating an early dinner with their husbands and wives. But after four delays in six days, and the hours strapped in on the flight deck that morning, nerves taut in the expectation of launch, the crew was agitated and jumpy: they couldn't sit still; conversation was anxious and muted. They were officially scheduled to go to sleep at 7:00 p.m., in order to rise early on Tuesday for another trip out to the pad. But the plummeting temperatures made a morning liftoff seem unlikely: Mike

Smith had already called Jane's mother and told her she might as well pack up and go home, where she had inventory to do at the department store where she worked; he warned Jane that the kids should be ready to get back to Houston before they missed any more school; it would be days before the cold let up. "There's no way we can launch," he said.

As cheerful as ever, Greg Jarvis did his best to keep his crewmates' spirits up, and Dick Scobee assured them that they could trust the Mission Management Team to make the right decision. "They've got people up there checking the weather now," he said. But he knew it was out of his hands: at this stage of the countdown, no astronaut—even the commander of the mission—had the authority to say whether or not they launched. After dinner, June Scobee took her husband into his room and closed the door behind them: behind the unflappable test pilot façade, she knew he felt tortured by every launch delay. But now he told her that a Tuesday takeoff was out of the question. When it was finally time to leave, in the doorway of the room the couple paused in a good-night embrace. "Well," June said, "you know you're not going tomorrow. So just sleep well."

Darkness fell, and each member of the crew made last-minute phone calls to friends and family around the country. Christa McAuliffe spent an hour talking to her friend Eileen O'Hara back in Concord: she described the agony of sitting on her back on *Challenger*'s middeck for five hours that morning, bound by safety harnesses, enmeshed in oxygen and communication lines. But she laughed about the failure of the bolt, and remained excited about the prospect of a launch in the morning. "I still can't wait," she said. Judy Resnik called Sylvia Salinas back in Houston, and the secretary made her friend promise to make some decisions about her future—and her romantic entanglements—as soon as she returned from the mission.

"When you get back, we're going to talk, OK?" Salinas said.

"OK."

And Scobee phoned his close friend and fellow astronaut Pinky Nelson: the commander admitted that dealing with the 51-L crew— the Teacher in Space publicity, the outsize personalities of Resnik and Onizuka—had tested the limits of his patience.

"I really can't wait to get this mission over with," Scobee said. "I can't wait to get this one done."

Four miles away in Firing Room 3, the glowing ruby digits of the Shuttle Count clock were once again ticking down toward zero.

It was 6:30 p.m. on the East Coast when reports of the morning's latest failed launch at the Cape made the headlines of all three national evening news shows. On CBS, the protracted fiasco on Pad 39B was the lead story, and anchor Dan Rather was pitiless. "Yet another costly, red-faces-all-around Space Shuttle launch delay. This time a bad bolt on a hatch and a bad-weather bolt from the blue are being blamed. What's more, a rescheduled launch for tomorrow doesn't look good, either," he said, before a mocking handover to the space reporter at Cape Canaveral: "Bruce Hall has the latest on today's high-tech low comedy." On ABC, John Quiñones was equally unforgiving: "Once again a flawless liftoff proved to be too much of a challenge for the *Challenger*," he said. Each of the network news programs carried NASA's footage from the White Room, of Johnny Corlew and the closeout crew struggling with the flat batteries and the failing drills—along with clips of the ignominious cascade of attempts to launch *Columbia* at the start of the month.

In his motel room, Stan Reinartz, the head of Marshall's Shuttle Projects Office, received a phone call from his deputy back in Huntsville, Jud Lovingood. A hardheaded engineer of the old school, Lovingood had been on the earlier conference call and was worried that, if the Thiokol engineers decided to oppose the launch, NASA would be forced to accept yet another delay. He told Reinartz that they should contact the agency's upper-level managers responsible for the final launch decision and let them know the way things were going—and prepare them for the worst. He also recommended that their own boss, Marshall Center Director Bill Lucas, sit in on the upcoming teleconference.

Reinartz put down the phone and walked over to the room where Lucas was staying; the head of the Science and Engineering Directorate from Marshall, Jim Kingsbury, was already there, and Larry Mulloy arrived soon after. Reinartz told his superiors about the impending tele-conference, but stopped short of suggesting Lucas attend. Mulloy ex-plained that the Thiokol engineers were concerned about the effect of the weather on the seals. But Lucas was dismissive.

"That sure is interesting," he said. "We get a little cold nip, and they want to shut the shuttle system down? I sure would like to see their rea-sons for that!"

And—in keeping with his boss's desire to keep Marshall problems sequestered from outside scrutiny—Reinartz also told Lucas he didn't see any reason to pass Thiokol's worries any further up the launch de-cision chain. As far as Reinartz was concerned, they need not involve either headquarters—Level I—or senior shuttle program managers at the Cape, on Level II: they could handle it themselves down on Level III. Lucas concurred.

"Fine," he said. "That's the thing to do. Keep me informed."

::::::::::::::::::

One hundred miles out in the Atlantic, NASA Retrieval Operations Manager Jim Devlin stood wedged into a corner on the bridge of the booster recovery ship *Liberty Star*, bracing himself on a grab rail as the vessel rose and fell beneath him, the deck plates groaning with the strain. The ship was in peril, and Devlin knew it. The same frigid wind that had rattled the walls of the White Room above Pad 39B that morning had now risen to gale force over the ocean, shrieking through the rigging as it gusted to seventy-five knots—upward of eighty miles an hour; around him in the dark, giant waves swelled to heights of more then twenty-five feet, hurling thick gobs of spume through the glow of the ship's running lights to smash across the windows of the bridge. Shallow-bottomed and less than 180 feet long, the *Liberty Star* had never been designed to deal with such weather. As her captain maintained his easterly course toward the booster recovery area, some 125 miles north of Grand Bahama Is-land, the ship struggled again and again to crest the steep faces of the mountainous waves, and rolled and wallowed in the black troughs be-tween them. The roar of the storm intensified, and through the windows overlooking the stern, Devlin could see that the huge winches used to tow in the boosters had been bent by the force of water breaking across the superstructure; the timbers intended to secure the recovered nose cones of the rockets on board were breaking loose from their moorings and sliding around on deck. The captain reduced speed, but now the raging sea threatened both the *Liberty Star* and her sister ship, the *Freedom Star*, with broach and capsize.

It was no good: both vessels would have to turn back.

Back in Utah, the Thiokol technicians were scrambling to organize their presentation. Standing at the whiteboard in the MIC room, Vice President for Engineering Bob Lund had sketched the outline of their argument, and handed assignments to each of his men to collate data in their areas of expertise. Roger Boisjoly returned to his office to work on a summary of the resiliency tests they had conducted on the O-rings; Brian Russell, a lanky thirty-one-year-old manager in the Solid Rocket Program Office, would handle the flight history of the boosters, including each incidence of seal damage and O-ring leaks; Arnie Thompson was assigned to describing the details of blow-by and compression tests. Meanwhile, the department specialists in heat transfer and gas dynamics were using the computers downstairs to model predictions for the way they expected the booster joints to behave at the temperatures forecast for the next morning's launch window.

The conference was so important that the company's two top executives—Cal Wiggins, Vice President for Wasatch Operations, and his boss, Senior Vice President Jerry Mason, General Manager of the Space Division—had already been called back to the plant from the Thiokol offices in Brigham City to attend. At fifty-nine, Mason was another twenty-five-year veteran at Thiokol, and reported directly to corporate headquarters in Chicago. Before his promotion to division chief, he had managed the 1,500 blue-collar staff on the shop floor of the Promontory plant with dogged resolve. Skinny and cadaverous, Mason had the etiolated look of a Dickensian schoolmaster, but was an avid outdoorsman whose home was filled with the glassy-eyed souvenirs of hunting and fishing trips. Wiggins, his reliable lieutenant, was mute and obedient; the consummate yes-man.

By this time, it was so late in the working day that the secretaries and other support staff were already heading home for the evening. With no one to type up the engineers' drafts into the formal charts required for the presentation, Bob Lund began to write them out himself, hand-lettering each Viewgraph slide in careful block capitals.

As Lund prepared the data, Wiggins and Mason sat and talked quietly with Joe Kilminster, chief of the company's solid rocket programs, at the head of the three large oak tables in the center of the MIC room; with NASA still threatening to open the lucrative shuttle booster contract to competition, and a crucial summit in Thiokol's ongoing negotiations with the agency scheduled for the next day, the stakes for the teleconference were high. If their engineers intended to cancel the launch, Mason and Wiggins needed to understand the rationale for doing so as precisely as they could.

By 6:00 p.m. in Utah, the first of the fourteen slides in the engineers' presentation were almost ready to send over to NASA—but there had been no time for them to rehearse, or even review, what they were going to say. As the handwritten pages began stuttering from the fax machines at Cape Canaveral and Marshall, many of the Thiokol team recognized that the data they had about the behavior of the joints in cold weather remained scant; some seemed contradictory. As a result, the recommendations they would be making were subjective and qualitative—it was far from a perfect case based in hard engineering fact.

Even so, it was one that they believed raised enough concerns about the flight safety of the *Challenger* mission to advocate a drastic course of action; the concluding chart Lund had drafted represented the unanimous recommendation of all fourteen managers and engineers who would eventually gather in the MIC room, including even Jerry Mason and Cal Wiggins:

Do not launch.

CHAPTER
TWENTY-TWO

MONDAY, JANUARY 27, 1986

7:55 p.m.

It was already freezing as Cecil Houston walked across the parking lot toward Trailer Complex C at the Kennedy Space Center. In the citrus groves nearby, the growers were bundled against the cold, sipping coffee and anxiously making the rounds of their trees and their temperature checkpoints, preparing to save what they could from the bitter weather. Something strange was happening on Merritt Island: an isolated cell of frigid air—ten degrees colder than it was less than twenty miles away in Cocoa Beach—had begun settling over the Space Center and the long rows of fruit trees around it. It threatened to destroy the entire crop of oranges and grapefruit. And the temperature was still dropping.

Inside the office, Houston made sure that the three-way transcontinental telephone hookup was working as planned. But before the teleconference could begin, he had a call from the Launch Director's office, relaying a radio report from the *Freedom Star* and the *Liberty Star* out in the Atlantic. The retrieval teams had notified the Cape that they'd been forced to abandon their mission, adding a further setback to plans to launch *Challenger* the following morning: the two ships wouldn't be in position to recover the shuttle boosters in the minutes after liftoff. So the burned-out booster rockets and their parachutes—sixty-four million dollars' worth of hardware—could simply be lost at sea.

Taking his seat at the master phone console at a square of tables in the conference room, Houston was ready to begin the meeting on time, at 8:15 p.m. He was tired, and impatient to get started: like many others summoned for the teleconference, he had been awake since the small hours of the morning, and could expect only a few hours' sleep later that night before rising early once more to do it all again the next day. The room around him was large enough to accommodate thirty, but by the time the voices from Alabama and Utah began announcing themselves over the network, it remained mostly empty: seated with Houston were just four men, including Larry Mulloy and Stan Reinartz—representing NASA—and Al McDonald from Thiokol. From here, it was Mulloy who would take the lead. By contrast, two hundred miles away in Huntsville, almost every one of the facility's available experts from the solid rocket booster program was assembled in Conference Room 411 at the Marshall Space Flight Center; there were fifteen of them in total, led by George Hardy, the white-haired Deputy Director of Science and Engineering who had overseen the rocket design from the outset, and Jud Lovingood from the Shuttle Projects Office.

Yet the meeting began in confusion: even as Thiokol Vice President Joe Kilminster started to speak, the members of his engineering team were still transmitting their data, one page at a time, to the other two groups across the country; the heat transfer and gas dynamics experts downstairs had not yet completed their calculations; and even when all the charts had arrived over the fax in Marshall, nothing had shown up at the Cape. And, in an unusual breach of protocol, Mason and Wiggins, the two senior executives from Morton Thiokol, did not announce their presence on the call; no one outside the MIC room knew they were even listening in.

It was 9:00 p.m. in Florida when the Utah engineers' presentation finally started, with a handwritten title chart: TEMPERATURE CONCERN ON SRM JOINTS, 27 JAN 1986. What followed was a point-by-point summary of everything they had learned to date about the solid rocket field joints and their projections for how they would respond to the freezing conditions at the Cape. First, Brian Russell laid out the bewildering thread of failure running through the flight history of the rockets. During the first nine Space Shuttle missions, Thiokol technicians had found only a single O-ring anomaly in the field joints of the boosters; but, beginning with the tenth flight in January 1984, they had witnessed an alarming escalation in problems. Since then, almost half of the field joint seals had seen some kind of damage—erosion or partial destruction of the O-rings; blow-by; leaks.

In the trailer at Cape Canaveral, Mulloy and the others sat back and listened; at Marshall, George Hardy scribbled notes on a pad, making sure he understood everything he was hearing.

From the Utah conference room, Roger Boisjoly and Arnie Thompson began to lay out their case. Boisjoly showed a sketch of the field joint in close-up, revealing its design flaws, and explained the importance of the split-second timing of the sealing process at launch. He said that if the primary O-ring didn't seal within the first six hundred milliseconds after ignition, then it was highly likely that the secondary seal would be destroyed before it had a chance to close the gap between the two halves of the joint. If it could bypass both seals, then hot gas would escape through the outside casing of the rocket; his audience knew only too well what this would mean for the spacecraft, and anyone on board.

In fine detail, Boisjoly explained how the O-ring material lost its elasticity in the cold, compromising the squeeze, and how low temperatures could critically delay almost everything about the way the joint was supposed to work; he displayed the frightening results of the resiliency tests kept secret for so long the previous year. "The colder temperatures predicted can change the O-ring material from a hard sponge to something more like a brick," he said. His charts showed that, at ambient temperatures of less than 50 degrees Fahrenheit, the O-rings could become so hard that not only might they fail to seal during the crucial ignition transient—but they might prove unable to do so at any point during the entire two-minute duration of the rockets' flight. Next, Brian Russell talked through Thiokol's database of temperature readings from previous solid rocket test firings and shuttle launches, along with comparable predictions of what to expect for *Challenger*. The numbers were bracing: they showed that the 51-L crew would be launching with rocket hardware at temperatures far outside the range of the engineers' experience—at least twenty degrees colder than the record lows faced by *Discovery* in January 1985.

As he wrapped up, Boisjoly felt sure that he had built a damning argument against launch. But there were complications: he knew his data was thin and, with no opportunity to review his colleagues' slides before he began, some of their information took him by surprise. And even before he had completed his part of the presentation, Marshall's Larry Mulloy began pressing him to provide further hard numbers to back his argument. Boisjoly grew flustered. He told Mulloy that he had no more data, but knew that, as the temperature dropped, the performance of the seal declined: "With colder temperatures, we're moving away from the direction of 'goodness,'" he said. Despite his unscientific language—and his sense that Mulloy's cross-examination was tougher even than those he had delivered in Flight Readiness Reviews—Boisjoly believed that good sense would still prevail among the NASA managers. After all, the safety of the astronauts remained the top priority of everyone involved in the shuttle program.

At last, Bob Lund stood to talk through the final pair of slides he had written out: CONCLUSIONS and RECOMMENDATIONS. The first showed that, based on the data they had presented, he and his staff would expect to see leaks in the rocket joints when their temperature at ignition fell to around 50 degrees Fahrenheit or less; the weather for the shuttle launch window at the Cape the next day was forecast to rise from 29 degrees at nine in the morning to just 38 degrees at two in the afternoon. The second slide made clear that the engineers regarded these conditions as profoundly dangerous. They advised that the solid rockets only be lit when the O-ring temperature was at or above that experienced after the coldest weather recorded to date: 53 degrees Fahrenheit. Until then, Lund said, the Morton Thiokol engineering department could not recommend that NASA proceed with firing the boosters. Larry Mulloy then asked Joe Kilminster for his position. He was firm: it was a "no-go" for launch.

For a moment, Roger Boisjoly felt his spirits lift; they had done it. At the last possible minute—and despite their limited evidence—he and his colleagues had ensured that the countdown for mission 51-L would be stopped. Elated, Boisjoly was certain that, after his months of frustration and anxiety, NASA's ponderous bureaucracy had finally worked as it should: the seven astronauts scheduled to fly aboard *Challenger* in just a few hours would now have to stand down, and launch in safety another day.

High above the steel latticework of the launch gantry on Cape Canaveral's Pad 39B, a gibbous moon had risen in the darkened sky. *Challenger* shone once more in the blue-white glow of the xenon searchlights. By 8:00 p.m., the pad technicians had completed the freeze protection preparations, and opened faucets up and down the Fixed Service Structure. But now the bitter wind sliced in over the Atlantic from the northeast, splashing water from the overflowing eyewash basins at the 235-foot level, the drains of the trickling showers, and the spigots of the rainbird system. As the temperature dropped, the spume and the puddles froze into a growing slick of ice along the catwalk between the orbiter and the crew's emergency escape baskets; more water dripped through the gratings onto the platforms and superstructure below, running down the girders and over communication equipment. Down on the Mobile Launcher Platform, the technicians had added more than 1,400 gallons of commercial antifreeze to the sound suppression troughs directly beneath the nozzles of the solid rockets, which would insulate the orbiter from acoustic shock at liftoff—turning the water green, and preventing it from icing over. But still the wind blew, the mercury plummeted, and the ineluctable cold crept down the full height of the gantry and across the steel deck of the platform. Gradually, even the emerald-colored liquid filling the giant troughs began to congeal and solidify.

"OK—is that it, Thiokol?" Mulloy asked.

"Yes," Bob Lund said.

"Well, I have a dissenting opinion. And I'd like to tell you why."

With that, the NASA manager now brought down the full force of his personality—and his technical expertise—on the engineers' opposition to launch. One by one, he pointed out the inconsistencies in their reasoning: that, of the two shuttle flights with the worst leaks to date, one had been launched when the seals were at room temperature; that four test motors had been fired at less than 50 degrees Fahrenheit with no ill effects; that the manufacturers of the O-rings themselves said that the seals would be within the acceptable range of operation all the way down to minus 20 degrees Fahrenheit; and, even on the rockets flown in January the previous year, six of the eight joints functioned perfectly well, when they had all been at the same temperature as the two that failed. Overall, Mulloy said, Boisjoly and the others had not convinced him that air temperature alone was a significant factor in causing blow-by. The data was simply inconclusive.

Although the Thiokol engineers did not fully realize it at the time, Mulloy's rebuttal marked a subtle shift in the tone and expectations in the meeting—a change that made it different from all previous Flight Readiness Reviews. In the past, if a contractor's data about the state of flight hardware had been inconclusive, the default position was not to fly: they were expected to prove that their equipment and components constituted an acceptable risk before launch. Now, it seemed, Mulloy was asking them to prove the opposite—to show him the data that proved conclusively it was *not* safe to launch.

Finally, Mulloy turned to the importance of the flight manifest. He told the team in Utah that their last-minute change of heart amounted to the creation of a new set of launch commit criteria—that the shuttle could only fly at temperatures above 53 degrees Fahrenheit. If accepted, such limitations would wreck the schedule of a spacecraft that rested on its ability to lift off from locations on both US coasts all year round. Sitting across the table from him in Trailer Complex C, Allan McDonald could see the fire burning in Mulloy's eyes; it was clear he had no intention of accepting the Utah engineers' no-go recommendation. "My God, Thiokol," Mulloy said, his voice rising. "When do you expect me to launch—next April?"

Now Mulloy asked for the views of the senior manager at Marshall. "What do you think about it, George?"

From the conference room in Alabama, George Hardy's voice came over the four-line connection.

"Well, I'm appalled that Thiokol would come up with this particular recommendation at this particular time," he said. Hardy didn't feel there was any correlation between temperature and the leaks; the Thiokol engineers were simply mistaken. Nonetheless, he recognized that NASA could not go on without them. "If they don't want to launch," he said, "we're not going to launch."

But the Thiokol team was stunned. They had never heard Hardy—a measured and judicious engineer who had worked on the program for twenty-six years—speak with such vehemence. For a few seconds, there was dead silence in the MIC room. Recognizing the pressure he and the Thiokol team were now under to change course, Roger Boisjoly felt his stomach churn. But to others listening in, the impasse was intractable: it seemed obvious NASA would be forced to scrub the launch.

Then Jerry Mason turned to Kilminster.

"Joe," he said softly, "ask them for a five-minute caucus."

"NASA, we need to go off-line and . . . talk this over some more in the light of what we've just heard," Kilminster said, and keyed the mute button on the phone. It was 10:30 p.m. in Florida; 8:30 p.m. in Utah.

As a humming quiet descended over the transcontinental telephone lines, in the trailer at Cape Canaveral Al McDonald turned to his four colleagues from NASA to continue the debate. The Thiokol engineer said that the agency's own flight qualification rules for the solid rockets would forbid launching in freezing weather; but Larry Mulloy disagreed.

McDonald suggested that they could just delay the launch until Tuesday afternoon, when temperatures might at least approach more acceptable limits. "We've considered this and rejected it," Mulloy said.

The five-minute caucus Kilminster had asked for now stretched into twenty minutes. Cecil Houston went down the hall for more coffee, and McDonald had time to think once again about the consequences of halting the launch. He knew that his bosses at Thiokol would be in a bind if they hindered NASA's flight schedule at such a vulnerable time for the company. Already behind schedule in production, they further risked jeopardizing their monopoly on their billion-dollar solid rocket contract; the critical summit meeting with the agency about the future of the deal was just hours away. But McDonald wasn't worried: he believed that the recommendations in Utah were coming directly from Bob Lund and the engineering department, and that the Thiokol managers trusted their expertise. If they were taking their time to return to the conference, it was probably because Roger Boisjoly was spending it on more calculations, or marshaling further evidence, to support their argument. Down in Alabama, some of the senior Marshall engineers were already discussing who they planned to notify to call off the launch.

Almost as soon as Joe Kilminster's finger had left the mute button, company Vice President Jerry Mason addressed the other senior executives clustered around him at the end of the table.

"We've got to make a management decision," he said in a low voice.

Nonetheless, for the next few minutes the four men listened as their engineers talked through the evidence once again. Boisjoly and Arnie Thompson reiterated their conviction that the cold posed a real threat to flight safety: it was a not a risk worth taking; the launch must be halted. But now Jerry Mason adopted the opposing view: yes, they might see erosion in the O-rings, but their tests showed that they could sustain extensive damage without a failure. And besides, if the primary seal failed, they could always rely on the secondary. The ten members of the engineering team in the room clearly did not agree; Boisjoly admitted that a lot of what they were saying could not be quantified, but they knew one thing for sure: the January 1985 flight had been bad, and this one would be worse.

Sensing that Mason was on the cusp of something unthinkable, Thompson went to the end of the table with his notepad and began sketching a diagram of the solid rocket joint to make clear the peril they were facing. But when he looked up from the drawing, the four managers were doing their best to ignore what he was saying. Mason simply glared at him until he returned to his seat. Then Boisjoly stepped in: snatching up one of the eight-by-ten color photographs taken inside Hangar AF a year before, he strode up the room and slapped it down on the table. Growing desperate, Boisjoly now lost all composure. His voice rose to a shout as he tried to make the executives see what he had seen that day in January 1985: the coal-black soot *here*; the terrifying evidence of heat on the secondary seal *here*; the arrows he had drawn, showing the path of superheated gas *here* and *here*. But it was no use. Spent and defeated, fearing he was moments away from losing his job, he sat down.

The caucus had extended to almost half an hour. There was no new data to consider. Regardless, it was clear Mason had made up his mind.

"It's time for a management decision," he said again. "Based on all the evidence that's presented to me, I think that we should go ahead and launch."

One at a time, Mason turned to his three fellow Vice Presidents for their decisions: first, to his reliable sidekick Cal Wiggins.

"Yes," Wiggins said, "I think we should launch."

Then to Joe Kilminster, who also agreed to reverse his position. Finally, he turned to face Bob Lund, the most senior engineer at the company. With his fellow executives all in agreement, Thiokol's position on the *Challenger* launch now came down to his vote alone.

A tormented look crossed Lund's face. For several seconds, he simply sat shaking his head. He looked imploringly down the table at Boisjoly and the other engineers. No one said a word.

Eventually, Mason broke the silence.

"Now, Bob, you take off your engineering hat and put on your management hat. We've done all we can from an engineering point, and now we've got to make a tough decision. And as a manager, you've got to do that."

The agony of the choice weighed heavily on the Vice President of Engineering, Wasatch Operations Division; no one had to tell him what would happen if the O-rings really did fail at launch; he knew that his own reputation, that of the company for which he'd worked for so many years, the lives of seven astronauts—perhaps even the fate of the space program itself—might rest on what he said in the next few moments.

Lund began to rise in his seat; he drew a palm across his face.

"Wishy-washy Lund," he said.

Then he raised both hands in submission.

"OK, I agree. Let's fire."

At around 11:00 p.m., Allan McDonald heard a click as Joe Kilminster came back on the line.

"We have reconsidered based on our former discussion with you," he said. Although Morton Thiokol remained concerned about low temperatures, Kilminster now agreed that the data his engineers had presented was inconclusive. "Therefore, MTI recommends launching."

McDonald was bewildered and dismayed by the sudden volte-face: he could only imagine that Bob Lund and his engineers had reviewed the data and reversed their recommendation—and Kilminster, so apparently resolute in his opinion before the caucus began, had followed their lead. Yet this time, there were no objections or arguments from NASA, either in Marshall or at the Cape.

"OK, Thiokol. We concur," George Hardy said. "We would like to have that in writing, and signed."

While Hardy spoke, Larry Mulloy stared hard across the table at McDonald; as Thiokol's senior manager at the Kennedy Space Center, it was his responsibility to give the company's formal "go" for launch. But before, the approval had always been in an oral poll. Never in the history of the program had a contractor been asked to provide written confirmation of the decision—and now, suspecting the agency managers were preparing to cover themselves in case something went wrong, McDonald refused to cooperate. "I won't sign that recommendation, Larry," he said. "It will have to come from the plant in Utah."

Back at the Promontory plant, Joe Kilminster and Bob Lund huddled together to draft the document, while the team of Thiokol engineers filed silently out of the conference room. Roger Boisjoly was somber. As they made their way downstairs, a colleague assured him that he had done everything he could: "You can't feel bad," he said. "We did, as engineering, what we were supposed to do . . . if there's a concern we bring that concern to management." Beyond that, it was out of their hands.

Down at the Cape, Al McDonald was making a final attempt to persuade Mulloy to change course, itemizing three separate reasons to cancel the launch: not just the O-ring problem; but the Atlantic storms that had overtaken the recovery ships, jeopardizing the retrieval of the boosters; and the possibility that ice formed on the launch gantry would damage the orbiter as it took off. But the NASA managers brushed him off. They told him only that they would pass along his concerns to those whose job it was to worry about them.

Still, McDonald insisted that he believed that by going ahead they were contravening their own flight qualification rules: "If anything happens," he said, "I wouldn't want to be the person that has to stand in front of a board of inquiry to explain why we launched . . ."

Mulloy said nothing.

It was approaching midnight when Joe Kilminster's signed recommendation for launch came over the fax machine in Trailer Complex C. McDonald walked back down the hall to deliver a copy to Mulloy, now in his final teleconference of the evening. As McDonald approached Cecil Houston's office, he could hear the two NASA managers—Mulloy and Stan Reinartz—on the phone with Arnie Aldrich, one of the Johnson Space Center chiefs whose final approval would be needed in order to launch in just a few hours. Aldrich said that the sea conditions out in the Atlantic might be enough to postpone the scheduled takeoff: he asked Mulloy what would happen if the two recovery ships couldn't make it back to the splashdown area in time to retrieve the boosters.

The Marshall manager told Aldrich not to worry: they might lose the parachutes and some supporting hardware, but the rockets themselves would float, and both were fitted with homing beacons; they could always locate and recover them later. The three men briefly discussed the danger of ice on the pad, and they agreed that the problem was under control. But neither Reinartz nor Mulloy told Aldrich of Thiokol's concern about the O-rings—and Cecil Houston sensed that they didn't want him to, either. So he stood and listened in silence. The call concluded with an agreement to resume the countdown: *Challenger* would launch at 9:38 a.m. the next day.

When they hung up, Al McDonald handed over the single sheet of paper testifying to Morton Thiokol's approval for the mission and waited while the men from Marshall read it. Then Stan Reinartz turned to Mulloy: perhaps, he thought, now was the time to notify Dr. Lucas about everything that had happened that night.

"Should we wake the old man," he asked, "and tell him about this?"

They agreed it was best to let him sleep.

CHAPTER
TWENTY-THREE

TUESDAY, JANUARY 28, 1986

2:00 a.m.

High up on the 235-foot level at the top of the launch gantry, it was dark, and cold; the wind sang in the girders, and Charlie Stevenson did not like what he saw. The leader of the Kennedy Space Center Ice Team, it was Stevenson's job to check the shuttle for ice that might prove dangerous during takeoff. His inspection was usually confined to the external tank, which, although insulated with thick blocks of polyurethane foam, became so cold when filled with liquid oxygen and hydrogen that moisture in the humid Florida air often formed a thick rime across its surface.

But when he had arrived at work in the Firing Room soon after midnight, Stevenson discovered a far more threatening problem developing out on Pad 39B. The cameras monitoring the gantry and the mobile platform supporting the shuttle revealed that the plan to protect the plumbing on the launchpad from the freezing weather had backfired. Although the pipes hadn't burst in the cold, the water trickling through the sprinkler system and the emergency showers and eyewash baths, spilling across the catwalks and cascading down the steel supports of the Fixed Service Structure, had left the pad encrusted in ice. What he found when he and his team arrived to inspect the scene in person was unlike anything he had ever seen before: sheets of ice an inch and a half thick glistened underfoot; the route Dick Scobee and his crew would take along the swing arm in an emergency—from the orbiter hatch to the slidewire escape baskets—was blocked by a dense slick of frozen water. Icicles a foot and a half long, like the pipes of a ghostly organ, dangled from handrails and walkways; they hung from the conduits, the cable trays, and the gratings.

Stevenson feared that, when shaken loose by the violent concussions of *Challenger*'s launch, the falling icicles could become devastating missiles if they struck the shuttle's delicate thermal protection tiles—causing damage that would make the spacecraft vulnerable to disaster on reentry. On the radio connecting him to Firing Room 2, he reported his concerns about the planned launch: the ice was dangerous, and growing worse; but if they stopped the water running, the pipes would freeze.

"Then what choices we got?" the Director of Engineering asked.

"Well, I'd say the only choice you got today," Stevenson said, "is not to go."

Back in the Launch Control Center, Launch Director Gene Thomas didn't want to hear it: "Boy, he's really stretching it," he said.

Thomas had already been delayed another hour by a computer glitch in the pad fire extinguishing system, moving the launch time back to 10:38 a.m.; if they were to have any chance of getting off the ground later that day, they would have to start filling the external tank with fuel—"tanking"—immediately. He told Stevenson to come back to the Firing Room to discuss their options, while he called in advice from the shuttle design team at Rockwell in California; they'd figure out some way of handling the ice.

The Launch Director gave the order for tanking to begin; the countdown continued.

※※※※※※※※※※※※※※※※※
※※※※※※※※※※※※※※※※※

At 5:00 a.m., Ellison Onizuka's older brother, Claude, awoke in the Quality Inn on International Drive in Orlando, and began preparing the large group of Ellison's friends and relatives for the hour-long drive over to the Cape. Claude had flown in from Kona the previous week to help shepherd more than sixty people—his mother, as well as aunts, uncles, cousins, nephews, and his brother's former scoutmaster Norman Sakata—from Hawaii to the Space Center to watch his brother leave for orbit. Now, for the second day in a row, he had to wrangle six vans full of guests out of bed before dawn for the long trip to the launch site.

At the Ocean Landings condo in Cocoa Beach, it was so early that Alison Smith pretended to be asleep when the phone rang, and her mother answered. It was Mike, calling from the crew dormitory: "We're not going to be able to launch today," he said, and asked what they were saying on TV about the weather. Although Jane and the other families would be picked up by NASA as planned and taken over to the Launch Control Center in just a few hours, he was almost certain they would simply be there to witness another postponement; it was just too cold. Mike told his wife he'd already called the Launch Director's secretary and arranged to change the children's plane tickets and get them to Houston that afternoon. They would return for another attempt later in the week. "We're going to come back down here on Thursday," he assured her.

It was still dark as Marshall Director Bill Lucas and his immediate number two drove through the security checkpoint at the Kennedy Space Center. The bright moon hung low in the sky, and the cold sent the men hurrying from the car, through the sliding doors, and into the marble-floored lobby of the Launch Control Center. Inside, they took their seats among the uppermost tiers of consoles in the double-height space of Firing Room 3, reserved for senior mandarins from NASA.

Confined within an air-conditioned glass box, from here Lucas could look down over the rows of engineers manning consoles beneath him—and, when the moment came to launch, swivel his chair around to take in the view of the pad through the laminated two-story windows behind him, louvered with steel panels to shield them from the sun. Nearby, a color TV raised on an alloy stand also displayed a continuous feed of pictures from NASA Select TV. As he sat down, Lucas was greeted by his managers Larry Mulloy and Stan Reinartz, who gave him a succinct summary of the previous night's teleconference: the rocket engineers at Thiokol had expressed some concerns about the weather, but had discussed their reservations at length. Eventually, they all agreed there would be no problem.

Mulloy handed Lucas a copy of the late-night fax from the Wasatch plant summarizing the company's official position—a single page of just sixteen lines, a dozen bullet points typed above the corporate logo of Morton Thiokol Inc. At the bottom was a simple, unequivocal statement of the kind the director of the Marshall Space Flight Center preferred: MTI RECOMMENDS STS-51-L LAUNCH PROCEED ON 28 JANUARY 1986.

Dr. Lucas, adhering to a strict interpretation of NASA's pyramidal reporting structure, saw no need to share word of Thiokol's initial qualms with his superiors. He had in his hand a signed launch recommendation; the matter was closed.

Over on the third floor of the Operations and Checkout Building, the seven *Challenger* astronauts gathered once again for a steak-and-eggs breakfast in the dining room of the crew quarters. In matching white polo shirts, for the second time they posed for prelaunch pictures around a table decorated with a centerpiece of red and white roses and a pair of small American flags, and a fresh cake frosted with the mission patch. Then, in a conference room down the hall, the crew received a final meteorology briefing from Houston: the cold weather was pushing the permissible limits for shuttle flights, so they could expect to wait for a while on the pad, but—for now, at least—the launch was still on. With a new time officially set for 10:38 a.m., they would likely lift off around noon. Back in their rooms, they changed into their flight overalls, packed up anything not destined for orbit into attaché cases and suit bags, and left it behind for NASA staff to return to Houston later in the day.

A little before 8:00 a.m., the astronauts rode the elevator downstairs, and walked out toward the waiting Astrovan. Led by Dick Scobee, they looked cheerful and excited, their faces washed by the orange light of the low morning sun. The last member of the crew in line was Greg Jarvis, and as he came down the ramp a small gang of his friends from Hughes cheered and clapped, waving a homemade banner: HAVE

A GREAT FLIGHT GREGO, it said in big black-and-red capitals, beside a cartoon shuttle blasting skyward on a plume consuming the numbers of all mission assignments he'd been given so far: 51-D–51-I–61-C–51-L. Behind Jarvis came John Young and, beside him, wearing a suit and tie, George Abbey. As he stepped outside into the glare of the camera lights, the Director of Flight Operations shot an apprehensive glance at the sky.

Out on the pad, temperatures had reached their lowest shortly after dawn, falling to 24 degrees Fahrenheit—eight degrees below freezing. The seething cold had crept into everything: the doors on the gantry elevator had become sluggish; several of the remote cameras had stopped working; an oxygen sensor aboard the shuttle had failed; one radio channel had been knocked out; and, down on the Mobile Launcher Platform, beneath the nozzles of the solid rockets—and in defiance of the hundreds of gallons of antifreeze—water in the sound suppression troughs had frozen solid. Sheer cascades of ice now descended from near the very top of the launch tower, bearding the walkways just sixty feet from *Challenger*'s left wingtip and encapsulating communications equipment. "Looks like, uh, something out of *Dr. Zhivago*," said the Rockwell liaison, surveying the closed-circuit TV images from the Launch Control Center; he had never seen anything like it. "There's sheets of icicles hanging everywhere," he said. Drowsy from lack of sleep, he sat back at his console and waited for the launch to be scrubbed.

In the meantime, Charlie Stevenson and his team had returned to the pad and begun smashing up the frozen water in the sound suppression troughs with long-handled nets and sweeping ice into the flame trench. As they took temperature readings around the shuttle with a handheld infrared thermometer, the instrument produced odd results: the surface of the left-hand solid rocket booster registered at around 25 degrees Fahrenheit, close to the air temperature at sunrise. But at the bottom of the right-hand booster, near the aft field joint, the reading was just 8 degrees—an astonishing 24 degrees below freezing; that just couldn't be right. Stevenson and his engineers assumed the thermometer was malfunctioning, and took note of the numbers, but kept the data to themselves. Inside the Launch Control Center, the ice remained the major concern: Rockwell technicians at the Cape and in California continued to discuss the threat it posed to the shuttle over the internal communications net. Ahead of a final NASA mission management meeting, they had to use their own computer models to decide whether to give their go-ahead for launch—but it seemed impossible to predict what would happen if thousands of fragments of ice were sent ricocheting from the gantry when the shuttle engines lit. "It's still a bit of Russian roulette," the chief engineer at Downey said. "You'll probably make it. Five out of six times you do, playing Russian roulette."

::::::::::::::::

Led by a white NASA security car topped with blue warning lights, it took the Astrovan a little more than twenty minutes to cover the nine miles to Pad 39B, stopping twice along the way: first to drop off John Young, who would make a final reconnaissance flight in the Shuttle Training Aircraft to check the weather over the Cape, and then George Abbey, who left at the Launch Control Center to join the other managers in Firing Room 3. The van sped northeast, straight up the deserted causeway toward the ocean, and made a sharp turn two miles from the pad, flashing through the deep shadows cast by the subtropical scrub. Then the driver turned into a long right-hand bend, and the seven astronauts saw their spacecraft revealed slowly before them, framed in the front windshield against a turquoise sky. Raised on a massive concrete ramp and anchored to the great, gray superstructure of the launch platform, *Challenger* lay dead ahead at the end of the road: fully fueled and ready to depart.

As the high-speed elevator carried them almost twenty stories up the launch tower, Commander Scobee and his crew could hear that their ship had come to life: the external tank heaved and groaned as its thin aluminum skin contracted in the cold, exhaling a stream of boiling liquid oxygen from beneath the conical "beanie cap" resting at its tip. When the doors slid open at their destination on Level 195, Scobee took a deep breath and smiled up at the sky; behind him, the Atlantic rollers glinted silently in the sunshine.

"This is a beautiful day to fly," he said.

"It's a little cold, though, Dick," said Johnny Corlew, once again leading the closeout crew for the day.

"Nah, that's good, that's great," Scobee said, shaking his head.

Even so, as they crossed the swing arm toward the White Room, the closeout technicians warned the crew about the treacherous slick of ice on the walkway. Judy Resnik shivered in her thin flight suit and, as he waited his turn to enter the orbiter, Mike Smith pulled out a folded white handkerchief from his top pocket to wipe his nose. But the mood was buoyant: as he entered the White Room to pull on his harness and helmet, Scobee beckoned to the chief Lockheed technician. "Here," he said, "you guys might need this today. Hang on to it."

It was a small alloy bolt, tied with a red ribbon.

The commander knelt to crawl into the orbiter, followed by Smith and Ellison Onizuka, who was teased by the technicians about his flight jacket, decorated with the patch of a previous shuttle mission. "Well," the astronaut said, "at least we're not on the Dan Rather show, like *somebody* I could name."

Resnik would take the flight engineer's position in the cockpit, behind the commander's and pilot's seats, and was the last of the upper-deck crew through the hatch. Before she crouched to cross the threshold, she turned to Christa McAuliffe. "The next time I see you, we'll be in space," she said.

Johnny Corlew had grown up in Indiana, and as a boy had picked apples for his teachers from the tree in his yard; this morning he had brought his own gift to the pad for McAuliffe: a Red Rome apple he'd had his wife pick up at the supermarket for the occasion. The Teacher in Space raised the fruit to her face with a smile, but then immediately returned it. "Save it for me," she said. "And I'll eat it when I get back."

Greg Jarvis, grinning and chatting before the technicians helped him into the tight-fitting clamshell helmet, was next.

"Well, we're really going to go today," Corlew said.

"Yeah," Jarvis replied. "I sure hope so."

Finally, Ron McNair shrugged into his equipment. He shook hands with the White Room team, stooped onto all fours, and crawled over the step to take his seat on the middeck.

Inside, the astronaut attached to the closeout team, Sonny Carter, moved from one member of the crew to another, tightening their harnesses, completing headset communication checks, and adjusting cables and hoses. As he hunched over McAuliffe to inspect her helmet one last time, he looked down into her face and saw that her Girl Scout pluck had deserted her at last. In her eyes he saw neither excitement nor anticipation, but recognized only one emotion: terror.

A few moments later, Carter crawled from the orbiter and Johnny Corlew swung the hatch shut behind him. It closed with a bang, and the latches fell into place.

Back in the Launch Control Center, the ground controllers broke into a round of applause.

It was 9:07 a.m.

::::::::::::::::::

Dawn was approaching in Southern California when Rocco Petrone, President of Rockwell's Space Transportation Division and a legendary former manager at NASA, picked up the phone to call his subordinates in the Firing Room on Merritt Island. A burly Italian American who had masterminded the construction of the facilities at Cape Canaveral and then overseen the Apollo program through the launchpad fire, the moon landing, and Apollo 13, Petrone ran the West Coast contractor's shuttle operations with an iron hand—and had left the Cape in disgust the previous afternoon, after witnessing the bolt fiasco unfold on Pad 39B. Now, watching the closed-circuit TV images of the icicles festooning the gantry and equipment around *Challenger* from the plant in Downey, he told his managers to make it clear to NASA that Rockwell could not approve the launch. "It is not safe," he said.

But the final decision now lay in the hands of the two most senior members of the Mission Management Team seated on the top tier of Firing Room 3: the chief of the shuttle program in Houston, Arnie

Aldrich, and his boss, Associate Administrator for Space Flight Jesse Moore. Aldrich, forty-nine years old, was another NASA veteran, who had started with the agency as a flight controller for the Mercury program at the end of the 1950s, and had worked on the shuttle since the beginning; he was confident he knew as much as almost anyone at the agency about the foibles of America's Space Transportation System.

Since taking the conference call in his hotel room at nearly midnight to talk to Larry Mulloy and Stan Reinartz about the recovery ships, Aldrich had managed to get little sleep: there had been another call at 3:00 a.m. to say that the count had slipped back by an hour, and that ice was building on the pad; at 4:30 a.m. he had arrived in the Firing Room to learn that the offshore winds had dropped, and the recovery ships would be in position to recover the spent boosters after all. But the ice on the pad could be a problem. Aldrich set the shuttle engineering teams at the Cape, and in Houston, Huntsville, and Downey, to analyze the risks and report back.

Soon after 9:00 a.m., the Launch Director put the countdown on hold, pushing liftoff back to no earlier than 11:08 a.m. In the meantime, Aldrich convened a meeting in the fourth-floor conference room of the Launch Control Center—attended by more than twenty senior managers and experts who had assessed the hazards of the freezing gantry. First, Aldrich heard the analysis from a team of NASA engineers: based on their calculations of wind speed, fragmentation, and debris trajectory, they felt good about the ice; it was unlikely to cause significant damage. They gave their go-ahead for launch, as soon as the air temperature rose above 31 degrees Fahrenheit. But when he turned to Rocco Petrone's men, they passed on the message from their boss: the situation was unpredictable, and unlike anything they'd seen before, said one. The other was more direct. "Rockwell cannot assure that it is safe to fly," he said.

And yet Aldrich had made up his mind: his own engineers were unanimous; a single dissenting voice—even from the contractors who had built the orbiter—was not enough to stop the launch. At around

10:00 a.m., he returned to the Firing Room, where he was immediately intercepted by George Abbey.

"What did you decide?" the Director of Flight Operations asked.

"We're still a go," Aldrich replied. He reminded Abbey that they were all on a tight schedule to get *Challenger* off the ground—and keep the rest of the year's missions on track.

"Did Rockwell say they were a go?" Abbey asked.

But the countdown had already resumed.

::::::::::::::::::

Standing on the aluminum bleachers more than three miles from the launchpad, the crowd shivered in the cold, wrapped in winter coats, hats, and sweatshirts, and sipped hot coffee or cocoa to ward off the chill. Christa McAuliffe's parents—Grace and Ed Corrigan—her sister, Lisa, and her brother, Christopher, were seated in the stands with other VIP guests, wearing buttons depicting Christa in her astronaut flight suit. Hundreds of other spectators had already left, chased away the previous day by the cold and the repeated delays: Ron McNair's older brother and his pregnant wife had already made the long drive back home to Atlanta, along with his youngest brother, Eric, and their father. In the nearby Space Center campground, so often filled with whooping onlookers watching the launchpad from the roofs of their vans and Winnebagos, only a few stragglers remained. The crowd on the VIP bleachers was so thin that, earlier that morning, NASA had brought all of Scott McAuliffe's third-grade classmates over by bus from the Banana River causeway to gather in their red, white, and blue baseball caps under a banner reading GO CHRISTA! At the request of photographers inside the media enclosure, the Corrigans had moved closer to the press—to bring them within range of their cameras at liftoff. "Christa would want to see what our faces look like," her mother said.

Up on the fourth floor of the Launch Control Center, Steven McAuliffe and his children had gathered once more with the other astronauts' families beneath the computer-printed banner in Gene Thomas's office.

There were more doughnuts from the nearby cafeteria, binoculars gathered on a table, and a videotape playing of the prelaunch coverage from earlier in the morning, including the crew breakfast and walkout. A separate TV set also continued to display live pictures from the pad, and the sheets of ice hanging from the gantry. Joy and Reggie McNair, once again in their miniature NASA jumpsuits, sat together on the floor: Joy played with Caroline McAuliffe, while Reggie pretended to fly a plastic Space Shuttle around the room. Alison Smith and her brother, Scott, took more naps on the chairs, and on the floor.

As the countdown entered yet another hold—while the NASA managers waited for the temperature to rise and the sun to melt some of the ice on the gantry—the adults sat quietly or stood around chatting. Jane Smith, still expecting the launch to be scrubbed at any minute, used the phone on a secretary's desk to rebook her children's plane tickets back to Houston. Lorna Onizuka, wearing a gray Mickey Mouse sweatshirt, drank coffee and cracked jokes; but her one-liners couldn't entirely smother an undercurrent of anxiety.

Back in New Hampshire, seniors at McAuliffe's high school in Concord had again packed the main auditorium, holding banners, balloons, and noisemakers, joining thousands of other students in schools across the country to witness the launch live on the feed from NASA TV. "All of America is watching and waiting," CNN's space correspondent Tom Mintier reported in a morning news update.

In Washington, DC, President Reagan was at the outset of a busy day in the White House: at 1:00 p.m. he was scheduled to meet a dozen national TV correspondents to give them a preview of the televised State of the Union address he would be delivering at nine that night. The speech had been the subject of intense last-minute wrangling among White House staff, tussling over how much tangible policy it should include, but NASA communications chiefs had suggested that Reagan mention the space program; the Teacher Flight seemed a perfect crowd-pleasing cue for the occasion. While the President met members of Congress in the Cabinet Room, the First Lady was upstairs in the

Executive Residence, planning to watch the *Challenger* launch on television as it happened.

:::::::::::::::::

At 10:30 a.m., Launch Director Gene Thomas watched on the remote cameras as Charlie Stevenson and his team drove back out to Pad 39B in their white government van. The countdown clocks inside the Launch Control Center were holding, as scheduled, at T-minus 20 minutes. High up in the cockpit of the orbiter, the *Challenger* crew bantered with one another over the intercom headsets, unheard by the public. They were still anticipating a scrub: "I hope we don't drive this down to the bitter end again today," Judy Resnik said.

Harnessed on their backs in the unyielding aluminum seats, the crew had little to do but wait for news from launch control: on the flight deck, Resnik, Onizuka, Smith, and Scobee were washed in sunlight and had a view to their left of the launch gantry and, above them, a cloudless cobalt sky. They attended to a handful of instrument checks and pressure readings and joked back and forth—about the cold, their breakfast, and the discomfort of yet more hours of supine inertia. But below on the middeck, Ron McNair, Greg Jarvis, and Christa McAuliffe saw only what sunshine struggled in through the porthole in the entry hatch and down the narrow flight deck gangway. Confined by their helmets and communications lines, the three astronauts had no tasks to perform during ascent, and would be scarcely more than cargo until they reached orbit. They would have almost nothing to look at beyond the wall of battered equipment lockers directly in front of them, and no source of information about their flight except their headset audio and the shuddering din from the solid rockets' burn, as six million pounds of thrust rattled through every nut, bolt, and fixture inside the spacecraft. As the hours of waiting had wound on, Jarvis interjected in the cross talk from the cockpit. But both McNair and McAuliffe sat silently beneath the wan fluorescent light, alone with their thoughts.

One hundred feet below them on the deck of the launch platform, Stevenson and his men were using their nets to fish more ice from the green waters of the sound suppression troughs, and still sweeping frozen debris away from the shuttle. The Ice Team leader noted that on the sunny side of the gantry melting icicles were coming loose and tumbling to the steel deck below. But in the shadows, right where the wind came in off the ocean and supercooled air drifted up from the base of the external tank, Stevenson could see that the lower part of *Challenger's* right-hand booster rocket was still freezing, glazed with a coating of ice an eighth of an inch thick that extended for thirty feet toward the strut holding it to the big orange fuel tank. Still, the air temperature had now risen above 34 degrees Fahrenheit: just within the formal limits NASA set for launch.

Inside Firing Room 3, the Director of Engineering was already making his final round of checks with the controllers at their rows of consoles.

"Any problems?"

"No problems."

"We're in good shape."

"Y'all are go."

"All our systems are go."

But he was still awaiting word from the Ice Team; at last, Charlie Stevenson came in over Channel 245.

"The vehicle looks good," he said.

In the cockpit, Dick Scobee heard the voice of launch control in his ear: the countdown was about to resume.

"Al-right!" the commander replied. "That's great."

In the windowless vault of Building 30 in Houston, Flight Director Jay Greene was making everything ready to assume control of *Challenger* on its path to orbit. Seated in the center of the hushed Flight Control Room, Greene was flanked by the eleven other managers and technicians who would monitor the launch, including the Flight Dynamics Officer—the FIDO—responsible for the spacecraft's trajectory, and the day's CapCom, astronaut and former Air Force pilot Dick Covey.

Responsibility for *Challenger* would pass to them as soon as the shuttle cleared the tower at the Cape. Now Greene polled each member of his team in the ritual chant of affirmation that proceeded a confirmed launch:

"FIDO?"

"Go!"

"GNC?"

"Go!"

"INCO?"

"Go!"

"Surgeon?"

"Go!"

"CapCom?"

"Go!"

In the top tier of seats in the Firing Room, Jesse Moore conferred with Arnie Aldrich and Gene Thomas: the final decision about whether to launch would be his. The three men talked quietly for a few moments, and Moore looked over some paperwork. Then he nodded.

Terminal count.

Nine minutes.

It was 11:29 a.m.

On the fourth floor of the Launch Control Center, a NASA official began leading the wives and children out of Thomas's office, down the hallway, past rows of cubicles, and out through a heavy self-closing door. One by one, they stepped out on to the roof and mounted the big steel staircase to the very top of the building: a broad expanse of pale concrete the size of a football field, surrounded by a white railing that cast a crisp shadow in the winter sunshine. Behind them, the black shapes of turkey vultures wheeled and soared in the thermals rising up the fifty-story cliff of the Vehicle Assembly Building. To the northeast, they had a perfect view: straight down the long path of the crawlerway toward the ocean and, to the left, of Pad 39B. They gathered at the rail; posed, smiling, for pictures.

From speakers on the roof, on the grandstands below, and all across Merritt Island, the voice of Cape Canaveral public affairs officer Hugh Harris picked up the count:

"The Ground Launch Sequencer has been initiated. T-minus 8 minutes, 30 seconds and counting. The flight instrument recorders are turned on."

At the Morton Thiokol plant in Utah, the time was approaching 9:30 a.m. Some of the solid rocket team had gathered in the MIC room at the Wasatch plant to watch the launch, but neither Arnie Thompson nor Roger Boisjoly were among them. Instead, Thompson was in his office one floor below, going over the details of the previous night's discussions with his colleagues. Boisjoly remained alone at his desk nearby.

But with only a few minutes of the countdown remaining, Boisjoly had left his office and was walking past the door of the conference room, when Bob Ebeling emerged to grab him by the arm, urging him to join the rest of the staff—and Ebeling's two daughters—to watch the live feed from the Cape.

Boisjoly said no. "I don't want to watch," he said.

But Ebeling insisted.

Inside the MIC room, there were no seats left, so Boisjoly sat on the floor directly in front of the screen of the big projection TV, his back resting against Ebeling's legs.

"T-minus 7 minutes, 30 seconds."

Running thousands of diagnostic tests and checks each second, the Ground Launch Sequencer began the process of severing *Challenger*'s last connections to Earth. At the computers' command, the crew access arm slowly retracted, swinging the White Room away from the orbiter hatchway with a robotic lurch. The solid rocket boosters were armed.

Four minutes.

Christa McAuliffe snapped down the visor of her helmet. She was breathing pure oxygen.

It was 11:35 a.m.

In the CNN studios in Atlanta, the producers switched over to broadcast live pictures from the Cape.

"T-minus two minutes and counting."

The vehicle began running on internal power. The vent hood lifted from the top of the external tank.

On the flight deck, the banter continued.

"OK, there goes the LOX arm," Smith said.

"Doesn't it go the other way?" Onizuka said, and laughed.

"God, I hope not, Ellison."

Ninety seconds.

In Firing Room 3, the console operators hunched over their screens. From his seat, Thiokol's Allan McDonald kept one eye on the monitor displaying the chamber pressures of the solid rockets, and the other on a TV screen showing the shuttle. Behind him, the senior NASA managers—Moore, Aldrich, Lucas, Abbey—turned to gaze out through the massive wall of glass toward the pad. Larry Mulloy was standing, his headset on.

In the cockpit, Scobee and Smith watched as the automatic sequencer worked through the final moments before launch: the propellant systems came up to pressure; all three engines were ready to fire.

"Thirty seconds down there," Scobee said.

"We are go for auto-sequence start."

The crew heard the distant whirring as the onboard computers verified the responses of the shuttle hydraulics.

"Fifteen," the commander said. Amber numbers blinked the final seconds of the countdown on the instrument panel.

Over the Kennedy Space Center loudspeaker system, the voice of Hugh Harris echoed the incantation for everyone to hear. The children in the grandstands joined in.

"T-minus ten."

"Nine."

"Eight."

"Seven."

"Six."

Inside the shuttle, the whine of the turbopumps rose to a roar. One by one, the three engines lit.

"We have main engine start."

At his console, Al McDonald broke into a cold sweat.

In Utah, Bob Ebeling and Roger Boisjoly held hands.

"Four."

"Three at a hundred," Scobee said; the trio of main engines had reached 100 percent thrust. The shuttle leaned away from the tower, straining against the hold-down bolts anchoring it to the Earth.

"Three."

"Two."

"One—"

The twin boosters lit: igniters simultaneously fired tongues of flame 150 feet long down the full length of the rockets' hollow cores, and more than seven hundred tons of aluminum perchlorate exploded into life. Within six hundred milliseconds the pressure inside their steel casings rose to nearly one thousand pounds per square inch. Almost invisibly, expanding gases pushed the walls of the half-inch steel casings outward, each of the six field joints flexed open, and the O-rings encircling them began to move into the widening gaps between the rocket segments.

The hold-down bolts blew. Sheets of ice more than three feet across tumbled from the launch gantry.

"—and lift off, *lift off* of the twenty-fifth Space Shuttle mission, and it has cleared the tower."

In the Thiokol conference room, Boisjoly looked up at Ebeling in relief. The seals had made it through ignition.

"We just dodged a bullet," he whispered.

:::::::::::::::::::

In the grandstands, Ed and Grace Corrigan stood side by side and watched their daughter ascend toward orbit, their faces lit with anxious smiles. They turned and embraced, linking hands with their daughter, Lisa. Standing on the roof of a small building nearby, Barbara Morgan, McAuliffe's backup, hollered and clapped: "Whooo! C'mon, go!" she shouted, and gave a gleeful wave toward the departing spacecraft. "Bye, Christa! Bye, crew!" Above her on the Launch Control Center roof, Judy Resnik's father, Marvin, stared intently skyward. Nearby, the three Smith children stood with their mother, witnessing their father's patience rewarded at last; Alison gazed up and lifted the viewfinder of her Kodak camera to her eye. At Concord High School, the students let loose with whoops and cheers.

But in the bottom-most field joint of the right-hand booster rocket, the cold had done its work: the synthetic rubber of the seals and the thick grease they were packed in had proved too inflexible to close the gap that opened in the case at ignition. Hot gas at more than 5,000 degrees Fahrenheit had blasted past the primary seal—and then broken through the second seal, too, instantly vaporizing portions of the O-rings as it went. Unseen by the crowd or the officials in the Launch Control Center, burning grease, insulation, and Viton rubber spurted from the ruptured joint in puffs of coal-black smoke.

Seven seconds into the flight, the shuttle's computers began to turn the orbiter onto its back as it thundered out over the Atlantic. Scobee opened his radio link with Mission Control.

"Houston, *Challenger:* roll program," he said.

"Go, you mother!" said Smith. The shuttle rattled and shook like a runaway train as it accelerated toward the speed of sound, making it hard for the pilot to read the instruments.

From her seat behind them, monitoring the laminated ascent check-list open on her knee, Judy Resnik gave an exuberant yell: "Shit hot!"

"Oooo-kay!" Scobee replied.

Now the same forces of combustion that had destroyed the booster seal momentarily conspired to heal it. Within twelve seconds of launch, molten aluminum oxides from the burning propellant built up in the fissure in the aft field joint of the rocket, sealing the breach around the ruptured O-rings and cutting off the leak.

In Houston, public affairs officer Steve Nesbitt had taken over the public commentary, and was watching the streams of black-and-white numbers appearing on his monitors, waiting to explain the next major event in the flight to the millions watching on TV. To his left, Jay Greene swept his gaze over the engine performance data appearing on his console. Everything looked good.

The next part of the ascent program would reduce the thrust of *Challenger*'s three main engines to take the shuttle through Max Q: the phase of maximum dynamic pressure at which the aerodynamic forces acting on the spacecraft would reach their most extreme. To reduce stress on the airframe as the shuttle shot through the atmosphere at Mach 1, the engines would throttle back to 65 percent of their rated power for fifteen seconds, before returning to full thrust on the other side of the pressure wave.

The Flight Dynamics Officer made the call:

"Throttle down . . . three at sixty-five."

"Sixty-five, FIDO," Greene replied.

On the broad projection screen at the front of the Flight Control Room, the red line marking the shuttle's trajectory tracked tightly to its nominal path. A perfect ascent.

Fifty-seven seconds into the flight, the maneuver was complete. In the cockpit, Scobee watched the thrust readings start to rise as *Challenger*'s computers began once more to increase engine power.

"Throttling up," he said.

But as *Challenger* shuddered through Max Q, it was also buffeted by the worst high-altitude wind shear yet encountered on a shuttle flight. The entire shuttle stack flexed and twisted in the turbulence, shattering the delicate glassy residues that had resealed the hemorrhaged rocket motor. At fifty-eight seconds, an orange flame flared through the field joint at the bottom of the right booster.

Still clearly visible to the spectators on the ground at the Cape, the shuttle was approaching an altitude of 35,000 feet, and a velocity of one and a half times the speed of sound, its engines firing at 104 percent of rated power.

"Feel that mother go!" said Smith. "Wooohooo!"

::::::::::::::::::

The flame grew in intensity, deflected down in the slipstream of the rising spacecraft until it made contact with the external fuel tank, close to one of the three steel struts securing the bottom of the booster to the spine of the shuttle stack. Yet neither the instruments on *Challenger*'s flight deck nor the readings on the consoles in Houston gave any indication that anything was wrong. The onboard computers, struggling to keep the orbiter flying true, swiveled the nozzle of the left-hand booster outward to compensate for the loss of pressure in its malfunctioning twin.

"*Challenger*, go at throttle up," the CapCom radioed from Mission Control.

"Roger, go at throttle up," said Scobee.

Burning at more than 6,000 degrees, in less than three seconds the errant flame escaping from the booster encircled the circumference of the giant external tank, incinerated its insulation, cut through its aluminum skin, and ruptured the welds of the pressurized fuel tank membrane within. A plume of liquid hydrogen burst into the slipstream of the rocket engines, where it ignited.

In Atlanta, CNN space correspondent Tom Mintier, watching the pictures of the spacecraft flying away into the empty sky, began to wrap up his live commentary. "So the twenty-fifth Space Shuttle mission is now on the way, after more delays than NASA cares to count. This morning it looked as though they were not going to be able to get off—"

He stopped abruptly.

At seventy-two seconds, the tank lost its structural integrity and tore apart, crumpling and disgorging the remaining liquid hydrogen—more than 300,000 gallons of it—which bloomed into a colossal fireball. Released from its aft anchors, the right-hand booster swiveled around its upper attachment point. Its nose smashed into the right wing of *Challenger*, and the liquid oxygen tank, tearing it open.

The orbiter was engulfed in a swelling cloud of combustible propellant, and the nozzles of its three main engines swiveled wildly as the onboard computers struggled to regain control of the disintegrating spacecraft; for the few fractions of a second it took for the engines to consume the fuel remaining in the feed lines, their high-pressure turbopumps continued to spin, until the computers shut them down one at a time. Then the booster rockets tore free from their mounts, and *Challenger*, still hurtling toward space at almost 1,500 miles per hour, tumbled from its precisely prescribed supersonic trajectory. Its airframe stressed far beyond its design limits, the most complicated machine in history began to come apart in flight: its stubby wings ripped away, the cargo bay bursting like a paper bag, the inrushing air pulling the fuselage asunder from the inside.

At seventy-three seconds, the transmission of telemetry from the shuttle suddenly ceased. On Jay Greene's console in Houston, on all the screens in Mission Control, the rapidly flickering lines of streaming data froze, and one column after another filled with the letter *S*.

Static.

CHAPTER
TWENTY-FOUR

TUESDAY, JANUARY 28, 1986

11:28 a.m.

On the ground, it was hard to see what had happened. Up on the concrete roof of the Launch Control Center, Alison Smith was taking pictures with her Kodak camera and snapped three frames in quick succession: first, the shuttle rising slowly from the pad on its plume of steam and smoke, and another as it streaked away, trailing a white ribbon of exhaust out over the ocean and across the endless sky. Then, through the viewfinder, she saw the single contrail suddenly flare with color, bloom, and separate in two. She clicked the shutter again, and turned to her brother in awe.

"Isn't it beautiful?" she said.

In the bleachers, the crowd broke into more cheers as the incandescent cloud above them grew slowly larger. There was uncertain applause as the trails of the solid rockets emerged, drew apart, and then crossed. Men craned their necks toward the sky, raising long lenses and binoculars. Grace and Edward Corrigan continued to stare upward, their anxious faces in full view of the nearby press and TV cameras. As the roar of the rockets rolled back across Merritt Island from miles overhead, Barbara Morgan waved and clapped. And the voice of NASA commentator Steve Nesbitt echoed once more from the speakers, with the reassuring incantations of nominal flight:

"One minute fifteen seconds. Velocity 2,900 feet per second," he said. "Altitude nine nautical miles. Downrange distance seven nautical miles."

But of *Challenger* and her crew, there was no longer any sign.

∷∷∷∷∷∷∷∷∷

Inside Firing Room 3, the relief and exhilaration of a textbook launch evaporated in an instant. Everyone watching the ascent through the massive eastern wall of windows understood the consequences of what had just happened. When he saw the two solid rockets break free from the cloud, for a moment Launch Director Gene Thomas hoped that, next, he would witness a miraculous escape; that he would catch sight of the orbiter emerging, intact, from the fireball, quick-thinking test pilot Dick Scobee at the controls, swooping around for a Return to Launch Site Abort and gliding safely to a landing on the two-and-a-half-mile concrete strip on Merritt Island. But his technical experience told him what his heart briefly refused to comprehend: such a maneuver was physically impossible. The orbiter had been destroyed; his friends were dead.

A heavy silence filled the room; beside Thomas in the top tier, the other senior managers sat stupefied. Some buried their heads in their hands. Others began to weep. "It's a bad day," murmured the Shuttle Operations Director, Bob Sieck. Like everyone else, George Abbey was poleaxed with shock. And then he thought of the families on the roof. First, he picked up the phone, calling drivers to take the relatives to crew quarters, away from the eyes of the press. Then he started toward the elevators.

A few hundred yards away on the press mound, CNN's local correspondent John Zarrella turned toward his cameraman.

"Steve, what the hell happened?"

Slowly, the cameraman raised his eyes from the viewfinder.

"The fucking thing blew up," he said.

∷∷∷∷∷∷∷∷∷

In Mission Control, Flight Controller Jay Greene looked to his left, where the big color television set was playing the live pictures of the launch from NASA Select TV: half a dozen white streamers of debris now cascaded silently from the roiling cloud, which hovered in the sky where *Challenger* had been only seconds before.

Greene called for an update from the Flight Dynamics Officer tracking the path of the shuttle.

"FIDO," he said crisply. "Trajectories."

"Flight, FIDO. Filters got discreting sources," the controller replied: the radar was tracking several objects; there should have been only one.

Then the Ground Control Engineer cut in.

"Flight—GC. We've had negative contact," he said. "Loss of downlink."

All telemetry from *Challenger* had broken off; they were no longer receiving any information from the spacecraft.

At the back of the room, Steve Nesbitt was staring at the columns of frozen numbers on his console screens, looking over at the TV set, waiting; struggling to decide how to fill the dead air.

"OK, all operators, watch your data carefully," Greene said. He continued to call around the room in a search for information.

"Procedures, any help?"

"Negative, Flight. No data."

Down at Cape Canaveral, the Air Force Range Safety Officer watched his own video feed as the two solid rockets veered across the open sky, out of control with almost a minute of fuel still left to burn. He reached for the self-destruct mechanism, pressed a button to send the encoded radio signal: ARM. He waited for a count of ten, then sent a second signal: FIRE. Seventeen miles out over the Atlantic, the two boosters were abruptly engulfed in black wreaths of flame and disintegrated in flight, pieces large and small adding to the shower of debris still tumbling earthward from tens of thousands of feet up.

::::::::::::::::::

On the roof of the Launch Control Center, Alison Smith glanced over her brother's shoulder and saw a group of NASA officials emerging onto the roof behind them.

Scott looked at his mother, and Jane knew.

"It's not good," she said.

"What do you mean?" he asked.

"They're lost," she said. "I have to go find Mr. Abbey."

Smith was the first one off the roof, down the stairs, and through the heavy door. She was heading along the hallway outside Gene Thomas's office when she saw Abbey hurrying toward her.

She stopped; looked him in the eye.

"Tell me," she said.

In the crowd below, the fog of confusion lifted only slowly. Near the low building where Barbara Morgan stood, other spectators were also staring up, trying to make sense of what they could see. Some had never witnessed a launch from the Cape before; among them were many who knew that the solid rockets were designed to fall away during ascent, but never this soon. And as the white trail of exhaust coiled and twisted in the sky above them, charting the path of the spacecraft to the point where it had suddenly vanished, disbelief gave way to chilling realization.

A voice said, "That's not right. That's not right." Another: "That's not right at all."

Barbara Morgan gazed silently at the sky; she grasped both hands beneath her chin.

"Oh no," she said softly.

At last, Steve Nesbitt's commentary sounded once more from the speakers strung out across the launch site. "Flight controllers here looking very carefully at the situation," he said, his voice a metallic echo reverberating among the grandstands, over the press mound, and along the causeways of Merritt Island. "Obviously a major malfunction."

A hush settled over the crowd in the bleachers. Christa McAuliffe's parents both looked about in bewilderment. Grace Corrigan wiped tears from her eyes and shook her head. "I don't believe this," she said. Around her, couples and families with small children began to pick up their belongings and climb to the ground.

"We have no downlink," said the impassive voice in the speakers.

In Houston, the engineers on the Flight Dynamics Console, still baffled by their radar data, had called up the Range Safety Officer at the Cape. Could he see anything?

"It all blew up," he replied.

"Say that again?"

"It all blew up."

"What did?"

"The shuttle."

Only then was Nesbitt able to state clearly what those straggling from the bleachers at the Cape had seen for themselves.

"We have a report from the Flight Dynamics Officer that the vehicle has exploded," he said, and his voice cracked. "The Flight Director confirms that. We are looking at—uh—checking with the recovery forces to see what can be done at this point. . . ."

A few moments later, a NASA official climbed into the bleachers to find the Corrigans, adrift amid the thinning crowd. "The vehicle has exploded," he told them.

"The vehicle has exploded?" Grace Corrigan asked, her voice rising in disbelief.

The official nodded, and led them away.

Back in Houston, Jay Greene had heard enough: "GC, all operators, contingency procedures in effect," he said. He gave instructions for everyone in Mission Control to gather their notebooks and preserve the data on their consoles; outgoing phone calls were forbidden: every log, every piece of paper would have to be impounded for later investigation.

Until then, NASA security personnel would ensure that no one could leave the room.

"OK," he said. "Lock the doors."

In the conference room at Morton Thiokol in Utah, the assembled rocket engineers watched in mute horror as the catastrophe unfolded: at first, some shared the vain hope that *Challenger* would somehow escape intact from the conflagration captured by NASA's high-magnification cameras; others were initially convinced that—because the boosters kept flying—they could not be to blame for what had happened.

It's not us, thought Roger Boisjoly. *It's the main engines.* But he said nothing; instead he rose silently from his seat on the floor, walked downstairs to his office, and closed the door. A few moments later, Boisjoly was joined by his colleague Brian Russell, the junior engineer who had been with him in the late teleconference the night before, and who had faxed the company's final approval for launch to the managers at Cape Canaveral.

The two men sat together and began to cry.

That afternoon, Boisjoly wrote one more entry in his logbook:

SRM-25 blew up approximately one minute into flight. Presently waiting for information on the cause of the disaster. I feel real sick about this but I did everything possible to convince them not to fly.

Based on what they had seen on television, there might have been a hundred different reasons for what happened. Yet as the phones in Building A-2 began to ring and engineers gathered in the hallways and cubicles of the Promontory plant to debate their theories, Boisjoly and Russell feared that they knew exactly what had killed the seven men and women aboard *Challenger.*

There was not yet any evidence, of course; but they knew it would come soon enough.

::::::::::::::::::

It was just after 11:40 a.m. in Washington, DC, when White House communications director Pat Buchanan pushed past the President's executive assistants and burst into the Oval Office, where Reagan was in the middle of a briefing.

"The Space Shuttle just blew up," Buchanan said.

For a moment, no one spoke; the President looked stricken.

"Isn't that the one with the teacher on it?"

"Yes, sir."

A few blocks away at NASA HQ, deposed Administrator James Beggs had watched everything happen on the agency's TV feed; in vain, he left his office to search the seventh floor for Bill Graham, his acting replacement.

"Where the hell is he?" Beggs asked.

But Graham was up on Capitol Hill in a meeting with the Republican head of the NASA oversight committee—who he had hoped would help him secure permanent appointment to the top job at the agency. It was half an hour before he returned to 600 Independence Avenue. When at last he came running back to his office, Beggs thrust a list of names for a board of inquiry into the Acting Administrator's hand. Then he told him to get on a plane to Cape Canaveral.

At the headquarters of CBS News in New York, Dan Rather was sitting in an editorial meeting when images of the accident came up on a monitor. "God almighty," he said.

Rather ran down the hall to the "flash" studio, always kept ready to broadcast breaking news, and took a seat before the cameras without makeup or his usual contact lenses. At 11:45 a.m., he cut into the morning's programs with an unscripted bulletin: "The Space Shuttle *Challenger*, apparently moments after takeoff, in a hard freeze from Florida—something went wrong," he said. "The vehicle, reportedly, has exploded."

Cutting back and forth between Steve Nesbitt's continuing commentary from Mission Control and with producers and researchers feeding him updates through his earpieces, Rather struggled to explain to his audience an event that no one yet understood. "What you have here," he said at one point, "is a reporter vamping for time." Starved of hard facts, the anchor introduced the brief clip of videotape showing the launch through the lenses of NASA's long-range cameras: the perfect takeoff; the crackle in the audio; the blazing cloud; the errant rockets; the streaming debris. Three minutes later, he did so once more: "Now, we're going to roll the tape again for you, of this morning's launch," he said. "Let's listen and watch."

The pattern was the same on the other national stations—on CNN, on NBC, and ABC—where the special bulletins would eventually broadcast continuously all afternoon; minute after minute, hour after hour, the footage played again, and again; and again.

On the fourth floor of the Launch Control Center, the relatives of the *Challenger* crew begged their NASA escorts for reassurance, clinging to hope that their loved ones had somehow survived. "Is he going to be OK?" Lorna Onizuka asked. "Is he going to be OK?" But the official at her side couldn't bear to tell her the truth.

"We'll have to see," he said.

George Abbey swept eight-year-old Erin, Mike Smith's youngest child, into his arms and carried her into the elevator. Jane Smith ran to retrieve their things from the Launch Director's office, where the welcome banner still hung from the ceiling, but agency staff now hurried to switch off the television sets. Erin screamed in anguish: "Daddy! I want you, Daddy! You always promised nothing would happen." Alison, although still holding on to the idea that her father might somehow have flown his spacecraft out of the fireball—and landed safely in Spain, or in Morocco, or Easter Island—was in tears. As the doors of the elevator closed on Abbey and the small group of astronauts' families, the cabin filled with the sound of sobbing.

"Stop crying," one of the other women told Alison. "You're upsetting everyone."

As the NASA vans carried them across the Space Center campus toward the Operations and Checkout Building, June Scobee looked from the windows at the strangers streaming from the viewing areas, or gathering on the shoulders of State Road 3; sagging into one another's arms or pointing out to the east, where the contrails of the lost spacecraft still hung in the sky. When the relatives finally reached their destination, it was clear that—however sophisticated NASA's contingency plans might have been—no one at the agency had prepared for a catastrophe like this.

The crew quarters, designed to accommodate a handful of astronauts for a few days before a launch into space, had not been built with comfort in mind—and the limited facilities were now overwhelmed by dozens of distraught women, children, and elderly relatives of the *Challenger* crew, ranging from Dick Scobee's eleven-month-old grandson, Justin, to Ron McNair's mother, Pearl, sixty. George Abbey planned to keep the families in a secure location while his deputy in Flight Operations, Rick Nygren, found aircraft to fly them home from the Cape airstrip; but first he had to have their luggage retrieved from the condos and motels where they had been staying up and down the coast, and the Space Center was now snarled in traffic as shattered spectators tried to leave the area by car. The local telephone exchange had jammed, making it all but impossible to dial into or out of Kennedy, so Nygren organized a team of astronauts—including Judy Resnik's boyfriend, Frank Culbertson, and Sonny Carter, who had been the last member of the ground support crew on the pad to climb from *Challenger* that morning—to act as runners, carrying messages from place to place. In the meantime, the crew quarters' staff, believing that their work was finished early that morning, had already left—so there was no one to provide food or drink; Nygren did what he could. The astronauts' weeping children were crammed together on a group of couches, where the crew flight surgeon tried her best to calm them and Sonny Carter sought to distract them with jokes and impressions. Carter's clowning made Alison Smith laugh, but her cheeks still burned from crying.

For the adults, many of whom had been awake since before dawn, time slowed to a crawl: Rick Nygren had made sure every TV set on the third floor of the Operations and Checkout Building had been turned off, and while they waited to be told when they could go home, no further word about the accident came in or out. The astronaut runners left and returned, bringing in more members of the crew's extended families—including both of Christa McAuliffe's parents, and Mike Smith's father-in-law, who had a heart condition—but no one had any information. Grace Corrigan kept repeating the same phrase again and again: *The vehicle has exploded . . . the vehicle has exploded.* In the kitchen, Chief Astronaut John Young, ninth man on the moon, wobbled from foot to foot, muttering to himself.

It was an hour before George Abbey gathered the adults in the conference room and confirmed what many of them already knew.

"We don't know all the details," he said. "But it looks like there has been an explosion. I don't believe there is any hope for the crew."

Jane Smith went down the hall to Mike's room, where his belongings remained just as he had left them that morning, marked with labels for return to Houston. Jane took her husband's plastic nameplate from the door, and asked Abbey if she could keep the terry-cloth robe he had worn that morning; it was one of the last things her husband had touched.

In Dick Scobee's room, June found his brown Samsonite attaché case, and in it flight manuals, star maps, NASA souvenir pins, an unsigned Valentine's Day card—"For My Wife"—and a scrap of paper on which he had copied out a few inspirational lines from science writer Ben Bova. Alone at last, Scobee broke down and sobbed.

No one had yet told the children.

"What happened?" Alison Smith asked her mother.

"If I tell you, you'll be upset," Jane said.

Alison began hyperventilating. The flight surgeon brought her a paper bag.

:::::::::::::::::::
:::::::::::::::::::

Back in the Launch Control Center, Gene Thomas and the other senior managers on the top tier of Firing Room 3 had sat in stunned silence for almost half an hour; only slowly did they begin to recover from the shock of what they had witnessed. At around 12:10 p.m., one of the armed NASA guards charged with securing the scene came to escort Thomas downstairs to a recording-playback room on the first floor. There, amid a group of twenty or more scientists and engineers, he stood quietly before the TV monitors and watched the video footage of *Challenger*'s brief flight to destruction. They played the recording a dozen times, over and over at different speeds, but it was no use: although Thomas could clearly see evidence of a bright point of light blooming near the external tank as the shuttle approached Mach 2, the grainy video would not give up any more information. Perhaps it was some kind of fuel leak from the tank, or from the solid rockets; it was just impossible to say. And whatever the video depicted, many senior managers remained certain that the accident had been caused by the elaborate and troublesome engineering of the Space Shuttle main engines: just as they had feared since the beginning of the program, one of the high-pressure turbopumps had thrown a blade. Both Larry Mulloy from the Marshall Space Flight Center and Al McDonald from Morton Thiokol felt certain that the cause of the accident couldn't have been their boosters; they knew that if you spring a leak in a solid rocket, it doesn't keep on flying—it explodes.

In a conference room in the Launch Control Center, Gene Thomas helped convene a group of senior managers to impound the console data from the Firing Room, and the movie and photographic film in the remote-controlled cameras out at Pad 39B. Then they appointed a NASA interim review board to begin the investigation into what had gone wrong. In the meantime, someone had to decide what to tell the press.

Fortunately, a contingency plan for how to handle the news of a disaster in the shuttle program had been completed less than two years earlier, by the then head of NASA's public affairs department. The aim of the plan was to avoid the spread of rumors and speculation in the press after an accident, and ensure what it described as "a full flow of accurate, timely and factual information" to the media. The document assessed accidents in categories of ascending severity: the first was a straightforward hardware malfunction jeopardizing the mission; the second, a failure that required the shuttle to ditch in the ocean—but that every member of the crew somehow survived; the third was a catastrophe that killed the astronauts. All took into account the possibility that, whatever happened, the American public would be watching it take place in real time.

"In the event of an emergency involving crew injury or fatality, the fact will be apparent to radio listeners and TV viewers as well as observers at launch and landing sites. Status of the crew will be the prime public consideration," the plan explained. "The facts, once confirmed, should be announced as promptly as possible." It recommended that, following any emergency, no more than twenty minutes should elapse before NASA officials issued a formal statement.

But in the two years since it had been devised, the frenetic pace of shuttle launches had left even the public affairs staff so busy that there had been no time to rehearse the contingency plan, and in mid-January the department chief who had written it left the agency for a job with the post office.

When the time came to put it into action, the plan fell apart almost immediately.

:::::::::::::::::

At the White House, the President met the gathered network reporters in the Roosevelt Room at around noon; he quickly explained that there would be no briefing about the State of the Union address. The journalists wanted to know how he felt.

"What can you say? It's a horrible thing. All of us have witnessed it, and actually seen it take place," Reagan said, standing in front of the fireplace, bracing himself against the back of a chair. "I just can't rid myself of the thought of the sacrifice, and of the families that have been watching this also, the families of those people on board and what they must be going through."

The President told the reporters that he knew nothing more than they did about what had happened—he and his staff had just watched it on TV, and NASA had not yet issued any comment. But he wanted to make sure that everything would be done to find out the causes of the accident and make sure it never happened again. "I'm confident that there will be no flight until they are absolutely as certain as a human being can be that it is safe," he said.

Yet he remained firm in his support for the space program, and—when asked if it had been a mistake to send a teacher into space—maintained that flying civilian passengers aboard the shuttle was a necessary part of conquering what he called "the last frontier." Reagan told the reporters that he intended to go ahead with the address, as planned, later that night—"you can't stop governing the nation because of a tragedy of this kind," he said—but that his remarks would include something directed at all the children who had been watching on TV when the accident took place. He wanted to explain to them that there had always been pioneers who had given their lives "out there," on the frontier—but for everyone back at home, life must continue. "You don't back up and quit some worthwhile endeavor because of tragedy," he said.

But, after consulting with his staff, within the hour Reagan agreed to call off the State of the Union. As the scope of the disaster became clear, he dispatched Vice President Bush and Acting Administrator Graham to the Cape to take his condolences to the families of the *Challenger* crew—and the White House announced that the President would give a live televised statement to the nation about the accident that afternoon.

In her office in the Old Executive Building adjacent to the White House, speechwriter Peggy Noonan had a phone call from the West Wing: they needed something quickly; keep it short—five minutes. Looking over a set of notes from the President's press conference with the reporters, Noonan built a speech around the spine of what Reagan had already said. As CNN played on a TV set beside her, she summoned the shared experience of an audience for whom the endless loop of *Challenger*'s final moments was already being ground into the national psyche. The words she wrote expressed the President's sorrow, but promised that NASA's mission of human exploration would continue.

Noonan had earned a reputation within the West Wing as the writer to turn to for patriotic and emotional rhetoric. Now she needed a lyrical twist for the end of the speech. As CNN replayed the footage of the crew cheerfully waving goodbye to the TV cameras on the way to the Astrovan that morning, she recalled the poem she had learned in seventh grade, a hymn to the joy of flying: "High Flight," by John Gillespie Magee. She worked lines from Magee's verses into her draft, but understood Reagan well enough to know that he would only deliver them if he was already familiar with the poem. It seemed a good bet—but, just in case, she wrote the quotation into the second of two discrete endings for the speech; if he chose to, he could wrap up before he reached the poetry. Noonan typed up her draft, made three copies, and rushed them over to Pat Buchanan. With no time for the intensive process of staff revisions to which Reagan's speeches were usually subjected, Noonan's words reached the President almost exactly as she had written them.

Down at Cape Canaveral, it was not until 4:30 p.m. that NASA Assistant Administrator Jesse Moore took a seat before the press to deliver the agency's first official statement about the disaster. In the five hours since it had happened, newspaper reporters and TV correspondents had poured into Florida from all over the country, as editors and producers scrambled to cover an event that had so abruptly transformed from high-spirited human interest to national catastrophe, the biggest news story in the world.

But NASA's usually slick public relations operation, renowned in the media for its responsiveness and transparency, had been paralyzed by the crisis: the newly appointed chief of public affairs was unfamiliar with the details of the emergency contingency plan, and the twenty-minute deadline to release information came and went. Hours passed, and the public affairs staff labored on in shock and grief: one member of the team was so shaken by the loss of the astronauts that he withdrew to his office inside the press dome, where he hid from reporters until nightfall. In the meantime, his colleagues misinterpreted the instructions to impound all data from the *Challenger* launch, in the apparent belief that not only should the information be preserved, but also kept secret. NASA instructed all employees and contractors not to speak to the press—and orders went out that no one in public affairs should talk without first clearing their statements with the agency's upper management. But those managers on the scene were busy trying to figure out what had gone wrong, and Acting Administrator Graham was again unreachable, now en route to Florida with the Vice President aboard Air Force Two.

In the growing information vacuum, NASA spokesmen appeared defensive, and the media grew suspicious; even journalists asking what the temperature had been that morning at Pad 39B were denied answers, and agency technicians seized film shot by news photographers of the launch. A press conference was announced for 3:00 p.m., and then for 4:00 p.m. By the time Jesse Moore finally appeared, so many reporters had gathered to hear him speak that the event had to be held outdoors, in front of a 350-seat grandstand. Swaddled against the cold in a black raincoat, his face shadowed and ashen, Moore began by reading his brief statement. "It is with deep, heartfelt sorrow that I address you here this afternoon," he said. "At 11:40 a.m. this morning, the space program experienced a national tragedy, with the explosion of the Space Shuttle *Challenger* approximately a minute and a half after launch from here at the Kennedy Space Center. I regret that I have to report that, based on very preliminary searches of the ocean where the *Challenger* impacted this morning, these searches have not revealed any evidence that the crew of *Challenger* survived."

Beyond that, Moore had little to offer: "We will not speculate as to the specific cause of the explosion," he said, and explained that the launch had seemed completely normal until the very instant the shuttle was destroyed. Further launches would be temporarily on hold, and a formal NASA review board would soon be appointed by Acting Administrator Graham; they would be responsible for determining exactly what had happened. Moore then fielded a handful of questions from the reporters, but many of his answers seemed evasive; when one journalist complained of a news blackout, Moore angrily corrected him. He was, however, unequivocal about some issues. No, he said, there had been nothing unusual about the weather that morning; and, no, the repeated delays over the previous three weeks had not created any pressure to launch. Flight safety had always been NASA's top priority.

"All of the people involved in this program, to my knowledge, felt that *Challenger* was quite ready to go," Moore said. "And I made the decision, based upon the recommendation of the team supporting me, that we launch."

:::::::::::::::::

At 5:00 p.m. Eastern time, each of the nation's TV networks broke into their continuing coverage of the accident to go live to the White House. The cameras revealed President Reagan seated in the Oval Office, his hands resting before him on the Resolute desk.

"Ladies and gentlemen, I'd planned to speak to you tonight to report on the state of the Union, but the events of earlier today have led me to change those plans," he said. "Today is a day for mourning and remembering." The President noted that it was almost nineteen years to the day since the deaths of the three astronauts in the Apollo launchpad fire, and then named each member of the *Challenger* crew: "We mourn seven heroes: Michael Smith, Dick Scobee, Judith Resnik, Ronald McNair, Ellison Onizuka, Gregory Jarvis, and Christa McAuliffe. We mourn their loss as a nation together."

The speech Peggy Noonan had written was brief—just four minutes in total—but anchored the disaster as a pivotal moment in American history. With gravity and sincerity, Reagan described not just grief at the deaths of seven men and women, but a loss of national innocence, experienced in real time by children and adults across the country: a shattering of confidence in the promise of high technology that had so far endured through two decades of reversals at home and abroad.

"On the day of the disaster, our nation held a vigil by our television sets," the President said. "In one cruel moment, our exhilaration turned to horror; we waited and watched and tried to make sense of what we had seen. . . . We've grown used to wonders in this century. It's hard to dazzle us. But for 25 years the United States space program has been doing just that. We've grown used to the idea of space, and perhaps we forget that we've only just begun. We're still pioneers. They, the members of the *Challenger* crew, were pioneers."

And then—after oblique jabs at the numerous unacknowledged fatalities in the Soviet space program—Reagan recommitted to the expansion and democratization of NASA's mission. "We'll continue our quest in space," he said. "There will be more shuttle flights and more shuttle crews and, yes, more volunteers, more civilians, more teachers in space. Nothing ends here; our hopes and our journeys continue."

As Reagan's address drew to a close, he alluded to another historical coincidence—the death of Elizabethan explorer Sir Francis Drake aboard his ship off the coast of Panama almost four centuries earlier. But when the President reached the final paragraph, he recognized the quotations from "High Flight," which had been inscribed on a plaque outside the door of his daughter's elementary school in Los Angeles. He kept reading; Magee's words would crystallize some of the most celebrated rhetoric of Reagan's presidency:

"The crew of the Space Shuttle *Challenger* honored us by the manner in which they lived their lives," he said. "We will never forget them, nor the last time we saw them, this morning, as they prepared for their journey and waved goodbye and slipped the surly bonds of earth to touch the face of God."

::::::::::::::::::

It was late afternoon when Air Force Two touched down on the shuttle landing strip on Merritt Island, carrying Vice President Bush, Acting Administrator Graham, astronaut-turned-Senator John Glenn, and Senator Jake Garn, who had flown into space aboard *Discovery* less than a year earlier.

During a brief exchange with the press as they arrived at the Kennedy Space Center, Bush spoke with tears in his eyes; he echoed the President's message of resolve to the children who had witnessed the catastrophe, and reiterated that the nation must press on with the pursuit of space exploration. Glenn, the outspoken critic of flying civilians on the shuttle, underscored how dangerous the space program had been from the start: "It's been nearly a quarter of a century that we thought this might happen sometime," he said, his voice cracking. "But we have delayed that day until today." Bill Graham refused to make any comment at all. "I can't right now," he told the reporters. "I hope to be able to in the next day or so."

By the time the politicians arrived at the Operations and Checkout Building, the bereaved relatives had been kept isolated on the third floor for hours—told only to expect a visit from someone from the White House, and with little idea of how events had been unfolding in the world outside. At one point earlier that afternoon Alison Smith had managed to reach a friend in Clear Lake on the telephone. Smith had been out of school for almost a week and wanted to talk to someone her own age. And when her classmate Jill answered, Smith thought she heard the sound of laughter tinkling in the receiver.

"No, no," Alison said, "the worst thing has happened."

But Jill was crying: she knew; everyone did. Smith was astonished. "It's all over the news," Jill explained. All the children in Clear Lake—the sons and daughters of the engineers and astronauts from the neighborhood—had been sent home early from school.

Numb with shock, exhausted by grief, and now desperate to return home, Lorna Onizuka, Cheryl McNair, Marcia Jarvis, Steve McAuliffe, Marvin Resnik, June Scobee, and Jane Smith sat silently in the conference room as George Bush and the delegation from Washington, DC, arrived. For a moment, no one said anything; then Smith and Scobee exchanged a glance, and the commander's widow stood up. "We want you to know one thing, that the husbands and wives love the space program and want to see it go on," she said. "We feel the same way," Bush replied.

Before leaving, the Vice President handed Scobee a piece of paper on which was written a telephone number, and a message: *Call us if you need us.*

::::::::::::::::::

Inside the Flight Control Room in Houston, each of the flight control engineers had gathered up their notes and binders, secured the remaining data on their consoles, and drifted back to their offices. It was hours before Steve Nesbitt finally stood up and realized that he was alone in the room; even the guard posted on the door had left. He made his way downstairs and out past the duck ponds behind Building 30, taking the back route to his office to avoid the reporters he felt sure would be waiting to pounce on him.

Outside, at the main gates of the Space Center, bouquets and wreaths of flowers were already being laid by Clear Lake residents in tribute to the lost astronauts; on NASA Road 1, flags hung at half-staff and the billboards outside hotels and businesses displayed messages of consolation: OUR SYMPATHY TO SHUTTLE TEAM AND FAMILIES; A TIME OF UTMOST SORROW AND MOURNING FOR ALL THE WORLD; and WE SALUTE AMERICAN SPACE PIONEERS.

Nesbitt drove home to his apartment and called his ex-wife, to ask if it would be OK not to pick up the kids that evening. "We've got a lot going on here," he said. Still struggling to comprehend what had happened, and desperate to anchor himself in something like normality, he got back in the car for the short ride over to Baybrook Mall in Friendswood. He bought himself an ice cream cone and for an hour wandered aimlessly among people wandering in and out of stores just as they would on any other day. Only then did he return home and there, for the first time, saw the explosion on TV.

:::::::::::::::::::

A steady rain was falling across Ellington Field as the first plane came in from the Cape that night, Chief Astronaut John Young at the controls. There was just a handful of passengers aboard, including the Smith family—Jane and the three children, and Mike's sister, Ellen—accompanied by George Abbey. They had passed the more than two-hour flight almost entirely in silence. By the time they landed in Houston, the local constable had corralled a group of reporters and photographers away from the arrival hangar, and doused the lights, to preserve the family's privacy. But when the door of the small plane fell open, Jane Smith saw a crowd stretching away into the darkness—hundreds of friends from Timber Cove, and from the Astronaut Office, waiting to help them return home.

At the end of the night, after the final aircraft had landed and the last of the dead astronauts' relatives had been swept away in NASA cars, the constable thought he was alone in the deserted airport. Then he turned to find the solitary figure of George Abbey standing outside the empty hangar, slumped with grief. Unsure what to do, the policeman enveloped Abbey in a hug, and then led him quietly to his car.

Judy Resnik during her first day in orbit aboard *Discovery* in 1984. Floating in the cockpit to her right is the clipboard she held up to the camera during a TV broadcast to Earth, showing a message to her father, Marvin: "HI DAD." *NASA*

A weightless Ron McNair playing the saxophone during his first journey into space, aboard *Challenger*, in February 1984. McNair recorded three songs, including "What the World Needs Now Is Love," in orbit—but all were accidentally deleted before the mission was over. *NASA*

Ellison Onizuka, photographed in his Air Force uniform soon after his promotion to lieutenant colonel, following his first journey into space as part of a three-day clandestine military mission aboard *Discovery*, in January 1985. *National Archives*

Dick Scobee in the cockpit of a two-seat biplane similar to the one he owned with fellow astronaut "Ox" van Hoften. *June Scobee Rodgers*

Mike Smith, photographed while lending a hand to his children's lawn-care business, near the family home in Timber Cove. *Smith Family Collection*

Greg Jarvis, training for weightlessness aboard the "Vomit Comet" in January 1985; Jarvis would join the crew of mission 51-L only after being bumped from two earlier flight assignments. *NASA*

The unofficial "gag" portrait taken of the 51-L crew in late 1985, in honor of Christa McAuliffe, the Teacher in Space, and her backup, Barbara Morgan; Greg Jarvis had not yet been assigned to the mission. *Clockwise from bottom left*: Onizuka, Smith, McNair, McAuliffe, Morgan, Scobee, and Resnik. *Smith Family Collection*

Challenger after its rollout from the Vehicle Assembly Building to Pad 39B at Cape Canaveral, photographed on December 22, 1985. *National Archives*

The 51-L crew in the White Room beside *Challenger* during a break in the Terminal Countdown Demonstration Test—a final full-scale rehearsal for launch—on January 9, 1986. *NASA*

The five wives of the *Challenger* crew pose in front of the NASA plane that took them to the Cape, on January 23, 1986. *Left to right*: Cheryl McNair, Jane Smith, Marcia Jarvis, June Scobee, and Lorna Onizuka. *Smith Family Collection*

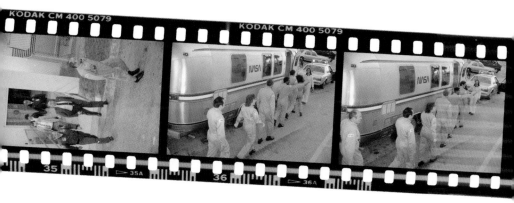

The crew of 51-L, accompanied by George Abbey and Chief of the Astronaut Office John Young, board the Astrovan for the short drive to the pad, on January 28, 1986. The final three frames shot by a NASA photographer outside the Operations and Checkout Building that morning, these are among the last photographs ever taken of the *Challenger* crew. *National Archives*

With less than nine minutes left in the countdown, family members of the crew gathered to watch the launch from the roof of the Launch Control Center. *Challenger* is visible in the distance, just over three miles away, on the far left of the photograph. *Left to right*: Scott Smith, Jane Smith, Erin Smith, Alison Smith, Marcia Jarvis, Kathie Scobee, June Scobee, and Rich Scobee. *Smith Family Collection*

From the ground, it was hard to tell immediately that anything had gone wrong. Alison Smith took this photograph around seventy-three seconds after launch. She turned to her brother and said, "Isn't it beautiful?" *Smith Family Collection*

Morton Thiokol engineer Roger Boisjoly had called for a redesign of the joints in the solid rockets and tried to stop the *Challenger* launch the night before it happened. Here, he holds up a sample of the Viton O-rings used in the joint. *Getty Images/ Denver Post/Duane Howell*

Allan McDonald being sworn in for his public testimony before the Rogers Commission on February 25, 1986. Thiokol's chief representative at Cape Canaveral, McDonald refused to sign the company's formal recommendation to launch. *Getty Images/Bettmann*

Larry Mulloy, NASA's manager of the solid rocket booster project at the Marshall Space Flight Center, points to the O-ring in a full-size cross-section of the rocket joint during his testimony before the Rogers Commission on February 13, 1986. *AP Images/Scott Stewart*

RICHARD FEYNMAN   NEIL ARMSTRONG

Caltech professor Richard Feynman alienated some of his colleagues on the commission with what they saw as his publicity-seeking antics; but his ice-water "experiment" helped crystallize the role of the O-rings in the accident with the media. *AP Images/Dennis Cook*

In January 1987, NASA workers began lowering the recovered wreckage of *Challenger* into a pair of disused underground missile silos at Cape Canaveral Air Force Station, which were then locked and sealed with concrete caps. *NASA*

CHAPTER TWENTY-FIVE

THE COMMISSION

$\mathbf{S}$urrendering slowly to gravity, falling from icy altitudes, tumbling and fluttering in the eddies of the jet stream, it would eventually take almost an hour before the final fragments of *Challenger* splashed into the ocean off Cape Canaveral. By then, much of the nation was already united in mourning. Whether gathered in classrooms, at home or at work, or transfixed by the images flickering across rows of TV screens in store windows, as midnight arrived on Tuesday an estimated 95 percent of American adults had seen the footage of the shuttle's final moments. And many felt the loss of the seven astronauts as a shattering blow—a national bereavement unlike any event since the assassination of John Kennedy more than twenty years before.

To men and women across the United States, the almost ideal diversity of the *Challenger* crew—"*one of everything*"—may have been enough to make the tragedy seem personal. Yet five of the seven were trained astronauts, and all but one were aerospace professionals, each of whom had understood and accepted the risks inherent in their mission. Christa McAuliffe was different. Not only had she captivated the attention of the nation's schoolchildren, but she had embodied the hopes of all those adults who believed, however remotely, that her journey was bringing the Walter Mitty fantasy of citizen spaceflight within their grasp. More than anything else, the poignant public death of the world's first Everyman astronaut made the loss of *Challenger* a shared tragedy: an instant-replay televised martyrdom in which everyone watching played a part.

Outside state capitols, military bases, stores, and hotels, and on porches up and down the country, the flags hung at half-staff; in Los Angeles, the Olympic torch at the Memorial Coliseum was reignited in tribute, and radio DJs dedicated songs to the lost astronauts; in Atlanta, hundreds of drivers turned on their headlamps in the bright sunshine, while in New York, the colored floodlights that lit the Empire State Building were extinguished. Over the next few days, gestures of sympathy large and small, both spontaneous and well-orchestrated, spread out across the world: seven black balloons were released over McAuliffe's alma mater in Framingham, Massachusetts, and schoolchildren began sending nickels and dimes to NASA to contribute to the cost of building a new orbiter; Congressman Bill Nelson proposed renaming seven of the recently discovered moons of Uranus after the members of the *Challenger* crew. Speaking from the Vatican, Pope John Paul II asked his audience in St. Peter's Square to pray with him for those who had died: "I lift up to God a fervent prayer so that he accepts in his embrace the souls of these courageous pioneers in progress in science and of man," he said; in Moscow, Soviet state television broadcast somber announcements describing the accident, while old Glenn Miller records played on the radio; even America's most despised enemy, Libyan despot Muammar Gaddafi, sent his own curdled message of sympathy. "Those who died were the victims of imperialist impatience," he said.

Back in Houston, the lights of the telephone switchboard in the NASA public affairs office once more flashed constantly, and the number of requests for press accreditation leapt from a handful to more than a thousand overnight; in the meantime, reporters and remote camera trucks laid siege to the homes of the *Challenger* widows. Although George Abbey had sent agency security staff armed with walkie-talkies to guard the front doors of the families' houses in Clear Lake, the press proved relentless, and had to be thrown off the scent: June Scobee's next-door neighbor, astronaut Fred Gregory, invited reporters into his own home so that they wouldn't be outside when the commander's widow first returned from Ellington Field. On Wednesday morning, Scobee woke to the familiar sights of her own bedroom—the winter light sifting through the sheer curtains, the dusty rose wallpaper, Dick's desk between the windows at the foot of the bed—and, still drowsy, for a moment she wondered where her husband could be.

Then it all came back.

A few streets away on Pebbleshire Drive, Cheryl McNair continued to believe that Ron might have escaped alive—that the orbiter had flown away from the blazing cloud and landed safely, unseen and unrecorded, somewhere halfway around the world at one of NASA's emergency landing strips; it would be days before she was able to accept that she would never see him again. Her parents stayed on to help look after Joy and Reggie, but the phone rang and rang, and she was haunted by a headache that seemed never to ebb or subside.

When Jane Smith had arrived back from the Cape, she found that the front door of the lakefront house in Timber Cove had been hung with a wreath, and already stood open. Inside, the place was filled with dozens of family friends from the neighborhood. Some stayed through the night so that Smith wouldn't have to spend it alone, and then remained for days afterward—joined by more astronauts and their wives—who fixed food, fetched drinks, answered the phone, and helped keep the media at bay.

Grace and Ed Corrigan did not reach their home in Framingham until late on Wednesday afternoon; they hadn't slept for two days. Escorted from the airport in Boston by a NASA press liaison, the parish priest, and a New Hampshire state trooper, the grieving couple arrived at their house in a small convoy of cars led by a cruiser from the local police department. But when they pulled up, they found the front lawn jammed with reporters and photographers, who immediately crowded around the car, pressing against the doors so tightly that no one could get out. Only after the priest and the public affairs officer had forced their way in and pushed the mob back were the Corrigans finally able to make it inside the house. The next day, a reporter from the *National Enquirer* knocked on the door, offering $100,000 in exchange for an exclusive interview.

:::::::::::::::::

Three days after the accident, on Friday January 31, more than six thousand mourners—NASA officials, astronauts, engineers, and contractors, and sixty members of Congress—gathered on the lawn outside Building 12 in Houston for a memorial to the fallen astronauts. The sky was gray, and the trees shivered in the winter wind that gusted over a lectern bearing the Presidential seal; an Air Force band played funeral hymns. Inside, Ronald and Nancy Reagan huddled with the families of the crew. "We'll all go out together in a few minutes," the President told them. "I wish there was something I could say to make it easier. But there just aren't any words."

The service lasted just half an hour. After an introduction from Acting Administrator Bill Graham and a reading from Psalm 46 by astronaut Charlie Bolden, the President delivered a eulogy that described how the catastrophe had brought the country together. "Across America, we are reaching out, holding hands, and finding comfort in one another," he said. He invoked once more the spirit of the High Frontier, comparing the astronauts to pioneers who had perished on the Oregon Trail, and addressed their families directly: "The sacrifice of your loved ones has stirred the soul of our nation," he told them. "We learned again that this America—which Abraham Lincoln called the last best hope of man on earth—was built on heroism and noble sacrifice. It was built by men and women like our seven star voyagers," he said.

Then, as the band played "God Bless America," the President embraced each of the widows and shook hands with Steven McAuliffe. The First Lady took eighteen-month-old Joy McNair in her arms; four T-38s roared overhead. June Scobee looked up to see one of the planes break away and rise steeply into the sky leaving three to fly on, in the missing man formation she had first witnessed years before above the desert airstrip at Edwards.

That morning, the entire ceremony was broadcast live on radio and television and, later, excerpts played on the evening news. The next day, newspapers printed Reagan's eulogy in full; eventually, more than three-quarters of all Americans—180 million people—watched some or all of the memorial on TV. And at first, the public saw the accident just as the President had framed it on the day it had happened: as a one-in-a-million accident, the inexplicable technical failure of a well-tested machine; the tragic but almost inescapable consequence of pressing the boundaries of technology. In the narrative of Reagan's rhetoric, the *Challenger* crew had laid down their lives as part of the virtuous quest to expand humankind's knowledge of the universe.

Initially, the proximate cause of what had killed the seven astronauts appeared to be a mystery, reported by the press in almost biblical terms, as if the spacecraft and its occupants had been swiped from the heavens by the hand of an Old Testament god: "The Shuttle Explodes" the *New York Times* declared. The accident was unlike the Apollo 1 fire, when technicians on the ground had fought back smoke and searing heat to reach the bodies of the stricken victims and begun almost immediately to search the wreckage for clues; or the crisis aboard Apollo 13, where the world had waited in suspense for days while ingenious engineers struggled to prize the astronauts from the jaws of death, and then gathered around their TV sets to witness a miraculous splashdown. Instead, the end of *Challenger* seemed sudden, remote, and inexplicable—the physical evidence of the tragedy scattered on the wind and drowned in the ocean. In the days after the accident, press conjecture about what had caused it ranged across seemingly every imaginable possibility: a shattered blade in an engine turbopump; cracks in the fuel of the solid rockets; the ice on the launchpad; terrorism; sabotage.

Despite all the information it had gathered from its scores of tracking cameras and hundreds of telemetry data sensors, NASA's continuing refusal in the days following the accident to speculate about what happened meant that, to the public, the mechanism of the shuttle's destruction at first seemed unclear: whether a fire had engulfed the spacecraft, or an explosion had torn it apart. Associate Administrator Jesse Moore's few statements to the press encouraged a belief that the crew had died instantly, blown to pieces or immolated without suspecting what had befallen them. Gradually, the grieving families came to accept that their bodies would never be found.

:::::::::::::::::::

But in the photo lab at the Johnson Space Center, some of the facts of the disaster had already become clear. In the confusion that had followed the accident, George Abbey had acted quickly to place astronauts in key roles in NASA's interim investigation—almost immediately dispatching Bob Crippen down to Florida to assist in the accident inquiry and lead the recovery of debris strewn in the ocean, and assigning others to study the millions of pieces of data gathered via telemetry during *Challenger's* final flight. Within twenty-four hours, the film from the 70-millimeter high-speed movie cameras—equipped with telescopic lenses, shooting forty frames a second, from angles not shown on TV—that had tracked the launch from Cape Canaveral had been processed and copies flown up to Houston from Florida. Spooling the footage across a light box, one frame at a time, the astronauts analyzing the images in the Space Center lab could see a tongue of bright flame flaring from the aft joint in the right-hand solid rocket booster about sixty seconds after launch. As soon as they saw it, they knew exactly what had happened.

Down at the Marshall Space Flight Center in Alabama, Al McDonald was working on the failure analysis team dedicated to the solid rockets. The team was based in a "war room" inside the Huntsville Operations Support Center, where they had covered a blackboard with an itemized list of every catastrophic scenario they could conceive. To gain access to the building, McDonald had been issued with a security pass card and a confidential keypad code number; all data pertaining to the accident was now apparently secret.

When he saw the film from the high-speed cameras on Wednesday evening, it confirmed his worst fears: that the accident had been caused by one of the Morton Thiokol boosters; worse still, it might have been caused by the very thing that McDonald had tried to warn NASA about the evening before the launch. That night he called his oldest daughter, who was attending college in Boston. "I feel like it's my fault that seven American astronauts have just been killed," he told her, his voice choking with tears. Even so, McDonald still held out hope that the team might yet discover another cause of the accident—or, at least, that something other than cold weather had caused a leak in the seals.

But evidence pointing to a failure of the solid rockets was growing: telemetry data revealed that the chamber pressure in the right-hand rocket had begun to fall at exactly the same time the flame became visible on the high-speed films, and that the nozzle of one rocket had swiveled during its last moments of guided flight, as if compensating for the failings of the other booster.

Still, NASA—insisting that its engineers would have to work systematically through an exhaustive fault tree diagnosis before reaching conclusions of any kind about the causes of the accident—chose not to release any of this information to the press.

But now the agency's reticence and apparent stonewalling about what it knew began to backfire. Veteran reporters on the space beat, accustomed to the agency's traditions of apparent transparency and reliant on their convivial relationship with its public affairs officers, now found themselves cut off from their chief sources of information and denied answers to questions about even the most basic facts surrounding the accident.

As editors and news producers scrambled for scoops on the story, the old guard was reinforced by a younger generation of reporters—who had not borne witness to the marvels of Apollo, but had instead been inspired to enter journalism by the revelations of Watergate and the downfall of Nixon. Alert to the government's capacity for incompetence and dishonesty, they were also primed for signs of conspiracy. With the official channels shut down, both groups of reporters began seeking answers to their questions from anonymous sources, working their connections inside NASA and with the agency's myriad contractors, talking to engineers and technicians in the bars of Clear Lake, Huntsville, and Cape Canaveral.

At around noon on Thursday, Jay Barbree—a fifty-two-year-old NBC radio reporter and sometime novelist who lived at the Cape and had covered NASA for almost thirty years—received a tip from a former agency official he trusted. Although he insisted on remaining anonymous, Barbree's source said that the NASA investigators had already narrowed the focus of their inquiry: their suspicion had fallen on the right-hand solid rocket booster, and a leak that may have developed between two of its segments. That afternoon, a second unnamed source confirmed the details for Barbree, and—while their network competitors continued to report that the cause of the accident remained unknown—at 6:30, NBC's Tom Brokaw led the network's evening news with the story. "Fresh developments in the shuttle disaster," he announced. "NASA is looking hard at the solid rockets." Meanwhile, not yet seventy-two hours after the catastrophe, the White House was already considering relieving NASA of responsibility for examining why it had happened.

Donald Regan, President Reagan's felicitously named chief of staff, had been arguing for days that the agency could no longer be entrusted with investigating itself, as NASA had been allowed to do in the aftermath of the Apollo 1 fire. Times had changed, he said: while the media had been willing cheerleaders in the race to the moon, both the space agency and the White House now faced an adversarial press, and a Congress made suspicious by years of scandal and duplicity in the executive branch. With a background in neither science nor technology, but in business, Regan believed that *Challenger* had likely been lost because of mistakes made by someone inside NASA, and an internal inquiry at the agency might present the public with a sanitized version of the truth; he urged the President to appoint an external panel to investigate what really went wrong. NASA's supporters in Washington, DC, fought the idea from the beginning, and some feared that Acting Administrator Bill Graham was so supine that he would agree to any form of inquiry as long as he kept his job; they felt certain that if James Beggs was still in charge, he would never have allowed outsiders to investigate his agency.

The President's team made their final decision on Friday afternoon as they flew back from the memorial in Houston aboard Air Force One.

Regan and National Security Advisor John Poindexter met the President in his private cabin, and the Chief of Staff was blunt. "We can't afford a charge of cover-up here," he told Reagan. Afterward, a cabinet official walked back through the plane to where Bill Graham was sitting and told him how it was going to be. That same day, in the war room in Huntsville, Allan McDonald sat for another screening of high-speed film footage flown in from the Cape—this time, from a camera with a perspective of Pad 39B he hadn't seen before.

With a view of *Challenger*'s right-hand solid rocket at the moment of ignition, this camera had captured a decisive piece of evidence: just over six-tenths of a second after the boosters had lit, McDonald could see a series of eight puffs of black smoke escape from near the aft field joint in the rocket. Looking more closely, it was clear to him that each puff blew upward toward the nose of the rocket, just as it would if it was being directed out of the joint by the clevis, following a failure of the O-rings at ignition. McDonald no longer had any doubt: his boosters had killed the astronauts, after all.

In the meantime, the press kept digging: reporters had begun asking about rumors that Rockwell executives had been so concerned by the closed-circuit TV pictures they had seen of ice on the pad during the countdown that they had called and asked to delay the launch. On Saturday, the *New York Times* quoted another anonymous source, who described the drop in pressure inside the right-hand booster a minute into flight, and said that NASA now believed the accident had been caused by a flame leaking from the rocket and burning a hole in the shuttle's external fuel tank. Even so, the agency wasn't ready to say anything publicly—perhaps, the paper suggested, because it feared who would be blamed for what happened. Asked that day if there had ever been a leak between the segments of a solid rocket booster, Jesse Moore said that no such fault had ever appeared in any one of nine static tests or twenty-four previous shuttle flights. "In 57 uses, there has never been a leak, not a single one. That's not an opinion, it's just a straight fact."

That night, at a press conference in Cape Canaveral, NASA finally released a fifteen-second sequence of videotape and three photographs taken from the high-speed film the astronauts had analyzed in Houston, showing the distinct bright orange glow near the aft joint of the rocket. But the agency's spokesman refused to describe the phenomenon as a flame. Instead, he insisted on calling it "an unusual plume," for which he offered no explanation beyond a statement from NASA's interim investigation board, which he read aloud: "The cause is unknown and neither the board nor NASA will speculate as to the cause or effects of this observation," he said. He made no mention whatever of the film showing the telltale black smoke at ignition.

::::::::::::::::::

On Monday, February 3, the President made the formal announcement that he was stripping NASA of the authority to continue investigating itself. Instead, he had appointed William Rogers—a Republican loyalist, Washington insider, and former Attorney General who had served as Secretary of State under Nixon—as the head of the independent Presidential Commission on the Space Shuttle Challenger Accident. "It's time now to assemble a group of distinguished Americans to take a hard look at the accident," Reagan said, "to make a calm and deliberate assessment of the facts and ways to avoid repetition."

The twelve members of the panel had been drawn from a list supplied by Acting Administrator Graham, and included a selection of prominent scientists and aerospace celebrities, among them Neil Armstrong; Sally Ride; lauded test pilot Chuck Yeager; Air Force General Don Kutyna, head of the Pentagon's side of the shuttle program; and the colorful Nobel Prize–winning physicist Richard Feynman. Empowering them to pursue any avenue of inquiry necessary to uncover what had gone wrong and assured of the full cooperation of NASA by Graham, Reagan gave the commission 120 days to investigate and deliver their official report. Until then, Rogers told the press, the agency's remaining fleet of three Space Shuttles would remain grounded.

While touting the independence of his newly assembled team, Rogers also made clear to the gathered reporters that he had no intention of fatally wounding a government agency that in his view had, for nearly thirty years, embodied all that was great about the United States. "We are not going to conduct this investigation in a manner which would be unfairly critical of NASA," he said, "because we think—I certainly think—NASA has done an excellent job, and I think the American people do."

::::::::::::::::::

Back in Utah, Roger Boisjoly had driven to his office at the Morton Thiokol plant the morning after the accident and picked up work where he had left off. He attended a meeting of the seal task force on time at 8:00 a.m., just as he had for months, while in the company lab technicians began their long-awaited temperature testing of O-ring samples as if nothing had happened. Although shocked and distraught, he tried to distract himself from his grief by concentrating on the effort to fix the faulty rocket joints, no matter how futile it might now seem.

But later that day, the company issued instructions for everyone who had worked on *Challenger*'s solid rocket motors to hand in any relevant documents they had—for safekeeping, pending an investigation. When word reached Boisjoly, his distrust in Thiokol's management got the better of him; fearing that it might be the first move in a corporate cover-up, an attempt to bury the long trail of decision-making that had led up to the *Challenger* launch, he decided to hand over only the handful of Viewgraph slides he had prepared for the teleconference the night before the disaster. Everything else—his notebooks and memos, including the desperate "loss of human life" message he had written to the Vice President of engineering in July the previous year—Boisjoly kept to himself. Before he left his office that day, he gathered all the documents and carefully locked them in a desk drawer.

When he arrived at work at the end of the week, Boisjoly was told by his manager that he'd been appointed to join Al McDonald and the failure analysis team in Huntsville. He hurried home to pack, ready for a long stay in Alabama. As he boarded the Thiokol corporate jet that afternoon, along with two suitcases full of clothes, the engineer also carried the package of what he would come to think of as his "smoking gun" memos.

Reporting to the NASA facility the next morning, Boisjoly received his security badges and got to work, but was struck by the lack of remorse he found among the Marshall engineers on the investigation. Apparently focused solely on finding and fixing problems with the aim of returning the remaining fleet of shuttles to flight as soon as possible, they seemed to him untouched by sorrow or regret at the deaths of seven astronauts. And, as they explored the possibility that *Challenger*'s right-hand booster had sprung a leak in flight, the leaders of the Marshall investigation team focused on what Boisjoly saw as the most far-fetched explanations for what might have happened.

The NASA managers directed the team to look for potential mistakes made by the Thiokol assembly teams who stacked the segments of the rockets at the Cape in the weeks before launch. They examined logbooks, processing records, and photographs of the assembled segments, and said there was evidence that the technicians in Florida had twisted one of the O-rings in the joint. They dispatched a team of engineers to Utah to perform tests to show how a twisted seal could cause a leak in the rocket. When that theory proved groundless, they tried to show that the rocket segments had been misshaped when put together, and the resulting distortion had led to the accident. But that, too, went nowhere. By the end of his first week in Alabama, Boisjoly had begun to suspect that the Marshall engineers were determined to look at every single possible explanation for the disaster but one: the freezing temperatures on the launchpad the night before the accident.

The first public hearings of the Rogers Commission began just before 10:00 a.m. on February 6, at the headquarters of the National Academy of Sciences in Washington, DC. Bill Graham was sworn in to introduce a day of briefings from NASA experts intended to bring the dozen members of the panel up-to-date on events since the accident; public fascination with the proceedings was such that CNN broadcast everything live. The tone was formal and genteel: NASA officials apologized for not being better prepared; Chairman Rogers thanked them for having appeared at such short notice. He explained to the audience that the agency had been cooperative, and that the commission planned to base their inquiry on investigative work that NASA itself conducted. And, at first, the agency managers might have been forgiven for not taking Rogers too seriously: when Jesse Moore had to explain to him that the initials STS stood for Space Transportation System, science reporters in the auditorium groaned in disbelief.

But whatever his ignorance of NASA jargon, the Chairman had been reading the papers, which just the day before had raised new questions about whether the accident had been caused by the record cold on Pad 39B: Morton Thiokol admitted to a *New York Times* reporter that the solid rocket propellant had not been intended for use below 40 degrees Fahrenheit. Rogers had also heard that Thiokol had warned NASA about the low temperatures at the Cape and, as Jesse Moore moved through his presentation about the background to the *Challenger* launch, the Chairman interrupted.

"I notice a press report that one of the contractors said that they gave a warning of some sort about the cold weather," he said. "Could you deal with that please?"

Moore promised he would get to it later.

"Fine," Rogers replied.

But Moore said nothing further about it, so Rogers followed up with the next witness, Arnie Aldrich.

"I do not recall such a warning," Aldrich said.

Next, Judson Lovingood, the Deputy Manager of the Shuttle Projects Office at the Marshall Space Flight Center, took to the stand to deliver a briefing dedicated to the shuttle main engines, solid rockets, and the external tank. In describing the nature of the rockets, he allowed that engineers had seen some damage to the O-ring seals between segments on previous shuttle flights, but assured the commission that there had never been any erosion of a secondary seal; although the charring in the O-rings was certainly an anomaly, the problem had been thoroughly addressed and documented.

Finally, as Lovingood drew his presentation on the rockets to a close, he returned to the question Rogers had asked at the beginning of the session. But—even though he had sat in the teleconference himself, and deemed it so important that he had suggested his boss, Center Director Lucas, should also attend—Lovingood remained vague about the details.

"We did have a meeting with Thiokol . . . we had a meeting where there was some concern about the cold temperatures," he said. "And Thiokol recommended to proceed on the launch."

He did not mention that the contractors had, before reversing themselves, vigorously opposed firing the rockets—and provided an initial written recommendation not to proceed. Instead, he pressed to wrap up briskly.

"Is there anything else on the booster?"

No one spoke.

"I guess not," Rogers replied.

Lovingood moved on to his next slide.

:::::::::::::::::::

But that weekend, the inquiry took an abrupt turn. On Sunday morning, the *New York Times* carried a shocking front-page splash: "NASA Had Warning of a Disaster Risk Posed by Booster." The story, based on internal memos leaked to the paper by a budget analyst at the space agency's headquarters, revealed the long history of damage to the seals stretching back to 1981, and quoted from internal documents citing the risks of O-ring failure: "loss of vehicle, mission and crew due to metal erosion, burn-through and probable case burst resulting in fire and deflagration."

The story described how copies of the memos had been sent to Jesse

Moore, and how the seal damage they documented seemed clearly to contradict what Judson Lovingood had told the commission under oath only a few days before. And all of it was news to William Rogers and the other members of the commission.

Rogers, embarrassed at having been caught flat-footed by the press—and made the victim of apparent dissembling by NASA—was furious. He immediately phoned Acting Administrator Graham and told him to get his act together: *Get in here and tell us everything you know about those memos.* Rogers demanded a full accounting of the history of problems with the O-rings, including copies of all the memos mentioned in the paper, any other relevant documents about the seals—and details of any issues related to cold weather.

That afternoon, Allan McDonald—taking the weekend off at home after almost two weeks of working twelve-hour days on the failure analysis team in Huntsville—received an urgent phone message from his boss at Thiokol. He needed McDonald to fly to Washington, DC, that night, to prepare for an emergency meeting of the commission on Monday afternoon; McDonald and two other engineers from the company were required to support NASA's presentation about the O-ring problems. The briefing would be delivered by Larry Mulloy from Marshall—but Mulloy wanted the Thiokol team to be on hand to provide answers to technical questions, if necessary. The hearing would be held in closed session, out of view of the media: Rogers wanted no further surprises before a planned public session the following day. And, though they would not be attending the meeting, also flying into Washington, DC, that night were three other company executives: Senior Vice President Jerry Mason and his sidekick Cal Wiggins, who had presided over Thiokol's teleconference the night before the launch, and an in-house counsel from the Space Division, Lee Dribin.

Chairman Rogers convened the hearing at 2:10 p.m. on Monday in Room 476 in the Old Executive Building on 17th Street NW in Washington, DC, directly across the street from the White House. The conference room was in the bowels of the building, and it was hot. Rogers sat at the head of a long wooden table, flanked by the other members of the commission, with space at the opposite end for those testifying; when McDonald and his fellow Thiokol engineers entered, they took seats with other observers along the back wall.

From the beginning, Rogers made his displeasure clear: he described the leaks in the press as unpleasant and unfortunate; he hoped that everyone had learned a lesson from the events of the weekend. He called on Bill Graham and his staff to honor the promises of cooperation they had made, and warned them against stonewalling: there was no point in trying to conceal what they knew; it would all come out in the end.

"This is not an adversarial procedure," he told them. "We would hope that NASA and NASA's officials will volunteer any information in a frank and forthright manner."

But as their testimony began, to Al McDonald it was obvious that the agency staff were not motivated by honesty. Graham began by issuing a lawyerly disclaimer that seemed designed to indemnify his men from perjuring themselves on the stand. Most of the testimony that followed came from Larry Mulloy, who steered the same careful line that his superiors had when asked about the teleconference on the eve of the launch. "We all concluded that there was no problem with the predicted temperatures," he told the panel, "and I received a document from the solid rocket motor project manager at Thiokol to that effect, that there was no adverse consequences expected due to the temperature on the night of the 27th."

In his seat at the back of the room, McDonald listened with rising disbelief until, at last, it dawned on him that Mulloy and the other NASA managers had no intention of admitting that Thiokol had tried to stop the launch. And he recognized, too, that if he was to make certain that a disaster like this never happened again, it wasn't enough to fix whatever technical faults had caused the accident; the decision-making process had to be changed as well. He made up his mind to do something, and raised his hand to speak.

No one paid any attention.

Instead, Mulloy kept talking, answering questions, moving along with his prepared presentation. Now McDonald stood and waved his arm in the air, but still to no avail. He heard Rogers take the floor, preparing to adjourn for the day. McDonald grew desperate; he left his seat and moved toward the end of the table.

As he approached, Mulloy noticed him at last.

"Mr. Chairman, Al McDonald from Morton Thiokol wanted to make a point."

"I wanted to say a point about the meeting," McDonald began, his hands trembling. He described how he had been at the Cape when he received the first call from Thiokol about the cold, and asked to set up the teleconference the night before the accident; he explained how he had sat in Trailer Complex C with Larry Mulloy, and received the data by fax from the engineers in Utah.

"The recommendation at that time," he said, "was not to launch below 53 degrees Fahrenheit."

General Kutyna asked what the actual temperature at launch had been.

"Twenty-nine degrees," Mulloy said.

McDonald sat down. The room fell silent.

"Could you stand up again and say that a little louder so we could hear it?" Rogers said. "I'm not sure we all understood what you said."

McDonald stood, and repeated himself—this time in more detail.

Now there was consternation among the commission members: Thiokol had recommended against the launch, but then *changed their minds*? Why? Richard Feynman seemed bewildered by the logic of the events McDonald was describing. For the first time, it was clear that the accident was not simply a technical failure—but also, perhaps, one of drastic human error. Rogers said he wanted to see hard evidence—documents and data—yet also established that McDonald himself had not been present in Utah when the crucial Thiokol caucus was held. To understand what had happened, other witnesses from Thiokol would have to testify.

"We really want to get the people before the Commission who made the decisions, and ask them, Why did they appear to change their minds?" he said.

Then he called for an immediate recess.

Still reeling from McDonald's revelations, the members of the commission gathered in a small anteroom to discuss what to do next.

"This is explosive," Rogers told them. "We've got to control this. If this gets out in the press, we're going to lose control of the investigation—and we'll lose credibility." The Chairman said it was imperative to find out who knew about the recommendation not to launch, and why it was reversed. "Who knew what, when?" he asked.

And until they established the facts for themselves, Rogers wanted to keep the news out of the public eye. Instead of taking testimony from the Thiokol engineers in the next day's public session, they agreed to avoid all discussion of the weather and the teleconference until a special secret hearing at the end of the week, in Cape Canaveral. Then they would put McDonald and his colleagues on the stand, and find out what really happened the night before *Challenger* launched.

But word of the drama in Room 476 traveled fast.

In the offices of Morton Thiokol a few miles away in Crystal City, Virginia, it was around 3:30 p.m. when Jerry Mason summoned the corporate counsel into a conference room.

The lawyer found Mason sitting with Cal Wiggins; he noticed that the Senior Vice President's long face had turned a ghostly white. "What's wrong?" he asked.

"Sit down," Mason said.

"I need to talk. The cat's out of the bag."

CHAPTER
TWENTY-SIX

THE TRUTH

By the time he took the call from Bill Graham on the first Sunday in February 1986, Dr. Richard Feynman was already a dying man. Seven years earlier, he had been diagnosed with a rare cancer for which the prognosis was bleak: inside his abdomen, doctors discovered a tumor the size of a melon that eventually enwreathed his intestines. Feynman had survived only after losing a kidney and his spleen and enduring two difficult surgeries; complications included a split aorta, and the transfusion of seventy-eight pints of blood. After that, he had been diagnosed with an equally rare and aggressive lymphoma, affecting his bone marrow. This time, the doctors were at a loss for solutions or explanations; Feynman refused to connect the development of two such unusual cancers with his work on the United States' atom bomb project as a young man. But when the Acting Administrator of NASA had reached him at home in California with the invitation to join the *Challenger* investigation, he knew that the work would devour much of what little time he had left. "You're ruining my life," he told Graham.

At sixty-seven, the Caltech professor was a long-haired iconoclast whose recent memoir *Surely You're Joking, Mr. Feynman!* had proved an unexpected bestseller and made him as famous for his countercultural antics—playing the bongos, visiting topless bars, experimenting with marijuana and sensory deprivation—as he was for his genius as a theoretical physicist. Disdaining the formality and plaudits of academia, Feynman was less proud of his scientific achievements than his ability to penetrate trickery and artifice; but he was also a shrewd curator of his own myth. And once his wife convinced him that what the *Challenger* commission really needed was one renegade investigator prepared to head out on his own in the pursuit of answers, he threw the full weight of his intellect into the inquiry.

Before flying to Washington, DC, Feynman—whose expertise lay in quantum mechanics—used his connections at NASA's Jet Propulsion Laboratory in Pasadena to arrange an intensive briefing on the technology of the shuttle. From the JPL engineers, he heard in detail about the troublesome history of the main engines and the difficulties with the seals in the solid rockets—about how bubbles in the zinc chromate putty packed in the joints helped create jets of hot gas, and how these could damage the O-rings. On the first page of notes he scribbled during his briefing that day, Feynman wrote: *O-rings show scorching in clevis check. . . . Once a small hole burns thru generates a large hole very fast! Few seconds catastrophic failure.*

But once the meetings of the commission began, Feynman was frustrated by the pace of their work and the torpid inefficiencies of the hearings: he felt he had learned more in that initial briefing in Pasadena than he had from hours of circumlocutions from NASA officials. He itched to go out and question individual engineers himself; he found the information he received through formal channels utterly useless. During the closed hearing in which Al McDonald revealed Thiokol's initial opposition to launch, Feynman asked the agency managers to provide him with detailed data about the resiliency of the O-rings at low temperature. Later that same afternoon, he received a thick stack of documents, reproducing the request at every level of NASA's meandering bureaucracy. Sandwiched in the middle was a piece of paper containing an answer—not to the vital question of how the rubber rings would spring back into shape during the first milliseconds of the rockets' ignition, but how they behaved over the duration of several hours.

But that night, during a dinner with Rogers and the other members of the commission inside the White House complex, Feynman looked down at the glass of ice water on the table before him and decided he didn't need NASA's engineers to help him understand the O-rings. He could do it himself, with his own experiment. All he needed was a sample of the rubber, and realized he could use a piece from inside the cutaway section of the solid rocket field joint the agency managers had been passing around during the hearings. When he took his seat on the panel the next morning, Feynman faced a packed auditorium lit by bright TV lights and a row of cameras. In the pockets of his jacket he had a screwdriver, a pair of pliers, and a metal clamp he had found in the medical center at NASA headquarters.

According to Chairman Rogers's strategy, the hearing would be devoted to testimony from Larry Mulloy—who would simply be repeating much of what he had said in the closed session the day before—and from the NASA budget analyst whose memos had appeared in the papers over the weekend. Rogers wanted to hold back all discussion of the cold temperatures at launch—and the Thiokol teleconference—until later in the week.

And, once Mulloy began to speak, the chairman told him to avoid dwelling on the weather. Instead, the Marshall engineer paced the floor with a wand-like microphone in his hand, working laboriously through the same presentation he had given once already: he described the Flight Readiness Review process and the rationale that lay behind acceptable risk; he detailed every possible scenario of O-ring damage in the solid rockets, and said that he had seen no evidence of erosion for a year, and they had no reason to believe that it was dangerous. For the first time, he admitted in public that cold temperatures could affect the sealing of the joints. But so far, he explained, it remained impossible to say whether the joints might be the cause of what he called the "51-L incident"; NASA's temperature tests on the O-ring material were still continuing: the results would not be ready until later in the week.

But Feynman had heard enough: he was impatient to cut through Mulloy's opaque jargon in a way the public could easily understand. At his seat, the physicist prepared the experiment: levering an inch-long piece of O-ring out of the steel channels in the cutaway model, he folded the rubber gasket over on itself and tightened it into the clamp; then he submerged the assembly in a Styrofoam cup filled with ice water, where the temperature hovered near freezing. During a recess, he approached Rogers to ask permission to conduct an on-camera demonstration. The chairman reluctantly agreed. When the meeting reconvened, Rogers gave Feynman the floor.

"This is a comment for Mr. Mulloy," Feynman began, and reached into the cup for the clamp.

"I took this stuff that I got out of your seal and I put it in ice water, and I discovered that when you put some pressure on it for a while and then undo it"—he leaned away from the table, giving the TV cameras an excellent view as he unwound the clamp and released the piece of O-ring—"it doesn't stretch back."

Like a seasoned illusionist, Feynman passed the sample to his left, as if Air Force General Donald Kutyna might affirm for the audience at home that seeing was believing.

"In other words, for a few seconds at least—and more seconds than that—there is no resilience in this particular material when it is at a temperature of 32 degrees.

"I believe that has some significance for our problem."

Two days after Feynman's televised experiment, the members of the commission flew down to Florida aboard a pair of NASA planes for their hearings at the Cape. By now, the causes of the accident were swimming into focus, for both Rogers's panel and the public: during testimony the next morning at the Kennedy Space Center, agency engineers revealed the photographs of black smoke escaping from the casing of Challenger's right-hand solid rocket as it left the pad; afterward, NASA finally released the pictures to the press. And the next day, for the first time the commission planned to cross-examine the Morton Thiokol executives who had presided over the teleconference the night before the launch: Jerry Mason and Cal Wiggins; Bob Lund; and Joe Kilminster, who had put his signature on their recommendation to launch. The company had also flown in Al McDonald and Roger Boisjoly, who had both asked to appear. McDonald was certain that NASA had deliberately misled the commission about the prelaunch debate with Thiokol—and if he and Boisjoly didn't stand up now and reveal the truth, it would never come out. McDonald had already made detailed notes about the teleconference, and believed them so important that he had left copies with a colleague in case his plane crashed on the way to the Cape. Meanwhile, in his motel room in Huntsville, Boisjoly had taken to sleeping with his packet of "smoking gun" memos beneath his pillow—and eventually grew so paranoid about what he knew that he feared his life might be in danger, and mailed an envelope of duplicates to his wife, with instructions to open it if anything should happen to him.

Following his unexpected intervention during the commission's closed session at the beginning of the week, both the NASA officials from the Marshall Space Flight Center and his bosses from Thiokol were wary of what McDonald might disclose at the Florida hearing. Before the engineer arrived at the Cape, the two groups had coordinated what they planned to tell Rogers, and Larry Mulloy had given the strategy his approval. "Looks good," he said, "if McDonald doesn't throw another turd on the table."

That night, Thiokol's corporate chief of aerospace, Ed Garrison, gathered his staff for a briefing from the company lawyers in a conference room at a Ramada Inn near Merritt Island. They suggested a strategy used by the defense in criminal prosecutions, advising their clients to say as little as possible in their testimony, and to reply to questions with simple "yes" or "no" answers. But McDonald made clear he had no intention of holding back; he told the lawyers he planned to tell Rogers every last detail he could recall about the night before the launch. "I agree with Al, and I'm going to do the same thing," Boisjoly said.

When the hearing began early the next morning, Chairman Rogers quickly stressed that he was only interested in hearing the complete truth. Before Thiokol Senior Vice President Jerry Mason began speaking, Rogers delivered a blunt warning: "Please disclose anything that you know about that may turn up," he said. "If you have documents that we don't know about that would be embarrassing to you, tell us about them now. We don't want to have to pry information out of you . . . tell us the whole story, if you will."

Almost immediately, the commissioners' questions began to lay bare the false assumptions and failures that had lingered in the solid rocket program since its inception: how the engineers had thought the booster design was so similar to the ones flown for years by the Air Force that its reliability was assured; that the way the joint worked in practice was a potentially dangerous kludge that they didn't fully understand; that no one had ever bothered to conduct meaningful tests at low temperatures until it was too late; that they had kept the shuttle flying regardless of a long series of ominous warnings. And, as the Thiokol executives described the progress of the prelaunch teleconference, their decision to change their recommendation from "no-go" to "go" came under scrutiny. Mason and Bob Lund told the panel that they guessed that there had been an even split among the engineering team about whether it would be safe to launch. But they admitted that they had never taken a poll to find out. At that point, Richard Feynman cut in to ask who among the Thiokol engineers were the most expert on the rockets' seals.

"Roger Boisjoly and Arnie Thompson," Bob Lund said.

"Mr. Boisjoly," Feynman asked, "were you in agreement with the result of this caucus that it was OK to fly?"

"No, I was not."

"Now, Mr. Thompson, were you in agreement, and so forth?"

"I was not," Thompson said.

The two engineers went on to testify in detail about the information they had presented before the launch; they showed the panel their Viewgraph slides; hours passed. Rogers was openly bewildered by Mason's convoluted account of how, after their recommendation to delay, he had changed his mind. "The impression is that you were directed to do it, that there was so much pressure to get this launch off the ground that you were directed to do it, and you did it," he said. "Now, if that is not the case, try to explain it in language that the public would understand."

Yet the Thiokol Vice President denied that any such direction had taken place. Neither he nor Larry Mulloy—or anyone else who had taken the stand that morning—would admit to having felt or applied pressure to alter the launch recommendation. The commissioners were astonished that, knowing everything they did about the history of the rockets and their own best-qualified engineers' stated concerns, the Thiokol Vice Presidents chose to recommend a launch regardless.

"I mean," Rogers said in exasperation, "that is just unbelievable to me."

After almost four hours of testimony, the chairman was once again concluding the proceedings for the day, when Al McDonald spoke up. "Since I caused this meeting to come about," he said, "I would like to testify, I guess."

McDonald was torn. He knew how damaging what he planned to say would be, both to NASA and to Morton Thiokol: to people he had respected, and worked with for years; his testimony would endanger their jobs, the future of NASA and the company, and his own career. The faces of men in the room he recognized were clouding with anger and distress. But he wanted the full truth to come out—not just to make sure such an accident never happened again, but to settle his own conscience.

Taking the stand, McDonald spoke briefly, but without interruption and, at last, the entire story of the night of January 27 emerged: he told the panel about how he had opposed the launch, about the repeated arguments he had with Larry Mulloy, and the belligerent attitude the NASA project manager had displayed during the conference; McDonald described how upset he had felt about the decision to proceed, and how he had tried again and again to make Mulloy and Stan Reinartz see reason; he explained how he had given them further grounds to delay the launch—the rough seas, the ice on the pad—but had been told that those were not his concerns. Finally, he repeated the warning he had given them: that he would not want to have to stand up in front of an inquiry and explain why they had decided to fire the rockets in violation of NASA's own launch parameters. His testimony made clear that Larry Mulloy had not been forthright with Rogers, and that the Thiokol executives were trying to protect their most valuable customer.

When McDonald finished, there was just one question—from Alton Keel, the commission's executive director, Rogers's right-hand man, and a trained engineer who had followed the shuttle program from the start.

"The inference, Mr. McDonald, from your testimony is that you were under pressure, perhaps unusual pressure, from NASA officials, to go ahead with the launch. Is that an accurate inference?"

"Yes," he replied.

"And did I understand, too, that you did not sign off on this one?" Rogers asked.

"No, I did not."

"Was that unusual?"

"I believe it was, yes."

Emotionally exhausted, ashamed at being forced to admit how NASA and Thiokol had failed, and still distraught by the unnecessary deaths of seven Americans, when he stepped down from the stand McDonald found himself fighting back tears. As he walked to the door, Sally Ride rose from her seat to embrace him.

"God, that took a lot of guts," she said.

::::::::::::::::::

That afternoon, McDonald and Roger Boisjoly boarded the Thiokol corporate jet in Titusville for the long flight home to Utah. The atmosphere in the cabin was venomous: neither Jerry Mason nor any of the other executives on board would speak to the two engineers. They rode in silence to Huntsville for refueling, where McDonald— unable to face his colleagues' enmity any longer—decided to stop for the night. Almost as soon as he had checked into his motel room, the phone began to ring as reporters from all over the country called him for comment. Mystified at how they had tracked him down—let alone how they had so swiftly learned of his testimony before a closed hearing—he refused to talk.

In the meantime, Chairman Rogers had telephoned the White House to tell the President that the accident had not simply been a technical failure after all; it was human error of the most shocking kind: he suspected that the entire launch approval process had failed. The next day, Rogers issued a public statement formally rerouting the course of the inquiry. From now on, none of the NASA officials involved in the final recommendation to launch mission STS-51-L would be permitted to play an active part in the investigation. "In recent days," the communiqué explained, "the Commission has been investigating all aspects of the decision-making process leading up to the launch of the *Challenger* and has found that the process may have been flawed."

It was another humiliating blow for the agency; staff in Washington, DC, were stunned. Jesse Moore stepped down from his position a few days later; James Beggs—still on a leave of absence to defend himself on federal charges—would resign soon afterward.

:::::::::::::::::::

Back on Merritt Island, the first shattered fragments of *Challenger* to emerge from the ocean had begun arriving at the Kennedy Space Center, concealed beneath olive-drab tarpaulins on the trailer of an 18-wheeled flatbed truck. Piece by piece, NASA engineers unloaded the wreckage and laid it out on a grid of four-foot squares marked with yellow tape on the floor of a three-story warehouse near the Vehicle Assembly Building—pieces of a colossal jigsaw that investigators hoped would provide the physical evidence of what had triggered the destruction of the most complicated machine in history. Supervising the effort to recover and analyze the debris was Bob Crippen, who had flown down from Houston in a T-38 within twenty-four hours of the accident and moved into a room in the same crew quarters building where the *Challenger* astronauts had spent their last night on Earth.

Devastated by the deaths of his friends, Crippen had spent hours after the accident in tears; he was as low as he had ever been. But he told himself that focusing on work was the best way of coping with his grief and—if he could help find and fix whatever had caused the accident so that the shuttle could fly again—of fulfilling the wishes of his lost colleagues. Taking an office in the Space Center headquarters building, Crippen had begun working twelve-hour days, seven days a week, rarely left the island, and wouldn't return to Houston for months. Officially, he was answerable to senior NASA managers appointed soon after the disaster to run the investigation and ocean recovery from Kennedy; unofficially, nothing there happened without his approval. "Crippen calls all the shots," one staff member told a reporter.

Work to recover the *Challenger* wreckage had begun as soon as the debris had ceased raining from the sky on January 28. That morning, the captain of the Coast Guard cutter *Point Roberts* had been keeping station seven miles offshore beneath the shuttle's flight path, just as he had many times before, as part of a routine to keep the area clear of maritime spectators during launch. He and his crew watched in awestruck silence, awaiting orders, as pieces of the spacecraft splashed into the water around them for almost sixty minutes. Eventually, at around 1:30 p.m., he received instructions to head into the center of a floating debris field that by then stretched for miles in every direction across the surface of the Atlantic, the sunlit waves glinting with fragments of aluminum and heat-resistant tiles torn from the skin of the shuttle. The Coast Guard crew collected what they could, and the next afternoon the *Point Roberts* docked inside the closed perimeter of the US Navy submarine base at Port Canaveral carrying six hundred pounds of wreckage—most of it in pieces so small that it was almost impossible to determine what they might once have been. It marked the beginning of what would become one of the largest maritime salvage operations in history.

Joined by a growing flotilla of ships from the Navy and the Coast Guard—and including NASA's two booster recovery vessels—within twenty-four hours the search had called in planes and helicopters to scour the surface of the sea from above, sweeping an area that ranged up the coast of Florida from Melbourne to St. Augustine, and eighty miles out to sea; Coast Guard jets equipped with surface-tracking radar took to the skies to scan 33,000 square miles of ocean, as far north as Savannah, Georgia. Meanwhile, the agency summoned help to the Cape from every corner of the United States: the Navy's Supervisor of Salvage, Charles "Black Bart" Bartholomew, flew in from Washington, DC, to coordinate a search for submerged wreckage, while an Explosive Ordnance

Disposal Team arrived from Eglin Air Force Base to handle the volatile remains of the shuttle's rocket propellants; and the head of the accident investigation bureau at the National Transportation Safety Board agreed to supervise a forensic examination of what they found.

Carried by wind and tide, the *Challenger* flotsam spread quickly; the crews of local scallop boats found fragments of wreckage in their nets, and pieces began washing up along the length of the Space Coast, where they were recovered by sightseers and local police officers assigned to help locate the debris. Meanwhile, a team of four astronauts was sent from Houston to comb the beaches for body parts. Late in the second day of the search, the crew of a Coast Guard ship recovered drifting objects apparently from *Challenger*'s crew cabin, including notebooks, a tape recorder—and a partially intact helmet, containing human remains. A large section of fuselage from the orbiter's nose was unloaded in Port Canaveral soon afterward. In their Miami operations center, Coast Guard officers used a sophisticated computer program to model the wind and current around where they had picked up the debris, and locate its likely impact point on the surface of the ocean. They soon believed they had found the area where the *Challenger* cabin had come down, within a circle roughly a mile in diameter near the Hetzel Shoal buoy, in shallow water some sixteen miles northeast of the Cape.

But NASA's official line remained that *Challenger* and its crew had been consumed by the explosion, their ashes scattered in the stratosphere, and the agency refused to release details of the salvage effort; after receiving a phone call from the Astronaut Office, a Coast Guard spokesman was convinced to keep the recovery of the cockpit debris secret, and lied about what he knew during a TV interview. Bob Crippen, since reviewing the film evidence of the spacecraft's last moments, refused to believe that the crew compartment could have survived the conflagration. He tried to stifle discussion of recovering the remains of the astronauts, fearing it would give false hope to their bereaved relatives. When fellow astronaut Jim Bagian—a flight surgeon and qualified diver assigned to the underwater search effort—told him he expected to find the cabin on the seafloor, Crippen became angry.

"I don't want to hear anybody say that again," he said. "We are not going to talk about it anymore."

By February 3, Crippen had determined the priority in which the *Challenger* debris would be recovered. First came the evidence of what had gone wrong: the lower parts of the right-hand solid rocket booster, where the black puffs of smoke and plume of flame had appeared; next, the same parts of the left-hand rocket, for comparison during the accident investigation, and the struts that had attached both boosters to the external fuel tank. Only after that would they focus on looking for whatever might be left of the crew compartment.

By February 7, the drifting debris from the shuttle had either sunk to the bottom or been carried away by waves and wind. Crippen gave his approval for the Coast Guard to call off their surface search, and Black Bart's underwater work began in earnest. Tall, blond, and mustachioed, Captain Bartholomew was a soft-spoken Navy diver who liked to tell people that he had quit flight school because it bored him. His unit arrived at the Cape with experience in some of the world's most complex deep-sea salvage expeditions—including the retrieval of a hydrogen bomb lost with a B-52 bomber that went down off the coast of Spain in 1966 and the search for the wreckage of the Korean airliner shot down by the Soviet air force over the Sea of Japan in 1983.

With the relentless coverage of the catastrophe in the press and TV news, and mounting pressure to uncover its cause, NASA was impatient for the more than 3 million pounds of *Challenger* debris to be found at once; but Black Bart's approach was as patient and methodical as it always was. Using film footage and flight-tracking radar data the agency had recorded of the shuttle as it broke up and tumbled into the ocean, he established an initial search area encompassing 250 square miles, in a part of the Atlantic that lay astride the powerful currents of the Gulf Stream, and the water was anywhere from 70 to 1,200 feet deep. To locate potential pieces of wreckage, Bartholomew sent out ships equipped with side-scanning sonar—similar to the equipment that had been used the year before to finally locate the wreck of the *Titanic*—which sailed north to south along parallel overlapping tracks, as if they were mowing a lawn.

The acoustic signals from the sonar "fish" towed behind each vessel were recorded on board as maroon-colored traces on eight-inch-wide rolls of strip chart paper, revealing the size and position of unidentified objects on the seafloor. At the end of every day, the strip charts were carried by a small boat back to the shore for analysis by a civilian sonar expert sitting in one of two beaten-up trailers Black Bart used as his headquarters, parked onshore near Port Canaveral. Fifteen hours a day, seven days a week, the technician pored over the sonar plots, relying on a combination of expertise and intuition to mark what seemed most likely to be pieces of *Challenger*—instead of coral formations, or the wreckage of the scores of missiles launched from the Cape over the previous thirty years, which also littered the search area. He assigned an individual number to each contact he picked out, entered its location into a computerized database, and then passed them on to Bartholomew for further investigation: those in shallow water would be examined by Navy divers; the deepest contacts went to teams deploying remote-controlled robot submarines or manned submersibles.

On Sunday, February 16, the crew of the *Johnson Sea Link II* were nosing their small battery-operated submersible through 1,200 feet of water 43 miles east of the Cape, in search of a piece of wreckage known only as Contact 21. As the vessel moved slowly across the bottom, its video cameras at first captured nothing but an undulating desert of yellow mud, flecks of white sediment caught in the floodlights as they drifted past the submersible in the half-knot current. But then, looming out of the dark water, the crew caught sight of a crescent of white carbon fiber, embedded in the mud and frayed at the edges; nearby, they spotted twisted lengths of piping and a small, spherical tank lying on the seafloor. Seizing a small piece of the wreckage with the submersible's manipulator arm, the crew carried it to the surface, where engineers from the Marshall Space Flight Center confirmed the details of what they had found: components from the aft skirt of *Challenger*'s right-hand solid rocket booster. The bottom-most section of the rocket, the piece provided no evidence of the cause of the accident; it had been forty feet from the joint where the telltale puff of black smoke appeared during liftoff. But it was a start.

At a press briefing at the Cape a few days later, Captain Bartholomew and the Air Force colonel coordinating the operation said they were confident they would eventually find everything they were looking for. "We're hot on the trail of the most important items," the colonel said. But they added that the recovery could take months: this was not an ordinary salvage expedition; they were treating the seabed like a crime scene.

Freed of its reliance on NASA for information, the work of the Rogers Commission was moving quickly. In the days following Allan McDonald's startling testimony in Florida, Rogers reorganized his team to begin conducting its own investigation and present to the public what they had discovered in private session. The dozen members of the panel split up and headed to Cape Canaveral, Huntsville, and Thiokol in Utah to gather more information, ahead of a series of open hearings in Washington, DC. Chairman Rogers planned to put NASA and senior Thiokol management on the stand beside Allan McDonald, Roger Boisjoly, and the other engineers who had tried to stop the launch, and make public how their recommendations had been ignored.

In the meantime, leaks about what McDonald had already told the panel propelled him into the national spotlight: after the details were reported on the ABC evening news, the telephone at home in Ogden began ringing and barely stopped. The following morning, when his story appeared on the front page of the *New York Times*—"Rocket Engineer Describes Arguing Against Launching"—furious company Vice President Jerry Mason called him to his office to explain himself; later that day, two members of Congress staged an abortive attempt to have him appear before them and the rest of the nation's press. To evade the journalists and camera trucks camped on the street outside the house, McDonald began sleeping at a neighbor's home instead. And, as the media focused with growing intensity on Thiokol's role in the accident, the reverberations could already be felt in Brigham City: on February 14, Jerry Mason informed staff at the Wasatch Division that he would be suspending manufacture of the Space Shuttle solid rockets indefinitely, leading to the first layoffs at the facility in almost twenty years. Two hundred employees would lose their jobs; others feared they would be next. The following week, demoralized men driving out to work at the Promontory plant noticed that someone had daubed graffiti in red fluorescent paint on the wall of a highway underpass.

In letters two feet high, it read MORTON THIOKOL MURDERERS.

Back in Houston, the families of the lost shuttle crew found themselves caught in the unrelenting grip of a personal trauma that had become a national spectacle. Weeks went by, but it remained almost impossible to turn on the TV or walk past a newsstand without a graphic reminder of the most harrowing moment of their lives: the images of *Challenger's* destruction now used to illustrate news stories about William Rogers's investigation, the future of the space program, or the search continuing in the waters off the coast of Florida. Each evening, the ships of the salvage flotilla returned to dock within the secure perimeter of the Trident submarine basin at Port Canaveral, where they were greeted by the Jetty Rats: a mob of reporters and photographers camped across the channel in a public park. In their pursuit of news about the progress of the deep-sea recovery—and, especially, any sign of the remains of the crew—the media's attempts to pierce the blanket of secrecy shrouding the operation had become increasingly sophisticated. Using radio scanners to eavesdrop on ship-to-shore communications and night-vision optics to examine any wreckage brought into port under cover of darkness, the journalists' hunt for a scoop grew ingenious and ruthless. Unconfirmed rumors and ghoulish details about what had been found on the seabed circulated in dockside gossip and in the bars of Cocoa Beach, and Black Bart was reported to hold court in his Holiday Inn hot tub at night, swigging beer and telling tales that would have left NASA officials aghast, had they heard them; but little ever made it into print.

The crew families traveled the country attending one commemorative event after another: George Abbey made sure that they could fly wherever they needed at government expense. But soon they found it hard to leave their homes without being recognized. The day after the televised memorial at the Johnson Space Center, Alison Smith and her sister went to buy new outfits for the next service and found a hush crept behind them through the mall, their unwelcome celebrity drawing silence along their path like a mournful train. Each of the families' numbers was listed in the Houston phone book, and strangers began calling Jane Smith at home, wanting to talk to her about Mike, until the astronaut assigned to her as a Casualty Assistance Officer persuaded her to install an answering machine to screen incoming messages. Before the accident, Smith had worked a part-time job in a friend's clothing store in nearby Seabrook, but afterward had to stop when her presence attracted so much attention that she couldn't get anything done. No longer able to go grocery shopping, Jane instead sent Alison to buy food for the family, rewarding her with the value of any coupons she could redeem at the checkout. June Scobee had stopped watching TV or reading the papers; she knew little of William Rogers's ongoing inquiry, and nothing of the machinations of Morton Thiokol executives. Heartbroken by the sudden loss of the partner who had steadied her through every stage of her life since high school, Scobee became too numb to support her two children. She stopped eating, and wandered the aisles of the supermarket in a forlorn daze: she would put a jar of Jif peanut butter—Dick's favorite—in her cart before recalling that there was no one at home to eat it; she'd collapse, sobbing, to the floor.

By the beginning of the final week in February, Commission Chairman Rogers was ready to go public with three days of new hearings dedicated to interrogating the decision to launch. Day one would be devoted to testimony from Morton Thiokol; the second, to the cross-examination of Larry Mulloy and other NASA Marshall engineers; and on the third day, Rockwell staff would take the stand. As the day of their hearing approached, Jerry Mason and the Thiokol lawyers planned carefully what the company's men would say as they faced questioning in front of the nation's TV cameras. Preparing yet another stack of colored Viewgraphs

and technical presentations with which to beat their audience into submission, by now Mason knew how damaging Al McDonald's testimony
could prove—and so organized the order of the presentations to dis-
credit him in advance. Mason himself planned to appear first, followed
by Bob Lund and Joe Kilminster. McDonald would speak last.

But Rogers had become convinced that some of those who had testified before him in the closed session were guilty of dissembling, or
worse—and certainly of misleading a Presidential inquiry. If anyone was
going to choreograph what happened next, it would be him. As Mason
and the Thiokol team gathered in the Crystal City Marriott to rehearse
their testimony the night before the hearing, they were interrupted by
the delivery of a press release from the commission. The document
listed the agenda for the following morning: the first speaker would be
Allan McDonald. Thunderstruck, Mason screamed at the lawyers to fix
it. But it was too late.

Summoning the Thiokol counsels to his hotel room, Rogers was frank:

"I'm not interested in anything but making a public scapegoat out
of Larry Mulloy and Thiokol management," he said. "The man has lied
under oath. I have an American hero in Allan McDonald. This will be a
public performance."

:::::::::::::::::

The culmination of everything that the commission had uncovered in
the weeks since the accident, the Washington, DC, hearings unfolded
almost exactly as Rogers promised. With calm authority, McDonald
opened the session by recounting his story in full from the beginning;
he described the pressure Mulloy and the other NASA managers had
exerted on Thiokol to approve the launch and his own attempts to stop
them. The commissioners asked only a handful of questions, and treated
him with more deference than they had even the most senior agency
officials at the outset of the inquiry.

But then Jerry Mason took his turn under the hot TV lights. With
most of his prepared remarks rendered redundant by McDonald's narrative, the session quickly turned prosecutorial. Questioning the logic

of Thiokol's final recommendation to launch, the commissioners at first treated Mason's answers with bewilderment, then mounting disbelief.

"The data was inconclusive, and so you said, 'Go ahead!'" Sally Ride said. Richard Feynman made clear that Mason's decision to overturn his engineers' advice to postpone the flight had no empirical foundation; it had, instead, been a reckless gamble.

As the cross-examination continued, Mason's composure faltered: he lost track of the questions; he waved his hands, struggled to answer, and leafed in desperation through his notes; camera shutters clattered like the rifle bolts of a firing squad. One after another, the members of the commission took aim at the condemned man; even Neil Armstrong proved merciless.

"Every launch has a risk, but you take that risk because something must be achieved," General Kutyna said at last. "What was driving here? What caused you to go?" Mason couldn't answer. Kutyna shook his head in exasperation and contempt. At the end of an agonizing hour, the Thiokol executive limped from the stand, his credibility in tatters.

That afternoon, Roger Boisjoly took his turn before the panel, and testified to his long months of attempts to fix the O-ring problems. He read aloud from his "smoking gun" memos, and recalled how, in the climactic teleconference, the Marshall engineers had said that they were "appalled" at his recommendation not to launch. Later, Thiokol engineering chief Bob Lund admitted that he had felt forced to invert NASA's usual priorities: instead of proving that Thiokol's rockets were safe to fly, he was being asked to show conclusively that they weren't. When asked why he couldn't simply have stuck to his original no-go recommendation and refused to bow to the pressure to make a "management decision," Lund was quiet for a long time.

An uncertain smile swam across his face.

"That's probably what I should have done," he said.

Larry Mulloy took the stand once again the next day. He remained pugnacious and self-assured: he said he hadn't been convinced by the Thiokol engineers' warnings because their data didn't seem coherent; and in the two years of problems leading up to the accident, issues with cold weather just hadn't been an overwhelming concern. But this time, almost no one on Rogers's panel appeared to believe him. And they seemed astonished that Mulloy had failed to make sure news of the debate reached the highest levels of the launch decision chain—or any of the astronauts. "Larry," General Kutyna told him, "if this were an airplane, an airliner, and I just had a two-hour argument with Boeing on whether the wing was going to fall off or not, I think I would tell the pilot."

The three days of hearings devoted to what Rogers called "possible human error" came to an end on February 27, almost exactly a month after *Challenger* disintegrated nearly nine miles above the Atlantic Ocean. By that time, the name of Morton Thiokol had become a byword for failure, and the painstakingly nurtured public image of the National Aeronautics and Space Administration had been demolished. The hours of often conflicting testimony and remorseless cross-examination had dispelled any lingering notion that the accident had been caused by an inexplicable technical failure—revealing instead a pattern of mismanagement and miscommunication at the highest levels of the agency. An organization that had, since its inception, boasted of its ability to manage extraordinary risk on the frontiers of technology and learn from its mistakes had instead overlooked a litany of clear warnings; the signals lost in the noise of a complacent can-do culture bred by repeatedly achieving the apparently impossible. Seduced by their own mythos, and blind to the subtleties of engineering complexity that none of them fully understood, the nation's smartest minds had unwittingly sent seven men and women to their deaths.

Arnie Aldrich and Jesse Moore, the two men responsible for giving the final approval to launch, came to face the commission once more near the very end of the third day of the hearings. As the questioning wrapped up, Chairman Rogers thanked Moore and Aldrich for their cooperation and, when he reached his conclusion, he spoke respectfully. His tone was warm, almost avuncular; but what he had to say was as damning for them as it was for the agency they represented.

"You will remember that I did say at one point that we thought the decision-making process may be flawed. I believe I am speaking for the whole commission when I say that it is flawed," Rogers told them. "Clearly flawed."

::::::::::::::::::

With their examination of the role of human error in the accident at an end, the members of the commission now turned to finding its immediate technical cause. They set out to investigate the exact mechanism by which the solid rocket boosters failed, and why it had come only after twenty-four apparently successful flights. They planned to explore how the seals were made, the way the rockets were assembled at the Cape, and whether the constant modifications made to the shuttle fleet since the first launch of *Columbia* in 1981 had placed additional strain on vital parts of the system. At the same time, a dedicated staff including serving astronauts, National Transportation Safety Board investigators, and FBI agents began interviewing those involved at every level of the *Challenger* launch process. Adopting the techniques of a criminal inquiry, the team gathered documents, took testimony, and collected signed affidavits from witnesses at NASA centers and contractors around the country; the work would continue for months.

As the tentacles of the investigation unwound into NASA's vast bureaucracy, in Washington, DC, President Reagan's staff began to wonder how badly damaged the space agency might eventually be by the commission's findings. What had begun at least in part as a political exercise in transparency—to provide dignified supervision of NASA's investigation into itself—had slipped from their control. Disquieting revelations were now spilling steadily into the media, with new and more pernicious details emerging all the time, just as it had during Watergate; there was no telling where it would end.

"It's like an artichoke," one White House official said. "Every day they peel something else."

CHAPTER
TWENTY-SEVEN

APOCALYPSE

It took almost exactly a month to find them, eighteen miles out and nearly ninety feet down, where the gray-green Atlantic water swept past the lighted buoy marking the end of the Hetzel Shoal.

By the middle of the afternoon on March 7, scuba divers Terry Bailey and Mike McAllister had only a few minutes of bottom time left in their search for Target 67. As he swam, the pulses of his handheld sonar gun echoed in Bailey's ears; McAllister carried an inflatable marker buoy. The water was cold and turbulent; visibility was down to ten feet. But the sonar contact was growing stronger: there was definitely something out there.

Bailey and McAllister had been attached to the search and salvage operation since the beginning, and knew the shallows of the Florida coast well. As part of the six-man civilian crew on the shabby Air Force landing craft they knew as the *Lucy*, over the years they had been on standby each time a Pentagon satellite launch went up from Cape Canaveral, ready to head out in the event of a malfunction and recover anything that crashed into the ocean. And McAllister was a local, intimate with each one of the Space Shuttles: back in 1980, he had spent months working among the Puzzle People gluing tiles to the skin of *Columbia* before her first launch, and then moved out to California to do the same on *Challenger* and *Discovery*. McAllister had watched from the ship's mooring in the Trident Basin as the spacecraft he had helped build disintegrated in the heavens, and in the hours afterward had cast off aboard the *Lucy*, with a team of electronic experts from the Cape Canaveral Air Force Station towing a side-scan sonar fish.

But it was weeks before Bob Crippen gave them clearance to begin diving off the Hetzel Shoal, where back in January the Coast Guard computers had determined that the remains of *Challenger*'s crew cabin would lie. On March 6, Bailey and McAllister went down for the first time, to identify a sonar contact designated Target 68. After just a few minutes in the water, the two men found a ragged twelve-foot piece of the orbiter's aft fuselage, already partly encrusted with barnacles, which the crew hauled up from the bottom with a crane. The next day, the *Lucy* anchored at coordinates where the sonar signature revealed what might be a large metal shape, resting in eighty-seven feet of water: Target 67.

The pinging sounds in Bailey's headset grew louder, and through the murky water he began to make out the edge of a debris field—wires and pieces of shredded metal embedded in the sand. Then he saw it: a pair of legs, bright white and pinned to the seabed beneath a piece of wreckage, the feet waving slowly in the current. The diver's eyes widened in horror, and he lurched toward the surface. When he reached the side of the ship, Bailey tore out his regulator. "I'm not diving in a graveyard!" he said.

Down on the bottom, it took McAllister a moment to realize that Bailey was gone. He had recognized at once that what he could see was the lower half of an Extravehicular Activity suit carried aboard the orbiter for space walks, not a dead body. He tied the nylon line of the marker buoy to a piece of debris nearby and looked at his watch; he was out of time. But he had already seen everything he needed to be certain: Target 67 was the crew compartment of the Space Shuttle *Challenger*.

Wary of eavesdropping reporters, the captain of the *Lucy* didn't want to transmit news of the divers' discovery over the radio. Instead, he dropped the ship's rubber Zodiac boat into the water, to take word personally to the US Navy salvage ship *Preserver*, anchored more than a mile away. When he arrived, the ship's senior officer, Lieutenant Commander John Devlin, was waiting for him at the rail.

"Is it?" Devlin asked.

"Yes, sir. It is."

::::::::::::::::::

At around eleven o'clock the following morning, Devlin pulled on scuba equipment and swam down to examine the wreckage himself. Beside him was flight surgeon Jim Bagian, one of the three astronauts sent out aboard the ship to assist in identifying anything they found on the sea-floor; the *Preserver's* complement of divers had also received a detailed briefing on the orbiter at the Cape before they left port. The agency's engineers seemed to think that the crew compartment would be dis-covered largely intact, and had told Devlin to make sure his men took note of the positions of dials and switches in the cockpit before raising it from the bottom, to help determine what had gone wrong. But during his two and a half years in Navy salvage, Devlin had recovered enough planes downed at sea to know what to expect. At the velocity with which most conventional aircraft struck the water, the surface of the ocean was as unyielding as concrete: metal fuselages shattered into thousands of pieces, and the bodies of those on board were instantly dismembered and decapitated by the forces of deceleration.

Devlin and Bagian approached the yellow tether of the marker buoy. Visibility had improved: circling the area from above, they could see wreckage spread across the sand in an oval, some sixty feet in diameter. But it barely resembled a part of *Challenger*. The aluminum alloy walls and floors of the compartment had been smashed with such force that it might have been blown apart by a bomb and swept by giant hands into a pile on the ocean bottom. Some pieces of debris protruded from the mass of wreckage as high as eight feet, and much of it was only held together by the brightly colored skeins of the orbiter's electrical wiring harness. And yet, amid the shambles of twisted metal, Devlin could make out parts of instrument panels; the lower half of the EVA suit; boots. There was no doubt.

Devlin's men began retrieving what was left of the seven mem-bers of the *Challenger* crew later that same day: the astronauts aboard wanted them brought up immediately, afraid that someone might be lost. What they found was no longer recognizable as human. Two scuba divers found parts of Dick Scobee, Mike Smith, and Christa McAuliffe still strapped into their seats, tangled among the debris, but identifiable by the name tags on their flight suits; another, winched down toward the seabed on a mechanical diving stage, discovered the remains of Judy Resnik, carried by the current and caught by the cable anchoring the platform to the bottom. Jim Bagian recovered one of Ron McNair's boots, and took it to the surface zipped inside his wetsuit. As a doctor, Bagian was able to ensure that the crew remains were handled carefully, and placed in plastic-lined bags for the journey back to shore, where military pathologists planned to conduct an autopsy. The *Preserver* docked in Port Canaveral just before nine o'clock that night, and the waiting Jetty Rats took note of bags and boxes being removed from the ship and driven away in the back of a white van.

Jane Smith was in the kitchen when the phone rang. "They've found the crew cabin," said the NASA official on the line. "They're bringing it up." The next morning, George Abbey gathered the Houston families at the Scobees' home to give them all the news. The agency couldn't yet confirm what exactly the divers had found, but Smith felt relieved: now, she thought, they could plan a funeral, and start to move on. For others, the discovery was harder to take: Judy Resnik's father had been assured that every member of the crew had been vaporized, and the family had already held a funeral service; if his daughter's remains now emerged from the ocean, it would do nothing to make her death any less final. Lorna Onizuka's friends felt that she had only just begun recovering from her grief—and the news would now rekindle the horror of January 28. June Scobee was filled once more with fury, disbelief, and pain. She had believed her husband was at peace, buried at sea—and now he would be disinterred from his resting place, forcing her to live through her public anguish all over again.

The news brought more reporters back to Scobee's door than ever—

calling on the phone, camping out on her lawn. This time there were so many that she couldn't even get out to her car, and called NASA in desperation; again, she sought refuge with her neighbors. Fred Gregory came to the back door, helped her pack a bag, and took her down the street to shelter at their house, putting her to bed and feeding her; she stayed for more than a month. Scobee yearned for the privacy and tranquility she had lost with the abrupt end of *Challenger*'s final mission. In public, the commander's wife continued to advocate for the future of the shuttle program, to fulfill America's destiny in space and honor the sacrifice made by her husband; in private, she pleaded with God to let her join him in death.

Although NASA waited until it had informed all the crew's immediate family about Target 67 before it released the news to the press, even then the agency refused to answer any questions about what it had found, citing its respect for the relatives of the dead. A spokeswoman for the Navy chose her words carefully: "We're talking debris, not a crew compartment, and we're talking remains, not bodies," she said.

Nevertheless, the discovery of the cabin wreckage raised further doubts about the official narrative of the accident. The nature of the debris indicated that it had not been blown apart or even severely damaged in an explosion, but remained almost entirely intact until it had hit the ocean. Now a further possibility emerged, remote but horrifying: if the crew compartment had escaped in one piece, then its occupants might also have survived the violent breakup of *Challenger* and—had they been equipped with any equipment to do so—could have escaped, alive, before the wreckage struck the ocean.

$$\cdot\cdot\cdot\cdot\cdot\cdot\cdot\cdot\cdot$$

While the salvage operation on the Hetzel Shoal seized headlines, and the accident investigators gradually assembled the jigsaw pieces of the shattered orbiter in the hangar at the Cape, the aftershocks of the disaster were opening new fissures at every level of NASA. When former astronaut Dick Truly returned to the agency to take up his post as the new head of spaceflight, he found the place in turmoil—unraveling so fast that it seemed it might soon cease to exist entirely.

Many of the astronauts in Houston had learned for the first time about the dangers of the solid rocket O-rings only from the public testimony of the NASA and Thiokol engineers—and were furious when they discovered the truth. Some felt betrayed by the organization to which they had entrusted their lives, and were aghast to discover how many of their own flights aboard the shuttle had come close to suffering the same fate as mission 51-L. Hank Hartsfield, the silver-haired former test pilot who had commanded Judy Resnik's first flight in 1984—on which two O-rings had shown signs of erosion, and blow-by—was astonished that no one had believed the astronauts should be informed about a problem so common, and so serious. In a startling breach of the astronauts' code of silence, Hartsfield told visiting reporters how angry he was at not being given the opportunity to make up his own mind about the risks he had been taking; as a result, some of his colleagues in Houston stopped talking to him. But, by then, his outrage had already been echoed at the top of the Space Center hierarchy: on the same day Hartsfield's comments were published, Chief Astronaut John Young wrote another of his famously unvarnished memos, to George Abbey, and sent a copy to every member of the astronaut corps.

The message described Young's dismay at watching the televised hearings of the Rogers Commission, and at the Marshall engineers' bland certainty that they had done nothing wrong; he explained that safety had been compromised by the NASA managers' pressure to keep launching; and he attached a list of potentially catastrophic mechanical faults that he insisted should be addressed before the shuttle ever flew again. Individually, the items on Young's list seemed insignificant. But together, they made clear that he and his fellow shuttle veterans were fortunate to have made it through twenty-four missions alive: "This list proves to me that there are some very lucky people around here," he wrote. When the Chief Astronaut's memo leaked to the *Houston Post* a few days later, it was front-page news across the country.

Elsewhere, the agency's glossy veneer of teamwork and brisk efficiency, decades in the making, continued to crack. In Cape Canaveral, the Director of the Kennedy Space Center lashed out in the press at

Rogers's investigation, accusing the commission of jumping to conclusions, and blaming the media's negative reporting about shuttle delays before the accident for pressuring NASA to launch. In Houston, Jesse Moore insisted that the true cause of the disaster had yet to be determined. "I'm very concerned we don't get myopic and say it's this O-ring thing," he said. In Huntsville, Marshall Space Flight Center Director Bill Lucas told reporters that his engineers' decision to launch *Challenger* had been "sound"; in the meantime, his senior staff continued to seek ways to place the blame for the accident elsewhere.

And around Lucas, even the wall of secrecy behind which Wernher von Braun had worked the miracles of propulsion to put men on the moon was crumbling. The offices and laboratories of the complex seethed with suspicion and anguish at the way the Huntsville engineers had been saddled with responsibility for the accident, even as hundreds of exhausted staff worked late into the night to isolate its cause. And in the first week of March, a three-page computer-printed letter arrived at the door of the NASA Inspector General's office, apparently written by a senior manager in Huntsville.

The author, who claimed to have attended the Flight Readiness Reviews before every one of the shuttle launches to date, said that he felt forced to come forward as a result of what he had witnessed over his years at Marshall—and especially in the weeks since the accident. The letter described how Bill Lucas presided over a dictatorial regime dominated by his handpicked "good old boy Mafia"; how all dissent was crushed, and how Lucas had made it known that under no circumstances was Marshall to be the cause for delaying a shuttle launch. The flight readiness process was nothing but a rubber stamp in which every engineer understood that they were expected to vote "go"; Lucas had repeatedly lied to the public, and deliberately attempted to conceal the true causes of the accident from the Rogers Commission.

Enclosed with the letter was a copy of the presentation given to Lucas the previous summer by a team of Thiokol engineers, laying out their fears of the O-ring problems, as evidence of the Marshall Space Flight Center Director's complicity in the tragedy. "Lucas proceeded at risk rather than have the MSFC delay the program," the author wrote. Explaining that he planned to retire soon, but wished to remain anonymous, the author called for Lucas to be removed, and thus bring to an end what he called "this Orwellian nightmare that many of us find ourselves trapped within."

Concluding with a desolate "GOD HELP US ALL," the letter was signed, "APOCALYPSE."

:::::::::::::::::

Out in the Atlantic, the winter weather had closed in. Hampered by high winds, heavy seas, poor visibility, and currents so strong that even divers wearing weighted suits struggled across the ocean bottom like men walking in a hurricane, the work to recover the wreckage of the *Challenger* crew compartment was slow. But at 9:45 p.m. on March 12, Commander Devlin brought the *Preserver* into dock at Port Canaveral. Debris from the spacecraft, hidden beneath canvas tarps, covered the fantail; an honor guard of sailors stood along the gunwale; and seven metal boxes bearing crew remains sat on the deck, draped with flags on Devlin's orders. With deliberate ceremony, sailors in dress uniform carried the boxes across the pier to three waiting ambulances, which left the dock thirty minutes later, blue lights flashing.

In the end, the divers found the remains of most of the *Challenger* crew still trapped inside the broken carcass of the cabin, along with their helmets, personal effects, clothing, and food from the lockers on the middeck, including condiments and an unbroken jar of Jif peanut butter. The Navy salvage teams also recovered many of the things the astronauts had carried with them, intending to return to friends and family as mementos flown in space: a ball from Janelle Onizuka's high school soccer team; a rib from the oldest glider in Australia, carried by Dick Scobee; two hundred embroidered STS-51-L mission emblems; Ron McNair's patch from the US Karate Association; and forty-seven copies of the US Constitution. Other small objects packed by the crew would never be found, among them the class ring from the Virginia Military Institute that Christa McAuliffe had taken aboard for her husband;

more than a dozen silver medallions packed by Greg Jarvis; a Dunhill cigarette lighter belonging to Judy Resnik's high school boyfriend; and the gold astronaut pin Mike Smith was expecting to be awarded when he got back home to Houston.

But amid the debris, the *Preserver* divers did manage to find the pieces of equipment that the crash investigators hoped would answer questions about how the accident had happened. These included not just the cockpit instrument panels, with gauges and switches registering activity on the flight deck during *Challenger's* final moments, but also every one of the five general-purpose computers that had controlled the spacecraft in flight, crushed flat by the impact with the ocean, yet otherwise intact. More important, the men of the *Preserver* brought up from the ocean floor the six tape machines that had recorded everything happening inside the orbiter during its seventy-three seconds of flight— including the two channels of the crew's conversation over the internal intercom system, which had not been transmitted to Mission Control.

Work inside the submerged wreckage was difficult, and dangerous— the tangle of serpentine wiring and jagged aluminum threatened to en- snare unwary divers, or cut the air and communication lines connecting them to the surface; sometimes visibility fell to less than twelve inches, and the men had to search by groping through the debris field with their fingertips. And although the salvage team believed they had successfully located the bodies of all seven members of the *Challenger* crew by the time they returned to port on March 12, there was one they had been unable to recover.

The remains of Greg Jarvis, who had been seated in the center of the middeck, between McAuliffe and McNair, were visible inside the heap of debris; but it proved impossible to reach them before the cabin was winched to the surface by crane. And as the wreckage emerged from the waves, the sailors and astronauts aboard the ship watched in horror as part of the compartment—weighed down by water and silt—tore loose, and Jarvis's blue-clad remains fell with it into the sea. A sailor standing on the fantail of the ship dived to the rescue, but it was too late: the engineer's body had been swallowed once more by the ocean. It would be more than a month before any trace of him would be found again; the Casualty Assistance Officer who called Marcia Jarvis with the news eventually told her that the largest part of her husband's body the divers recovered was a fragment of his pelvis.

By that time, at the end of almost eight weeks of searching the turbulent waters off the Cape, Commander Devlin and the crew of the *Preserver* had been released from the salvage operation, and the last remnants of the debris field off the Hetzel Shoal had been scoured from the bottom. Directing the process from his office at the Kennedy Space Center, Bob Crippen insisted on recovering 100 percent of the wrecked crew compartment and its contents from the ocean, and eventually NASA chartered a Port Canaveral scallop boat to dredge the seafloor with nets, to trawl for any fragments that may have been left behind or buried in the sand.

:::::::::::::::::::

By the middle of March, Rogers Commission investigators had gathered the beginnings of what would eventually encompass hundreds of hours of testimony from managers, engineers, and technicians at NASA and its contractors. They explored potential causes of the accident that ranged from poor quality control in the assembly of the solid rockets at the Cape to fatigue and drug use among console operators in Mission Control; they collected signed affidavits from agency managers who swore that they had felt no pressure from outside the agency to launch *Challenger* on the morning of January 28; and questioned every one of the thirty-four men who had sat in on the teleconference the night before.

Meanwhile, the dozen members of the commission panel had divided into four specialized teams, each devoted to one aspect of the inquiry, to conduct their own interviews at NASA facilities around the country. While continuing his work on the investigation, Richard Feynman made several visits to a hospital in Washington, DC, for blood tests and medication to treat his worsening cancer, and consulted his doctor in California over the phone; he scribbled notes on the progress of the treatment on the back of his minutes of meetings about the accident.

As part of the group headed by General Donald Kutyna—dedicated to accident analysis—Feynman flew to Huntsville to help supervise tests intended to establish exactly how the solid rocket seals had failed during *Challenger*'s brief final flight. But Feynman, already satisfied that O-rings made inflexible by extreme cold had burned through and destroyed the shuttle, had little interest in pursuing the details any further. Instead, he embarked on a wider mission. He suspected that the wishful thinking and incompetence that the investigation had uncovered so far was not confined to the handling of the solid rockets, but could be found elsewhere in the shuttle program. To the growing irritation of some of his colleagues—who saw him as a publicity-seeking loose cannon whose antics were a distraction from the stated purpose of the commission—Feynman set out to prove that the management rot ran deep inside NASA. He chose the infamously precarious technology of the Space Shuttle main engines as his example, and asked for a presentation on the system from the engineers at the Marshall Space Flight Center.

When Feynman had first arrived in Huntsville, one of his initial meetings had been with a Range Safety Officer, who wanted to tell him about NASA's estimates of the risk of failure in the solid rockets before the accident. The officer explained that his own analysis suggested such rockets would explode or otherwise misfire in around 4 percent of launches—or one in twenty-five. This was in keeping with the independent probabilistic studies NASA officials had chosen to disregard at the outset of the program. But, since then, the agency mandarins had apparently concocted a new estimate, even more optimistic than ever; years before the loss of *Challenger*, they told the range safety officer that, in fact, he should consider the chance of failure closer to one in one hundred thousand.

Feynman was incredulous.

"That means you could fly the shuttle *every day* for an average of *300 years* between accidents—every day, one flight, for 300 years—which is obviously crazy!"

"Yes, I know," the officer said.

Two hours into the briefing on the Space Shuttle main engines,

Feynman interrupted to conduct another experiment. Tearing a sheet of paper into several pieces, he handed one to each of the engineers and the senior manager in the room, and asked them to write down their estimate of how reliable they believed the shuttle's engines were. The engineers each offered up similar evaluations: they calculated the risk of failure at around one in three hundred, or perhaps as high as one in two hundred. But the manager didn't want to be pinned down: instead, he wrote on the scrap of paper a list of criteria for determining reliability; there were no numbers.

"You weaseled," Feynman told him. Pressed to quantify an answer, the manager at first insisted that the engines were 100 percent reliable; only when his engineers' jaws fell open in astonishment at this assertion did he provide his final assessment. It was the same impossible statistic that NASA management had attributed to the solid rockets: one in one hundred thousand.

Yet, that afternoon, the engineers told Feynman in detail about the engines' history of test stand failures, of the fractures in the turbine blades, and the potential for disaster in the high-pressure turbopumps. They told him that they faced at least a dozen serious problems; only half had been fixed. By the end of the day, Feynman was convinced that he had found evidence of the same syndrome in the main engine program that had led to catastrophe with the solid rockets: NASA managers who prioritized magical thinking over technical realities, while the alarm bells rung by their own engineers went unheard.

The whole agency, Feynman decided, had a very serious problem.

It was not until March 21 that NASA officials finally appeared before an open hearing of the Rogers Commission panel in Washington, DC, and testified that they now knew conclusively what had caused the *Challenger* accident: the failure of the aft field joint in the shuttle's right-hand solid rocket booster. "I believe we have eliminated all of the other possibilities," said the Marshall engineer supervising their inquiry.

In the deep water forty miles out into the Atlantic, the search for the physical evidence of the rocket's demise ground on, dogged by bad weather and the mammoth scale of the task; for all the public attention garnered by the discovery of the crew cabin, the wreckage of the right-hand booster remained the primary objective of Black Bart Bartholomew's salvage operation. As the end of March drew near, the recovery teams were scouring an area of 420 square nautical miles of ocean bottom, and had found almost six hundred sonar contacts to investigate. But so far, Bartholomew's flotilla of ships, submarines, robot submersibles, and divers had managed to examine less than a fifth of those discoveries: just twenty-nine of them had turned out to be pieces of shuttle wreckage; only a handful were parts of *Challenger's* right-hand solid rocket; and not one was from the joint that the Marshall engineers believed had burned through, around sixty seconds into flight.

With time running out to wrap up the accident investigation, Bartholomew called in help from the Navy's small experimental nuclear submarine, the NR-1. Diving on the final Monday in March, carrying a crew of twelve—including a NASA rocket engineer sent along to help visually identify pieces of solid rocket—the vessel embarked on a mission to clear the entire backlog of unexamined debris in a single voyage.

The NR-1 would eventually stay submerged for almost two weeks, traveling from one suspect location to another until the crew had examined 281 different targets. Among them was Contact 131, a low-priority object first discovered more than a month earlier, a broken hunk of steel wedged into the sand in more than five hundred feet of water, thirty-five miles from Cape Canaveral, and which the rocket engineer found difficult to make out through the submarine's five-inch porthole. He thought it looked like a piece of wreckage from a sunken ship's boiler—just one more piece of seabed trash, among the dozens of fifty-five-gallon drums and experimental missile debris; the paint cans; the filing cabinet; the kitchen sink; and the duffel bag containing $13 million worth of cocaine that the salvage teams would find elsewhere. But Contact 131 was captured more clearly by the nuclear submarine's video cameras, and when the tapes reached NASA, the agency told Black Bart to investigate further: whatever it was, the debris was marked with signs of scorching.

In the second week of April, the head of the National Transportation Safety Board investigation team working at the Cape finally allowed reporters to inspect some of the wreckage recovered to date. Inside a three-story warehouse surrounded by an eight-foot fence topped with barbed wire, journalists and photographers gathered around dozens of fragments of the orbiter resting on the floor, or set up on sawhorses— including a large chunk of the right wing, a piece of the nose that had enveloped the crew compartment, and a twelve by six foot section of the fuselage bearing a scratched and gouged Stars and Stripes, scrubbed clean of barnacles and brought into port by the crew of the *Lucy* just a few days earlier. Two of the main engines, pieces of the external tank, and the nose cones of both solid rockets were held in a separate building; but there was no sign of the crew compartment, which remained concealed in a closely guarded area nearby.

After weeks lying on the bottom of the ocean, much of the debris was still fouled with seaweed and decaying marine life, but the NTSB investigator explained that—contrary to what viewers of the endlessly repeated video clips of the accident might believe—almost none of it showed any signs of the incineration he would expect from an explosion. Although a large section of its tail was badly scorched by the exhaust of the flailing solid rockets, the orbiter had otherwise escaped unscathed from the fireball that had blossomed from the external tank; instead, careening out of control at twice the speed of sound, it had broken up in flight moments later.

The worst damage to the spacecraft had been caused by the same aerodynamic forces that tore it apart, and collision with other pieces of debris as it disintegrated. The accident investigator said that the conical nose section of the orbiter had apparently separated cleanly from the fuselage, taking the crew compartment with it; other experts believed it did so with such violence that it was unlikely anyone aboard could have survived the initial breakup, the effects of sudden cabin depressurization, or the extreme gravitational forces of its subsequent descent. Yet it was also clear that the compartment had remained intact as it plummeted toward the surface of the Atlantic, where the investigator estimated it made impact with the water at a speed of between 130 and 180 miles per hour.

"It did not float down slowly," he said.

⁘⁘⁘⁘⁘⁘⁘⁘⁘

Contact 131 finally came up from the ocean bottom in the twilight hours before dawn on April 13, raised by the heavyweight crane of an offshore oil rig support ship hired by NASA for the purpose. Later that day, a pair of agency engineers arrived on board to examine it. They found a curved section of hardened steel eight-and-a-half feet long and weighing four thousand pounds: part of the casing of *Challenger's* right-hand booster, broken away from close to the point where it had been attached to the shuttle's external tank—and where the puffs of black smoke had emerged from the rocket at ignition. The white paint that had once covered the booster segment had been blackened by a fire or explosion, and in some places seared away entirely. At the bottom were the remains of the tang where the segment had been joined to the one below. And, in the middle, the engineers could see a ragged hole in the casing two feet across, where the steel had been burned right through, as if cut with a blowtorch.

The salvage operation would continue for four more months—eventually employing sixteen aircraft, twenty-four ships, three submarines, and ten thousand personnel from the Navy, Coast Guard, NASA, and several civilian contractors. At one point costing as much as $1 million a day, by the time it concluded in August 1986 it had become the largest and most ambitious salvage operation in US Navy history. But as the two NASA engineers stood beside Contact 131 on the rolling deck of the recovery ship that day, they confirmed that at last they had found what they were looking for; the most important part of Black Bart's work was complete—and the mystery of *Challenger's* destruction one step closer to resolution.

⁘⁘⁘⁘⁘⁘⁘⁘⁘

Conducted by two military doctors from the Armed Forces Institute of Pathology in Maryland, the autopsies and final identification of the remains of the *Challenger* crew were shrouded in secrecy from the beginning. After being brought ashore in Port Canaveral and transferred to Patrick Air Force Base under cover of darkness, the astronauts' body parts were transported by NASA helicopter to a makeshift morgue set up inside Hangar L—the Life Sciences Building—at the Kennedy Space Center. George Abbey and other agency officials pulled political strings to shut out the local medical examiner—who had jurisdiction over all deaths at the Cape—from the pathologists' work, apparently to prevent any information about what they discovered reaching the public. As a result, further waves of rumor and suspicion swept through the ranks of the national media following the story in Florida.

The formal identification of the remains was complicated both by the condition in which they were found and NASA's lack of preparation for such a devastating accident. The agency had collected no fingerprint or footprint records of the *Challenger* crew and—with DNA profiling still in its infancy—the pathologists resorted to dental history, X-rays of bone fragments, and anatomical comparison to arrive at their final conclusions. Although delayed while divers searched for the lost remains of Greg Jarvis, by the end of the third week in April the doctors had identified parts of all seven astronauts. Joe Kerwin, the veteran astronaut and flight surgeon who headed NASA's Life Sciences Laboratory in Houston, wrote a brief letter to each of the crew families, itemizing what had been recovered.

The pair of forensic pathologists overseeing the autopsy were specialists in air crash fatalities—Institute of Pathology doctors had conducted postmortems in previous astronaut deaths, including the victims of the Apollo 1 fire—and a key part of their work lay in determining exactly how those aboard *Challenger* had died. By examining the tissues of the crew, they hoped to be able to deduce what conditions had been like inside the shuttle cabin in the minutes before its final disintegration: whether the astronauts had been killed almost instantly by an explosive high-altitude depressurization; if the skin of the compartment had been ruptured, admitting toxic hypergolic propellant from the spacecraft's Orbital Maneuvering System; or if a slow leak of the cabin atmosphere had rendered them slowly unconscious in their seats as the wreckage fell back toward the Earth.

While the forensic work continued, the families of the astronauts received word from its contacts at NASA that the bodies of their loved ones would be released for burial at the end of the month. At the same time, the agency issued its first public statement acknowledging that crew remains had been recovered from the ocean; yet it refused to provide any details of the autopsy, or of what the pathologists might have discovered. When reporters asked whether the agency's technicians had recovered any audio from the shuttle's cockpit voice recorder, NASA public affairs officials would not say when—or if—it might be possible to do so.

But inside the Astronaut Office, those who had seen computer-enhanced imagery of *Challenger's* breakup, taken from the tracking cameras at the Cape, had already begun to speculate about the true fate of the crew. To some of the engineers and astronauts in Houston, the film presented firm evidence that the disintegration of the orbiter was not as violent as the crash investigators believed. Only after several news organizations filed Freedom of Information requests with the agency did NASA officials release the material to the public, and a new picture of the final moments of the shuttle emerged: a sequence of ten color photographs showed the distinctive bullet shape of the crew compartment, shooting out of the cloud of blazing rocket fuel, arcing upward amid the shower of smoking debris. The dark frames around the windows of the cabin were clearly discernible; morning sunlight flashed on the glass as it spun slowly in flight. Sonny Carter told Jane Smith that the g-forces inside the compartment would have been no greater than those Mike experienced when his plane had been catapulted off the deck of a carrier at sea.

Tumbling gently in free fall, it took two minutes and forty-five seconds for the broken section of *Challenger* to hit the Atlantic, and it now seemed possible that the seven members of the crew might have been alive the whole way down.

At 9:30 a.m. on April 29, 1986—almost three months to the day since the countdown clocks of Space Shuttle mission 51-L had reached zero— the crew of *Challenger* left Cape Canaveral for the last time. Earlier that morning, a cortege of seven black hearses had carried the caskets containing the astronauts' remains up the long causeway toward the shuttle landing strip, where a C-141 Starlifter waited to fly them to Dover Air Force Base, to be returned to their families. Hundreds of Space Center employees turned out along the route; the headlights of the big black limousines glowed yellow in the flat spring light. Beside the runway, a line of senior NASA officials including John Young and new shuttle chief Dick Truly stood in solemn silence as the cars came to a stop behind the giant aircraft.

At the end of the line of dignitaries stood George Abbey. The Director of Flight Operations held his hand over his heart as twenty-eight Air Force pallbearers in gleaming boots, black berets, and white gloves carried the flag-draped coffins slowly past a color guard bearing the standards of the United States and NASA, and up the ramp of the transport plane. There were no speeches, and no music; no sound except the shrill chatter of the birds and the tramp of boots on the concrete. As the cars drew off one by one and the color guard led the column of pallbearers away across the flight line, Abbey and Young climbed aboard the aircraft, and its four jet engines began to whine. The plane lumbered down the runway, rose steeply into the air above the black water marshes, and then banked east toward the sun, roaring over the dunes and the breakers, in a final pass over the deserted gantry of Launch Complex 39B.

CHAPTER
TWENTY-EIGHT

THE LONG FALL

William P. Rogers formally delivered the *Report of the Presidential Commission on the Space Shuttle Challenger Accident* to the White House on the sunny afternoon of June 9, 1986. After a briefing on its contents in the Cabinet Room, President Reagan stood before the press at a lectern in the Rose Garden to thank Rogers and the panel—and to cast their findings as a conclusive step toward restoring the nation's pride in NASA, and America's prominence in space exploration.

"We've suffered a tragedy and a setback," Reagan said, "but we'll forge ahead, wiser this time, and undaunted—as undaunted as the spirit of the *Challenger* and her seven heroes."

But the conclusions of the commission's 230-page report were damning. In meticulous detail, Rogers and his team exposed the concatenation of cost cutting, faulty design, management blunders, and institutional hubris that had led to the catastrophic end of Space Shuttle mission 51-L. In clear, simple language, the report described the development of the shuttle program and the background of *Challenger's* final flight, from the crew's training schedule to the record freeze that swept across Merritt Island on the night of January 27. It examined the most likely causes of the accident, provided a forensic description of the physical evidence hauled out of the Atlantic by the search and recovery teams, and concluded that—without doubt—the seven astronauts had been doomed by the failure of a single Viton pressure seal in the aft field joint of their spacecraft's right-hand solid rocket booster. "We know exactly how this accident occurred," Rogers told reporters at a press conference later. "I certainly hope there will be no nagging questions."

The report—the first of five volumes of material, which would include reproductions of many key documents and full transcripts of both the open and closed sessions of the committee hearings—was exhaustive and comprehensive. It outlined the history of the solid rockets, and how budget cuts had shaped NASA's decision to employ the technology in the shuttle program and select Morton Thiokol as a prime contractor; it revealed that the design of the field joints had been faulty from the beginning; that Thiokol had taken little appropriate action to fix them, and it opposed the agency engineers' suggestions to do so in the years before the shuttle first flew. When the launches began and initial signs of damage to the O-rings emerged, the contractors had proved slow to plan or execute a program to find a solution.

But the commissioners reserved their harshest criticism for the managers of the Marshall Space Flight Center: they found that Larry Mulloy had failed for years to alert his superiors to the seriousness of the failings in the O-rings and that, in his later testimony before the commission, he had misled Rogers and his panel about the extent of the warnings he had provided. They reported that they had found not a single mention of O-ring problems in any of the paperwork—a stack of documents several inches thick—prepared during the flight readiness process of the 51-L mission.

Quoting extensively from the Thiokol engineers' sworn testimony, the report documented the attempts of Al McDonald, Roger Boisjoly, Bob Ebeling, and Arnie Thompson to have the launch stopped, and the pressure Mulloy had applied to have their no-go recommendation reversed. The commissioners concluded that, had Mulloy passed on the seriousness of the engineers' concerns about the seals and the effect of the cold weather to those at the top of the decision chain, the launch would have been postponed and the accident would never have happened.

Nor did those in the upper reaches of NASA management escape censure: the commission found that the agency had become increasingly blind to the hazards inherent in the shuttle program, had fatally weakened its own safety and quality assurance structures, and had insisted on conducting ambitious spur-of-the-moment experimental missions while simulaneously trying to run an airline in space. For more than a decade it had been trying to do too much with too little.

And even as the staff at NASA headquarters pushed harder to fulfill the political goals of a launch schedule that many of them knew was chimerical, long before *Challenger* left the ground they had received ample warning that something was amiss in the shuttle program. The commissioners concluded that, by the summer of 1985, Jesse Moore and other senior officials were in possession of enough information about the failings of the solid rocket seals that they should have grounded the entire fleet of spacecraft until the problems could be fixed.

The report concluded with a series of nine recommendations to correct the failings of the program before the shuttle returned to flight—including a redesign of the joints in the solid rockets, an overhaul of NASA management, the creation of an independent safety review board, installation of a new abort-and-escape mechanism to save astronauts in the event of an emergency, and a reduction of the flight rate within realistic goals. Nonetheless, at Bill Rogers's insistence, the document ended on an optimistic note, with a final statement urging that the agency continue to receive the nation's full support. "The Commission applauds NASA's spectacular achievements of the past," it read, "and anticipates impressive achievements to come."

. . . .
. . . .

The fallout from the findings, which began before the commission had started drafting the report, had only intensified as its publication drew closer. In Washington, DC, Bill Graham left the agency to take up a job inside the White House as Science Adviser to the President. The position Graham had wanted so badly was now taken by sixty-six-year-old former administrator James Fletcher—the same man who had persuaded President Nixon to endorse the shuttle program in the first place—who reluctantly returned to the job he had occupied in the early seventies.

In Huntsville, Bill Lucas dismissed Larry Mulloy and other senior managers at Marshall from their positions on the shuttle program, before announcing, just a few days before Rogers delivered his report, that he would also retire as Director of the Marshall Space Flight Center. In Utah, Joe Kilminster—who had signed Thiokol's formal approval to launch *Challenger*—was transferred to the company's automotive products division. His boss, Jerry Mason—who had presided over the decision to overturn the engineer's recommendations—was forced into early retirement. Mason's faithful yes-man, Cal Wiggins, and Bob Lund, the head of engineering who had cast the deciding vote in the fateful teleconference, were both relieved of their responsibilities and reassigned.

Bob Ebeling, the hot-tempered seals expert at Thiokol who had wanted to shut down the company's production of the shuttle solid rockets until the joint was redesigned, took each dismissal as a personal victory, and crowed in triumph when he heard the news. But Ebeling and the other whistleblowers at the Promontory plant had paid a steep price for their role in bringing the truth to light. In the weeks after Al McDonald's and Roger Boisjoly's shocking initial testimony before the commission, Thiokol's upper management had acknowledged their employees' apparent lack of loyalty with a show of corporate displeasure. The two men were first sidelined from the accident investigation, and the intensifying effort to redesign the booster rockets to make them safe for flight, and then isolated from further contact with NASA; McDonald was stripped of his responsibilities and effectively demoted.

Together with the other men who had testified to their efforts to stop the *Challenger* launch—Ebeling, Brian Russell, and Arnie Thompson—when they returned to work at the plant the engineers found themselves frozen out: not just by management, but by colleagues who feared that any show of solidarity might endanger their own future at the company, or who—ironically—blamed them for job losses caused by the accident. One afternoon in early May, Boisjoly was standing in Ebeling's office when he heard a furious voice booming from a nearby cubicle: "If I lose my job because you guys couldn't resist testifying, then I'm going to dump my kids on your doorstep!"

Presently, the outcast engineers began calling themselves the "Five Lepers."

⸭⸭⸭⸭⸭⸭⸭⸭⸭⸭⸭⸭⸭⸭⸭⸭

When the commission's report reached the press and the public on June 9, there were many who thought it hadn't gone far enough. Richard Feynman had threatened to withhold his signature from the document over the concluding statement, which he regarded as a Pollyanna sop to an institution that had grown complacent and, perhaps, incompetent—and lost touch with the rigorous engineering principles on which it had been founded. Only after negotiations with Rogers, brokered by Feynman's new friend General Kutyna, did the physicist and the chairman reach a compromise. Rogers softened the language of the report's final endorsement of NASA, and allowed Feynman to publish his own, more critical, opinion as one of fifteen appendices to the main document.

Although it would not be released officially until weeks later, Feynman handed out copies of *Appendix F: Personal Observations on Reliability of Shuttle* at a press conference he gave at Caltech the day after attending the Rose Garden ceremony in Washington, DC. Outlining the NASA bureaucrats' magical thinking about the odds of catastrophic failure in the shuttle's engines that he had uncovered in Huntsville, Feynman's *Appendix F* presented an unsparing analysis of the agency's failings. "It would appear that, for whatever purpose, be it for internal or external consumption, the management of NASA exaggerates the reliability of its product to the point of fantasy," he wrote. He suggested that agency officials had, from the very beginning, oversold the shuttle in order to maintain their access to federal coffers; and he compared the Marshall engineers' insistence on flying after the discovery of O-ring damage to Russian roulette: "If you hold a gun to your head and shoot and don't kill yourself," he said, "a man would be foolish to say, Let's spin it and fire again."

At the same time Feynman was talking to reporters in California, William Rogers and Neil Armstrong were defending their work on Capitol Hill as Congress embarked on its own investigation into the accident. Generally, the report was well-received by the politicians. But in the last of three back-to-back hearings that day, Chairman Rogers came under attack from Ernest Hollings, a publicity-hungry Democrat on the Commerce, Science, and Transportation subcommittee. Hollings wanted to know why the commission had not more fully examined the rumors suggesting that the White House had put pressure on NASA to launch *Challenger*, so that the President could have a live satellite linkup with the orbiting Teacher in Space during his State of the Union address. Rogers explained that the commission had taken the suspicion very seriously, questioning the relevant personnel under oath and taking depositions—and that a team of FBI agents found the story was groundless. When Hollings refused to listen, Rogers became incensed, his face flushing with color, and both men began shouting at each other across the packed room. "It did not happen!" Rogers said. "If you can prove it, I will come back here and apologize!"

Like some members of the chairman's own staff, Hollings was also angry that Rogers had not sought criminal prosecution for those NASA officials and contractors they believed had caused the deaths of the *Challenger* astronauts. "I want the individuals involved to know they are going to be held responsible," Hollings said, insisting that Marshall Director Bill Lucas had shown neither understanding nor remorse about his role in the accident, and that Larry Mulloy was guilty of gross negligence. "We ought to take action against the Mulloys, Lucases and some of those people at Thiokol who changed their minds and recommended the launch."

But, as a former prosecutor, Rogers argued that there was insufficient evidence to build a case of willful negligence. He suggested that the accident was, instead, a failure of the system—and of the nation; now both NASA and the country needed to learn from what had happened, and move on. "People involved have suffered a lot," he said. "I'm not sure picking out any scapegoat and prosecuting would serve the national interest. I hope it doesn't happen."

Down in Huntsville the next day, Larry Mulloy read these words aloud from a transcript of the chairman's remarks, and laughed his good old boy laugh.

"I heartily agree with Mr. Rogers," he said.

In a series of public and private ceremonies that spanned the breadth of the United States, during the weeks it took the Rogers Commission to finalize their report, the seven members of the *Challenger* crew had gradually been laid to rest. Christa McAuliffe was the first to be buried, with neither flowers nor headstone to mark the plot, in a secluded Catholic service in Concord at the beginning of May. A few days later, Jane Smith and her three children walked slowly through Arlington National Cemetery behind a gun caisson pulled by six white horses; she watched as a navy honor guard fired three rifle volleys over the grave and a bugler played taps; then her husband's casket was lowered into the ground beneath a large maple tree. On May 10, Marcia Jarvis and a handful of friends took a sailboat out into the ocean off Long Beach, where she read aloud a poem she had written in memory of Greg, and scattered his ashes into the Pacific. At the same time, Judy Resnik's father, Marvin, who had held a full Jewish memorial service for his daughter in the weeks after the accident, was hounded by calls from reporters demanding details about his plans to bury the remains returned from the Cape. But Resnik refused to be drawn, and begged for privacy. "I don't think that's anyone's business," he said. "It's so hard to go through a tragedy like this. Why should you make it harder?"

On a hot Saturday morning in Lake City, South Carolina, a funeral cortege half a mile long passed down Deep River Street, and turned left at the vacant lot where Ron McNair's boyhood home had once stood, carrying his flag-covered casket on toward a half-acre plot in the Rest Lawn Cemetery. In his eulogy, McNair's pastor remembered the astronaut as an inspiration to Black children. "I can hear Jesus greeting him, saying, 'Servant of God, well done,'" he said. "Rest yourself. The voyage is over. The mission is complete . . . someone will stand and tell the story." Two days later, on what would have been her husband's forty-seventh birthday, it was June Scobee's turn to visit Arlington, where she saw Dick's cremated remains buried in a simple wooden box, as he had requested. It wasn't until June 2 that Ellison Onizuka became the final member of the crew to be interred, his coffin flown home to Hawaii aboard a government plane, for an Air Force and Buddhist ceremony at the National Memorial Cemetery of the Pacific, on Puowaina—the "Hill of Sacrifice"—overlooking Honolulu.

:::::::::::::::::

Even before the publication of Rogers's report, the relatives of the *Challenger* crew had considered seeking damages against NASA and Morton Thiokol over their role in the accident. In the months following the disaster, Ronald Krist, a Houston attorney with offices practically across the street from the Johnson Space Center, heard indirectly from several of the families asking if he would be willing to represent them. "You bet," he replied.

Krist, who had acted for Gus Grissom's widow, Betty, in the aftermath of her husband's death in the Apollo launchpad fire, regarded the lawsuits as inevitable. Apart from the psychological trauma and loss they had all endured, the widows and children of five of the astronauts were suddenly bereft of their principle means of financial support; NASA's responsibility seemed obvious. But the degree of the agency's liability—and, perhaps, that of Morton Thiokol—was something that Krist couldn't determine until he'd read the findings of the presidential commission.

The families were among the first to learn the contents of the report, invited by Rogers to a private briefing in a Washington, DC, hotel the morning before it was delivered to the White House. The commission team tried to keep the location secret, but the hotel was besieged by reporters, who climbed the fire escapes to gain access to the building. Over coffee and doughnuts, Rogers, Sally Ride, and Bob Crippen gave the relatives a detailed outline of the commission's findings and recommendations. Despite the months of televised hearings and exhaustive newspaper coverage, it was the first June Scobee and her family had heard about the teleconference the night before the launch, or how the NASA managers had ignored repeated warnings about the cold.

"Those idiots!" twenty-two-year-old Rich Scobee said, and pounded his fist on the table.

Although drawn closer by the tragedy, and sucked into the eye of a hurricane of relentless media scrutiny, the seven families were divided in their public response to the report. June Scobee maintained the steely resolution expected of the commander's widow, expressing continued support for the shuttle program and NASA, but refusing to comment on the commission's conclusions. "If words could bring my husband back, I would speak volumes," she said. "If money could bring him back, I would sue for millions. All that was important to me was him—and nothing will bring him back."

Greg Jarvis's mother praised the commission, and said that her son had understood the risks he was taking—but then broke down when asked whether NASA should be blamed for the accident. "I think something should be done," she said. "They shouldn't've let this happen." Jane Smith was left angry and disappointed by what she had learned. Particularly sickened by Larry Mulloy's comments during the teleconference— *My God, Thiokol, when do you expect me to launch—next April?*—Smith prepared a blistering statement, which she read over the phone to a reporter. "The report reflects incredibly terrible judgments, shockingly sparse concern for human life, instances of officials lacking the courage to exercise the responsibilities of their high office and some very bewildering thought processes," she said. Gathering details of the investigation from friends inside the Astronaut Office—especially Sonny Carter, who had worked on the recovery of the crew cabin—Smith suspected that NASA officials had not been completely open about what had happened. She wanted the whole truth, and Mike's old flying friend—attorney Bill Maready—told her that a damages suit might be the best way to get it. The legal discovery process, mandated in civil litigation, could force the agency to reveal everything it knew.

In the meantime, NASA's doctors and technicians had been probing the evidence gathered from the wreckage of the cabin, seeking to unlock the secrets of the astronauts' final moments. At first, NASA engineers had believed the six cockpit data recorders found by divers back in March had been too badly damaged by their long immersion in seawater to reveal anything at all. Even after being dried out in a vacuum chamber, the tapes flaked apart when technicians attempted to unspool them. But a team at the laboratories of IBM in Tucson, Arizona, eventually devised a technique to undo the chemical degeneration, unwound the tapes by hand, and made copies that they flew back to Houston. In late June, NASA engineers began trying to decode audio data from the recording of the astronauts' voices in the cabin during the seventy-three seconds after launch: if the crew were aware of the calamity enveloping them, they may have given some indication to one another, even if they had no time to radio an alarm to Mission Control.

The results of the forensic pathologists' work at the Cape had proved less encouraging. By the time NASA had released the bodies of the crew for burial, the military doctors examining the remains had proved unable to determine how exactly the seven astronauts had died. So the head of the shuttle program instructed Joe Kerwin and his team in the Life Sciences lab in Houston to widen their analysis of the wreckage. The destruction was so complete that they could find no conclusive signs of an explosive decompression of the crew compartment, which would have killed its occupants almost instantly; they examined every piece of paper dredged up from the Atlantic for notes possibly scribbled by the astronauts in their final moments, but found nothing. Then, after almost a month of investigation, Kerwin reported that they had found something unusual: in the crew's Personal Egress Air Packs, recovered from the ocean floor.

The packs, compact metal boxes containing an individual air supply connected to each astronaut's helmet, were only ever intended for use in the event of an emergency on the ground. If there was a fire or a release of toxic gas inside the shuttle on the pad, each crew member would be able to use their pack to breathe as they made their escape. But Kerwin's team found signs that three of the seven packs on board *Challenger* had been used during flight—and one of these was identifiable by a serial number as that assigned to Mike Smith. Further tests established that the valve activating Smith's air supply could not have opened by accident, nor could it have been reached by the pilot himself, because the equipment was mounted on the back of his seat.

This evidence suggested that either Judy Resnik or Ellison Onizuka—both sitting behind Smith on the flight deck—had been conscious in the moments after the orbiter's disintegration, and had reached over to activate the pilot's air pack. The volume of air remaining in Smith's pack also revealed that someone had been breathing from the supply for around two and a half minutes: almost exactly the length of time it took for the sundered crew cabin to fall the twelve miles from its apogee to the surface of the Atlantic.

After learning what Kerwin had found, Jane Smith tried to discover whatever else she could about what had happened in the cockpit during the last seconds of her husband's life. In the simulator in Building 5 in Houston, Smith sat in Resnik's seat on the flight deck, and deduced that it would have been impossible for her to activate Mike's air pack, because it was beyond her reach; Onizuka must have been the one to do it. She flew down to Cape Canaveral and visited the hangar where the wreckage was stored, and the refrigerated area, redolent with the greasy scent of decay, where the fragments of the crew compartment had been laid out. She laid her hands on what remained of her husband's seat, and the harness that had strapped him into it; and Sonny Carter told her where the spring-loaded switches on the pilot's console had apparently been moved from their ascent settings. To Jane, this discovery confirmed not only that Mike had been conscious when he began his long descent toward the ocean, but that he had been busy, working through every procedure he could think of as he fell. He was nobody's fool; he knew he was going to die. But he never stopped trying to live.

On July 2, 1986, Jane Smith filed a personal injury and wrongful death claim against NASA and Larry Mulloy, seeking more than $15 million in damages. The filing alleged that Mulloy and the agency "directed, allowed and participated in the launch of *Challenger* when they knew or should have known that the segments of the right hand solid rocket booster would not properly seal and that a catastrophic accident would likely occur as a result thereof." The filing cited as witnesses "several thousand people at the Kennedy Space Center, and untold thousands of others watching on television."

The decision to take action had been difficult: Smith knew that she and the other families of military personnel or NASA employees had limited legal recourse against the government, and bringing a suit against the agency would violate the old unwritten code of the astronauts. When Betty Grissom had eventually chosen to seek damages from North American Rockwell, twenty-five years earlier, she had been cast out of Togethersville and showered with hate mail from strangers across the country. But Smith still believed that the agency knew more about the accident than it would admit.

That suspicion was apparently borne out a few weeks later, when the first public statement about the last moments of the crew emerged from Houston. With a brief announcement on July 17, NASA officials at the Space Center acknowledged that the preliminary work to recover data from the cockpit voice recorder had been successful, but had revealed no anomalies of any kind; the crew had received no warning of the catastrophe. They provided no details, and did not release a transcript. Another eleven days passed before the agency changed its story. On the same day it released the news about Joe Kerwin's analysis of the emergency air packs, a NASA spokesman explained that engineers listening to a garbled portion at the end of the recording had picked out something else from the noise and distortion.

It was the voice of Jane Smith's husband:

"Uh-oh."

The final words uttered in the shuttle cockpit before the cabin data recorder and the intercom system were severed from their onboard power supply, these two murmured syllables were proof that—in the fractions of a second before his spacecraft was torn apart—the pilot of *Challenger* understood that something had gone terribly wrong.

＊＊＊＊＊＊＊＊＊＊＊＊＊＊＊＊＊

Eventually, all seven families of the crew of mission 51-L would receive financial settlements in compensation from the government, or Morton

Thiokol, or both, for the loss of their loved ones. Jane Smith's claim was followed a few months later by a lawsuit brought against Thiokol by Cheryl McNair, and another that the attorney Ronald Krist filed on behalf of Judy Resnik's mother. "To do nothing would be a tacit acquiescence or stamp of approval of the type of conduct that took my husband's life," McNair said. "And this I am unwilling to do." When NASA refused to settle Smith's claim, she brought a separate action against the rocket contractor, this time seeking $15 billion in damages. Together, the three women's cases would wind on for years, and eventually be settled out of court, the details kept confidential.

June Scobee, Lorna Onizuka, Marcia Jarvis, and Steven McAuliffe did not hire legal representation, but instead settled with the government and Thiokol after taking informal advice from McAuliffe's law partner in Concord. In exchange for promising never to discuss the details publicly—or sue NASA in the future—the four families were jointly awarded more than $7.7 million in cash and annuities, 40 percent of which would be paid by the US government, and the rest by Morton Thiokol. On learning of the agreement, Ronald Krist described the settlement as "woefully inadequate." At $4,641,000, the rocket company's share of the compensation was little more than a rounding error for the corporate accountants: since signing the deal to manufacture the boosters in November 1973, the NASA contract had earned Thiokol revenues of some $1.5 billion.

The company's treatment of the Five Lepers out at the Promontory plant had proved more problematic. When the Rogers Commission discovered that Morton Thiokol had apparently punished Roger Boisjoly and Allan McDonald for their damaging testimony, the chairman publicly rebuked the contractor's senior management, and NASA's Office of Inspector General opened an investigation into its behavior. At the same time, Democratic Congressman Ed Markey wrote a furious letter to Thiokol's CEO explaining that he had called on NASA to cancel a $200 million contract recently awarded to the company to redesign the solid rockets and make them safe for flight; more ominous still, he had introduced legislation to the House floor to disqualify Morton Thiokol from receiving any future work at all from the agency until Boisjoly and McDonald had been reinstated, or until the company had proved that the two engineers had not been reassigned because of their testimony before the commission.

After Markey's letter reached the Thiokol CEO's desk in Chicago, Roger Boisjoly was reassigned to his previous responsibilities, and Allan McDonald was appointed to a new position. As director of the Wasatch Division's Solid Rocket Motor Verification Task Force, McDonald would now be at the head of the intensifying effort to redesign the boosters; Boisjoly's expertise would be central to the testing and verification process. But while both men were feted as heroes in the national media, and McDonald was swept up in a frenzy of meetings, consultations, and equipment tests, Boisjoly remained tormented by thoughts of the accident he had fought so hard to prevent.

Blaming himself for not doing more to stop the launch, he was profoundly distressed by suggestions that, if he had been more determined to do so, he should have telephoned NASA's shuttle program chiefs himself—or even called the White House. Plagued by sleeplessness, nightmares, headaches, and double vision, Boisjoly began overeating and piled on weight. His doctor had prescribed Xanax for his anxiety, but taking it made him feel as if he were in a trance. And unlike McDonald, who felt supported by those in his local community, Boisjoly eventually discovered that the same people who had once elected him mayor of Willard had now turned against him: neighbors ostracized him in church and hounded his wife. One day, he found that a dead rabbit had been left in his mailbox; twice, while taking walks out of town beside the highway, passing drivers tried to run him off the road with their cars.

In July 1986, Boisjoly took two weeks' vacation, but remained consumed by anger and depression. He returned to work only for long enough to request permission to take extended medical leave, and was eventually diagnosed with post-traumatic stress disorder. At the end of January 1987, he returned to the Promontory plant, packed fourteen boxes containing every document and scrap of paper he had kept in his office into his Toyota Celica, and drove out of the parking lot for the last time. A few days later, on the anniversary of the loss of *Challenger*, Boisjoly filed a suit against Morton Thiokol, accusing the company of fraud, negligence, racketeering, and manslaughter, and seeking more than $1 billion in damages. The case was dismissed by a federal judge the following year. Although feted as a hero by his fellow engineers and presented with awards for upholding the ethical standards of his profession, Boisjoly never worked in the aerospace industry again.

In the end, it would be almost three years before the shuttle flew once more. Late on the bright, almost windless, morning of September 29, 1988, *Discovery* cleared the tower at Cape Canaveral carrying a small veteran crew, equipped with a rudimentary new escape system, and riding on Thiokol's redesigned solid rocket boosters. When the shuttle touched down safely at Edwards Air Force Base four days later, the astronauts were met by an audience of 380,000 people and a band playing the national anthem, as the CapCom in Mission Control heralded their triumphant return: "Welcome back, *Discovery*," he said. "A great ending to a new beginning."

By then, the members of the *Challenger* crew had been memorialized in the names of schools and grants, playgrounds, buildings, streets, asteroids, and craters on the moon. The seven families they left behind had also come together to found a new charity dedicated to their memory: the Challenger Center for Space Science Education. Chaired by an indomitable June Scobee, the center became the foundation of an ambitious initiative that would build a network of sophisticated space flight simulation facilities across the United States, allowing children to role-play many aspects of space exploration—and continue the educational mission of flight 51-L.

The changes undertaken by NASA in the wake of the Rogers Commission report included not only the redesigned hardware—the booster rocket seals, the bailout system, rebuilt main engines and, for the astronauts, a return to wearing full pressure suits, fitted with built-in parachutes and oxygen supplies—but also a wholesale reorganization. In the years since the accident, almost every senior- and mid-level manager involved in approving the launch of *Challenger* had resigned or been reassigned; senior staff who had worked on the space program in the glory days of Apollo had been lured back to the agency, and the autonomy of the centers in Huntsville and Houston diminished in favor of control from headquarters in Washington, DC. There was a new, independent safety and quality assurance organization reporting directly to the Administrator—and astronauts had been given powerful positions in the launch decision chain; promoted to Deputy Director for Space Shuttle Operations at the Cape, Bob Crippen had headed the team giving final approval for *Discovery* to leave the ground. Congressman Bill Nelson compared the healing he had seen since the accident to the redoubling of effort to reach the moon that had followed the deadly fire on Pad 34, back in 1967.

But some of those who had served on Rogers's panel were not so sure. Soon after the launch of *Discovery*, former *Aviation Week* editor Robert Holz told a reporter that he and his fellow commissioners had long feared that, once the glare of the Presidential inquiry's scrutiny had receded, NASA would simply go back to business as usual.

"We have the feeling that nothing much has changed," Holz said. "The problem is not that they'll get one off. They'll get one off and maybe two or three, but they're heading for trouble down the road."

In the meantime, on Crippen's instructions, most of the debris of *Challenger* had been photographed, cataloged, packed into 102 wooden crates, and lowered into a pair of disused missile silos on the grounds of Cape Canaveral Air Force Station, which were then locked and capped with concrete. The shuttle was dead, the veteran astronaut had decided; the time had come to bury it. Slowly, the warren of subterranean rooms began to flood with brackish water.

And for more than fourteen years after *Discovery*'s successful return to flight, the Space Shuttle flew routinely and successfully—completing a total of eighty-seven further missions by the end of 2002. As the world was swept by one seismic change after another—the end of the Cold War, and the collapse of the Soviet Union; the 9/11 attacks—the harrowing memory of the *Challenger* disaster softened and faded, carried away by the river of history.

In that time, the diverse roles that had once defined the ambitions of

America's Spaceliner had fallen away one by one, disavowed in the interests of safety and national security: first, President Reagan halted commercial satellite launches aboard the shuttle, declaring that he was instead opening the field for private enterprise in rocketry and space travel; then, NASA quietly curtailed the Space Flight Participant Program: it seemed that, despite the President's promises, there would be no journalists or artists in space, after all. Finally, after years of mothballing its facilities and winding down involvement with the agency, the Pentagon announced that it would stop flying its astronauts and sensitive payloads aboard the shuttle altogether; it was just too dangerous.

And yet manned spaceflight was at last focused on a permanent presence in Earth orbit—the International Space Station, a joint project between five agencies from nations around the globe, including the United States and Russia. And, just as Wernher von Braun had once envisioned, the shuttle fleet played a key role in its construction and supply. Space travel seemed to become almost quotidian. A new generation of NASA management grew ever more confident in its abilities. Slowly, insidiously, some of the old ways and attitudes became reestablished in the offices in Washington, DC, Houston, Huntsville, and Cape Canaveral; others had never really gone away.

::::::::::::::::::

By the time it embarked on its twenty-eighth mission on January 16, 2003, the long-serving Space Shuttle *Columbia*—the first to reach orbit—had been flying for more than two decades. The crew of seven was taking the venerable old lady of the shuttle fleet up on only its second voyage in three years. The launch of STS-107 was exceptional due to the long break between missions—and also because the astronauts would not be docking at the International Space Station, but instead carrying out a series of scientific experiments in orbit. Even so, few Americans were even aware the flight was taking place.

When a piece of the thick foam insulation covering part of the external fuel tank broke free during the orbiter's ascent, striking the leading edge of its left wing at around 500 miles per hour, there was little immediate alarm in Mission Control. Falling insulation had been a concern for NASA engineers since the earliest days of the program, but it had happened so frequently that they now regarded it as an acceptable risk. And while the potential for damage to the fragile tiles covering the shuttle remained as great as it had on *Columbia*'s first hair-raising test flight back in 1981, since then, there had apparently never been any tile loss significant enough to endanger the safety of the astronauts on reentry. Nevertheless, the size of this piece of foam—one foot by two feet, the size of a briefcase—was enough to trigger the creation of a debris assessment team to analyze the problem in Houston.

Once *Columbia* reached orbit, mid-level engineers in Houston made three requests to repeat the kind of remote inspection performed during John Young and Bob Crippen's test flight in 1981, using Pentagon spy satellites to capture images of the suspect area of the left wing. But two of the requests were denied by senior managers, and the third never came to the attention of the correct officials due to a breakdown in communication. There were repeated email exchanges between concerned structural engineers in the various parts of NASA's sprawling technical bureaucracy, but their worries were also brushed aside by upper management. No pictures of the wing were taken.

On the shuttle's eighth day in space, Mission Control in Houston sent a message to *Columbia*'s commander, forty-five-year-old former test pilot Rick Husband, informing him that the piece of foam had struck the leading edge of the left wing on the ascent, but it was a familiar scenario it had seen before. There was, the message said, "absolutely no concern for reentry." Five days later, on January 28, the crew suspended their orbital routines to mark the anniversaries of the Apollo launchpad fire and the loss of *Challenger* with a moment of silence. "They made the ultimate sacrifice, giving their lives for their country and mankind. Their dedication was an inspiration to each of us," Husband said.

On the morning of Saturday, February 1, 2003, the crew prepared to return *Columbia* to Earth. Touchdown on the Shuttle Landing Facility runway at Cape Canaveral was scheduled for 9:16 Eastern time.

The shuttle entered the atmosphere over the Pacific at an altitude of 400,000 feet, in the same steep nose-up attitude Maxime Faget had once demonstrated with his balsa and paper model. The shuttle was enveloped in a sheath of superhot plasma, which grew hotter and hotter as it plunged farther into the atmosphere; around the nose and the leading edges of the orbiter's wings, temperatures rose to 3,000 degrees Fahrenheit.

Fifteen minutes later, soon after *Columbia* crossed the California coast, the shuttle's trajectory remained perfectly on track. But by 8:54 a.m., one of the engineers in Mission Control had noticed a cascade of anomalous telemetry readings, sent from inside the orbiter's left wing: among them, the hydraulic lines and tires of the landing gear seemed to be losing pressure. Yet the data was scattered, offering no consistent pattern: the most logical explanation appeared to lie in faulty sensors.

What the flight controllers did not know was that, more than a week earlier, the piece of foam debris—weighing less than two pounds— had struck the orbiter with the force of a rifle bullet, smashing a hole six inches long through the brittle carbon-carbon heat shielding on the edge of the wing. In the quarter of an hour the shuttle had spent falling through the atmosphere, first nose-up and then riding the series of steep, banked turns designed to shed speed high in the atmosphere, a jet of superhot gas had surged through this gap—and, undetected by Mission Control or the crew—begun melting the aluminum framework of the left wing, and destroying mechanical and electrical components one by one.

Ground staff in the Flight Control Room radioed through a message to *Columbia*'s commander about the falling tire pressures, but the shuttle remained in an anticipated period of intermittent communication blackout. Only a fragment of an acknowledgment from Rick Husband came in over the headsets in Clear Lake.

"And, uh, Hou—" the commander began; he was cut off in the middle of the word "Houston."

A few seconds later, just before 9:00 a.m., he said, "Roger, uh, buh—" Then the signal broke off once more.

It would be the final voice transmission from *Columbia*.

As it flew over north-central Texas at an altitude of around forty miles, witnesses on the ground watched as the single bright plasma trail of the descending orbiter began to separate, birthing a handful of glittering points of light streaking across the cerulean sky.

The shuttle was coming apart in flight.

Inside the Flight Control Room, the astronaut in charge of communications with the crew continued to follow the normal procedure for landing approach, trying to reestablish contact via UHF radio.

"Columbia—Houston. UHF comm check," he said.

"Columbia—Houston. UHF comm check."

But there was only silence.

At 9:12 a.m., Flight Director LeRoy Cain turned from his console to the back of the room, where he received word of a Dallas TV station broadcasting live images of multiple contrails high overhead, on *Columbia*'s flightpath. Thousands of pieces of debris were already hitting the ground, falling into ponds and backyards, on highways and car windshields, in a track of terrible rain ten miles wide and three hundred miles long, stretching from East Texas into Louisiana.

Cain knew he could wait no longer. It was time to give the instruction that no one in Mission Control wanted to hear; formal acknowledgment that another seven astronauts—

Rick Husband

William McCool

Michael Anderson

Kalpana Chawla

David Brown

Laurel Clark

Ilan Ramon

—were dead.

Cain straightened, and stepped forward to face his team. A single tear ran down his cheek.

"GC—Flight," he said.

"Flight—GC," the Ground Controller replied.

"Lock the doors."

EPILOGUE

The Columbia Accident Investigation Board delivered its report on August 26, 2003, and concluded that many of the lessons of the *Challenger* disaster had gone unheeded. The agency once again acknowledged its faults, and vowed more technological and management changes.

Shuttle missions would eventually resume, orbiters docking each time with the International Space Station, in part to provide a safe haven in the event of an emergency. In 2007, Barbara Morgan, the Idaho schoolteacher who had been Christa McAuliffe's backup on *Challenger*—having quit her job, moved to Houston, and spent nine years in preparation to become a fully trained astronaut—at last reached space, as a Mission Specialist on STS-118. The final mission—the 135th shuttle voyage—aboard *Atlantis*, would not take place until July 2011. But NASA's forty-year dream of the reusable spaceplane—of creating routine access to orbit on an airline schedule, of taking the astronomical costs out of astronautics—had perished long before, along with the crew of *Columbia*, in the clear winter skies over Texas.

By then, Richard Feynman, too, was dead, but his blunt conclusion in *Appendix F* had became famous. His words had proved prophetic, and provided an epitaph not only for NASA's heroic goals with the National Space Transportation System, but for many of the historic catastrophes caused by mankind's overconfidence in his own ingenuity.

For a successful technology, reality must take precedence over public relations, Feynman wrote, *for nature cannot be fooled.*

:::::::::::::::::::

George Abbey was among those dismissed from their positions following the *Challenger* accident, and exiled to NASA headquarters in Washington, DC. The launch of *Discovery* in 1988 was the first at the Cape since 1975 at which he did not accompany the astronauts out to the pad. He eventually returned to the Johnson Space Center in 1994 and became Center Director two years later. With John Young once more at his side, Abbey went on to oversee the selection of more than one hundred further astronauts and the successful construction of the International Space Station. One of his first acts as Director was to create a memorial grove for astronauts in Clear Lake, with a tree and plaque dedicated to each deceased member of the corps. He retired, at seventy, in January 2003, less than a month before the loss of *Columbia*.

Former NASA Administrator **James Beggs** continued to fight the legal case against him for more than a year, and in June 1987 prosecutors dropped all of the charges against him. Attorney General Ed Meese wrote a letter of "profound apology" to Beggs for the prosecution, and two years later a federal judge ordered all records in the case destroyed. Although reluctant to discuss NASA's responsibility for *Challenger*, Beggs suggested that the accident may never have happened if he had remained in charge. "Whether I would have done anything different at the time, I've thought about that," he said. "I think I would have, but that's pure conjecture." He died in 2020 at the age of ninety-four.

In 1988, the American Association for the Advancement of Science awarded **Roger Boisjoly** the Award for Scientific Freedom and Responsibility. He died in Nephi, Utah, on January 6, 2012. He was seventy-three.

Astronaut **Charlie Bolden** would fly into space four times aboard the Space Shuttle, twice as a pilot and twice in the left-hand seat, as commander. He left NASA in 1994 to return to active service with the Marine Corps, where he rose to major general. In 2009, President Barack Obama appointed Bolden the first Black administrator of NASA. He served until 2019, when he stepped down at the age of seventy.

Bob Crippen continued to ascend the ranks of NASA management, first taking charge of running the shuttle program from Washington, DC, and later becoming Director of the Kennedy Space Center, where he would oversee twenty-two successful shuttle launches. Following further budget cuts to the program, he left NASA in 1995, and moved to Brigham City, Utah, to become president of the Thiokol Propulsion Group. As the head of three divisions of the company, in 1999 Crippen announced an extension of the rocket manufacturer's contract to provide shuttle boosters to NASA, worth $1.7 billion over six years. In 2001, he retired to Florida with his second wife, Pandora, a former shuttle engineer he met while at work on the *Challenger* recovery.

Unable to face the possibility of another malfunction for which he would feel responsible, **Bob Ebeling** retired from Morton Thiokol in the summer of 1986. He spent much of the rest of his life working at a bird refuge near his home in Utah; in 2013, he was named Volunteer of the Year for the National Wildlife Refuge System. But for three decades Ebeling remained tormented by his failure to stop the launch of STS-51-L. In an interview he gave to NPR to mark the anniversary of the accident in 2016, Ebeling said, "I think that was one of the mistakes God made. He shouldn't have picked me for that job . . . next time I talk to him, I'm going to ask him, 'Why? You picked a loser.'" After the broadcast, for the first time since the disaster, Ebeling heard directly from those involved in the launch decision, including George Hardy and Bob Lund, assuring him that he bore no responsibility for the deaths of the *Challenger* crew. "You did all that you could do," Lund told him. Ebeling succumbed to cancer two months later. He was eighty-nine.

After the disintegration of *Challenger*, **Jean-Michel Jarre** decided to cancel the outdoor concert at which he'd planned to feature Ron McNair playing the song they had practiced together. Only after receiving a series of phone calls from McNair's colleagues and friends in the Astronaut Office asking him to continue was Jarre convinced to proceed. On the night of April 5, 1986, the show lit up downtown Houston with a spectacle of lasers, fireworks, and images projected onto massive sheets of canvas hung from the city's skyscrapers. Intended to celebrate the 150th anniversary of Texas, and 25 years since NASA had arrived in Houston, the concert featured McNair's friend, saxophonist Kirk Whalum, performing the part the astronaut had intended to play in space—as the centerpiece of "Last Rendez-Vous," a song Jarre later renamed "Ron's Piece." A record-breaking live audience of 1.3 million people in Houston watched the show, which was also screened on television around the world.

Marcia Jarvis never returned to her job at the dental practice in Los Angeles. In the spring of 1989, she moved to Mammoth Lakes in the Sierra Nevada, where she spent her time riding horses and doing volunteer work. She helped out at a probationary school for boys, made educational appearances supporting NASA, and worked for a hot-air ballooning company. While there, she met Ron Tinsley, a local contractor and a former member of the ski patrol: in 2001, the couple moved to Colorado and were married at Thanksgiving. Each year, on January 28, she tries to rise at dawn and finds a place for quiet contemplation. Occasionally, she thinks of nothing at all; often, she recalls the final conversation she had with Greg on the beach that afternoon at Cape Canaveral: *But if something happens, just be happy.* At first, she was angry about the way he died. "But anger isn't constructive," she said recently. "People make mistakes. And other people pay for it sometimes."

After retiring from his position as Director of the Marshall Space Flight Center, **William R. Lucas** took part-time work with the University of Alabama at Huntsville, where he led an initiative to examine the school's space education and research programs. He took to gardening and hunting, and spent free moments reflecting on his career. But more than thirty years after the accident, he still refused to accept the judgment of William Rogers and his commission that the decision-making process leading to the *Challenger* launch had been flawed. "I did what I thought was right in the light of the information I had. And if I was going over it, with the same information I had at the time, I'd make the same decision—because I thought it was right," he said in a TV interview aired in 2020. "I didn't do anything that I thought was wrong then, and I didn't do anything that I think was wrong in retrospect." In March 2022, Lucas celebrated his one hundredth birthday.

In the months that followed the death of his wife, **Steven McAuliffe** and his two children withdrew almost entirely from the public eye. Although a member of the board of the Challenger Center for Space Science Education from the beginning, McAuliffe agreed to appear at events promoting the initiative only if he could avoid speaking to the press. In 1992, he married Kathy Thomas, a reading teacher for the Concord School District and mother of two, in a private ceremony, and in November that year was sworn in as a federal judge for the US District Court for the District of New Hampshire. In January 2016, Judge McAuliffe issued a statement that began, "The passage of thirty years since the *Challenger* accident is not of great personal significance to our family. For us, *Challenger* will always be an event that occurred just recently."

After overseeing the redesign of the Space Shuttle solid rocket boosters, in 1988 **Allan McDonald** was awarded the NASA Public Service Medal. In a letter supporting his nomination for an American Institute of Aeronautics and Astronautics prize, William Rogers wrote that "I think the Nation owes a debt of gratitude to Mr. McDonald." But the engineer continued to face the enmity of his colleagues at Thiokol and their families and, gradually, again found himself sidelined at the company. He retired in 2001 and became a successful speaker, lecturing on engineering ethics and decision-making, and published a memoir, *Truth, Lies,* *and O-Rings,* describing his experience. In March 2021, he suffered a fall while helping neighbors in Utah clear snow from their property and died a few days later; he was eighty-three.

Cheryl McNair stayed in Houston after the accident to raise her children, dedicating time to the Challenger Center—and to a local organization providing day care to teenaged mothers in the city. McNair established an educational foundation in her husband's name, and in 1989, Congress created the McNair Scholars Program, to help first-generation students, and those from low-income and underrepresented backgrounds, attend graduate school. In 2011, Cheryl attended the opening of the Ronald E. McNair Life History Center in Lake City, South Carolina, dedicated to celebrating her husband's life and mission—and housed in the same building that had once been home to the library he had stubbornly integrated as a boy in 1959. Joy McNair, Ron's daughter, became an attorney specializing in estate planning, and remained close to her mother in Houston; Reggie eventually moved to Atlanta and began work in finance. Cheryl never remarried, and for years kept Ron's workroom at home just as he had left it; although many of the mementos her husband flew into space would later be damaged or destroyed in a house fire, his saxophones survived. In the 2023 academic year, the McNair Scholars Program provided funding to students at more than two hundred schools and colleges across the United States.

In July 1986, **Larry Mulloy** left NASA for good and began work as a management consultant doing business with the agency and its government contractors in Washington, DC. For the rest of his life, he continued to insist that he had been made a scapegoat for the wider failings of NASA and the shuttle program. He said that he had followed the rules of the Flight Readiness Review process to the letter, and from the start kept his senior managers in the launch decision chain informed of problems with the O-rings. In an email to Allan McDonald in 2014, Mulloy wrote that he had provided the Rogers Commission with video evidence that proved his case, and afterward Executive Director Alton Keel had telephoned him to apologize for the way he had been treated—but by that time the commission's final report had been published; it was too late to alter the record. In one of his final messages to McDonald, Mulloy signed off, "Sleep well, Al. I do." Almost a decade later, Keel remained adamant that no such call or conversation had ever taken place. Mulloy died in Tennessee in October 2020, aged eighty-six.

In 1990, Representative **Bill Nelson** resigned his seat in Congress to run unsuccessfully for governor of Florida. Returning to Washington, DC, as a Senator in 2001, he was instrumental in passing legislation that codified and expanded the commercialization of US spaceflight. In 2021, he was sworn in as the fourteenth Administrator of NASA.

Steve Nesbitt withdrew from his media duties at the Johnson Space Center at the end of January 1986, and took the first of a series of management positions within the NASA public affairs office. Yet he remained eager to show the resilience and continuity of the agency—and reassure those who associated his voice with disaster—and received permission to provide the launch commentary of the shuttle's successful return to flight two years later. Nesbitt continued to serve occasionally as the voice of Mission Control for another dozen years and retired, at the age of fifty-nine, in December 2010.

In the months following the 51-L disaster, **Lorna Onizuka** remained in Houston, and began work as a liaison for JAXA—the Japanese counterpart of NASA. The accident proved particularly hard for Onizuka's two daughters, who each year greeted its anniversary with dread: long afterward, Darien would pick up the phone and imagine she was speaking to her father, asking him when he was going to come home; Janelle blamed herself for not telling him more forcefully that she didn't want him to fly. In an alcove in the house, Lorna maintained a shrine to Ellison, including a baseball signed by Buzz Aldrin and a can of his favorite beer, and eventually became a grandmother seven times over. She never remarried, but retained close bonds to the cause—and the community—for which her husband had given his life: through her work, the Challenger Center, and the Houston Livestock Show and Rodeo; on the day that George Abbey finally cleared out his office at the Johnson Space Center, Onizuka was there to help him pack.

Marvin Resnik left Akron and retired to California, where he died at the age of ninety, in March 2010.

Sally Ride remained in Washington, DC, after the publication of the Rogers Commission report, to lead work on a NASA study considering the future of the space program. She retired from the agency the following year, divorced her husband, and returned to academia, as a professor of physics at the University of California at San Diego. In 1992, as part of President Clinton's transition team, Ride provided guidance on how to reform NASA, and afterward Clinton offered her the job of agency Administrator; she declined. Together with her partner, Tam O'Shaughnessy, Ride wrote six science books for children and founded a company, Sally Ride Science, to "make science and engineering cool again"—particularly for girls. In 2003, she reluctantly accepted a position on the Columbia Accident Investigation Board, becoming the only individual to serve on the panels of both shuttle accident inquiries. She died of cancer in 2012, at sixty-one.

As the chair of the Challenger Center for Space Science Education, **June Scobee** was present in Houston for the opening of the first of the series of learning centers in 1988. That same year, she moved to Washington, DC, and in June 1989 married US Army general Don Rodgers, whom she had met after an Easter service at Arlington. On January 28, 1996, **Rich Scobee**, then an Air Force fighter pilot, marked the anniversary of his father's death by flying his F-16 in the missing man formation over the stadium hosting the thirtieth Super Bowl in Tempe, Arizona. By 2021, the educational nonprofit founded by the families of the 51-L crew was responsible for thirty-seven separate Challenger Center facilities around the world.

Jane Smith remained in Timber Cove for almost four years after the loss of her husband, a part of her always waiting for him to come home, or expecting to see him around the next corner. In 1990, she married Hugh Dixon Walcott—an old family friend, a doctor, and classmate of Mike's from Annapolis—and the next year returned to Virginia Beach and the familiar sound of jets from Naval Air Station Oceana roaring overhead. Her son, Scott, moved with his family to San Diego, where he became a building contractor; Alison married, became a fourth-grade schoolteacher, and raised two girls; Erin settled on a farm in rural Virginia. In the closet at home, Jane kept the attaché case that she had found in her husband's room at crew quarters on the afternoon of the launch, still labeled RETURN TO HOUSTON. And each time a hurricane struck the coast, she would step outside. "Hello, Mike," she'd say, as the weather howled around her. "Are you out here with me?"

Despite a thorough investigation, the author of the **"Apocalypse" letter** was never identified.

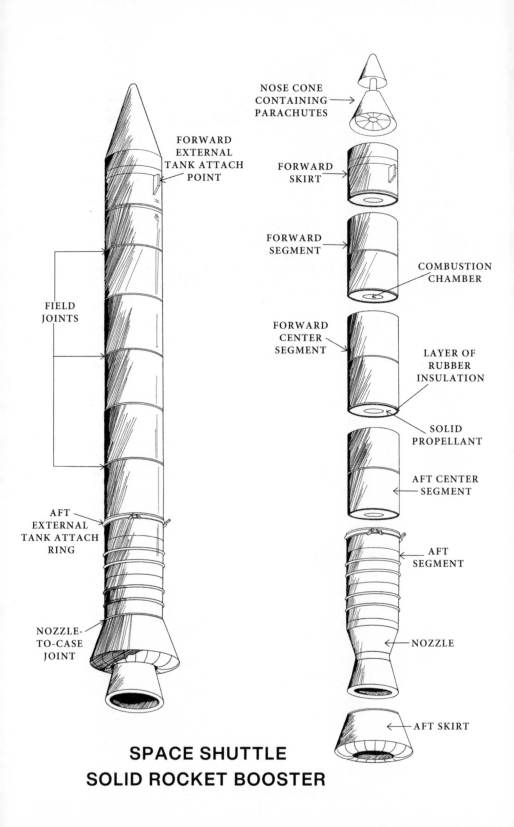

FORWARD
EXTERNAL
TANK ATTACH
POINT

FIELD
JOINTS

AFT
EXTERNAL
TANK ATTACH
RING

NOZZLE-
TO-CASE
JOINT

NOSE CONE
CONTAINING
PARACHUTES

FORWARD
SKIRT

FORWARD
SEGMENT

COMBUSTION
CHAMBER

FORWARD
CENTER
SEGMENT

LAYER OF
RUBBER
INSULATION

SOLID
PROPELLANT

AFT CENTER
SEGMENT

AFT
SEGMENT

NOZZLE

AFT SKIRT

SPACE SHUTTLE
SOLID ROCKET BOOSTER

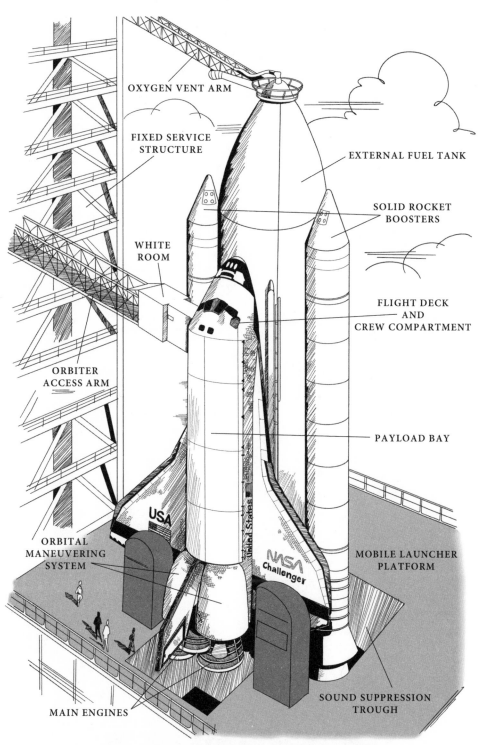

OXYGEN VENT ARM

FIXED SERVICE
STRUCTURE

EXTERNAL FUEL TANK

SOLID ROCKET
BOOSTERS

WHITE
ROOM

FLIGHT DECK
AND
CREW COMPARTMENT

ORBITER
ACCESS ARM

PAYLOAD BAY

USA

United States

NASA
Challenger

ORBITAL
MANEUVERING
SYSTEM

MOBILE LAUNCHER
PLATFORM

MAIN ENGINES

SOUND SUPPRESSION
TROUGH

**THE SPACE SHUTTLE *CHALLENGER*
ON THE LAUNCHPAD**

ACKNOWLEDGMENTS

While this book draws on the memories of many former astronauts, engineers, contractors, and NASA staff, it would not have been possible without the trust and cooperation of the families of the crew of STS-51-L. Through many hours of interviews, emails, and telephone calls, and through sharing their collections of photographs, documents, letters, and journals, Alison Balch, Marcia Jarvis, Cheryl McNair, June Scobee Rodgers, and Jane Smith-Wolcott helped me reconstruct the experience of the *Challenger* astronauts—and those closest to them—from long before they arrived at NASA to the present day. I'm grateful to them for their patience and generosity, and for revisiting their memories of the seven men and women whose extraordinary lives form the heart of this story.

I'd also like to thank those whose recollections were not only crucial to my understanding of the space agency and the shuttle program, but also helped connect me with key sources at the outset of my research: Alan Ladwig; Mike Mullane; Steve Hawley; Rick Hauck; Chet Vaughan; Tom and Mary Snitch; and Estella Gillette of the NASA Alumni League. I'm especially grateful to those who agreed to meet face-to-face during that period when doing so seemed uncomfortable, or potentially hazardous: my first scheduled in-person interview for the book—with Ambassador Alton Keel, outdoors, on a hilltop in Virginia in October—was also the first occasion I've ever had to postpone an interview because of the weather forecast.

When the archives eventually reopened, I received superb guidance from a handful of experts—Gene Morris of the Textual Reference Branch, Kaitlyn Crain Enriquez and Todd Crumley of the Still Pictures Branch of the National Archives, and Elizabeth Borja at the National Air and Space Museum—who helped me locate material essential to the narrative of the book. At NASA I received enthusiastic support not just from Chief Historian Brian Odom, but also from public affairs staff across the country who indulged my odd-seeming requests to see inside deserted conference rooms or long-disused facilities, and allowed me access to places few other reporters have visited. Thanks to Valerie Buckingham, Kelly Humphries, Adeline Morgan, Cindy Shoemate—and to Laura Rochon, who hosted me on one marathon dawn-to-dusk tour of what may have been every NASA location in Houston. I'm especially grateful to John Tribe and Steve Nesbitt, whose help with the project extended to guiding me through the Kennedy and Johnson Space Centers with their own invaluable recollections of what each place was like at the height of the shuttle program. A visit to the Columbia Room in the company of Mike Ciannilli was a forceful reminder of why the lessons of NASA's three most infamous accidents should not be forgotten.

Over the four years of my work on the book, several writers and researchers proved extremely generous with their advice and expertise in the history of the shuttle program and the fate of *Challenger*; I'm grateful to Jennifer Ross-Nazzal, Mark Maier, Martin Ditkoff, Robert Pearlman, David Hitt, Andrew Chaikin, Eric Knops, and Charles Fishman for their help and their insights. Former Thiokol engineer Brian Russell made time to read draft sections of the manuscript charting the troubled engineering history of the solid rocket boosters, saving me from undue technical embarrassment; any errors or oversimplifications that remain are my own.

I'm once again indebted to Melanie Locay and the staff of the Vartan Gregorian Center for Research in the Humanities at the New York Public Library, who provided me with a beautiful and silent place to work, and access to rare and otherwise hard-to-find materials; without the resources of the NYPL system—and especially the George Bruce Library on 125th Street—the project would have proved much more difficult. I'm grateful, too, for the kind assistance of Connie Moore at NASA headquarters, Elisa Picchio at the Caltech Archive, and the wonderful Kate Suiter, who tracked down the more obscure photographs in the photo inserts.

From the beginning of the process, I've continued to rely upon the good humor and wisdom of my agent, Edward Orloff. At Avid Reader, the imperturbable Ben Loehnen has been a steady source of encouragement—and judicious notes and edits; I'm also grateful for the support and hard work of Jofie Ferrari-Adler, Meredith Vilarello, David Kass, Caroline McGregor, Allison Green, Nick Rooney, and of Susan Hobson at McCormick Literary and Kristina Moore at UTA. I'd like to thank the staff at Viking Penguin UK—Daniel Crewe, my editor Greg Clowes, and Annie Lucas—as well as my British agent Sophie Lambert, who helped the book find a home in London. Carolyn Kelly and Jonathan Evans have guided me seamlessly through the production process, and Rob Sternitzky rescued me from a wilderness of typos and grammatical idiosyncrasies. Alison Forner has—for the second time running—produced a wonderful jacket, and Lewelin Polanco has matched it with striking interior design; Alexis Seabrook provided the beautiful illustrations. Sean Lavery was responsible for fact-checking.

For providing welcome respite during the gloomiest months of working through a pandemic, I'm grateful to my fellow attendees at the semiregular virtual cocktail hours of 2020 and 2021: Clark Collis; Trevor Giles; Colin Greenwood; Andrew Harrison; Jane McDonald; Ben Mitchell; and Kate and John; and for two years of often freezing but always entertaining all-season alfresco lunches, Chris Heath. For their advice and friendship along the way: Toby Amies; Dan Crane; Ross Jones; David Keeps; Brendan Koerner; Andrew Marshall; Matt McAllester; Ian Parker; Evan Ratliff; Julie Satow; Gabe Soria and Felix; and Greg Williams; I owe you all a drink (not you, Felix).

Finally, for their unwavering love and support through reporting setbacks and unexpected breakthroughs; unrelenting deadlines; abrupt disappearances to Florida, or Texas, or Arizona, or Maine; and the year spent confined in a one-size-fits-all apartment/office/school/vacation home amid growing drifts of government reports, press clippings, and interview transcripts, I cannot thank Vanessa Mobley, and our daughter, Isla, enough. Without them, none of this would be worthwhile.

—New York, January 2024

NOTE ON SOURCES

The published literature on the history of the Space Shuttle is extensive, and often overwhelmingly technical. To create a narrative nonfiction account of the *Challenger* story told from a human perspective, I drew upon my own interviews with eyewitnesses, existing first-person descriptions and memoirs, archival documents, and a comprehensive collection of oral histories that NASA began assembling more than fifty years ago.

The recollections of the men and women I spoke to during my research were essential to establishing the world of NASA in the 1970s and '80s, and the personalities and idiosyncrasies of those at the center of this story. The dialogue and direct speech in the narrative are taken from my interviews, from contemporary audio and video recordings, news reports, verbatim transcripts, books, or other sources; similarly, any descriptions of the thoughts and feelings of the protagonists are drawn from personal testimony. Those occasions where memories conflict, or accounts differ, are addressed in the notes.

In addition to these individuals' recollections, I made use of a wealth of primary materials, often preserved by NASA from the period, to support and corroborate their testimony—from the technical papers cataloged on the agency's Scientific and Technical Reports Server and specific incident and mission reports to the archives of Johnson Space Center press releases. Those key documents that proved more difficult to locate I obtained through FOIA requests. The photographic and video record also available through NASA, the National Archives, and elsewhere—not just of shuttle missions and launches, but of public events, astronaut interviews, and news coverage—is extensive, and proved an invaluable resource in describing scenes and events in the story. I'm grateful to Brian Odom for making it possible for me to visit each of the

NASA facilities that played a major role in the narrative, following in the footsteps of astronauts and their families from the rocket engine test stands of Mississippi to the roof of the Launch Control Center.

My understanding of the broader background and history of the shuttle program was informed by the work of many outstanding academics and historians. For the political and administrative story of the shuttle program, and its roots in the public disenchantment with Apollo, I consulted T. A. Heppenheimer's two comprehensive studies, *The Space Shuttle Decision, 1965–1972*, and *Development of the Space Shuttle, 1972–1981*, as well as John Logsdon's *After Apollo?*. Heppenheimer's work, and former NASA engineer Dennis R. Jenkins's definitive three-volume *Space Shuttle: Developing An Icon*, provided an encyclopedic grounding in the technical history of the Space Transportation System. Joseph Trento's *Prescription for Disaster* gave me a clear overview of the political and administrative path leading through NASA headquarters toward the accident; I'm indebted to Joe for his early encouragement with this project, and for directing me to the Joseph and Susan Trento NASA Safety Investigation Collection at the National Air and Space Museum. This invaluable archive of transcripts and primary materials assembled by Joe and his wife, Susan—including Roger Boisjoly's engineering notebooks, his "smoking gun" memos, letters, affidavits, and personal documents—provided insights into the thoughts and actions of many key protagonists in the story, which would otherwise have remained out of reach.

For more detail on the relationships and machinations at the Johnson Space Center, and within the Astronaut Office, I relied especially upon Michael Cassutt's excellent biography of George Abbey, *The Astronaut Maker*; the frank memoirs of two members of Astronaut Group 8, Mike Mullane's *Riding Rockets*, and Rhea Seddon's *Go For Orbit*; and Lynn Sherr's biography of Sally Ride. The drama surrounding the launch of STS-1 is grippingly recounted in Rowland White's *Into the Black*; and former *Concord Monitor* columnist Bob Hohler's *I Touch the Future* remains the most complete account of the Space Flight Participant Program through the lens of Christa McAuliffe's experience.

To help reconstruct the story from inside Morton Thiokol, I drew upon two detailed memoirs: Allan McDonald's *Truth, Lies, and O-Rings,* and one completed by Roger Boisjoly in 1999, but never published. My discovery of a draft of Boisjoly's book, on an unmarked CD buried in a file box at the National Air and Space Museum, was a turning point in my research. I was more fortunate still that Dr. Mark Maier at Chapman University, a leading authority on the *Challenger* tragedy, provided me with a full copy of the memoir—with which Boisjoly had entrusted him before his death in 2012—as well as copies of interview transcripts and documents Mark has gathered over his own many years of research into the subject. Former Thiokol engineers Brian Russell and Jerry Burn shared with me internal reports, records, and photographs that helped tell the story of the solid rocket seals; Jenny Stein was equally generous with her own collection of documents charting the troubled history of the Space Shuttle Main Engines. Diane Vaughan's *The* Challenger *Launch Decision* proved an essential guide to the nature of the Flight Readiness Review process and the bureaucracy of the Marshall Space Flight Center, and Andrew Dunar and Stephen P. Waring's landmark *Power to Explore* added a richly sourced counterpoint to the prevailing narrative of institutional failure in Huntsville.

Above all, the records of the Rogers Commission provided vital detail and analysis of the long trail of history leading to the accident—not only in the five published volumes of its final report, but in the vast trove of materials gathered during the investigation, all of which is held at the National Archives. The transcripts of the Commission investigators' many hours of interviews with NASA and Thiokol staff, conducted in the early months of 1986 and running to more than five thousand pages, were essential in understanding the development of the rockets, and the interplay of personalities and interests in the final hours before launch—especially during the fateful teleconference between Marshall and Morton Thiokol engineers. These documents made it possible to build a minute-by-minute, and sometimes second-by-second, account of what happened, and what was said, from dozens of different eyewitness perspectives.

With the disintegration of *Challenger* just over a minute into flight, the story became one of the most intensively covered news events in history, with journalists from all over the world contributing to a detailed contemporaneous record in print, on radio and TV. The work of the *New York Times* alone, which would later win the newspaper a Pulitzer Prize, fills hundreds of pages on the subject. The book *Challenger: A Major Malfunction*, based on *Reader's Digest* space correspondent Malcolm McConnell's reporting from the scene, is an excellent hour-by-hour description of events immediately surrounding the explosion. The work of reporters from these and other institutions, which proceeded over months and sometimes years, now preserved on microfilm and in digital archives, enabled me to construct a detailed narrative of January 28 and beyond, encompassing multiple points of view. I'm very grateful to three of the journalists assigned to the story from the beginning—Bill Broad, Charles Fishman, and David Sanger—for sharing recollections that were essential to my understanding of how the accident changed NASA's relationship with the media, and how the management of that relationship affected the course of the Presidential inquiry and its eventual conclusions.

NOTES

ABBREVIATIONS:

JSTC: Joseph and Susan Trento NASA Safety Investigation Collection, National Air and Space Museum, Air and Space Archives, Chantilly, VA

OHI-J: Johnson Space Center Oral History Project Interview

OHI-H: NASA Headquarters History Office Oral History Collection Interview

OHI-M: Marshall Space Flight Center Oral History Interview

LAT: Los Angeles Times

NYT: New York Times

RCR: Report of the Presidential Commission on the Space Shuttle Challenger Accident (The Rogers Commission Report), vols. I–V

R-NARA: Transcript of Rogers Commission investigators' interview held in Record Group 220 (Records of Temporary Committees, Commissions, and Boards) at the National Archives and Records Administration, College Park, MA

WP: Washington Post

PROLOGUE

1   *The coffee:* Steve Nesbitt, author interview, April 7, 2021; author visit to Building 30 with Nesbitt, January 27, 2023; William Harwood, "Timeline of *Challenger's* Final Flight," CBS News, https://www.cbsnews.com/network/news/space/home/memo rial/51l.html; "STS-51L—Launch Flight Directors Loop," YouTube video, January 28, 1986, https://www.youtube.com/watch?v=H3SYYrKxHD8.

2   *The hushed:* Dick Covey, author interview, December 13, 2021; Covey, OHI-J, by Jennifer Ross-Nazzal, November 15, 2006, 27.

2   *Nesbitt rarely:* Nesbitt, author interview; "STS-51L—Launch Flight Directors Loop."

CHAPTER ONE: FIRE ON PAD 34

7   *Martha Chaffee was in her kitchen:* Dora Jane Hamblin, "The Fire and Fate Have Left Eight Widows," *Life* 64, no. 4, January 26, 1968; Grissom and Still, *Starfall,* 188; Cernan and Davis, *The Last Man on the Moon,* 10–11, and interview with Martha Chaffee in the film of the same name, 36m; Collins, *Carrying the Fire,* 271.

7   *At thirty-one, Roger Chaffee was:* Chrysler and Chaffee, *On Course to the Stars,* 41–61; Grissom and Still, *Starfall,* 178.

8   *At the moment in May 1961:* Kennedy called for a manned moon landing in his speech to Congress on May 25, 1961. Alan Shepard had flown into space on May 5, but Gagarin had beaten him by three weeks—and Shepard's flight was a brief

suborbital mission of just fifteen minutes; Gagarin orbited the Earth for almost two hours.

9 *In June 1966:* Paul D. Spudis, "Surveyor 1, America's First Lunar Landing," *Air & Space,* June 2, 2016, https://www.smithsonianmag.com/air-space-magazine/surveyor-1-americas-first-lunar-landing-180959289/.

9 *As the summer:* Murray and Cox, *Apollo,* 120–23, 183–84, 186.

10 *"The fact that":* This line has been variously attributed to Scott Carpenter, John Glenn, and Shepard, in anecdotes dating back to 1963. This version is from Kranz, *Failure Is Not an Option,* 200–201.

10 *Gus Grissom rarely:* Leopold, *Calculated Risk,* 168, 226.

11 *But some:* Ibid., 170–74; Eisele, *Apollo Pilot,* 34.

11 *between forty and seventy: Report of Apollo 204 Review Board,* D-13-11.

11 *Leading the:* Murray and Cox, *Apollo,* 184; Andrew Chaikin, "Apollo's Worst Day," *Air and Space Magazine,* November 2016, https://www.airspacemag.com/history-of-flight/apollo-fire-50-years-180960972/.

12 *"You are the contractor":* Chaikin, "Apollo's Worst Day."

13 *"acceptable risk":* There seemed to be sound reasons for this. By the time of the fire, NASA had collectively logged more than twenty thousand hours both in flight and tests using a pure-oxygen atmosphere. Leopold, *Calculated Risk,* 170.

13 *The regulations:* Ibid., 172.

13 *So when:* Murray and Cox, *Apollo,* 185; Chaikin, "Apollo's Worst Day."

13 *Finally, as the review:* Murray and Cox, *Apollo,* 184.

14 *It had arrived:* Wally Schirra, OHI-J, interview by Roy Neal, December 1, 1998, 12.26–12.27.

14 *In early October:* Chaikin, "Apollo's Worst Day."

14 *Grissom had become:* Leopold, *Calculated Risk,* 175; Grissom and Still, *Starfall,* 179–82.

15 *it would have been shot:* Walt Williams, NASA Director of Flight Operations during the Mercury program, quoted in Leopold, *Calculated Risk,* 176.

15 *Yet Grissom:* He felt like "a wolf howling in the wilderness," he said. Ibid., 178.

15 *" 'Go' Fever":* Schirra, OHI-J, 12.27.

15 *"If I say anything":* Leopold, *Calculated Risk,* 172.

15 *At the end:* Chrysler and Chaffee, *On Course to the Stars,* 117–18.

15 *"As far as we're concerned":* Leopold, *Calculated Risk,* 178.

15 *With Christmas approaching:* Chrysler and Chaffee, *On Course to the Stars,* 118–19.

16 *domed concrete launch control blockhouse:* Murray and Cox, *Apollo,* 221.

16 *NASA officials had chosen the site:* Lipartito and Butler, *A History of the Kennedy Space Center,* 57–58, 100–101.

17 *the subtropical weather:* Testimony of Colonel E. F. Kolczynski, US Air Force meteorologist, RCR, vol. V, 984.

17 *the weather was perfect:* NOAA Online Weather Data, Climatological Data for Melbourne Area, Florida, January 1967.

17 *Sealed in:* Leopold, *Calculated Risk,* 185; NASA photograph of crew crossing gantry, Chaikin, "Apollo's Worst Day."

17 *As they had suited up:* Murray and Cox, *Apollo,* 187; Leopold, *Calculated Risk,* 183.

17 *The plugs-out test:* Ibid.; Chrysler and Chaffee, *On Course for the Stars,* 123.

18 *It took more than an hour:* Leopold, *Calculated Risk,* 187–88; Murray and Cox, *Apollo,* 191–93; John Tribe, author interview, July 12, 2021.

20 *Up on Level A8:* Leopold, *Calculated Risk,* 192, 194–96.

21   *a single handprint:* Ibid., 190, and Murray and Cox, *Apollo,* 197. Contrary to the description in NASA's official Apollo 204 report, Leopold explains the true position of Grissom's body: Leopold, *Calculated Risk,* 196.

CHAPTER TWO: WHITEY ON THE MOON

22   *The call came in:* Collins, *Carrying the Fire,* 269–71.

22   *Later that night:* "Bulletins on Astronauts Interrupt TV Schedule," *NYT,* January 28, 1967.

23   *Johnson issued a statement:* AP, "Johnson Voices Sorrow at Loss of 'Three Valiant Young Men,' " *NYT,* January 28, 1967.

23   *pulled back the curtain:* John Noble Wilford, "The Apollo Spacecraft Is Due for a Long, Hard Look," *NYT,* February 5, 1967.

23   *Four days after:* "Burials of Astronauts Will Be Shown on TV," *NYT,* January 31, 1967; "Burial of Apollo 1 Astronauts," Getty Archive video, https://www.gettyim ages.com/detail/video/packed-arlington-cemetery-for-the-burials-of-nasa-news -footage/453230238; "3 Astronauts Buried with Honors," *NYT,* February 1, 1967.

23   *Escorting the shattered family:* Cernan and Davis, *The Last Man on the Moon,* 14.

23   *Back at Cape Canaveral:* Murray and Cox, *Apollo,* 210–11, 216.

24   *determined to protect:* Wayne Biddle, "Two Faces of Catastrophe," *Air & Space,* August–September 1990.

24   *Their findings revealed: Report of the Apollo 204 Review Board,* 5–12.

24   *In retrospect:* Murray and Cox, *Apollo,* 214–15.

24   *When engineers:* The investigation failed, however, to emphasize that the intended mission for this new spacecraft had been changed from manned to unmanned when the capsule was half built—and that North American had installed an additional fifteen miles of wiring to fly the craft on automatic systems. Andrew Nahum, "Review: Storming the Moon," *New Scientist,* June 12, 1993, https://www.newscien tist.com/article/mg13818774-900-review-storming-the-moon/#ixzz6lXW3BBvi.

25   *The intensity of the fire:* Murray and Cox, *Apollo,* 189–90.

25   *"criminal negligence":* John Noble Wilford, "Criticism Sharp; Faulty Wire Is Termed the Probable Cause of Blaze Fatal to 3," *NYT,* April 10, 1967.

25   *Administrator Webb:* "Space: Blind Spot," *Time,* April 27, 1967, http://content .time.com/time/subscriber/article/0,33009,843575-1,00.html; John Noble Wilford, "Webb Holds Firm on Apollo Goals at House Inquiry," *NYT,* April 11, 1967.

26   *it was not until August:* The mission was Apollo 7; the spacecraft was delivered to the pad on August 9, 1968: "About Apollo 7, the First Crewed Apollo Space Mission," NASA, July 8, 2015, https://www.nasa.gov/mission_pages/apollo/missions /apollo7.html.

26   *By then the capsule:* Collins, *Carrying the Fire,* 274.

26   *The management:* Murray and Cox, *Apollo,* 222–23.

27   *"You lose crew":* John Hodge, quoted in ibid., 225.

27   *Joseph Shea, whose bosses:* Murray and Cox, *Apollo,* 220.

27   *Gene Cernan stood:* Cernan interview, *The Last Man on the Moon* film, 1h17m.

28   *"I'd like to just say":* Jeffrey Kluger, "Farewell to Eugene Cernan, the Last Man to Walk on the Moon," *Time,* January 16, 2017, https://time.com/4635852/cernan-eu gene-gene-last-man-moon-nasa-astronaut/.

28   *three further landings:* Kenneth Silber, "Down to Earth: The Apollo Missions That

Never Were," *Scientific American,* July 19, 2009, https://www.scientificamerican.com/article/canceled-apollo-missions/.

28 *milking the cows:* The Last Man on the Moon film, 5m.

28 *"OK, Jack":* Cernan and fellow Apollo astronaut Walt Cunningham maintained afterward that these were among his final words on the lunar surface—but they do not appear in the official transcript. Cernan and Davis, *The Last Man on the Moon,* 338; Cunningham, *The All-American Boys,* 292; Megan Garber, "What Were the Last Words Spoken on the Moon?," *Atlantic,* December 15, 2012, https://www.theatlantic.com/technology/archive/2012/12/what-were-the-last-words-spoken-on-the-moon/266287/.

28 *This time, however:* John J. O'Connor, "Apollo 17 Coverage Gets Little Viewer Response," *NYT,* December 14, 1972.

29 *wasteful and quixotic:* In a paper examining public attitudes toward manned spaceflight in the period, space historian Roger Launius quoted a *Newsweek* story from 1966: "The US space program is in decline. The Vietnam war and the desperate conditions of the nation's poor and its cities—which make space flight seem, in comparison, like an embarrassing national self-indulgence—have combined to drag down a program where the sky was no longer the limit." Roger D. Launius, "Public Opinion Polls and Perceptions of US Human Spaceflight," *Space Policy* 19 (2003): 163–75.

29 *an angry mob:* Trento, *Prescription for Disaster,* 76.

29 *a distant, fragile Earth:* Dennis Overbuy, "Apollo 8's Earthrise: The Shot Seen around the World," *NYT,* December 21, 2018, https://www.nytimes.com/2018/12/21/science/earthrise-moon-apollo-nasa.html.

29 *600 million people:* Tiffany Hsu, "The Apollo 11 Mission Was Also a Global Media Sensation," *NYT,* July 15, 2019; "First Moon Landing Fast Facts," CNN, July 23, 2023, https://www.cnn.com/2013/09/15/us/moon-landing-fast-facts/index.html.

29 *Two months later:* "Space: The Price of Mars," *Time,* September 26, 1969, https://content.time.com/time/subscriber/article/0,33009,844910,00.html.

29 *But even as:* Trento, *Prescription for Disaster,* 73–74.

30 *equipment failures and narrow escapes:* Thomas O'Toole, "Apollo Space Program Has Woes: Missions Cut for Survival," *Anniston Star,* September 27, 1970; Staff of the *Washington Post, Challengers,* 38–39.

30 *And with each subsequent:* Bryan Greene, "While NASA Was Landing on the Moon, Many African Americans Sought Economic Justice Instead," *Smithsonian,* July 11, 2019, https://www.smithsonianmag.com/history/nasa-landing-moon-many-african-americans-sought-economic-justice-instead-180972622/.

31 *The following day:* Imani Perry, "For the Poor People's Campaign, the Moonshot Was Less Than a Triumph," *NYT,* July 16, 2019, https://www.nytimes.com/2019/07/16/us/for-the-poor-peoples-campaign-the-moonshot-was-less-than-a-triumph.html.

31 *Nor were the astronauts:* Harold M. Schmeck Jr., "Apollo 15 Crew Is Reprimanded," *NYT,* July 12, 1972; Staff of the *Washington Post, Challengers,* 42.

31 *no plans to screen:* Trento, *Prescription for Disaster,* 116; Jennifer Levasseur, "Leaving the Moon, Watching at Home," National Air and Space Museum, December 30, 2011, https://airandspace.si.edu/stories/editorial/leaving-moon-watching-home.

32 *Back in Houston:* White, *Into the Black,* 91.

32 *In January 1972:* "Challenger's Last Flight," *LAT,* February 9, 1986; Logsdon, *After Apollo?,* 269, 287. The statement was based on a draft prepared by Apollo 8 astronaut Bill Anders.

33 *Early drafts:* White, *Into the Black,* 92; Logsdon, *After Apollo?,* 265; "Name to describe program to develop new space transportation system," memo from Peter Flanagan to President Nixon, January 4, 1972, reproduced in Rebecca Onion, "I Say Space Shuttle, You Say Space Clipper," *Slate,* November 15, 2012, https://slate.com /human-interest/2012/11/the-space-shuttle-nixon-s-aides-asked-him-to-reconsid er-the-program-s-name.html.

CHAPTER THREE: THE SPACEPLANE

34 *At around 10:00 a.m.:* Ivy Hooks, OHI-J, interview by Jennifer Ross-Nazzal, March 5, 2009, 22; White, *Into the Black,* 3–4; time-lapse image of Faget dropping model, reproduced in Jennifer Ross-Nazzal, "The Right Place," *Houston History,* Fall 2008.

34 *At forty-seven:* Murray and Cox, *Apollo,* 39–42; Dorothy B. Lee, OHI-J, interview by Rebecca Wright, November 10, 1999, 12.10.

35 *Born in British Honduras:* Max Faget, OHI-J by Jim Slade, June 18–19, 1997, 12.1–12.6; OHI-H, NASA Culture Study interview by Howard McCurdy, November 9, 1987, 2–15; Murray and Cox, *Apollo,* 39–40.

35 *It was at Langley:* Heppenheimer, *The Space Shuttle Decision,* 26; Caldwell Johnson, OHI-J, by Michelle Kelly, April 1, 1998, 12.5.

36 *reaching temperatures of up to 12,000 degrees:* David K. Stumpf, "Reentry Vehicle Development Leading to the Minuteman Avco Mark 5 and 11," *Air Power History* 64, no. 3 (Fall 2017): 13.

36 *Using wind tunnel:* Murray and Cox, *Apollo,* 39fn.

36 *It was this work:* Faget, OHI-J, 1997, 12.34, 12.57.

37 *a new task:* Chester Vaughan, author interview, Clear Lake, May 5, 2021.

37 *Early in 1969:* White, *Into the Black,* 46; Milton A. Silveira, OHI-J, by Sandra Johnson, April 18, 2006, 7–8; James Oberg, "Max Faget: Master Builder," *Omni* 17, no. 7 (April 1995), 64.

37 *The prototype Space Shuttle:* Hooks, OHI-J, 22; White, *Into the Black,* 7–8. Faget had been building shuttle-like models even before Mercury flew: Robert O. Piland, OHI-J, by Summer Chick Bergen, August 21, 1998, 12.11.

38 *The monumental feats:* The shuttle was "an order of magnitude more complex" than the Apollo spacecraft, according to Bob Sieck; author interview, October 26, 2020. "People don't appreciate that the shuttle, as a technical goal, is much more ambitious than the moon program," MIT professor Eugene Covert told the *Washington Monthly* in 1980: Gregg Easterbrook, "The Spruce Goose of Outer Space," *Washington Monthly,* April 1980.

38 *But by the time:* Ibid.; Tom Moser, "Lecture 5: Orbiter Structure + Thermal Protection System," Aircraft Systems Engineering lecture video, https://ocw.mit.edu /courses/16-885j-aircraft-systems-engineering-fall-2005/resources/lecture-5/; Sieck, author interview, October 27, 2020.

40 *Regardless of the obstacles:* Amy Shira Teitel, "Eugen Sanger: Germany's Other Rocket Genius," *Popular Science,* June 10, 2014, https://www.popsci.com/blog-net work/vintage-space/eugen-sanger-germanys-other-rocket-genius/; White, *Into the*

Black, 141; Jenkins, *Space Shuttle: Developing an Icon,* vol. I, 2–5 and 52; Reuter, *The V2 and the German, Russian, and American Rocket Program,* 96–99.

40 *As the Third Reich:* Alejandro de la Garza, "How Historians Are Reckoning with the Former Nazi Who Launched America's Space Program," *Time,* July 18, 2019, https://time.com/5627637/nasa-nazi-von-braun/.

41 *Dornberger once complained:* Thomas O'Toole, "W. R. Dornberger Dies, German Rocket Expert," *WP,* July 2, 1980.

41 *Dornberger suggested:* Krafft Ehricke, OHI-H interview by John Sloop, April 26, 1974, 30; White, *Into the Black,* 141–42.

41 *In October 1957, the USSR:* Clarence J. Geiger, *History of the X-20A Dyna-Soar,* Historical Division Information Office, Aeronautical Systems Division, Air Force Systems Command, October 1963, ix.

41 *Dyna-Soar was a flat-bottomed:* Jenkins, *Developing an Icon,* vol. I, 55.

41 *The Air Force spent years and hundreds:* It had cost $410 million by 1963. Launius and Jenkins, *Coming Home,* 179; Jenkins, *Developing an Icon,* vol. I, 68.

42 *engineers understood little:* Jenkins, *Developing an Icon,* vol. I, 47–48; Jenkins, *Space Shuttle: The History of the National Space Transportation System,* 20.

42 *Instead of the heavy:* Launius and Jenkins, *Coming Home,* 176; Richard Hallion, *The Hypersonic Revolution: Case Studies in the History of Hypersonic Technology,* vol. 2, 12.

42 *a tricycle undercarriage:* Launius and Jenkins, *Coming Home,* 178.

42 *In March 1962:* AP, "6 Pilots Assigned to Dyna-Soar Tests," *NYT,* March 16, 1962; "Dyna-Soar Space Glider Is Displayed in Mock-Up," *NYT,* September 21, 1962.

43 *the United Nations ratified:* "Treaty Banning Nuclear Weapon Tests in the Atmosphere, in Outer Space and Under Water," United Nations Treaty Collection, October 10, 1963, https://treaties.un.org/pages/showDetails.aspx?objid=080000028 01313d9.

43 *Scientists and Air Force generals:* Chris Bergin, "The Story of the Dyna-Soar," NASA Space Flight, January 7, 2006, https://www.nasaspaceflight.com/2006/01/the-story-of-the-dyna-soar/; George Abbey, author interview, September 25, 2020.

43 *"The MOL is conceived":* The official was Air Force missile chief General Bernard A. Schriever. "Orbiting Lab, Space Station, Separate," *Charlotte Observer,* January 2, 1964; Ted Lewis, "Why We Want Eyes in Space," New York *Daily News,* August 26, 1965. Schriever was also quoted referring to the "development of means against possible enemy actions in space," and anti-satellite weapons, but he drew a distinction between these concepts and MOL.

44 *the Pentagon announced:* UPI, "Military Project Picks Astronauts," *NYT,* November 13, 1965.

44 *That day:* White, *Into the Black,* 30.

44 *In October 1954:* Heppenheimer, *The Space Shuttle Decision,* 29–32. The effort was inspired in part by Walter Dornberger's fascination with the Silbervogel concept. Hansen, *Engineer in Charge,* 350; Thompson, *At the Edge of Space,* 33–35.

45 *Each pilot:* Thompson, *At the Edge of Space,* 38.

45 *"the imagined problems":* Dr. Hugh L. Dryden, quoted in Gorn, *Expanding the Envelope,* 3.

46 *"We produced research":* Thompson, *At the Edge of Space,* 270.

46 *Keen to extend:* Evans, *The X-15 Rocket Plane,* 394–95.

46 *Their exploration:* "NASA Armstrong Fact Sheet: X-15 Hypersonic Research Program," NASA, February 28, 2014, https://www.nasa.gov/centers/armstrong/news

/FactSheets/FS-052-DFRC.html. The experience was more terrifying still for those on the ground. When heart rate monitors were attached to both pilots and flight controllers, the controllers' recorded pulse rates during the flights were even higher than those of the men in the aircraft. Thompson, *At the Edge of Space*, 248.

CHAPTER FOUR: THE MOST COMPLICATED MACHINE IN HISTORY

47   *By the time Air Force major Mike Adams:* Evans, *The X-15 Rocket Plane*, 393.

47   *Earlier generations:* Even wives had been discouraged from watching earlier X-plane test flights; all data concerning the flight of the Bell X-1 in which Chuck Yeager broke the sound barrier was classified as top secret within hours of his landing. Thompson, *At the Edge of Space*, 258; Christian Gelzer, "First Generation X-1," NASA, February 28, 2014, https://www.nasa.gov/aeronautics/first-generation-x-1/.

47   *the rate of attrition:* Evans, *The X-15 Rocket Plane*, 393–95.

47   *It was a little frightening:* Ibid., 362.

47   *Stoic and reserved:* Ibid., 367.

48   *In October 1966:* Ibid., 376.

48   *As her husband:* Thompson, *At the Edge of Space*, 257–58; AP and UPI dispatches, "X-15 Crashes in Desert, Killing Pilot," *Courier Journal* (Louisville, KY), November 16, 1967.

49   *For a few seconds:* Thompson, *At the Edge of Space*, 263; Evans, *The X-15 Rocket Plane*, 381–87.

50   *The wreckage:* Evans, *The X-15 Rocket Plane*, 369, 394.

50   *The lack of:* AP, "Pilot Killed in Crash of an X-15; First Death in Research Program," *NYT*, November 16, 1967; AP, "Capital Flier Dies in Crash of X-15," *Sacramento Bee*, November 15, 1967.

51   *But among:* Evans, *The X-15 Rocket Plane*, 398.

51   *the first American:* In recognition of this tragic distinction, Adams's name was eventually added to the Space Mirror astronauts' memorial at the Kennedy Space Center in 1991.

51   *Yet the X-15:* Thompson, *At the Edge of Space*, 265–71; Pete Knight achieved a speed of Mach 6.7 on his X-15 flight of October 3, 1967, an unofficial record for a piloted aircraft, which still stood in 2024. James Young, "William J. Knight," X-15 Biographies, NASA, October 1999, https://history.nasa.gov/x15/knight.html.

52   *NASA Administrator:* Logsdon, *After Apollo?*, 33; Heppenheimer, *The Space Shuttle Decision*, 95–98.

53   *it was no coincidence:* Dethloff, *Suddenly Tomorrow Came*, 41–42.

53   *In Huntsville:* Dunar and Waring, *Power to Explore*, 63, ix.

53   *Under his direction:* Author visit to Marshall Space Flight Center, Huntsville, January 23, 2023; Dunar and Waring, *Power to Explore*, 3; Tim Radford, "Space Racers," *Guardian*, June 30, 1994; C. S. Pyun, "Slowdown in Space Programs: Its Impact on the Southeast," *Monthly Review* 54, no. 5 (May 1969): 58–64, https://fraser.stlouis fed.org/title/economic-review-federal-reserve-bank-atlanta-884/may-1969-35173 /slowdown-space-programs-impact-southeast-414922.

53   *But once:* Dunar and Waring, *Power to Explore*, 151; Heppenheimer, *The Space Shuttle Decision*, 102–3.

54   *Fortunately, Thomas Paine:* Logsdon, *After Apollo?*, 35; Heppenheimer, *The Space Shuttle Decision*, 111; Thomas O. Paine, NASA Biography, October 22, 2004, https:// history.nasa.gov/Biographies/paine.html.

54   *Paine pressed:* Heppenheimer, *The Space Shuttle Decision,* 147–48.

54   *At the center:* Ibid., 136–50; Mahaffey, *Atomic Adventures,* 135–64.

55   *scant appetite:* Heppenheimer, *The Space Shuttle Decision,* 96, 178–79.

55   *When Armstrong, Aldrin, and Collins were feted:* Neufield, *Von Braun,* 436–37.

55   *As the President:* Heppenheimer, *The Space Shuttle Decision,* 92–93.

56   *Dottie Lee got the call:* Dorothy B. Lee, OHI-J, by Rebecca Wright, November 10, 1999, 12.38.

56   *Lee, forty-two:* Lee, OHI-J, by Wright, 12.30; Dorothy Lee NASA biography; Dorothy Lee, obituary and comments, Dignity Memorial, https://www.dignitymemorial .com/obituaries/webster-tx/dorothy-lee-9886222.

56   *After the meeting:* Lee, OHI-J, by Wright, 12.2–12.3; Warren Brasher, author interview, May 5, 2021.

57   *two separate winged vehicles:* Heppenheimer, *The Space Shuttle Decision,* 221. These specifications had been settled upon by the end of 1969.

57   *At first:* McConnell, *Challenger,* 33–35.

57   *But long before:* Heppenheimer, *The Space Shuttle Decision,* 288–89.

58   *Despite the budget:* Ibid., 245.

58   *To make this argument:* Ibid., 278–81.

59   *"They start at a number":* Ibid., 284.

59   *did not yet officially exist:* White, *Into the Black,* 36.

59   *Code-named Hexagon:* McConnell, *Challenger,* 38; White, *Into the Black,* 73–74.

59   *sixty* miles *of film:* Thom Patterson, "1970s Spy Satellite 'Better Than Google Earth,'" CNN, September 1, 2016, https://www.cnn.com/2016/09/01/us/declassif ied-spy-satellite-hexagon/index.html.

59   *The Air Force also:* Heppenheimer, *The Space Shuttle Decision,* 215.

60   *Yet such long-distance:* Ibid., 234. Faget's Design MSC-040C became the basis of the design: Jenkins, *Space Shuttle,* 79, 142–49, 201. Chris Kraft later explained the decision-making process in detail during a lecture at MIT. "Apollo and the Space Shuttle: Part II," MIT Open Learning Library, https://openlearninglibrary .mit.edu/courses/course-v1:MITx+16.885x+3T2019/courseware/cb948628e7e 44720ae04161dfde794a8/53e12af64ba34d40ac5528f365d1448e/?activate_block _id=block-v1%3AMITx%2B16.885x%2B3T2019%2Btype%40sequential%2B block%4053e12af64ba34d40ac5528f365d1448e.

61   *when in November:* Memo from Faget to Deputy Director MSF, November 30, 1970; Memo to Deputy Director from AF STS Group, December 3, 1970, in Scott Pace, "Engineering Design and Political Choice: The Space Shuttle, 1969–1972," Massachusetts Institute of Technology, May 1982, 115–16, cited in Heppenheimer, *The Space Shuttle Decision,* 232.

61   *"a pact with the devil":* Joe Allen, author interview, June 9, 2021.

61   *veteran engineers at NASA:* Neufeld, *Von Braun,* 454.

62   *four main contracts:* Trento, *Prescription for Disaster,* 129; Heppenheimer, *The Space Shuttle Decision,* 174.

62   *Structural assembly:* Orbiter Vehicles: Enterprise (OV-101), Background, October 3, 2010, archived at https://web.archive.org/web/20160514085926/http:// www.pao.ksc.nasa.gov/shuttle/resources/orbiters/enterprise.html.

CHAPTER FIVE: THE FUTURE BLACK SPACEMAN

63 *Standing near:* Dethloff, *Suddenly Tomorrow Came,* 42–53. Contracts were issued for the first eleven buildings on the site in December 1962, and the center formally opened less than two years later, in June 1964.

63 *From the windows:* Author visit to Center Director's office, Building 1, January 27, 2023.

63 *And it was here:* Atkinson and Shafritz, *The Real Stuff,* 145, 175; Carolyn Huntoon, author interview, May 19, 2021.

64 *conference room:* Shayler and Burgess, *NASA's First Space Shuttle Astronaut Selection,* 166; author visit to Room 966, Johnson Space Center, January 27, 2023.

64 *Abbey had been:* Cassutt, *The Astronaut Maker,* xix, 166, 154; George Abbey, author interview, September 25, 2020; Steve Hawley, author interview, October 15, 2020; Charlie Bolden, author interview, August 11, 2021.

64 *under instructions:* Atkinson and Shafritz explain that the orders came down from above, but others suggest that Abbey was crucial to its success. Later, former Johnson Space Center Director of Human Resources Jack R. Lister said: "[I think], the product that George [Abbey] produces is always excellent. . . . He's done a magnificent job in [locating and] selecting women and minorities. And I can tell you, it's his leadership that did that. Nobody else. . . ." Jack R. Lister, OHI-J, by Sandra Johnson, March 19, 2002, 37.

65 *It would be the first:* Cassutt, *The Astronaut Maker,* 167–74.

65 *the Equal Employment Opportunity Act:* Sherr, *Sally Ride,* 74.

65 *"with full consideration":* Fletcher spoke at the NASA center directors' retreat at Peaks of Otter Lodge, September 11–12, 1972. Atkinson and Shafritz, *The Real Stuff,* 138.

65 *Public Law 94-106:* Cassutt, *The Astronaut Maker,* 172.

65 *Not everyone:* Carolyn Huntoon, author interview, May 19, 2021; Cassutt, *The Astronaut Maker,* 175–76.

66 *Slayton concealed:* The accident destroyed a government helicopter and Cernan barely escaped with his life from beneath a slick of burning aviation fuel. But Kraft did not learn how he had been deceived until Cernan sent him the galleys of his memoir decades later—by which time Slayton was dead. Kraft, *Flight,* 346–48.

66 *In September 1959:* Katharine Q. Seelye, "Geraldyn M. Cobb, 88, Who Found a Glass Ceiling in Space, Dies," *NYT,* April 19, 2019; Margaret A. Weitekamp, "Lovelace's Woman in Space Program," NASA, https://history.nasa.gov/flats.html; Richard Haitch, "Follow-Up on the News: Foiled Astronaut," *NYT,* June 26, 1983; William McGaffin, "Women's Rights Battle Explodes in Space," *Press & Sun-Bulletin* (Binghamton, NY), March 18, 1962; AP, "Woman Pilot First of Sex to Pass Rugged Tests for Space Flights," *Fort Worth Star-Telegram,* August 20, 1960.

67 *As Cobb began:* Atkinson and Shafritz, *The Real Stuff,* 88–89.

68 *Yet the truth:* Shayler and Moule, *Women in Space,* 80–81.

68 *"Despite the manifest":* Letter to Jerrie Cobb from Hiden T. Cox, quoted in Atkinson and Shafritz, *The Real Stuff,* 91.

68 *the two women sat down:* Weitekamp, *Right Stuff, Wrong Sex,* 136–37; Robert Branson, "Want a Date in Space . . . ," *Battle Creek Enquirer,* March 16, 1962.

69 *"We seek only":* Atkinson and Shafritz, *The Real Stuff,* 99.

69 *But this effort:* Ibid., 92–94; Sherr, *Sally Ride,* 68–69.

70   *Of this half dozen:* Sherr, *Sally Ride,* 70.

70   *Nine months later:* "First Woman in Space," *NYT,* June 17, 1963.

71   *Even as the drama:* Atkinson and Shafritz, *The Real Stuff,* 98–99.

72   *But when that year:* Paul and Moss, *We Could Not Fail,* 90–91; Emily Ludolph, "Ed Dwight Was Set to Be the First Black Astronaut. Here's Why That Never Happened," *NYT,* July 16, 2019.

72   *The public relations risks:* Ibid.

73   *a B-57 bomber pilot:* Paul and Moss, *We Could Not Fail,* 93.

73   *Ed Dwight was born:* Ludolph, "Ed Dwight Was Set to Be the First Black Astronaut"; Jen Chen, "KCK Native on Being the First African-American to Train for NASA," NPR, January 19, 2017, https://www.kcur.org/show/central-standard/2017-01-19 /kck-native-on-being-the-first-african-american-to-train-for-nasa; "Lincolnite Damages Russian MIGs in Strafing Fight," *Lincoln Clarion,* November 30, 1951; Paul and Moss, *We Could Not Fail,* 92; Bill Becker, "Negro Astronaut Aiming for Moon," *NYT,* April 2, 1963.

74   *He was elated:* Ludolph, "Ed Dwight Was Set to Be the First Black Astronaut"; "Select Negro for Aerospace School: May Become First of Race in Space," *Cleveland Call and Post,* July 7, 1962; "Tagged as First Tan Astronaut," *Baltimore African American,* June 7, 1962; Chen, "KCK Native on Being the First African-American to Train for NASA."

74   *by being admitted to ARPS:* Dwight enrolled in Phase One of the course in August 1962: Atkinson and Shafritz, *The Real Stuff,* 101. Details of ARPS curriculum from Edwards Air Force Base Fact Sheet, archived at: https://web.archive.org/web/20160303231218 /http://www.edwards.af.mil/library/factsheets/factsheet.asp?id=6586. Until October 12, 1961, the school was known as the Air Force Test Pilot School, but from that point, the course was increased in length to a year and the school redesignated as the U.S. Air Force Aerospace Research Pilot School (ARPS).

74   *Dwight's day-to-day reality:* Paul and Moss, *We Could Not Fail,* 92–93.

75   *With additional funding:* According to Yeager's account, described by Paul and Moss, the intake was expanded only to fifteen. Paul and Moss, *We Could Not Fail,* 92. But other sources agree the class had sixteen—Atkinson and Shafritz, *The Real Stuff,* 101.

75   *did little to correct:* Indeed, NASA used footage of Dwight in a filmstrip shown to children as part of its "Spacemobile" initiative, a traveling educational program directed at "inner city" schools. Paul and Moss, *We Could Not Fail,* 94.

76   *"the future black spaceman":* Gene Seymour, "Dwight's Career Lost Altitude When JFK Died," *Detroit Free Press,* August 18, 1983.

76   *In one year:* Ludolph, "Ed Dwight Was Set to Be the First Black Astronaut."

76   *in late 1963:* In his autobiography, Dwight is not specific about the date on which he completed the course—but, unlike most other sources, he implies that he only qualified after Kennedy's assassination in November. Dwight, *Soaring on the Wings of a Dream,* 285–87.

76   *The agency announced:* Atkinson and Shafritz, *The Real Stuff,* 101; "NASA Selects 14 to Be Astronauts," *NYT,* October 19, 1963.

76   *"there was not":* Ludolph, "Ed Dwight Was Set to Be the First Black Astronaut." For more detail about Slayton's deliberations, see Charles A. Berry, OHI-J, by Carol Butler, April 29, 1999.

76   *"I respect their decision":* Paul and Moss, *We Could Not Fail,* 96.

76 *But the following:* Ibid., 96–102; Atkinson and Shafritz, *The Real Stuff,* 102–3.

77 *The first African American:* James Oberg, "The Unsung Astronaut," NBC News, February 23, 2005, https://www.nbcnews.com/id/wbna7018497#.WPgWSqK1uM8; John Uri, "Robert Lawrence: First African-American Astronaut," NASA, February 21, 2018, https://www.nasa.gov/feature/robert-lawrence-first-african-american -astronaut. "Accident report, Summary": archived at https://web.archive.org/web /20190708104028/http://raahistory.com/military/airforce/lawrence/report1.jpg. (The first Black person in space would eventually be a Cuban of African descent, Brigadier General Arnaldo Tamayo Méndez, a member of a joint Soviet-Cuban mission to the Salyut 6 space station, in 1980.)

CHAPTER SIX: THE FNGS

78 *When it finally came:* UPI, "Astronauts Wanted; Women, Minorities Are Urged to Apply," *NYT,* July 8, 1976.

78 *an eight-page pamphlet:* "Opportunities as Candidates for Mission Specialist Astronaut," Appendix A, Atkinson and Shafritz, *The Real Stuff,* 211.

79 *electrified the entire brotherhood:* For example, nearly everyone at EAFB reportedly applied. Staff of the *Washington Post, Challengers,* 123.

79 *In his office:* June Scobee Rodgers, author interviews, October 15, 2021, and September 29, 2023.

79 *It was a day June Scobee had dreaded:* Scobee Rodgers, author interview, September 29, 2023.

79 *When she and her husband:* Scobee Rodgers, *Silver Linings,* 49.

79 *When the couple married:* Scobee Rodgers, author interviews, September 29 and October 15, 2021; Scobee Rodgers, *Silver Linings,* 64.

80 *June finally relented:* Scobee Rodgers, author interviews; Scobee Rodgers, *Silver Linings,* 89–90. Astronaut Dick Covey said that, in his class of sixty pilots, only the top graduate was eligible to choose a fighter assignment. Covey, OHI-J, by Jennifer Ross-Nazzal, November 1, 2006, 5.

80 *June was in the spring semester:* Scobee Rodgers, author interview, September 29, 2021.

81 *But Dick's flying—testing new aircraft:* Scobee Rodgers, author interview, October 15, 2021; flight engineer Thomas Gray describes the circumstances of this 747 brake test in Amanda Gardner, "Engineer Thomas H. Gray (BSEE '61) Helped Boeing Aircraft Fly Right," *Mirage Magazine,* Spring 2021, https://mirage.unm.edu/a-ca reer-in-flight/.

81 *in October 1975:* Scobee's first flight was October 21, 1975; his second was November 19, 1975: X-24B flight log, Appendix B, Johnny G. Armstrong, *Flight Planning and Conduct of the X-24B Research Aircraft Test Program,* 93, https://apps.dtic.mil /sti/pdfs/ADB029224.pdf.

81 *Scobee was well trained:* Armstrong, *Flight Planning and Conduct of the X-24B Research Aircraft Test Program,* 84; Scobee Rodgers, author interviews, October 15, 2021, and September 29, 2023.

81 *Her fears:* Scobee Rodgers, author interview, October 15, 2021. Mike Love died on March 1, 1976; AP, "X24B Space Launcher Pilot Killed in Crash," *Shreveport Times,* March 2, 1976.

82 *Eventually, it would seem:* Staff of the *Washington Post, Challengers,* 115–16, 122–23; Ogawa and Grant, *Ellison S. Onizuka,* 75–77.

82 *Onizuka's grandparents:* Ogawa and Grant, *Ellison S. Onizuka,* 17, 39–40, 42, 46.

82 *As a teenager:* Ogawa and Grant, *Ellison S. Onizuka,* 41–42; Staff of the *Washington Post, Challengers,* 123. Ogawa and Grant suggest that Onizuka told his wife about the application soon after it was submitted but led her to believe it was a last-minute shot in the dark—not the culmination of a long-gestating plan. Ogawa and Grant, *Elison S. Onizuka,* 81–85.

83 *Six months after:* Atkinson and Shafritz, *The Real Stuff,* 150–55; Huntoon, author interview, May 19, 2021; Sherr, *Sally Ride,* 79.

84 *a five-minute recruitment film:* "Nichelle Nichols—NASA Recruitment Film (1977)," YouTube, https://www.youtube.com/watch?v=Lca9_EDMcX0.

84 *When Ron McNair found:* McNair, *In the Spirit of Ronald E. McNair,* 109; Staff of the *Washington Post, Challengers,* 109; Cheryl McNair, author interview, May 3, 2021.

84 *McNair had already come:* McNair, *In the Spirit of Ronald E. McNair,* 9–12.

85 *One night in 1956:* Ibid., 16; "Pellets Pepper Home of NAACP Official," *Orlando Sentinel,* March 5, 1956.

85 *The two older McNair boys:* McNair, *In the Spirit of Ronald E. McNair,* 30–47.

85 *As a teenager:* Ibid., 44–72; Jon Kirby, "The Wonder of It All," *Oxford American,* no. 107 (Winter 2019): https://oxfordamerican.org/magazine/issue-107/the-wonder-of-it-all.

86 *Although he received:* Ibid. McNair, *In the Spirit of Ronald E. McNair,* 77.

86 *During his first year at college:* Ibid., 85.

86 *magna cum laude:* "McNair Family Starts Scholarship Fund Drive," *Jet,* February 24, 1986.

87 *"you can't give it to anyone":* McNair, *In the Spirit of Ronald E. McNair,* 96.

87 *he would publish a paper:* Michael S. Feld and Ronald E. McNair, "The Physics of Karate," *Scientific American,* April 1, 1979.

87 *McNair also continued:* McNair, *In the Spirit of Ronald E. McNair,* 99–100.

87 *There, in the fellowship hall:* Ibid.,100.

87 *In Boston:* Ibid., 97–98.

87 *McNair was still in his first year:* Cheryl McNair, author interview, May 3, 2021.

88 *Back in Cambridge, Ron took Cheryl to his lab:* Ibid.; McNair, *In the Spirit of Ronald E. McNair,* 101–3.

88 *Refusing to quit:* Cheryl McNair, author interview, May 3, 2021; "Remembering Ronald E. McNair, PhD, '76," MIT Physics, December 1, 2003, https://physics.mit.edu/news/remembering-ronald-mcnair/.

89 *McNair completed:* McNair, *In the Spirit of Ronald E. McNair,* 110; Staff of the *Washington Post, Challengers,* 109.

89 *Judy Resnik, a talented:* Elizabeth Kolbert, "Two Paths to the Stars, Turnings and Triumph; Judith Resnik," *NYT,* February 9, 1986; Scott Spencer with Chris Spolar, "The Epic Flight of Judith Resnik," *Esquire,* December 1986; Staff of the *Washington Post, Challengers,* 84–93; Barbara Galloway, "A Private Astronaut," *Akron Beacon Journal,* June 17, 1984; Katherine Foran, "Specialist Aimed High All Her Life," *Kansas City Times,* February 7, 1986.

90 *In the end, nearly 25,000:* Sherr, *Sally Ride,* 83–85; Cassutt, *The Astronaut Maker,* 181; Shayler and Moule, *Women in Space,* 171; Kolbert, "Two Paths to the Stars."

92 *"artists as well as technicians":* Cassutt, *The Astronaut Maker,* 207.

92 *One candidate:* Ibid., 184; George "Pinky" Nelson, author interview, January 5, 2022; Nelson, OHI-J, by Jennifer Ross-Nazzal, May 6, 2004, 4–5.

92 *When the day:* Cassutt, *The Astronaut Maker,* 190–91; Guion Bluford, author interview, January 13, 2021.

92 *Just before:* Nicholas C. Chriss, "6 Women, 3 Blacks Named Shuttle Flight Astronauts," *LAT,* January 17, 1978, Kolbert, "Two Paths to the Stars"; Staff of the *Washington Post, Challengers,* 93; Anna Fisher, author interview, January 13, 2022.

92 *on the bus:* Staff of the *Washington Post, Challengers,* 124.

93 *The six successful:* Seddon, *Go for Orbit,* 27.

93 *the dead center:* NASA photo S78-26569, January 31,1978.

93 *But the white men:* Steve Hawley, author interview, October 1, 2020.

93 *Over the course of a week:* Shayler, *NASA's First Space Shuttle Astronaut Selection,* 164.

94 *But the acronym:* In his memoir, *Half Way Down the Trail to Hell: A Wartime Remembrance in Three Parts,* Vietnam veteran Stephen E. Kirkland wrote of the term "FNG," "This was not an insult, merely a defense mechanism against getting too friendly with green troops since the attrition rate among replacements was outlandishly high." Stephen E. Kirkland, *Half Way Down the Trail to Hell: A War Time Remembrance in Three Parts,* 14, http://international.loc.gov/master/vhp/0567/056741 /pd0001.pdf.

94 *As part of their induction:* Mullane, *Riding Rockets,* 59.

CHAPTER SEVEN: NEVER A STRAIGHT ANSWER

95 *It was a blazing afternoon:* Robert Lindsey, "Part Plane and Part Boxcar, New Space Shuttle Unveiled," *NYT,* September 18, 1976; White, *Into the Black,* 155; Stephen Fox, "Space Shuttle 'Enterprise' Unveiled," *Lexington Herald,* September 18, 1976. The United States celebrates Constitution Day on September 17 every year, marking the day the Constitution was signed in Philadelphia in 1787.

95 *"This day":* Lindsey, "Part Plane and Part Boxcar, New Space Shuttle Unveiled."

95 *In reality, the Enterprise:* Jenkins, *Developing an Icon,* vol. I, 422–23.

96 *an incomplete hulk:* Final assembly of *Columbia* did not begin until November 7, 1977; rollout from Plant 42 was not until March 8, 1979: Jenkins, *Space Shuttle,* 242.

96 *And although:* William K. Stevens, "New Generation of Astronauts Poised for Shuttle Era," *NYT,* April 6, 1981.

96 *initial dreams of a fleet of a dozen:* Jenkins, *Developing an Icon,* vol. I, 455.

97 *one hundred missions:* Lane, ed., *Wings in Orbit,* 60, 162.

97 *designed to carry:* External Tank Isometric and description, Library of Congress, https://www.loc.gov/resource/hhh.tx1116.sheet/?sp=2.

97 *weighing almost:* "The Space Shuttle," NASA Encyclopedia, https://www.nasa.gov /reference/the-space-shuttle/. The External Tank weighed roughly 800 tons; a Union Pacific locomotive weighs 216 tons.

98 *Drawing the fuel:* "Space Shuttle Propulsion Trivia," NASA Facts, https://www .nasa.gov/sites/default/files/113069main_shuttle_trivia.pdf; William Vietinghoff, "Space Shuttle Main Engines: Just the Stats," *Air & Space,* September 2013, https://www.smithsonianmag.com/air-space-magazine/space-shuttle-engines

-just-the-stats-239729/. Vietinghoff, in 1977 a chemical engineer with the SSME Development Group at Rocketdyne, created the pump power comparison based on the volume of the backyard pool of company secretary Dorothy Rowlands.

98 *a jet of liquid hydrogen:* Steven Siceloff, "Space Shuttle, The Thrust of the Matter," archived at https://web.archive.org/web/20170618010254/https://www.nasa.gov/mission_pages/shuttle/behindscenes/ssmephotoessay.html.

98 *Yet at the Rocketdyne plant:* Author interviews with Maury Vander and Jeff Henderson and retired Stennis engineers, January 25, 2023; Trento, *Prescription for Disaster,* 139–40; Robert E. Lindstrom, OHI-J, by Jennifer Ross-Nazzal, July 19, 2010, 9.

98 *"all-up testing":* Trento, *Prescription for Disaster,* 151. This method had been used in the initial testing of the Minuteman ICBM to speed development, meaning that the missile "fired all three stages and flew to its full range" on its maiden flight, and then again with the first test flight of the Saturn V. Heppenheimer, *The Space Shuttle Decision,* 61.

98 *at least $40 million:* Jeff Foust, "Aerojet Rocketdyne defends SLS engine contract costs," *SpaceNews,* May 7, 2020, https://spacenews.com/aerojet-rocketdyne-defends-sls-engine-contract-costs/; J. F. Thompson, "MIT Aircraft Systems Engineering, Lecture 6: Propulsion—Space Shuttle Main Engines," https://ocw.mit.edu/courses/16-885j-aircraft-systems-engineering-fall-2005/resources/lecture-6/1:03.53.

98 *Early in the shuttle design process:* Maxime A. Faget, OHI-J, by Carol Butler, August 19, 1998, 13.16; Heppenheimer, *Development of the Space Shuttle,* 212–13.

99 *purified Minnesota sand:* White, *Into the Black,* 135.

99 *The shuttle designers:* Heppenheimer, *Development of the Space Shuttle,* 226.

100 *"a horrendous job":* Ibid.

100 *The hard outer coating:* Tom Moser, "Lecture 5: Orbiter Structure + Thermal Protection System." At one point, the tiles also fell victim to deliberate sabotage. NASA had once ensured that all contracted employees working on what it called "life-dependent systems" of the manned program—the failure of which would kill astronauts—were subject to full background checks. But budget cuts meant that the agency could no longer afford such thorough vetting. As a result, NASA did not know that Rockwell had begun running a small, part-time program employing formerly incarcerated men to glue the insulation tiles on the orbiter. To their horror, random quality-control inspection at the Rockwell plant in California revealed that many of the tiles installed on the spacecraft had been punctured with sharp tools. An agency investigation discovered that the technicians had been sabotaging the tiles to make more work, in the hope of having their contracts extended. Correcting the damage took months. Trento, *Prescription for Disaster,* 164–65.

100 *furious to discover:* Trento, *Prescription for Disaster,* 169; Eric Berger, "A Cold War Mystery: Why Did Jimmy Carter Save the Space Shuttle?" *Ars Technica,* July 14, 2016, https://arstechnica.com/science/2016/07/a-cold-war-mystery-why-did-jimmy-carter-save-the-space-shuttle/?comments=1.

100 *Carter's new NASA chief:* Robert Frosch, interview by David DeVorkin on July 10, 1981, Niels Bohr Library & Archives, American Institute of Physics, College Park, MD, USA, Session V, 65–66.

101 *A blunt New Yorker:* Robert Frosch, author interview, September 14, 2020.

101 *Frosch flew down:* Berger, "A Cold War Mystery."

101 *Eventually Frosch:* Mark, *The Space Station,* 72; Trento, *Prescription for Disaster,* 159,

162, 169; Frosch, personal communication with author, October 5, 2020; Berger, "A Cold War Mystery"; Heppenheimer, *Development of the Space Shuttle*, 344.

102 *due to commence:* Shayler and Burgess, *NASA's First Space Shuttle Astronaut Selection*, 165.

102 *the first to begin:* Rick Hauck, author interview, October 26, 2021; NASA photograph of Resnik, NASA ID: S-78-29319, May 12, 1978.

102 *Dick Scobee arrived:* June Scobee Rodgers, author interview, October 15, 2021.

103 *perks unavailable to most test pilots:* Koppel, *The Astronaut Wives Club*, 44, 90–92; the seven Mercury astronauts signed a deal with *Life* magazine that paid them a share of half a million dollars a year, offering each as much as ten times their military salaries at the time: *The Astronaut Wives Club*, 19.

103 *But the men and women:* Anna Fisher, author interview, January 13, 2022, and June Scobee Rodgers, author interview, October 12, 2021; Cassutt, *The Astronaut Maker*, 183fn.

103 *Driving them around:* Mike Mullane, author interview, November 11, 2020; Dick Covey, author interview, December 6, 2021; Mullane, OHI-J, by Rebecca Wright, January 24, 2003, 8, and Covey, OHI-J, by Jennifer Ross-Nazzal, November 1, 2006, 17.

103 *At the Johnson:* Milton Reim, "NASA Selects 35 Astronaut Candidates," JSC NASA News release 78-03, January 16, 1978. Details of Astronaut Offices are from Mullane, *Riding Rockets*, 45; Rick Hauck, author interview, December 14, 2021, and author visit to Building 4, January 27, 2023. Other details are from Mullane, OHI-J, by Rebecca Wright, January 24, 2003, 12, and Mullane, *Riding Rockets*, 36. Fred Haise, Don Lind, and Joe Engle were all members of Astronaut Group 5, inducted in April 1966, and had been trained for Apollo missions—but by 1978 had still not flown in space.

104 *At eight o'clock:* Anna Fisher, author interview, January 13, 2022.

104 *Slight and dark-haired:* Mullane, *Riding Rockets*, 47–48; "Commissioners Reverse Action of Zoning Board," *Orlando Evening Star*, November 7, 1967. Photo of Young smoking his corncob pipe: https://www.tapatalk.com/groups/drgrabows/astronaut-john-young-t10096.html.

104 *The focus:* Mullane, *Riding Rockets*, 48. Mission named as OFT-1 in Milton Reim, "NASA Names Astronaut Crews for Early Shuttle Flights," JSC NASA News release 78-15, March 16, 1978. *NYT* story on tentative launch window, May 18, 1978, cited in NASA, *Astronautics and Aeronautics, 1978, A Chronology, The NASA History Office, 1986*, 115.

105 *Crippen, a dark-haired Texan:* Crippen, author interview, November 18, 2021; Crippen OHI-J, by Rebecca Wright, May 26, 2006, 15.

105 *Although Young and Crippen had spent months:* Crippen, author interview, November 18, 2021; Mullane, *Riding Rockets*, 48.

105 *Standing on the deck of the landing craft:* Seddon, *Go for Orbit*, 47–52; AP/NASA film: "UPITN 26 9 78 Space Shuttle Candidates Water Survival Training in the USA," AP Archive, YouTube, https://www.youtube.com/watch?v=lrJrIu2KTs8.

106 *A key part:* Milton Reim, "Astronaut Candidates Learn Water Survival Techniques," JSC NASA News release 78-32, July 24, 1978.

107 *NASA's public affairs officers:* Shayler and Burgess, *NASA's First Space Shuttle Astronaut Selection*, 172, 189; AP, "Training for Space Starts in the Seas," *Fort Lauderdale News*, August 11, 1978; "People" column, *Time*, August 14, 1978.

107 *Although none of the women discussed it:* Seddon, *Go for Orbit*, 48.

107 *Back in Houston:* Jay Honeycutt, quoted in John Noble Wilford, "35 New Astronauts to Start Tomorrow," *NYT,* July 7, 1978; Seddon, *Go for Orbit,* 67–68; Hauck, author interview, October 26, 2021.

108 *The new intake:* Hauck, author interview, October 26, 2021, and June Scobee Rodgers, author interview, October 15, 2021.

108 *For months:* John Creighton, quoted in Shayler and Burgess, *NASA's First Space Shuttle Astronaut Selection,* 180; Hauck, OHI-J, by Jennifer Ross-Nazzal, November 20, 2003, 10; Hauck, author interview, October 26, 2021; Seddon, *Go for Orbit,* 46, Dan Brandenstein, OHI-J, by Carol Butler, January 19, 1999, 12.5–12.6.

108 *"Your test":* Mike Coats, quoted in Shayler and Burgess, *NASA's First Space Shuttle Astronaut Selection,* 179.

108 *The candidates':* Seddon, *Go for Orbit,* 61; Nelson, author interview, January 5, 2022; James "Ox" van Hoften, author interview, January 11, 2022.

109 *The cockpit of the:* William K. Stevens, "New Generation of Astronauts Poised for Shuttle Era," *NYT,* April 6, 1981.

109 *the windowless:* Author visit to Building 4, January 27, 2023.

109 *Inside the harshly:* Cooper, *Before Lift-Off,* 35; Jerry Swain, author interview, and visit to Building 5, January 27, 2023.

109 *Alongside individual:* Mullane, *Riding Rockets,* 71–74; author visit to Building 5, January 27, 2023.

110 *These simulations:* Sherr, *Sally Ride,* 136. Although the rehearsals were designed to be as realistic as possible, the astronauts were not expected to remain inside the simulator—which lacked the bathroom and sleeping facilities of the real orbiter, which would only function in microgravity—for thirty-six hours continuously and would leave periodically to go home and rest; Jerry Swain, author interview, January 27, 2023.

110 *The supervisors:* Cooper, *Before Lift-Off,* 118–19; Hohler, *I Touch the Future,* 160.

110 *Nearby, in Building 9:* Author visit to Buildings 9, 5, and 29, January 27, 2023; Mullane, *Riding Rockets,* 73.

111 *On Friday evenings:* Author interviews with Guy Bluford, January 13, 2021; Estella Gillette, April 6, 2021; Sylvia Salinas Stottlemyer April 7, 2021; Pinky Nelson, January 5, 2022; and Ox van Hoften, January 11, 2022; Nick Powell, "Patrons of Long-Closed NASA Watering Hole Outpost Tavern Cling to Memories," *Houston Chronicle,* May 17, 2019; Mullane, *Riding Rockets,* 56–57; Bagby, *The New Guys,* 70; Hohler, *I Touch the Future,* 116–17; Lester Peter "Pe-Te" Johnson, Froberg Funeral Home Obituary, https://www.frobergfuneralhome.com/obituary/Lester-Johnson.

111 *At the end of summer 1978:* Mullane, *Riding Rockets,* 59–60.

CHAPTER EIGHT: THE GREAT TILE CAPER

112 *Ordered down:* Lawrence Kuznetz, author interview, April 21, 2022; Lawrence Kuznetz, *Save the Shuttle,* ebook loc. 481.

112 *eleven-story:* "Space Florida Facilities," Space Florida, https://www.spaceflorida.gov /facilities/c3pf/.

113 *"You'll find out!":* Kuznetz, *Save the Shuttle,* loc. 481.

113 *the first week:* March 8, 1979. John Uri, "40 Years Ago: Space Shuttle Columbia Arrives at Kennedy Space Center," NASA, March 25, 2019, https://www.nasa.gov/his tory/40-years-ago-space-shuttle-columbia-arrives-at-kennedy-space-center/.

113 *November that year:* Heppenheimer, in *Development of the Shuttle,* 243, writes that a date of May or June 1980 was fixed in the spring of 1979, citing an *Aviation Week & Space Technology* report in September 1979.

113 *seven thousand:* Jenkins, *Space Shuttle,* 239.

113 *NASA had ordered:* White, *Into the Black,* 210–11; Rick Nygren, author interview, February 23, 2021; Jenkins, *Space Shuttle,* 239.

113 *But the strategy:* "Troubled Space Shuttle Arrives at Space Center," *NYT,* March 25, 1979; John Schwartz, "Loss of the Shuttle: Heat Shields; Protective Tiles Have Been Major Concern from the Start," *NYT,* February 3, 2003; Heppenheimer, *Development of the Shuttle,* 234–35; Trento, *Prescription for Disaster,* 163.

114 *Showing his pass:* Kuznetz, *Save the Shuttle,* loc. 663. He writes that 20,000 tiles were attached when he arrived: Kuznetz, *Save the Shuttle,* loc. 665.

114 *A few weeks later:* NASA announced Kleinknecht's appointment on June 4, 1979: Robert Gordon, "Kleinknecht Assumes New Role," JSC NASA News Release 79-39.

114 *torn away:* Kenny Kleinknecht, OHI-J, by Carol Butler, September 10, 1998, 12.2, 12.25.

115 *named Pipeline:* Kuznetz, *Save the Shuttle,* loc. 564.

115 *forced to admit:* AP, "NASA Revising Budget for 1980; Proposes $220 Million Additional," *NYT,* May 20, 1979.

115 *By July 1979:* Heppenheimer, *Development of the Space Shuttle,* 236.

115 *At the time:* Moser, author interview, February 22, 2022; Moser, OHI-J, by Rebecca Wright, April 9, 2010, 40; White, *Into the Black,* 211–12; Heppenheimer, *Development of the Space Shuttle,* 240–42.

115 *Back in Clear Lake:* Mullane, *Riding Rockets,* 88.

116 *"OK, that's it":* Anna Fisher, author interview, January 13, 2022; Anna Fisher, OH-J, by Jennifer Ross-Nazzal, February 17, 2009, 20.

116 *The tasks ranged:* Rick Hauck, author interview, December 14, 2021; for a full list of assignments, see Shayler and Burgess, *NASA's First Space Shuttle Astronaut Selection,* 198–200.

116 *In September 1979:* Anna Fisher, author interview, January 13, 2022; NASA Release 79-120 and JSC Release 79-58, September 20, 1979, cited in Bette R. Janson and Eleanor H. Ritchie, *Astronautics and Aeronautics, A Chronology, 1979–1984,* NASA History Office, 1990, 65.

117 *In the meantime:* John Noble Wilford, "Day and Night, 1,100 Workers Fight Shuttle's Problems," *NYT,* July 22, 1980.

117 *According to their:* White, *Into the Black,* 211–12.

117 *The remedy:* Heppenheimer, *Development of the Space Shuttle,* 237, 241.

118 *twenty construction trailers:* Ibid., 239.

118 *a Rube Goldberg–looking device:* Mike McAllister, author interview, July 14, 2022.

118 *The skin of* Columbia: Heppenheimer, *Development of the Space Shuttle,* 238–41; Kuznetz, *Save the Shuttle,* loc. 598.

119 *In Houston:* Tom Moser, author interview, January 23, 2022; Moser, OHI-J, by Rebecca Wright, April 9, 2010, 44.

119 *In July, as the tenth:* "Skylab's Fiery Fall," *Time,* July 16, 1979; Richard D. Lyons, "Skylab Debris Hits Australian Desert; No Harm Reported," *NYT,* July 12, 1979.

120 *Carter delivered:* Jimmy Carter, "July 15, 1979: 'Crisis of Confidence' Speech," Miller Center, University of Virginia, https://millercenter.org/the-presidency/president ial-speeches/july-15-1979-crisis-confidence-speech.

121 *His appeal went:* Hendrik Hertzberg, "A Very Merry Malaise," *New Yorker,* July 17, 2009; Bernard Gwertzman, "US Attempt to Rescue Iran Hostages Fails," *NYT,* April 25, 1980; " 'Desert One': Inside the Failed 1980 Hostage Rescue in Iran," CBS News, August 16, 2020, https://www.cbsnews.com/news/desert-one-inside-the-fail ed-1980-hostage-rescue-in-iran/.

121 *first to September 1980:* Janson and Ritchie, *Astronautics and Aeronautics, A Chronology, 1979–1984,* NASA History Office, 1990. January 4, January 29, February 22, and July 31, 1980, 103, 110–11, 122–23, 187–88.

121 *In an editorial:* Joel S. Hirschhorn, "Aluminum Dumbo," *WP,* March 28, 1980; Wayne Biddell, "The Endless Countdown," *NYT,* June 22, 1980; Gregg Easterbrook, "The Spruce Goose of Outer Space," *Washington Monthly,* April 1980; Robert Lee Hotz, "Decoding Columbia: A Detective Story," *LAT,* December 21, 2003.

123 *By the beginning:* Janson and Ritchie, *Astronautics and Aeronautics, A Chronology, 1979–1984,* June 3 and July 31, 1980, 165, 187–88; "Around the Nation," *NYT,* June 16, 1980; "Space Shuttle Wins a Couple," Ideas & Trends, *NYT,* June 22, 1980.

123 *Meanwhile, down at the Cape:* Heppenheimer, *Development of the Space Shuttle,* 243–44; Kuznetz, *Save the Shuttle,* loc. 729, loc. 732.

123 *Back at NASA:* Trento, *Prescription for Disaster,* 172–73; Robert Frosch, author interview, September 14, 2020.

CHAPTER NINE: WHEN THOUGHTS TURN INWARD

127 *Up on the third floor:* Young with Hansen, *Forever Young,* 223, 229; John Young interview, and footage in *The World about Us: The Ultimate Explorer,* BBC, 1983, 37:49, https://www.youtube.com/watch?v=Jb_vsyeL-HM; Henry Casselli, *When Thoughts Turn Inward,* painting, 1981; author visit to O&C Building, January 20, 2023.

128 *In the months since:* Rick Gore, "When the Space Shuttle Finally Flies," *National Geographic,* March 1982; Steve Nesbitt, author interview, April 7, 2021.

128 *10 million readers:* Philip Shenon, "Geographic's Troubled World," *NYT,* December 19, 1982.

129 *At the end of March:* John Uri, "40 Years Ago: Three Weeks before the Launch of STS-1," Roundup Reads, NASA, March 19, 2021, https://roundupreads.jsc.nasa .gov/pages.ashx/1648/40%20Years%20Ago%20Three%20Weeks%20Before%20 the%20Launch%20of%20STS1; Terry Berlin, "Columbia's First Victims," Baen, https://www.baen.com/columbia; John Noble Wilford, "Shuttle Passes First Test; A Worker Is Killed," *NYT,* March 20, 1981.

129 *"Smooth sailin' ":* "STS-1 Mission Commentary Tape, Tapes 1 thru 6," NASA transcript, April 12, 1981, 52, https://historydms.hq.nasa.gov/sites/default/files/DMS/e 000018445.pdf.

129 *"Hey," he said:* Crippen, quoted in "The Boldest Test Flight in History," April 2006, https://www.nasa.gov/mission_pages/shuttle/sts1/sts125.html.

129 *As the final moments:* Rick Nygren, author interview, February 23, 2021; "STS-1 Mission Commentary Tape, Tapes 1 thru 6," 52–54; White, *Into the Black,* 267; Judy Resnik interviewed by Tom Brokaw, *Today* show, April 9, 1981, YouTube video, https://www.youtube.com/watch?v=EaafRyuwA8w.

130 *Pinky Nelson gazed:* Nelson, OHI-J, by Jennifer Ross-Nazzal, May 6, 2004, 48.

130 *flight director Neil Hutchinson:* Hutchinson, OHI-J, by Kevin M. Rusnak, July 28, 2000, 27.

130 *Around them:* MCC Mission Operations Room (MOCR) Positions Room 231, STS-1 First Space Shuttle Mission Press Kit, April 1981, 37; author visit to Building 30, Johnson Space Center, January 27, 2023.

131 *As the shuttle ascended:* "STS-1 Mission Commentary Tape, Tapes 1 thru 6," 54; STS-1-The Launch-Complete Day 1 (40th Anniversary) video, https://www.you tube.com/watch?v=SS7MNPWES-E; Neil Hutchinson, OHI-J, by Kevin M. Rusnak, July 28, 2000, 26–27; Hutchinson, author interview, July 20, 2021.

131 *pitilessly narrow parameters:* The margins were such that if the computers failed for as little as 120 milliseconds, the shuttle could be destroyed: Steven Siceloff, "Shuttle Computers Navigate Record of Reliability," June 28, 2010, https://www.nasa.gov /mission_pages/shuttle/flyout/flyfeature_shuttlecomputers.html (archived at https:// web.archive.org/web/20230619085841/https://www.nasa.gov/mission_pages/shuttle /flyout/flyfeature_shuttlecomputers.html).

132 *yellow-white flames:* Crippen, OHI-J, by Rebecca Wright, May 26, 2006, 67; Jenkins, *Developing an Icon,* vol. II, 287–88.

132 *Soon afterward:* White, *Into the Black,* 270.

132 *Unbeknownst to:* Ibid., 268–69; AP, "Damage to Launching Pad Is Called Only 'Minimal,'" *NYT,* April 17, 1981.

133 *The astronauts were not quite:* White, *Into the Black,* 276–78; Young with Hansen, *Forever Young,* 232.

134 *Down in Houston:* White, *Into the Black,* 278; "STS-1 Mission Commentary Tape, Tapes 1 thru 6," 89–92; Wilford, "Shuttle Rockets into Orbit on First Flight."

134 *But when:* White, *Into the Black,* 300; Dorothy B. Lee, OHI-J, by Rebecca Wright, November 10, 1999, 23–24.

136 *"We have no usable":* White, *Into the Black,* 328.

136 *What Kranz couldn't:* Rowland White, "The Spysat and the Shuttle," *Air & Space Magazine,* April 12, 2017, https://www.smithsonianmag.com/air-space-magazine /spysat-and-shuttle-180962872/; Eric Adams, "Everything We Know About America's Secret KH-11 Recon Satellites," *Popular Mechanics,* September 6, 2019, https:// www.popularmechanics.com/military/research/a28937898/kh-11-satellites/.

136 *But the KH-11's orbit:* White, *Into the Black,* 332, 336; Joe Allen, author interview, June 9, 2021.

137 *Two days, five hours:* White, *Into the Black,* 349.

137 *It was not quite:* "STS-1 Flight Sequence of Events," STS-1 First Space Shuttle Mission Press Kit, April 1981, 15; Brian Welch, "The Amazing All-Electric Flying Machine," *Space News Roundup,* April 12, 1991; Rick Hauck, author interview, October 13, 2020.

137 *For all the hazards:* Welch, "The Amazing All-Electric Flying Machine."

138 *Steeply descending:* Allen with Martin, *Entering Space,* 195; Guy Gugliotta, "In Reentry, Shuttle Was Insulated Ball of Fire," *Washington Post,* February 2, 2003.

138 *For the most treacherous:* White, *Into the Black,* 355; "The Greatest Test Flight—STS-1 (Full Mission 16)," YouTube video, https://www.youtube.com/watch?v=zV iFdJwBpGM; Rick Hauck, author interview, October 13, 2020.

138 *As the sun climbed:* Welch, "The Amazing All-Electric Flying Machine"; Joseph B. Treaster, "Nation Reacts with Cheers and 'We Are No. 1 Again,'" *NYT,* April 15, 1981.

139 *In the Flight Control Room:* "The Greatest Test Flight—STS-1 (Full Mission 16)."

139 *In the glassed-in VIP:* Neil Hutchinson, online panel to mark fortieth anniversary of STS-1, recorded by NASA Alumni League, April 1, 2021.

139   *Out in the desert:* Treaster, "Nation Reacts with Cheers and 'We Are No. 1 Again.' "

140   *plummeting from an altitude:* "Shuttle Trajectory (Final), STS-1," STS-1 First Space Shuttle Mission Press Kit, April 1981, 27.

140   *When at last John Young:* "The Greatest Test Flight—STS-1 (Full Mission 16)"; Pinky Nelson, author interview, January 5, 2022.

141   *Crippen was recognized:* Bob Crippen, author interview, November 18, 2021.

141   *The same newspapers: Time* story quoted in Brian Welch, "The Amazing All-Electric Flying Machine"; "We're in Space to Stay," *Newsweek,* April 27, 1981, 24; William Boot, "NASA and the Spellbound Press," *Columbia Journalism Review* 25, Iss. 2 (July 1, 1986), 23.

141   *The man expected:* UPI, "New Head of NASA Selected by Reagan," *LAT,* April 24, 1981; Beggs, interviewed by Joseph Trento, August 7, 1986, transcript and notes, 1, JSTC, Box 1; Ed Campion, author interview, May 11, 2021; Wolfgang Saxon, "From Navy to Space Agency, Beggs Has Worn Many Hats," *NYT,* December 3, 1985.

142   *"NASA is already":* Peter Larson, "Defense Contractor New NASA Boss," *Orlando Sentinel Star,* March 19, 1981.

142   *Beggs now presided:* Trento, *Prescription for Disaster,* 198.

142   *On July 4, 1982:* Howell Raines, "Reagan Affirms Support for US Space Program," *NYT,* July 5, 1982.

143   *140 miles:* Historic Engineering Record, Space Transportation System, Motor Vessels Liberty Star and Freedom Star, HAER No. TX-116-M, 10.

143   *Alarmed, the Thiokol inspectors:* Dunar and Waring, *Power to Explore,* 351, 357.

CHAPTER TEN: THE FIRST AMERICAN WOMAN IN SPACE

144   *Early one morning:* Sherr, *Sally Ride,* 128–30; Bob Crippen, author interview, November 18, 2021; Cassutt, *The Astronaut Maker,* 160, 239–41; Mullane, *Riding Rockets,* 101. Estella Gillette, author interview, April 6, 2021. Sherr writes that Sally Ride was told of her assignment just a few hours before the announcement; Cassutt and Mike Mullane suggest the crew first learned the news a full week earlier.

144   *Until that moment:* Mullane, *Riding Rockets,* 61–62; Sylvia Salinas Stottlemyer, author interview, April 7, 2021; Dick Covey, author interview, December 6, 2021; June Scobee Rodgers, author interview, October 15, 2021; Barbara Galloway, "Nameless, Faceless Members of the Team," *Akron Beacon Journal,* June 17, 1984.

145   *Many of the astronaut:* Addresses in personal archive of Rick Nygren; Scobee Rodgers, *Silver Linings,* 102, 104; Scobee Rodgers, author interview, October 15, 2021; Ox van Hoften, author interview, January 11, 2022; author visit to Brookpoint Drive, January 28, 2023.

146   *At first, the handful of Black:* Author interviews with Cheryl McNair, May 3, 2021; Guy Bluford, January 13, 2021; and Charles Bolden, August 11, 2021. Later, Bolden was almost equally startled to discover that he wasn't welcome at every Astronaut Office happy hour venue: when one evening he walked up to the bar at Bill and Marie's—a dimly lit juke joint off the coastal highway down near the water—he was greeted by a wide-eyed Bob Crippen. "What are you doing here?" whispered Crippen, who had grown up in the bar his parents ran in Porter, Texas, at a time when drinking in the state was still segregated. "You are out of your mind. They've never seen anybody like you in here . . . just stick close to me." Bolden, author interview, August 11, 2021; Crippen, author interview, October 4, 2023.

147   *But the first:* Mullane, *Riding Rockets,* 52–53; Anna Fisher, author interview, January 13, 2022; Rick Hauck, author interview, October 26, 2021.

147   *"keep their hands clean":* Tom Wolfe, *The Right Stuff* (New York: Bantam, 1980), 145.

148   *Yet the singles:* Sherr, *Sally Ride,* 139–40; Seddon, *Go for Orbit,* 140.

148   *Resnik had been:* Spencer with Spolar, "The Epic Flight of Judith Resnik"; Barbara Galloway, "A Private Astronaut," *Akron Beacon Journal,* June 17, 1984; Sylvia Salinas Stottlemyer, author interview, April 7, 2021.

149   *Physically, too:* Ibid.; Seddon, *Go for Orbit,* 83.

149   *Four years after:* Spencer with Spolar, "The Epic Flight of Judith Resnik"; Staff of the *Washington Post, Challengers,* 95; Sylvia Salinas Stottlemyer, author interview, April 7, 2021.

150   *seen as the most likely:* "Everybody just thought they were the top of the heap": Dick Truly, quoted in Sherr, *Sally Ride,* 127. The location of NASA headquarters during this period is shown in NASA photo 83-HC-542, August 29, 1983, http://heroicrelics.org/info/nasa-hq/nasa-hq-circa-1983.html.

150   *Resnik designed:* Anna Fisher, author interview, January 13, 2022.

150   *But for all:* Jay Honeycutt, quoted by Estella Gillette, personal communication with author, Houston, January 14, 2022; Ox van Hoften, author interview, January 11, 2022; Seddon, *Go for Orbit,* 145–47; Mullane, *Riding Rockets,* 163; Sherr, *Sally Ride,* 140; Anna Fisher, author interview, January 13, 2022; Sylvia Salinas Stottlemyer, author interview, April 7, 2021; Pinky Nelson, author interview, January 5, 2022; NASA photograph of Resnik in her office at JSC, September 5, 1985, NASA ID S85-40171.

150   *Officially, the final:* Anna Fisher, author interview, January 13, 2022; "Don Jorge": Estella Gillette, personal communication with author, April 6, 2021; Cassutt, *The Astronaut Maker,* 239–40; Shayler and Moule, *Women in Space,* 198–99; Guion Bluford, OHI-J, by Jennifer Ross-Nazzal, August 2, 2004, 2.

151   *On April 19:* Anna Fisher, author interview, January 13, 2022; author visit to Building 4 at the Johnson Space Center, January 27, 2023; Mullane, *Riding Rockets,* 100. Astronauts Bonnie Dunbar and Mary Cleave joined in 1980, bringing the total number of women in the office to eight.

152   *"Firsts are only":* Lon Rains, "Former College Park Dean Honored," *WP,* September 25, 1983.

152   *But others believed:* Mullane, *Riding Rockets,* 100.

152   *One began:* Ibid., 108.

152   *on the day Abbey turned:* Ibid., 92; Cassutt, *The Astronaut Maker,* 230–31.

153   *The seething:* Mullane, *Riding Rockets,* 94; Burrough, *Dragonfly,* 19–22. Estella Gillette, author interview, April 6, 2021; Story Musgrave, author interview, September 28, 2023.

153   *a media frenzy:* Sherr, *Sally Ride,* 147.

153   *At the same time:* Cassutt, *The Astronaut Maker,* 236; John Lawrence, "Astronaut Recruitment," JSC NASA News release no. 83-015, May 16, 1983, 2.

154   *Yet Abbey:* Cassutt, *The Astronaut Maker,* 162–63; Joseph P. Allen, OHI-J, by Jennifer Ross-Nazzal, March 18, 2004, 30–31.

154   *But even as the shuttle fleet:* These efforts were summarized by Warren North in "Shuttle Contingency Abort and Crew Escape," NASA Memo CA-WJN-84, W. J. North, May 4, 1984. "After several weeks of delay the JSC request was shelved apparently because of adverse public response that might evolve from overt NASA concern for crew safety," he wrote, in describing how the August 1983 Langley initiative was stifled.

155 *"ten consecutive miracles"*: Cassutt, *The Astronaut Maker,* 163.

155 *If, for some reason:* Rick Hauck, author interview, October 26, 2021; Mullane, *Riding Rockets,* 164.

155 *Finally, if:* Mullane, *Riding Rockets,* 212; W. J. North, NASA Memo CA-WJN-84.

CHAPTER ELEVEN: THE SQUEEZE

156 *The young engineers:* Margalit Fox, "Maxime Faget, 83, Pioneering Aerospace Engineer, Dies," *NYT,* October 12, 2004; Bruce Nichols, "Chris Kraft: 'A Giant among People . . .' Space Center Chief Retiring," UPI, May 1, 1982.

156 *From his aerie on the eighth floor:* George Abbey, OHI-M, NASA Culture Study interview by Howard McCurdy, November 10, 1987, 13–14.

156 *Down at the Cape:* Rick Nygren, author interview, February 23, 2021.

156 *At the top:* Trento, *Prescription for Disaster,* 209.

157 *No part:* Dunar and Waring, *Power to Explore,* 147.

157 *Bill Lucas had come:* William R. Lucas, OHI-M, by Stephen P. Waring, April 4, 1994, 36–38; Bill Lucas, OHI-M, by Stephen Waring and Andrew Dunar, November 3, 1992, 31; Dunar and Waring, *Power to Explore,* 43.

158 *Now, as Center Director:* Dunar and Waring, *Power to Explore,* 153–54.

158 *Dr. Lucas had excellent credentials:* Neufeld, *Von Braun,* 93–94, 393–94; McConnell, *Challenger,* 105–111; Eric Lipton and Robert S. Capers, "Corporate Changes, NASA Cutbacks Hit Project," *Hartford Courant,* April 1, 1991; Boisjoly, *Challenger Book Material,* 509; "Apocalypse" letter, JSTC, Box Two, RB Writings. Famously, von Braun had once rewarded a rocket engineer who admitted a catastrophic mistake by giving him a bottle of champagne, to encourage openness about any future errors: Neufeld, *Von Braun,* 94.

158 *von Braun would last:* Neufeld, *Von Braun,* 449–53.

158 *Although Max Faget:* Heppenheimer, *The Space Shuttle Decision,* 174–75; George Hardy, OHI-H, by Joe Guilmartin, July 13, 1984, 15.

159 *When fully assembled:* Heppenheimer, *The Space Shuttle Decision,* 189; Jenkins, *Developing an Icon,* vol. II, 279.

159 *The solid rocket contractor:* Eric G. Swedin, "Thiokol in Utah," *Utah Historical Quarterly* 75, no. 1 (Spring 2007): 64–78, https://issuu.com/utah10/docs/uhq_volume75_2007 _number1/s/10239216; "Biophysical Description of Golden Spike National Historic Site," National Park Service, June 5, 2018, https://www.nps.gov/im/ncpn/bpd-gosp .htm; McDonald with Hansen, *Truth, Lies, and O-Rings,* 19; Duane Morley Cox, author interview, August 5, 2021; Edward T. Thompson, "The Rocketing Fortunes of Thiokol," *Fortune,* June 1958, 106–14; "Home of the Minuteman," *Time,* January 25, 1960.

160 *The plant was more:* McDonald with Hansen, *Truth, Lies, and O-Rings,* 28; RCR, vol. I, 121; Brian Russell, author interview, April 6, 2022.

161 *In 1982:* McDonald with Hansen, *Truth, Lies, and O-Rings,* 9–10.

161 *Thiokol's Wasatch operation:* Cecil Houston, R-NARA, April 10, 1986, 83–86; Boisjoly, *Challenger Book Material,* 11; Brian Russell, author interview, April 6, 2022. For example, Russell's father worked in research and development at the plant and was still there when his son started his job in engineering; Bob Ebeling and one of his daughters were also both employed at the company at the same time.

161 *In keeping with:* RCR, vol. I, 122; Brian Russell, author interview, March 2, 2022.

161 *The main body of the rocket:* RCR, vol. I, 9.

162   *Inside Thiokol's Promontory plant:* Brian Russell, personal communication with author, October 24, 2023; Heppenheimer, *Development of the Space Shuttle,* 186–88.

162   *The astronauts' lack:* Bob Crippen, author interview, October 4, 2023; Jenny Stein, personal communication with author, November 20, 2023. The only information about the rockets relayed directly into the cockpit came as burnout approached, when a computer message warning that separation was imminent flashed on the CRT screens on the flight deck: Mullane, *Riding Rockets,* 102.

163   *Baked in:* Author visit to former Thiokol facility with Duane Morley Cox, December 8, 2021; Thompson, "The Rocketing Fortunes of Thiokol"; Heppenheimer, *The Space Shuttle Decision,* 175–76; photographs in the collection of Brian Russell.

163   *The four fueled:* Statement before Congressional Subcommittee on Science, Technology and Space by Morton Thiokol vice president Ed Dorsey, January 22, 1987, 159; RCR, vol. I, 99.

164   *flameproof putty:* The exact function of the putty was disputed by engineers working on the rocket—to the extent that some referred to it as the "lucky putty." One discussion of its function can be found in the letter to Larry Mulloy from George Morefield, "Zinc Chromate Putty in SRM Joints," March 9, 1984, RCR, vol. V, 1551.

164   *Designed to fit:* Jerry Burn, author interview, April 7, 2022; Vaughan, *The Challenger Launch Decision,* 96–97; Heppenheimer, *The Space Shuttle Decision,* 422–23; RCR, vol. I, 8–9, 120–22; Dunar and Waring, *Power to Explore,* 341.

165   *The first full-scale:* Brian Russell, author interview, April 6, 2022; Heppenheimer, *The Space Shuttle Decision,* 186.

166   *But when Thiokol engineers:* RCR, vol. I, 85, 122–23. The test took place in September 1977: Thiokol Case Burst Test Report dated December 21, 1977, TWR-11664, cited in Dunar and Waring, *Power to Explore,* 343–44.

166   *leather-covered mallets:* Jerry Burn, author interview, December 6, 2023.

166   *a pair of memos:* John Q. Miller, "Restatement of Position on SRM Clevis Joint O-Ring Acceptance Criteria and Clevis Joint Shim Requirements," Memo EP25 (78-1) January 9, 1978; John Q. Miller, "Evaluation of SRM Clevis Joint Behavior," Memo EP-25 (79-13), January 19, 1979, reproduced in RCR, vol. I, 234–36.

167   *The Marshall laboratory engineers:* RCR, vol. I, 122; Dunar and Waring, *Power to Explore,* 346, 350; Brian Malloy, "Parker Says Shuttle O-Rings Met Standards," UPI, February 12, 1986, https://www.upi.com/Archives/1986/02/12/Parker-says-shuttle-O-rings-met-standards/2791508568400/.

167   *Meanwhile, static firing:* Report, "Analytical Evaluation of the Space Shuttle SRM Tang/Clevis Joint Behavior," Thiokol, October 17, 1978, PC 102302; RCR, vol. I, 122.

167   *presented their findings:* Dunar and Waring, *Power to Explore,* 346–47.

168   *Apparently satisfied:* Heppenheimer, *The Space Shuttle Decision,* 278.

168   *On the Critical Items List:* The shuttle hardware was deliberately designed to incorporate as many redundant systems as possible, and some components were duplicated several times to provide backups in case of failure. For example, the shuttle was equipped with a total of five IBM avionics computers, so that if one malfunctioned, the crew could continue their mission as if nothing had happened; if two or three failed, the orbiter could keep flying without obvious difficulty, but the crew might have to return to Earth earlier than planned. These components, with multiple redundancies, NASA designated Criticality 3, or 2R; those with a single backup were designated Criticality 1R. But those parts on the Critical Items List that were reviewed with the most care before launch were the Criticality 1 components, with

no redundancy at all; if any one of these failed in flight, the result would spell disaster: "Loss of vehicle, mission and crew." Heppenheimer, *The Space Shuttle Decision*, 278–81, 290–92.

168 *At around the same time:* T. E. Bell and K. Esch, "The Space Shuttle: A Case of Subjective Engineering," *IEEE Spectrum*, June 1989; Heppenheimer, *Development of the Space Shuttle*, 283–84.

169 *Still, when the Thiokol:* Brian Russell, author interview, April 6, 2022; Malloy, "Parker Says Shuttle O-Rings Met Standards"; "Challenger's Last Flight," *LAT,* February 9, 1986; Mark A. Vigil, R-NARA, March 27, 1986, 16; Curtis J. Newsome, R-NARA, March 27, 1986, 25; Kenneth Koby, R-NARA, March 27, 1986, 12.

170 *The meticulous:* Dunar and Waring, *Power to Explore*, 351; McDonald with Hansen, *Truth, Lies, and O-Rings*, 35; Brian Russell, author interview, April 6, 2022; Russell, personal communication with author, November 29, 2023.

171 *which had never been designed:* According to former Thiokol engineer Jerry Burn, the Viton material began to ablate and vaporize at 800 Fahrenheit; the temperature of the gas inside the rocket approached 6,500 Fahrenheit. Burn, author interview, April 7, 2022.

171 *The sterile:* Martin Ditkof, "Space Shuttle Challenger January 28, 1986, Tragedy 36 Years Later: A Retrospective on Causation and Moral Injuries," Thesis, University of Colorado at Colorado Springs, January 28, 2022, 36; Ben Powers, R-NARA, March 12, 1986, 6.

171 *Back in the labs in Huntsville:* Vaughan, *The* Challenger *Launch Decision*, 120–25; Dunar and Waring, *Power to Explore*, 351.

172 *NASA's solid rockets chief:* Vaughan, *The* Challenger *Launch Decision*, 123.

172 *In November 1982:* Testimony by Larry Mulloy, RCR, vol. V, 826; Charles Fishman, "Questions Surround NASA Official," *WP,* February 24, 1986; Fishman, author interview, October 11, 2023; Mulloy, R-NARA, April 2, 1986, 6–7; McDonald with Hansen, *Truth, Lies, and O-Rings*, 80–81; "Lawrence Benjamin Mulloy, April 13, 1934–October 2, 2020," Tribute Archive, https://www.tributearchive.com/obituaries/21103270/lawrence-benjamin-mulloy.

172 *America's National Spaceline:* Kaplan, *Space Shuttle*, 38.

173 *Despite his decades:* Larry Mulloy, R-NARA, April 2, 1986, 7–8.

173 *Astronauts and senior NASA managers:* Dick Covey, author interview, December 13, 2021; Rick Hauck, author interview, October 13, 2020; Mike Weeks, quoted in Dunar and Waring, *Power to Explore*, 353.

CHAPTER TWELVE: THE BLACK CAT

174 *"What the shuttle really is":* General James Abrahamson, quoted in Frank Yacenda, "Shuttle Success Makes US No.1—Abrahamson," *Florida Today,* April 11, 1983.

174 *Afterward, Bluford:* Guion Bluford, OHI-J, by Jennifer Ross-Nazzal, August 2, 2004, 27–28; "Here and There . . . ," New York *Daily News*, October 25, 1983.

175 *"the cleanest mission yet":* Mike Toner, "Shuttle Flight 'Cleanest Mission Yet,'" *Austin American-Statesman,* September 6, 1983.

175 *But when:* McDonald with Hansen, *Truth, Lies, and O-Rings*, 14.

175 *some six weeks:* Ibid., 26.

176 *risk could never be eliminated:* Bill Lucas, OHI-M, by Stephen Waring and Andrew

Dunar, November 3, 1992, 31. "In a high-risk business that NASA is in, the only way that you can preclude a flight problem is not fly," Lucas said.

176   *It was necessarily:* Vaughan, The Challenger *Launch Decision,* 82.

176   *The Flight Readiness Review:* RCR, vol. I, 15.

177   *Here, Administrator Beggs:* Vaughan, The Challenger *Launch Decision,* 82–84; Weeks, R-NARA, April 7, 1986, 23.

177   *By that time:* McDonald with Hansen, *Truth, Lies, and O-Rings,* 26.

178   *One senior engineer:* Jay Honeycutt, author interview, July 15, 2022.

178   *Within this exacting:* Vaughan, The Challenger *Launch Decision,* 86–87. According to Roger Boisjoly, Marshall Directorate engineer Leon Ray was seen as a "pain in the ass" at Thiokol. "Joseph Trento interview with Roger Boisjoly at home in Mesa, Arizona, April 12, 1991," 41, JSTC, Box Three, Roger Boisjoly.

179   *On the morning of March 2:* McDonald with Hansen, *Truth, Lies, and O-Rings,* 18–19.

180   *impressive wooden desk:* Lucas sat at von Braun's former desk, which was later placed on display in the U.S. Space & Rocket Center, Huntsville. Author visit to exhibit on January 23, 2023.

180   *The three men raced:* McDonald with Hansen, *Truth, Lies, and O-Rings,* 19–21.

181   *Al McDonald had worked:* James R. Hansen, "Big Sky Values, A Biography of Allan J McDonald," *Truth, Lies, and O-Rings,* 580–82.

181   *Coolheaded and capable:* Brian Russell, author interview, April 6, 2022; McConnell, *Challenger,* 182.

181   *Until a few months:* McDonald with Hansen, *Truth, Lies, and O-Rings,* 10–22.

183   *Officially, the change was designed:* Bob Crippen, author interview, October 4, 2023; Ben Evans, *Tragedy and Triumph in Orbit,* 211.

183   *Using the new method:* Scobee Rodgers, author interview, October 15, 2021, and Crippen, author interview, October 4, 2023; an image of the patch Scobee designed can be found at http://www.spacepatches.nl/sts_mis/sts13.html. STS-41-C was originally scheduled in February 1983 as the thirteenth mission to fly, and thus given the STS-13 designation; when the flight manifests changed, it ultimately became the eleventh shuttle launch. The thirteenth mission was also aboard *Challenger:* STS-41-G in October 1984. Betty Johnson, "Crewmembers Named for STS-13, Spacelab 2 and 3," NASA News Release No. 83-004, February 18, 1983; John Lawrence, "Shuttle Crews Selected," NASA News Release No. 83-036, September 21, 1983.

183   *From the spot reserved:* Scobee Rodgers, author interview, September 29, 2023; "STS-41-C Launch—1st Shuttle Direct Ascent Trajectory, Challenger, 1984, 60fps," YouTube video, https://www.youtube.com/watch?v=tY65RkJOeG4.

184   *After bad weather:* Sarah Blakeslee, "Shuttle Lands in Desert on Coast, Diverted by Bad Weather on Cape," *NYT,* April 14, 1984; Scobee Rodgers, author interview, September 29, 2023; Scobee Rodgers, *Silver Linings,* 115.

CHAPTER THIRTEEN: THE HUMAN SATELLITE

185   *One afternoon:* Kurt Heisig, author interview, July 25, 2023; Kirby, "The Wonder of It All"; Kurt Heisig, "Sax in Space," Kurt Heisig Music, 1986, https://www.kurt heisigmusic.com/sax-in-space/.

185   *Ron McNair had received:* The date of the public announcement was February 4, 1983: AP, "2nd Woman Astronaut Selected for Space Trip," *Florida Today,*

February 5, 1983; McNair, *In the Spirit of Ronald E. McNair,* 126; Cheryl McNair, author interview, May 3, 2021.

186 *Now he had:* Kirby, "The Wonder of It All"; Heisig, "Sax in Space"; Heisig, author interview, July 25, 2023.

187 *"Buck Rogers jetpacks":* McCandless, *Wonders All Around,* 129.

187 *the shuttle toilet:* Ibid., 148.

187 *The total bill:* Leonard Sloane, "A Specialized Market for Satellite Insurance," *NYT,* February 7, 1984.

188 *They were not:* McCandless, *Wonders All Around,* 150; John Noble Wilford, "2 Astronauts Float Free in Space, 170 Miles Up," *NYT,* February 8, 1984.

188 *Moving backward:* Hoot Gibson, OHI-J, by Jennifer Ross-Nazzal, January 22, 2016, 31; McCandless, *Wonders All Around,* 151; Andrew Chaikin, "Untethered," *Air & Space,* October 2014, https://www.smithsonianmag.com/air-space-magazine/untethered-180952792/.

188 *Still gazing:* McCandless, *Wonders All Around,* 149, 162, 242fn; Wilford, "2 Astronauts Float Free in Space."

189 *Looking out:* Hoot Gibson, OHI-J, by Jennifer Ross-Nazzal, January 22, 2016, 31; Wilford, "2 Astronauts Float Free in Space"; Chaikin, "Untethered."

189 *Meanwhile, Captain McCandless:* McCandless, *Wonders All Around,* 154.

189 *"the saddest moment":* White returned only after receiving a personal instruction from Chris Kraft—who, to do so, violated his own directive that no one but the CapCom should communicate directly with the astronauts. Burgess and French, *In the Shadow of the Moon,* 32.

189 *The experimental plan:* McCandless, *Wonders All Around,* 156, 161.

189 *Almost six hours:* Wilford, "2 Astronauts Float Free in Space"; Seddon, *Go for Orbit,* 180.

190 *"one of the great":* Chaikin, "Untethered."

190 *And although unreported:* Heisig, "Sax in Space"; Heisig, author interview, July 25, 2023.

190 *captured floating:* NASA photo of McNair, S84-27211, February 8, 1984.

190 *When McNair discovered:* Cheryl McNair, author interview, May 3, 2021; Kurt Heisig, author interview, July 25, 2023.

191 *A week or so after:* Heisig, "Sax in Space"; Ron McNair to Kurt Heisig, undated letter, in the collection of Heisig.

191 *In the months that followed, the astronaut:* McNair, *In the Spirit of Ronald E. McNair,* 20, 23, 26 (Montgomery was sixty-five years old on Lake City High Commencement Day in 1971: McNair, *In the Spirit of Ronald E. McNair,* 147–48). Parade and other details: Ibid., 144–49; Michael Lewis and Margaret Sprott, "Astronaut McNair Lands in Lake City," *The State* (Columbia, SC), April 13, 1984.

192 *Over the course of:* McNair, *In the Spirit of Ronald E. McNair,* 149, 157; Lewis and Sprott, "Astronaut McNair Lands in Lake City."

192 *Later, as he awaited:* McNair, *In The Spirit of Ronald E. McNair,* 156–58.

193 *The notice pinned:* Croft and Youskauskas, *Come Fly with Us,* 341–42.

193 *Hughes planned:* Original flyer reproduced in pages of Hughes's *Signal* magazine, Jack Fisher, "The Hughes Astronauts," June 4, 2013, https://hughesscgheritage.wordpress.com/2013/06/04/hc-payload-specialists-signal-hughes-communications-newsletter/.

193 *Greg Jarvis, thirty-nine:* Marcia Jarvis, author interview, February 28, 2022; Staff

of the *Washington Post, Challengers,* 134–36; Armstrong and Aldrin's *Eagle* LM touched down at 20:17:40 UTC, Sunday, July 20.

195 *In 1973, he began work at Hughes:* Marcia Jarvis, author interview, March 14, 2022; Staff of the *Washington Post, Challengers,* 131, 136.

195 *From the very start:* Staff of the *Washington Post, Challengers,* 142; Marcia Jarvis, author interview, February 28, 2022; Hughes flyer from *Signal* newsletter.

196 *It was still dark:* Mullane, *Riding Rockets,* 150–53; Hohler, *I Touch the Future,* 82.

197 *Just a few days earlier:* Sylvia Salinas Stottlemyer, author interview, April 7, 2021.

197 *Inside the cabin:* Mullane, *Riding Rockets,* 153–55; John Noble Wilford, "Shuttle Launching Delayed 24 Hours by the Failure of a Backup Computer," *NYT,* June 26, 1984.

198 *"Discovery," the Launch Director announced:* Mullane, *Riding Rockets,* 155–58.

200 *four hundred yards:* Diagram of FSS Slidewire System, Shuttle Operations Manual, JSC-12770, Volume 12, Crew Systems, August 14, 1987, 718, fig. 4.4-1.

200 *On the roof:* Mullane, *Riding Rockets,* 156–60; Sylvia Salinas Stottlemyer, author interview, April 7, 2021.

200 *It was more than forty minutes:* Edward Kolcum, "NASA Assesses Effects of Failure in Launch of Discovery," *Aviation Week & Space Technology,* July 2, 1984; Hohler, *I Touch the Future,* 82.

200 *"I was disappointed":* Staff of the *Washington Post, Challengers,* 96.

200 *Much later:* Mullane, *Riding Rockets,* 160; Mullane, author interview, November 10, 2020.

200 *It would be another two months:* John Noble Wilford, "Countdown Begins for Space Shuttle," *NYT,* June 24, 1984.

200 *Once in orbit: Solar Array Flight Experiment Final Report,* Lockheed Missiles and Space Company, April 1, 1986, https://ntrs.nasa.gov/archive/nasa/casi.ntrs.nasa.gov/19890004113.pdf.

201 *during one live broadcast:* View of Mission Specialist Judith Resnik on the middeck, August 30, 1984, and caption. NASA photo, ID: 41d-13-025; "Shuttle Trying 3 for 3," *The Stuart News,* September 1, 1984.

CHAPTER FOURTEEN: ACCEPTABLE RISK

202 *Less than two weeks:* Details of the refurbishment inspection and reporting process described by Brian Russell, author interview, April 6, 2022. The date of the inspection was no later than September 12, 1984: Presentation chart from Thiokol SRM Preboard FRR, Vaughan, *The Challenger Launch Decision,* 145.

202 *Since reluctantly agreeing:* McDonald with Hansen, *Truth, Lies, and O-Rings,* 28–30.

203 *McDonald's new tasks:* Ibid., 30–31.

203 *Even so:* Ibid., 25.

204 *In the months since:* Weeks, R-NARA, April 7, 1986, 24–26.

204 *Besides, the tweaks:* Vaughan, *The Challenger Launch Decision,* 123–25. Engineers found signs of heat reaching the rings on STS-6, but no erosion: RCR, vol. II, H-1.

205 *So when the Thiokol:* Boisjoly, *Challenger Book Material,* 139–40; Boisjoly Work Assignments (MTI), 30, JSTC, Box 2; Vaughan, *The Challenger Launch Decision,* 137.

205 *Even so, the Thiokol engineers:* Ibid., 137–39.

206 *Hans Mark instructed:* Hans Mark to MSFC/Larry Mulloy, "STS-41-C Programmatic Action Item," March 30, 1984, reproduced as Chart 20 in RCR, vol. II, Appendix H, 13.

207 *Mulloy passed the message:* Larry Wear to Joe Kilminster, "Review of SRM Case-to-Case and Case-to-Nozzle Joint Sealing Procedures," April 11, 1984, reproduced as Chart 20 in RCR, vol. II, Appendix H, 13–14; Vaughan, *The* Challenger *Launch Decision,* 140.

207 *In the meantime, Mulloy:* Ibid., 130; McDonald with Hansen, *Truth, Lies, and O-Rings,* 217–18; McDonald, R-NARA, March 19, 1986, 49–50; RCR, vol. I, 156–58; SRB Critical Items List, "Solid Rocket Booster: Criticality 1," December 17, 1982, reproduced in RCR, vol. I, 157.

207 *Five months later:* Brian Russell, author interview, April 6, 2022; McDonald with Hansen, *Truth, Lies, and O-Rings,* 39–41; Thiokol, "SRM Preboard Problem Summary Chart," September 12, 1984, in Vaughan, *The* Challenger *Launch Decision,* 145.

207 *At a meeting on September 12: Investigation of the Challenger Accident: Hearings before the Committee on Science and Technology,* U.S. House of Representatives, Ninety-Ninth Congress, second session, October 29, 1986, 55.

208 *The analysis passed into the decision chain:* Ibid.

208 *The package:* McDonald with Hansen, *Truth, Lies, and O-Rings,* 41–42; Vaughan, *The* Challenger *Launch Decision,* 143; Problem Summary, Shuttle Projects Office Board, September 19, 1984, reproduced as Chart 29 in RCR, vol. II, Appendix H, H-17–H-18.

208 *And yet:* Vaughan, *The* Challenger *Launch Decision,* 141–48.

209 *Instead, for the first time: Investigation of the Challenger Accident,* 55; Problem Summary, Shuttle Projects Office Board, September 19, 1984, RCR, vol. II, Chart 30, H-18. In March, Mulloy had previously used the term "acceptable erosion"; in September this became "allowable." Problem Summary, Shuttle Projects Office Board, March 8, 1984; Chart 9: STS-11 (SRM-10) O-Ring Erosion Assessment: RCR, vol. II, Chart 8, H-7.

209 *But he was reassured:* McDonald with Hansen, *Truth, Lies, and O-Rings,* 45–46; "Ox" van Hoften, for instance, eventually learned that erosion had occurred on both his flights—but he did not hear anything about it until after the accident. (Van Hoften, author interview, January 11, 2022.) Following the flight of STS-51-C in January 1985, Bob Crippen sat in on a Flight Readiness Review in Huntsville where O-ring blow-by was mentioned, but he later recalled the failure being presented as if "it wasn't considered that much of a big deal." (Crippen testimony, RCR, vol. V, 1418.)

209 *In the middle of September 1984:* John Konrad, author interview, March 16, 2022; Staff of the *Washington Post, Challengers,* 142–43; *Hughes Signal* newsletter quoted at https://hughesscgheritage.wordpress.com/2013/06/04/hc-payload-specialists-sig nal-hughes-communications-newsletter/3/.

209 *Posing for:* Staff of the *Washington Post, Challengers,* 143; photograph from *Signal* newsletter, August 1984, https://hughesscgheritage.files.wordpress.com/2023/03/ba a8f-astronauts.jpg.

210 *At the Cape:* Staff of the *Washington Post, Challengers,* 144; John Konrad, author interview, March 16, 2022.

211 *two months later:* Marcia Jarvis, author interview, March 14, 2022; "STS-51A Launch CNN Coverage," video, https://www.youtube.com/watch?v=GwmFcH8lrCE.

211 *"the most challenging flight":* Jay Greene, quoted in "Space Shuttle Is Launched on Satellite Salvage Flight Friday," November 9, 1984, *NYT.*

211 *What had originally been: Lost in Space,* BBC documentary; Rick Hauck, author interview, October 13, 2020; Anna Fisher, author interview, January 13, 2022; Joe Allen, author interview, June 9, 2021.

212 *Moser, who had only:* Tom Moser, author interview, February 23, 2022.

212 *The astronauts, too:* Joe Allen, OHI-J, by Jennifer Ross-Nazzal, November 18, 2004, 11–12.

212 *Not only were proximity operations:* Rick Hauck, OHI-J, by Jennifer Ross-Nazzal, November 20, 2003, 32–33; Hauck, author interview, October 13, 2020; Vance Brand, OHI-J, by Rebecca Wright, April 12, 2002, 13; Larry Kuznetz, author interview, April 21, 2022. Although the timing of a rescue would have been tight, in *Wonders All Around,* Bruce McCandless III quotes a Martin Marietta engineer saying that sending the shuttle after the MMU would be like "a Ferrari going after a Volkswagen." McCandless, *Wonders All Around,* 242fn.

213 *On the fifth day:* Allen, *Entering Space,* 228–36; *Lost in Space;* John Noble Wilford, "Space Salvage: Lost Satellites Are the Prize," *NYT,* November 6, 1984.

214 *"I just want to tell you":* "Remarks by Telephone with Crewmembers On Board the Space Shuttle Discovery," November 15, 1984, Ronald Reagan Presidential Library & Museum, https://www.reaganlibrary.gov/archives/speech/remarks-telepho ne-crewmembers-board-space-shuttle-discovery.

214 *The* Discovery *crew's triumphant:* John Noble Wilford, "Astronauts Snare Errant Satellite for the First Time Tuesday," *NYT,* November 13, 1984.

214 *From the VIP room:* Tom Moser, author interview, February 22, 2022.

214 *On the day Hauck:* Chris Kraft to Rick Hauck, November 16, 1984; copy of letter in the collection of Rick Hauck.

215 *Yet in their postflight:* "Space Shuttle Flight 14 (STS-51A) Post Flight Press Conference Video," NSS, November 8, 1984, https://space.nss.org/space-shuttle-flight-14 -sts-51a-post-flight-press-conference-video/.

215 *After receiving:* Rick Hauck, author interview, October 26, 2021; Joe Allen, OHI-J, by Jennifer Ross-Nazzal, November 18, 2004, 22; Hauck, quoted in James Reston Jr., "The Astronauts After Challenger," *NYT,* January 25, 1987.

CHAPTER FIFTEEN: THE TEACHERNAUTS

216 *On the last Wednesday of July: The Tonight Show,* July 31, 1985, Library of the Paley Center for Media, Catalog ID B:44276; Hohler, *I Touch the Future,* 141–44.

217 *Ordinary Americans had begun petitioning:* Ibid., 56–57.

217 *By the time the initial test flights:* Ibid., 57; Alan Ladwig, *See You in Orbit?,* 191.

218 *Yet Beggs soon concluded:* Ibid., 192.

218 *Former agency Administrator Robert Frosch:* Ibid., 195–98.

218 *Meanwhile, former Apollo astronaut:* Ibid., 198.

219 *"an unemployed, undereducated, ne'er-do-well":* George Feld letter to Carl Praktish, January 3, 1983, quoted in Ibid., 199.

219 *The final report:* Hohler, *I Touch the Future,* 58; Ladwig, *See You in Orbit?,* 203–4; Walter Pincus, "NASA's Push to Put Citizen in Space Overtook Fully 'Operational' Shuttle," *WP,* March 5, 1986.

219 *By early the next year:* Hohler, *I Touch the Future,* 57–58; Logsdon, *Reagan and the Space Frontier,* 220–21. NASA officials gave serious consideration to sending *Sesame Street* puppeteer Caroll Spinney—in his costume as Big Bird—into orbit, and the agency sent a letter to Spinney soliciting his opinion. He agreed to go, but the logistics of storing the puppet costume aboard the orbiter proved tricky, and the plan was never approved. Alan Boyle, "NASA Confirms Talks to Fly Big Bird on

Doomed Shuttle *Challenger,*" NBC News, May 4, 2015, https://www.nbcnews.com/science/weird-science/nasa-confirms-talks-fly-big-bird-doomed-shuttle-challenger-n353521; Jerry Rice, "Spinney Cycle: A Look at the Muppeteer Who Plays Big Bird, Oscar the Grouch," *Variety,* May 16, 2006, https://variety.com/2006/scene/awards/spinney-cycle-1117939890/.

220 *In July 1984, Beggs:* Logsdon, *Reagan and the Space Frontier,* 221–22; "Remarks at a Ceremony Honoring the 1983–1984 Winners in the Secondary School Recognition Program," Ronald Reagan Presidential Library & Museum, August 27, 1984, https://www.reaganlibrary.gov/archives/speech/remarks-ceremony-honoring-1983-1984-winners-secondary-school-recognition-program; Ladwig, *See You in Orbit?,* 223; Philip M. Boffey, "First Shuttle Ride by Private Citizen to Go to Teacher," *NYT,* August 28, 1984.

220 *That same day:* Hohler, *I Touch the Future,* 54–55.

221 *The next morning:* Ibid., 59; "Reagan Wants Teacher in Space," Monitor Wire Reports, *Concord Monitor,* August 28, 1984, 1.

221 *Meanwhile, citing his lack:* Tom Mirga, "Reagan Says Teacher to Be First Space Passenger," *Education Week,* September 5, 1984, https://www.edweek.org/education/reagan-says-teacher-to-be-first-space-passenger/1984/09.

221 *On Capitol Hill:* Hohler, *I Touch the Future,* 84. Once the final selection of the teacher had been made, Glenn softened his tone but remained opposed in principle to flying anyone but professional astronauts aboard the shuttle: "I think it's OK that we're sending a teacher into space," he said in July 1985, "but if we start trivializing space—sending every butcher, baker and candlestick maker—the American people will start seeing this as a space bus to take people on joy rides." Tom Bayerlein, "Former Astronauts Disagree on Worth of Mission to Mars," *Dayton Daily News,* July 21, 1985.

221 *Even so, many teachers:* Hohler, *I Touch the Future,* 60.

222 *Sharon Christa Corrigan:* Ibid., 20–38; Corrigan, *A Journal for Christa,* 40–41. Although Hohler writes that reservations for flights to the moon were sold by Eastern Airlines, the marketing stunt was in fact staged by Pan Am. Jon Swan, "The Moon as Tourist Trap? That's Not Really So Far Out," *NYT,* January 31, 1971.

223 *McAuliffe began work:* Matthew L. Wauld, "2 Space Novices with a Love of Knowledge; Christa McAuliffe," *NYT,* February 10, 1986; Hohler, *I Touch the Future,* 44–46.

223 *At Concord High:* Ibid., 48–53.

224 *McAuliffe might have forgotten:* Ibid., 60–67.

225 *Greg and Marcia Jarvis:* Marcia Jarvis, author interview, March 14, 2022; Marcia Jarvis, journal entries for January 5–31, 1985; author visit to the former site of Peachtree Apartments, January 27, 2023; "January 5, 1985, Weather History in Houston," Weather Spark, https://weatherspark.com/h/d/9247/1985/1/5/Historical-Weather-on-Saturday-January-5-1985-in-Houston-Texas-United-States.

225 *Jarvis was scheduled to fly:* Marcia Jarvis, author interview March 14, 2022; Marcia Jarvis journal entries for January and February 1985. Image of original 51-D patch, and crew portrait: http://www.collectspace.com/ubb/Forum18/HTML/000129.html.

225 *Amid the acrimonious:* Cassutt, *The Astronaut Maker,* 211; *Come Fly with Us* contains an account of how in the late seventies Kraft and astronauts including John Young predicted the failure of the Spacelab One program. Croft and Youskauskas, *Come Fly with Us,* 25–31.

226 *But the ambitious:* John Noble Wilford, "Fear of Ice Delays Shuttle Launching," *NYT,* January 23, 1985; AP, "Senator's Shuttle Flight Is Delayed for a Week," *NYT,*

February 6, 1985. While previous shuttle flights had carried individual classified pay-loads, STS-51-C was the first NASA mission that was almost entirely clandestine.

227   *The second mission of the year:* Ladwig, *See You in Orbit?*, 237–41; Mullane, *Riding Rockets*, 202–5; AP, "Senator's Shuttle Flight Is Delayed for a Week"; AP, "Delayed Space Flight Is Now Set for March 3," *NYT*, February 8, 1985; "Launching of Shuttle Rescheduled for March 4," *NYT*, February 22, 1985.

227   *On March 1, 1985:* John Lawrence, "Note to Editors," JSC NASA News Release, No. 85-101, February 22, 1985; Charles Anzalone, "Space Flight of UB Grad Is Delayed," *Buffalo News*, March 14, 1985; Marcia Jarvis, author interview, March 14, 2022; NASA press release 85-34 cited in "Astronautics and Aeronautics 1985, A Chronology," NASA SP-4025, entry for March 6, 279; Staff of the *Washington Post, Challengers*, 145; Marcia Jarvis journal entries, March 1985. "What a crock!" the ordinarily mild-mannered Marcia wrote in her diary on March 5. "I hope Garn gets *a lot* of flack."

227   *a function room:* Marcia Jarvis, personal communication with author, November 29, 2023.

228   *In her room at the Hyatt:* Hohler, *I Touch the Future*, 73–76.

228   *At the climax of it all:* Alan Ladwig, OHI-J, by Sandra Johnson, April 11, 2017, 14; Hohler, *I Touch the Future*, 86–87.

229   *The Teacher in Space mission:* Ibid., 79. At the time the final interview process was taking place, the STS-51-L mission was still slated for November that year. But by the time the winner was selected, the launch date had slipped back to January 1986. Charles Redmond and Terry White, "NASA Names Crews to Deploy Satellites in Year-end Flights," JSC NASA News Release 85-005, January 29, 1985; John Noble Wilford, "Teacher Is Picked for Shuttle Trip," *NYT*, July 20, 1985.

229   *Over the next few days:* Hohler, *I Touch the Future*, 80–81.

CHAPTER SIXTEEN: ABORT TO ORBIT

231   *offering a quotation:* Hohler, *I Touch the Future*, 73.

231   *But Reagan's second-term:* Joseph J. Trento and Susan B. Trento, "Why Challenger Was Doomed," *Los Angeles Times Magazine*, January 18, 1987; Trento, *Prescription for Disaster*, 184.

232   *But when NASA officials:* John Noble Wilford, "In Harsh Light of Reality, the Shuttle Is Being Re-evaluated," *NYT*, May 14, 1985.

232   *The upstart European Space Agency:* McConnell, *Challenger*, 62–63; Hohler, *I Touch the Future*, 80.

232   *The only way to:* Craig Couvalt, "Shuttle/Station Cost Challenges Key to Future Space Operations," *Aviation Week & Space Technology*, March 18, 1985; McConnell, *Challenger*, 72. The JSC public affairs office announced in April that the maiden voyage of *Atlantis* would be a military mission, launching in September 1985. John Lawrence, "Note to Editors," JSC NASA News Release 85-014, April 8, 1985.

233   *Instead of the returning spacecraft being subjected to:* McConnell, *Challenger*, 68–69.

233   *When Discovery returned:* Jenkins, *Developing an Icon*, vol. III, 58; John Noble Wilford, "One Space Shuttle Checked; Another Readied," *NYT*, April 21, 1985; Thomas O'Toole, "Shuttle Survives Its Roughest Landing," *WP*, April 20, 1985.

233   *Meanwhile, cost cutting meant:* Bob Sieck, author interview, July 12, 2021; Bob

Drogin, "Cuts in NASA Oversight of Shuttle Program Hit: Quality Control Allegedly Suffered; Workers Complain That Speed-ups Have Led to Accidents," *LAT,* February 5, 1986.

234 *In the meantime, the three shuttles:* Wilford, "In Harsh Light of Reality, the Shuttle Is Being Re-evaluated."

234 *In the same week the competing:* Hohler, *I Touch the Future,* 78; Ox van Hoften, author interview, January 11, 2022. Lynn Sherr told van Hoften and Pinky Nelson this before their launch in April 1984, when she explained that the thirteenth mission would be the last one ABC covered live. James D. A. van Hoften, OHI-J, by Jennifer Ross-Nazzal, December 5, 2007, 27.

234 *Christa McAuliffe had prepared:* Hohler, *I Touch the Future,* 87, 91–99, 102–3, 106–7, 109, 111, 119–20.

237 *The name of "the first private citizen passenger":* "George Bush Announces Christa McAuliffe as the First Teacher to Go to Space," Space Archive video, YouTube, https://www.youtube.com/watch?v=7McTr25xJuA.

237 *By that time, the story was:* Hohler, *I Touch the Future,* 7–11.

237 *Only when Beggs leaned:* "George Bush Announces Christa McAuliffe as the First Teacher to Go to Space"; Hohler, *I Touch the Future,* 13–14.

238 *Back in Concord:* Richard Blais, "The Phone, Doorbell Never Stopped Ringing," *Concord Monitor,* July 20, 1985.

238 *Two weeks after:* Loren Acton, quoted in "In the Age of Spaceplanes," *Air & Space Magazine,* November 18, 2010, https://www.smithsonianmag.com/air-space-magazine/in-the-age-of-spaceplanes-72344760/.

239 *But down on the "booster":* Jenny M. Stein, author interview, May 5, 2021; Eugene Krantz, OHI-J, by Jennifer Ross-Nazzal, December 7, 2011, 21.

241 *"Booster—Flight":* "STS-51F Abort to Orbit," YouTube video, https://www.youtube.com/watch?v=9Luwm_gjLCs; Hohler, *I Touch the Future,* 140.

241 *Inside the shuttle cockpit, flight engineer:* Story Musgrave, author interview, September 28, 2023; Musgrave, quoted in Reichhardt, *Space Shuttle: The First 20 Years,* 153; AP, "Karl G. Henize, NASA Scientist, Dies at 66 Climbing Mount Everest," *NYT,* October 10, 1993.

242 *But down in Mission Control:* Stein, author interview, May 5, 2021; "STS-51F Abort to Orbit."

242 *The rocket engineer would:* John O'Neill, " 'Doing My Job': Hoosier's Quick Decision Saved Shuttle Challenger from Aborting in Flight," *Indianapolis Star,* August 1, 1985; "National Space Club Honors JSC Employees," *JSC Space News Roundup* 25, no. 6 (April 18, 1986), 3.

242 *Two days later:* Tonight Show video, Paley Center Library.

CHAPTER SEVENTEEN: THE MYSTERY OF HANGAR AF

244 *It wasn't until 1:30 a.m.:* Staff of the *Washington Post, Challengers,* 115; "The Onizuka home in Houston, Texas," photograph in Ogawa and Grant, *Ellison S. Onizuka,* 90; "January 28, 1985, Weather History in Houston," Weather Spark, https://weatherspark.com/h/d/9247/1985/1/28/Historical-Weather-on-Monday-January-28-1985-in-Houston-Texas-United-States#metar-02-00. It was a cloudless night, and there was a quarter moon in the sky: "Moon Phases 1985—Lunar Calendar for Houston,

Texas, USA," Time and Date, https://www.timeanddate.com/moon/phases/usa/houston?year=1985.

244 *But Ellison Onizuka was no longer:* Staff of the *Washington Post, Challengers,* 115–16; Ogawa and Grant, *Ellison S. Onizuka,* 114, 122.

245 *Two days later, Morton Thiokol engineer:* Thiokol inspection photograph, dated January 30, 1985, documenting gas path on subject A68, in the collection of Jerry Burn; Boisjoly, *Challenger Book Material,* 237–38; Historic American Engineering Record, "Space Transportation System, Motor Vessels Liberty Star and Freedom Star," HAER No. TX-16-M; "NASA Facts: Solid Rocket Boosters and Post-Launch Processing," FS-2004-07-012-KSC (Rev. 2006).

246 *A big, quiet man who:* Trento interview with Boisjoly, April 12, 1991, 41; author interviews with Brian Russell and Jerry Burn, Utah, April 6 and 7, 2022.

246 *This trip, however:* Trento interview with Boisjoly, April 12, 1991, 54.

247 *the coldest weather in Florida history:* Jon Nordheimer, "Arctic Chill Grips South as Cold Ebbs in North," AP; " 'Freeze of the Century' Damages 90% of Citrus Crop in Florida," *NYT,* January 23, 1985; Russell Baker, "Observer: Sacrificial Digits Saved," *NYT,* January 23, 1985; "January 1985 Record-Breaking Cold," National Weather Service, https://www.weather.gov/ilm/January1985cold.

247 *At Cape Canaveral, the water:* Trento interview with Boisjoly, April 12, 1991, 3–4; Boisjoly, *Challenger Book Material,* 236; McDonald with Hansen, *Truth, Lies, and O-Rings,* 47; Vaughan, *The Challenger Launch Decision,* 154–55; Arnold Aldrich testimony, RCR, vol. V, 1018. The ambient temperature on the launchpad at lift-off was 18 degrees Centigrade; it had been lower on two previous occasions, for STS-41-B and STS-41-C: G. L. Jasper et al., "Atmospheric Environment for Space Shuttle (STS-51C) Launch," NASA Technical Memorandum TM-86508, April 1985, 10.

247 *Although an experienced engineer:* Trento interview with Boisjoly, April 12, 1991, 12–13, 23–24.

248 *One Saturday in 1974:* Boisjoly, *Challenger Book Material,* 297; Trento interview with Boisjoly, April 12, 1991, 12–13. Details of Flight 981: *Turkish Airlines DC-10 TC-JAV in the Ermenonville Forest, France on 3 March 1974,* Accidents Investigation Branch, Department of Trade, HMSO, February 1976, https://assets.publishing.service.gov.uk/media/5422eedde5274a1317000247/8-1976_TC-JAV.pdf; Perrow, *Normal Accidents,* 141–42.

249 *the deadliest accident:* The death toll would be surpassed in March 1977, when 583 people were killed in a collision between two jumbo jets on the runway at Tenerife Los Rodeos International Airport; a grim record that stood well into the twenty-first century. "100 Worst Accidents," Aviation Safety Network statistics, https://aviation-safety.net/statistics/worst/worst.php?list=worstcoll.

250 *Boisjoly began:* William C. Rempel, "Engineer Who Opposed Launch Known for Integrity, Intensity," *LAT,* February 26, 1986; "Aurora to Willard: Here's Election Results from Utah's Smaller Towns," *Salt Lake Tribune,* October 7, 1981.

250 *At the same time:* Boisjoly, interview with Trento, April 12, 1991, 21; Brian Russell, author interview, April 26, 2023.

251 *a weakness the new executive:* Larry Sayer, R-NARA, March 20, 1986, 23.

251 *Down in his office:* Boisjoly, *Challenger Book Material,* 170–71; Boisjoly, Biographical Summary, JSTC, Box One, Roger Boisjoly/Fox Film.

251   *since the very start:* Roger Boisjoly, Work Assignments (MTI), 30, JSTC, Box Two, Roger Boisjoly Material.

252   *When the erosion to the seals:* Boisjoly, R-NARA, April 2, 1986, 7–14; Boisjoly, *Challenger Book Material,* 140.

252   *By the time he stepped:* Boisjoly, *Challenger Book Material,* 159–61. Boisjoly, interview with Trento, April 12, 1991, 42, 54–55; Mark Hayhurst, "I Knew What Was about to Happen," *Guardian,* January 23, 2001, https://www.theguardian.com/science/2001/jan/23/spaceexploration.g2; Morton Thiokol photographs documenting gas path on subject A68, dated January 30, 1985, from the collection of Jerry Burn; Boisjoly, R-NARA, April 2, 1986, 14; *Investigation of the Challenger Accident,* 55.

252   *almost a third:* Blackened grease extended across a one-hundred-and-ten-degree arc of the rocket joint. Boisjoly, *Challenger Book Material,* 237.

254   *Back in Utah, Boisjoly and a team:* Jerry Burn, author interview, April 7, 2022; Brian Russell, author interview, April 6, 2022; McDonald with Hansen, *Truth, Lies, and O-Rings,* 47.

254   *That left the one aspect:* Ibid., 47; Vaughan, *The* Challenger *Launch Decision,* 154–55.

255   *In the past, Boisjoly had been:* Ibid., 27; Boisjoly, R-NARA, April 2, 1986, 38–40.

256   *"Come on—this is a hundred-year event":* Boisjoly, interview with Trento, April 12, 1991, 55.

256   *So while the Thiokol:* Vaughan, *The* Challenger *Launch Decision,* 161; Boisjoly, R-NARA, April 2, 1986, 46–48.

256   CONDITION IS NOT: Vaughan, *The* Challenger *Launch Decision,* 156.

256   *no mention of low temperatures:* Ibid., 161; transcript of Mulloy's Level I presentation, February 21, 1985, RCR, vol. II, H-42.

256   *Yet while the twinned:* Brian Russell, author interview, April 26, 2023; "O-Ring Resiliency Test Procedures" (undated document extract) and "SRM Joint History Apr 1984–Jan 1986," TWR-15501, from the collection of Russell; Boisjoly, R-NARA, April 2, 1986, 34. The test review meeting was held on February 28, 1985: "Boisjoly Field Notes—Field Notes 3/19/84–5/8/85," JSTC, Box Two, Roger Boisjoly Material.

257   *his office overlooking:* Personal communication with Brian Russell, December 12, 2023; author visit to Promontory plant, December 8, 2021.

257   *McDonald's attention:* McDonald with Hansen, *Truth, Lies, and O-Rings,* 50–54.

257   *another catastrophic:* The fire was the result of a rare summer storm passing over the plant, during which a bolt of lightning struck Mix House M-24, causing an electrical malfunction in the blades of a remote-controlled six hundred-gallon mixing bowl filled with seven thousand pounds of shuttle propellant: Ibid., 57–58; "Rocket-Fuel Mixer Explodes at Morton-Thiokol," *Salt Lake Tribune,* June 4, 1985; "Lightning May Have Caused Blaze at Thiokol," *Daily Herald* (Provo), June 5, 1985; "Telephonic Report for a NASA/NASA Contractor Mishap," June 3, 1985, reproduced in *NASA's Quality Assurance Program: Hearing before the Subcommittee on Space Science and Applications of the Committee on Science and Technology House of Representatives,* Ninety-Ninth Congress, Second Session, May 21, 1986, 77.

257   *In the meantime, the Thiokol engineers':* McDonald with Hansen, *Truth, Lies, and O-Rings,* 51–52; Boisjoly, R-NARA, April 2, 1986, 18; STS-51-G launched June 17, 1985; Boisjoly, *Challenger Book Material,* 282, gives date of disassembly as June 25.

258 *as little as three-tenths:* Don Lind, OHI-J, by Rebecca Wright, May 27, 2005, 49–50.

259 *My God, he thought:* Boisjoly, R-NARA, April 2, 1986, 18.

259 *The new damage report:* Boisjoly, *Challenger Book Material*, 282–84. STS-51-F was scheduled for launch on July 12.

259 *In Washington, DC, Mike Weeks:* Mike Weeks, R-NARA, April 7, 1986, 27–28. Weeks's account, suggesting he summoned the Thiokol team to Washington in response to the O-ring problem, differs from that given by Al McDonald in testimony to the Rogers Commission: McDonald recalled that the August 19 meeting was principally about the June mixer fire, and Weeks requested that McDonald include material about the O-rings as an additional issue. RCR, vol. V, 1591–92.

259 *In the meantime, the members:* McDonald with Hansen, *Truth, Lies, and O-Rings*, 59–60; extracts from Thiokol memos and presentations July 2, 1985, to July 9, 1985, in Boisjoly, *Challenger Book Material*, 288–93; Dunar and Waring, *Power to Explore*, 362.

260 *Back in Utah:* Date cited in Boisjoly to Lund, Thiokol interoffice memo, 31 July 1985, "SRM O-ring Erosion/Potential Failure Criticality," RCR, vol. V, 249–50; Boisjoly, R-NARA, April 2, 1986, 18–19; Vaughan, *The* Challenger *Launch Decision*, 174.

260 *At the end of a long meeting:* McDonald with Hansen, *Truth, Lies, and O-Rings*, 63–65.

261 *At the Thiokol plant, Roger Boisjoly:* Boisjoly, *Challenger Book Material*, 296–300; Boisjoly, R-NARA, April 2, 1986, 19–21; Boisjoly to Lund, "SRM O-ring Erosion/Potential Failure Criticality."

262 *now he decided:* McDonald, R-NARA, March 19, 1986, 46–48; McDonald with Hansen, *Truth, Lies, and O-Rings*, 65.

262 *But before leaving:* McDonald with Hansen, *Truth, Lies, and O-Rings*, 66; Boisjoly, *Challenger Book Material*, 301–3; RCR, vol. V, 1592. McDonald writes that he regrets agreeing to this change, but in her account Diane Vaughan argues that there were three sound reasons for eliminating the statement from the presentation. Vaughan, *The* Challenger *Launch Decision*, 178.

CHAPTER EIGHTEEN: THE BACK GATE

263 *Wreathed in the scent:* Author interviews with Alison Balch, January 10, 2023, and with Jane Smith-Wolcott and Alison Balch, November 15, 2022; photograph of Mike Smith on the Cannons' lawn in the Smith family collection; author visit to Timber Cove, January 28, 2023.

263 *At forty, Mike Smith:* Staff of the *Washington Post, Challengers*, 68–77; Smith-Wolcott and Balch, author interview, November 15, 2022; NASA Biographical Data, Michael J. Smith, JSC, September 1984, National Archives, NAID 198457: https://catalog.archives.gov/id/198457; photos in the Smith family collection.

266 *"dynamic, aggressive":* USN Document P1611-1, Report on the Fitness of Officers, Michael J. Smith, April 30, 1978, and May 1, 1979.

266 *"OK, kids":* Smith-Wolcott and Balch, author interview, November 15, 2022.

266 *The neighborhood was:* Rebecca Wright, "A Home for Heroes: Timber Cove," *Houston History* 6, no. 1 (Fall 2008): 48–54, https://historycollection.jsc.nasa.gov /JSCHistoryPortal/history/HouHistory/HoustonHistory-Fall08.pdf.

266 *When they arrived:* Smith-Wolcott and Balch, author interview, November 15, 2022; Anna Fisher, author interview, January 13, 2022; Balch, author interview,

January 10, 2023; author visit to Timber Cove, January 28, 2023; photographs in the Smith family collection.

267 *As a farm boy:* Smith-Wolcott and Balch, author interview, November 15, 2022.

267 *Although he wasn't the type:* Ibid.; Cassutt, *The Astronaut Maker,* 270–71; the tankard is on display as part of the *Forever Remembered* memorial exhibition at the Kennedy Space Center; James A. Michener, *The Bridges at Toko-Ri* (Greenwich, CT: Fawcett Crest, 1953), 126: In both the novel and the 1954 film adaptation, the admiral's words are "Where *did* we get such men?" However, when he delivered the speech in which he quoted the dialogue in 1981, Ronald Reagan altered the line to "Where do we get such men?"—and this is the version that Abbey had engraved on the tankard.

268 *But Smith's closeness:* Smith-Wolcott and Balch, author interview, November 15, 2022; Ben Evans, " 'I Went Flying, Alright!': Remembering Challenger Pilot Mike Smith on his 70th Birthday," AmericaSpace, https://www.americaspace.com/2015/04/30/i-went-flying-all-right-remembering-challenger-pilot-mike-smith-on-his-70th-birthday/; Steve Nesbitt, "Crews for 51-H and 61-E Announced," *Space News Roundup* 23, no. 11 (June 15, 1984).

268 *four more months:* Simulator training for the mission began thirty-seven weeks before launch. RCR, vol. I, 14–15, states that simulator training specific to flight didn't start until T-9 weeks.

269 *"What do you think":* Balch, author interview, January 10, 2023.

269 *more than a dozen officials:* "List of attendees at 850819 Headquarters briefing, provided to investigator Emily Trapnell by CA Speak, March 3, 1986," National Archives Textual Reference, The Presidential Commission on the Space Shuttle Challenger Accident, 1986; PC Numbered Documents, Preliminary Control (PC) Numbered Document: NAID: 262413301, 4, https://catalog.archives.gov/id/262413301?objectPage=4.

269 *the engineer was intent:* McDonald with Hansen, *Truth, Lies, and O-Rings,* 66.

269 *the plan had changed:* Jesse Moore, R-NARA, April 8, 1986, 20–21.

269 *no DC bureaucrat:* Weeks, R-NARA, April 7, 1986, 3–4; Jessi Strawn, "Space Pioneers, Iowa Natives, Receive State Aerospace Engineering Alumni Awards," *Iowa State University College of Engineering News,* April 11, 2012, https://news.engineering.iastate.edu/2012/04/11/space-pioneers-iowa-natives-receive-iowa-state-aerospace-engineering-alumni-awards/.

269 *the most candid report:* RCR, vol. II, H-69; Vaughan, *The* Challenger *Launch Decision,* 176.

270 *The Thiokol engineer talked:* McDonald with Hansen, *Truth, Lies, and O-Rings,* 66–67; Vaughan, *The* Challenger *Launch Decision,* 176–78.

271 *But Mike Weeks felt:* Weeks, R-NARA, April 7, 1986, 33–35; Weeks testimony, RCR, vol. V, 1666; McDonald with Hansen, *Truth, Lies, and O-Rings,* 395; the testing was scheduled for January 1986, according to Boisjoly's notebook entry for September 12, 1985: Boisjoly, Notebook #1, JSTC, Box Two, Roger Boisjoly Material; the date was eventually fixed for February 13, 1986: Larry Mulloy testimony, RCR, vol. IV, 304.

271 *satisfied that everything:* Weeks, R-NARA, April 7, 1986, 35. In describing his exchange with George Hardy, Weeks later told investigators, "And maybe even the ablest man I knew at the agency could err." Separately, Weeks told Richard Feynman: "In most calamities, nature has given us a warning," foreshadowing the physicist's

own conclusions. (Feynman personal notes dated "Sunday March 29, 1985," http://natedsanders.com/Richard_Feynman_Handwritten_Diary_Style_Document__-LOT60358.aspx.)

271   *Back in Utah*: Boisjoly, *Challenger Book Material,* 304–5, 307.

272   *"Hooray for the management"*: Boisjoly, "Notebook #1", September 11, 1985, JSTC, Box Two, Roger Boisjoly Material.

272   *But Boisjoly was soon*: Boisjoly, *Challenger Book Material,* 309–22; Vaughan, *The Challenger Launch Decision,* 179–81; Memo, "Weekly Activity Report," October 1, 1985, reproduced in RCR, vol. V, 286.

272   *Ebeling, who had been assigned*: Ebeling, R-NARA, March 19, 1986, 1–3; Duane Morley Cox, author interview, August 5, 2021; Brian Russell, author interview, April 6, 2022.

272   *But, like Boisjoly*: Views of Ebeling from Brian Russell, author interview, April 6, 2022, and Jerry Burn, author interview, April 7, 2022.

272   *a scarlet blazer*: Boisjoly, *Challenger Book Material,* 318–19.

273   *"HELP!"*: Memo, Weekly activity report, October 1, 1985.

273   *they watched the countdown*: Russell, author interview, April 6, 2022; Boisjoly, *Challenger Book Material,* 296; the anxiety was shared by at least one member of the engineering team at the Marshall Space Flight Center, Luther Ben Powers: Powers, R-NARA, March 12, 1986, 34–35.

273   *As the weather grew colder*: Boisjoly, interview with Trento, 60.

273   *"stop sending this shit"*: Brian Russell, author interview, April 6, 2022.

CHAPTER NINETEEN: THE ULTIMATE FIELD TRIP

274   *Christa McAuliffe arrived*: Hohler, *I Touch the Future,* 153–54.

274   *checked into a motel*: Christa McAuliffe, letter to parents, September 9, 1985, Christa McAuliffe online collections, Framingham State University, https://digitalcommons.framingham.edu/uncategorized/IO_a6ccd658-8ac6-4f4f-aeea-474d893dedde/.

274   *mobbed by press*: AP, "Teacher Prepares for Spaceflight," *Del Rio News Herald,* September 9, 1985.

274   *"That's the apple"*: "51-L Teacher Training/Meeting with Crew; Food Evaluation; Clothing and Gear Fitting," NASA 253 STS113.778 Reel 3, US National Archives video, https://www.youtube.com/watch?v=RpV1vKPRyJ8.

274   *Like many of the veteran astronauts*: Cook, *The Burning Blue,* 54. At one point, Resnik had asked a friend in exasperation, "What are we going to *do* with these people?" Staff of the *Washington Post, Challengers,* 95; Mullane, *Riding Rockets,* 197–200.

275   *McAuliffe's own astronaut training*: Hohler, *I Touch the Future,* 154–55; Gayle Golden, "Teacher Considers Shuttle 'The Ultimate Field Trip,' " *Dallas Morning News,* December 15, 1985.

275   *But over the weeks McAuliffe spent*: Cook, *The Burning Blue,* 54–55; Barbara Morgan, author interview, December 12, 2023.

275   *Greg Jarvis had spent*: Staff of the *Washington Post, Challengers,* 145; Marcia Jarvis, author interview, March 14, 2022.

276   *At last, in September*: Croft and Youskauskas, *Come Fly with Us,* 314, 317.

276   *This time*: Nelson with Buckingham, *Mission,* 52; "Florida Lawmaker Assigned to Flight on Space Shuttle," *NYT,* October 5, 1985.

276   *Anxious to avoid*: Nelson with Buckingham, *Mission,* 186–89.

276   *Behind his back:* Mullane, *Riding Rockets,* 205.

277   *On November 11:* RCR, vol. II, J-7.

277   *checklists and thick binders:* NASA photo, ID:S85-46207, December 17, 1985.

277   *Onizuka and Ron McNair:* Ben Evans, " 'Moderately Complex': The Mission 51-L Should Have Been," AmericaSpace, https://www.americaspace.com/2014/01/31/moderately-complex-the-mission-that-51l-should-have-been; "Space Shuttle Mission STS-51L," NASA Press Kit, January 1986; Hugh Harris, NASA countdown commentary: "The Challenger Disaster: STS-51-L Countdown and Launch," video, https://www.youtube.com/watch?v=WqDxYFzETCk.

278   *At the same time:* Jean-Michel Jarre, author interview, September 30, 2021; Aubrey Powell, dir., *Making the Steamroller Fly,* documentary, 1997, 22:23, https://www.youtube.com/watch?v=jspTodS0AOI.

279   *And yet, whatever:* Cook, *The Burning Blue,* 33; Hohler, *I Touch the Future,* 180–81; Samira Asma-Sadeque, "Debris from Challenger Shuttle Disaster Found off Florida Coast," *Guardian,* November 10, 2022, https://www.theguardian.com/science/2022/nov/10/space-shuttle-challenger-debris-found-florida-coast.

279   *Deluged with:* Hohler, *I Touch the Future,* 137–40, 180–84.

280   *"What do I do?":* Ibid., 177.

280   *By mid-November:* Ibid., 160, 192; June Scobee Rodgers, author interview, September 29, 2023; Scobee Rodgers, *Silver Linings,* 120; Barbara Morgan, author interview, December 12, 2023. Scobee's grandson Justin was born on January 30, 1985.

281   *The walls of:* Cook, *The Burning Blue,* 66–67.

281   *Michael Collins had:* Collins, *Carrying the Fire,* 243–44.

282   *her own small selection:* Hohler, *I Touch the Future,* 210.

282   *Behind the green doors:* Beggs, interview by Trento, August 7, 1986, JSTC, Box One, 4–5; details of Beggs's office from Ed Campion, author interview, May 11, 2021. The US attorney was Robert C. Bonner, who, during his tenure, embarked on a crusade against white-collar crime. Kim Murphy, "Aloof U.S. Attorney Broadens Scope and Influence of Office," *LAT,* March 14, 1988.

282   *But by that time:* Trento, *Prescription for Disaster,* 272; Joseph and Susan Trento, "Why *Challenger* Was Doomed," *LAT,* January 18, 1987; Beggs, interview by Trento, August 7, 1986, 10; AP, "Presidential Arms Adviser to be Named to NASA Post, *LAT,* September 13, 1985.

283   *Graham arrived:* Philip Shenon, "NASA Chief Takes Leave to Answer Fraud Charges," *NYT,* December 5, 1985; Beggs, interview by Trento, August 7, 1986, 5, JSTC, Box One. Graham was sworn in on November 25, 1985. "William R. Graham," NASA, https://history.nasa.gov/Biographies/graham.html.

283   *"He is a disaster":* Beggs, interview by Trento, August 7, 1986, 7, JSTC, Box One.

284   *Graham had little intention:* The reporter was Joe Trento. Trento, *Prescription for Disaster,* 272.

284   *More disruption:* Sam Marshall, "NASA after *Challenger:* The Public Affairs Perspective," *Public Relations Journal,* August 1986, 18.

284   *On December 12:* Hohler, *I Touch the Future,* 205–6.

285   *There was more:* Ibid., 207.

285   *Ellison Onizuka told:* Ibid., 209.

285   *overran the allotted:* Ibid., 206–7.

285   *"a potted plant":* Staff of the *Washington Post, Challengers,* 172.

285   *She rolled her eyes:* Hohler, *I Touch the Future,* 202.

286 *After almost eight years:* Staff of the *Washington Post, Challengers,* 83; author interviews with Sylvia Salinas Stottlemyer, April 7, 2021, and June Scobee Rodgers, April 7, 2021; Tribe, *My Story,* 387. The rumor was that the senior astronaut was Bob Crippen, whose first marriage would end in divorce. "I've heard that before," Crippen said in 2023. "That's not true." Crippen, author interview, October 4, 2023.
286 *"I'm just a person":* Hohler, *I Touch the Future,* 203.
286 *"Let's put it this way":* Ibid., 208.
286 *It took eight hours:* Ibid., 216.

CHAPTER TWENTY: FRIDAY, JANUARY 10, 1986, 7:00 A.M.

289 *7:00 a.m.:* William J. Broad, "Space Shuttle Lift Off Delayed for 7th Time, Setting Record," *NYT,* January 11, 1986; Nelson with Buckingham, *Mission,* 99.
289 *Strapped into the left-hand seat:* Nelson with Buckingham, *Mission,* 101; Croft and Youskauskas, *Come Fly with Us,* 322–23; "Space Transportation System," HAER No. TX-116, 311: 535,000 gallons of fuel and oxidizer, https://www.nasa.gov/wp-con tent/uploads/2015/12/4pfd.pdf.
289 *Over in Firing Room 1:* Thomas, *Some Trust in Chariots,* 122–23.
290 *It was the* Columbia *crew's fourth trip:* Broad, "Space Shuttle Lift Off Delayed for 7th Time, Setting Record."
290 *fifteen shuttle flights:* There were fourteen missions manifested for the calendar year of 1986, plus the delayed *Columbia* flight from December 1985. "Space Shuttle Payload Flight Assignments," NASA, June 1985, https://ntrs.nasa.gov/api/cita tions/19850021678/downloads/19850021678.pdf.
291 *would carry a journalist:* McConnell, *Challenger,* 87; Brinkley, *Cronkite,* 589–91.
291 *But the agency's:* William J. Broad, "Shuttle in Space After 7 Delays; Astronauts Launch RCA Satellite," January 13, 1986; McDonald with Hansen, *Truth, Lies, and O-Rings,* 80, 203.
291 *Even so:* Trento, *Prescription for Disaster,* 280.
291 *The technicians:* McConnell, *Challenger,* 73.
292 *On January 6:* Michael Wines, "Shuttle Nearly Lofted with Too Little Fuel," *LAT,* August 6, 1986, https://www.latimes.com/archives/la-xpm-1986-08-06-mn-1443 -story.html.
292 *on duty for eleven hours:* McConnell, *Challenger,* 73.
292 *"mechanical failure":* Trento, *Prescription for Disaster,* 280–81.
292 *bad weather:* Chet Lunner, "Weather Delays Shuttle," *Florida Today,* January 8, 1986.
292 *a lucky escape:* John Young's testimony to this effect is cited in *NASA's Quality Assurance Program,* 62. AP, "Space Shuttle Repaired; Launching Set for Today," *NYT,* January 10, 1986; J. Rudolph, "STS-32 LOX Replenish Valve Scrub," Lockheed, February 19, 1986, reproduced in *NASA's Quality Assurance Program,* 111–28; Nelson with Buckingham, *Mission,* 100–101. Details of both scrubs are discussed in Arnold D. Aldrich to Mike Weeks et al., "STS 61-C Launch, Jan. 14 1986," attachment 1, "Arnold D. Aldrich, 'Challenger,' " August 27, 2008.
293 *The path:* "The History of Halley's Comet," NASA, https://solarsystem.nasa.gov/as teroids-comets-and-meteors/comets/1p-halley/in-depth/; "A Brief History of Halley's Comet," Ian Ridpath, http://www.ianridpath.com/halley/halley5a.html.
293 *When shuttle mission:* Broad, "Shuttle in Space after 7 Delays; Astronauts Launch RCA Satellite"; Chet Lunner, "Columbia a Delight at Dawn," *Florida Today,*

January 13, 1986. "They were all glad to see *Columbia* leaving town," said Jim Ball, the NASA commentator at the time.

294　*"It's going to be very tight":* Bob Sieck, quoted in Chet Lunner, "Columbia Gets Joyful Sendoff," *Fort Meyers News-Press*, January 13, 1986.

294　*Late in the morning:* Hohler, *I Touch the Future*, 231.

294　*They were joined:* Author interviews with Jane Smith-Wolcott, November 15, 2022, and Marcia Jarvis, March 14, 2022; photographs in the collections of Smith-Wolcott and Jarvis; the plane was scheduled to depart at 12:20 p.m.: John Lawrence, "Note to editors," JSC NASA News Release 86-002, January 21, 1986.

295　*It was late afternoon:* "Challenger Resource Tape," NASA STI Program, YouTube video, https://www.youtube.com/watch?v=IviOm71Iml0&t=498s; McConnell, *Challenger*, 93; Chet Lunner, "Teachernaut, Crewmates, Wow Crowd," *Florida Today*, January 24, 1986.

296　*"Well, I am so excited":* "Challenger Resource Tape."

296　*three thousand people:* Hohler, *I Touch the Future*, 224.

296　*an austere NASA invitation:* Original launch invitation in the collection of Marcia Jarvis.

296　*shipments of kalua pig:* Ogawa and Grant, *Ellison S. Onizuka*, 149.

296　*Ron McNair's grandmother:* Cheryl McNair, author interview, May 3, 2021; McNair with Brewer, *In the Spirit of Ronald E. McNair*, 163.

296　*yet another postponement:* Michael Lafferty, "Dust Storm Holds Shuttle until Sunday," *Florida Today*, January 23, 1986.

297　*Mike Smith had warned:* The Staff of the *Washington Post*, *Challengers*, 178.

297　*McAuliffe's relatives:* Corrigan, *A Journal for Christa*, 1.

297　*Steven McAuliffe and the children:* "Space-Bound Teacher Draws a Crowd," *Florida Today*, January 23, 1986; Hohler, *I Touch the Future*, 229; Michael Kranish, "Off They Go on a Florida Field Trip to See Scott's Mom Head for Space," *Boston Globe*, January 23, 1986.

298　*In addition to the astronauts':* Trento, *Prescription for Disaster*, 282; Hohler, *I Touch the Future*, 236; Michele Candace, "Bush Sets Shuttle Stop," *Florida Today*, January 25, 1986. The vice president's attendance at the inauguration of the Honduran premier would be a pivotal moment in the developing Iran-Contra conspiracy: "Bush to Attend Inaugural Ceremonies in Honduras," *LAT*, January 10, 1986, https://www.latimes.com/archives/la-xpm-1986-01-10-mn-850-story.html.

298　*Elsewhere, the carnival:* William J. Broad, "Teacher Is Focus of Space Mission," *NYT*, January 25, 1986.

298　*The trinket stores:* Hohler, *I Touch the Future*, 230.

299　*On Friday afternoon:* "Teacher in Space: Mission Watch Report, NASA TV, January 24, 1986," Cape Canaveral Space Force Museum video, https://www.youtube.com/watch?v=PcUd17sSoIY; Thomas, *Some Trust in Chariots*, 91; Hohler, *I Touch the Future*, 233–34; photographs in the collection of Marcia Jarvis; "Space Shuttle Mission Information: Countdown 101," archived at https://web.archive.org/web/20200813094007/https://www.nasa.gov/mission_pages/shuttle/launch/countdown101.html.

299　*Steve McAuliffe, who had become:* Hohler, *I Touch the Future*, 234; beach house photographs in the collection of Marcia Jarvis; Corrigan, *A Journal for Christa*, 126–27; Bill Butterworth, quoted in Staff of the *Washington Post*, *Challengers*, 180; photograph of Resnik in sweatsuit in the Smith family collection.

300   *But not every member:* Marcia Jarvis, author interview, March 14, 2022.

300   *The sky was bright:* Photographs in the collection of Marcia Jarvis; "Teacher in Space: Mission Watch Report, NASA TV"; weather summary in the *Orlando Sentinel,* Saturday, January 25, 1986.

300   *At around 10:30 a.m. on Saturday:* McConnell, *Challenger,* 105–6.

300   *That same bad weather:* Hohler, *I Touch the Future,* 237; McConnell, *Challenger,* 111–12, 123. The meeting finished a little before 1:00 p.m. It was ninety minutes long, and Bill Lucas sat almost silent through the whole process; no one mentioned the O-rings. In his public testimony before the Rogers Commission, Mulloy claimed to have mentioned the O-ring erosion from the previous flight at L-1. But it seems he was mistaken—it was L-2. McDonald with Hansen, *Truth, Lies, and O-Rings,* 93.

302   *The crew:* Staff of the *Washington Post, Challengers,* 180; McConnell, *Challenger,* 123–24; Hohler, *I Touch the Future,* 238; "The Challenger Disaster Part 1 (Pre Mission News Coverage)," YouTube video, https://www.youtube.com/watch?v=gJKy wSdnzfw.

302   *Jane Smith threw a party:* Smith-Walcott and Alison Balch, author interview, November 15, 2022; McConnell, *Challenger,* 123–24.

302   *Then, as darkness gathered:* Corrigan, *A Journal for Christa,* 129. Although Corrigan recalled that it was already cold, the weather report suggests it was warm and windless: "January 25, 1986, Weather History in Cape Canaveral," Weather Spark, https://weatherspark.com/h/d/18782/1986/1/25/Historical-Weather-on-Saturday-January-25-1986-in-Cape-Canaveral-Florida-United-States#metar-20-00.

303   *At 9:00 p.m., Jesse Moore:* McConnell, *Challenger,* 129–37.

304   *"We'll sit on the ground":* Staff of the *Washington Post, Challengers,* 180.

304   *Linda Long called McAuliffe:* Hohler, *I Touch the Future,* 239.

304   *On Sunday morning:* Thomas, *Some Trust in Chariots,* 161–62; author visit to LCC roof, January 20, 2023.

305   *Mike Smith had woken:* McConnell, *Challenger,* 137; author visit to crew quarters, January 20, 2023.

305   *The scrub was:* Bob Hohler and Ralph Jimenez, "The Wind Keeps Christa Grounded," *Concord Monitor,* January 27, 1986; Trento, *Prescription for Disaster,* 282.

306   *That afternoon:* McConnell, *Challenger,* 141–42.

306   *"We're in quarantine!":* Ibid., 142.

307   *Over in the Launch Control Center:* Ibid., 127–29, 141.

CHAPTER TWENTY-ONE: MONDAY, JANUARY 27, 1986, 8:00 A.M.

308   *From the big:* Balch and Smith-Wolcott, author interview, November 15, 2022; photographs in the collection of the Smith family and Marcia Jarvis.

308   *Outside:* McConnell, *Challenger,* 151–55; "Space Shuttle Mission Information: Countdown 101."

309   *Alison Smith grew tired:* Balch and Smith-Wolcott, author interview, November 15, 2022; photographs in the Smith family collection.

309   *It took almost another hour:* McConnell, *Challenger,* 156–58.

310   *At last, more than two hours:* Ibid., 153–59; Jay Hamburg, "Jan. 28, 1986, Etched Forever: An Account of the Final Hours of Space Shuttle Challenger," *Orlando*

Sentinel, January 28, 1987; Thomas, *Some Trust in Chariots,* 167–68; Cecil Houston, R-NARA, April 10, 1986, 9. Timings given in RCR, vol. I, 85.

310 *"Hey, forget it":* McConnell, *Challenger,* 160.

310 *Yet over in the Launch Control Center:* Arnie Aldrich, OHI-J, by Rebecca Wright, April 28, 2008; Richard Kohrs, testimony, RCR, vol. IV, 795; Thomas, *Some Trust in Chariots,* 168–69. In the meeting, Thomas mentioned the dangers of rainwater freezing inside the external tank, but the tank team reassured him that the problem had been taken care of. The solid rocket boosters contained batteries that might prove sensitive to the cold, but the engineers didn't feel this would pose any significant threat to the mission.

311 *Back at the Marshall:* Wear, R-NARA, March 12, 1986, 5–6; Brinton, R-NARA, March 13, 1986, 4.

311 *It was not yet noon:* Arnie Thompson said that Brinton called at "about 10:00" a.m. MST: Thompson, testimony before Presidential Commission, RCR, vol. V, 804. Hamburg, "Jan. 28, 1986, Etched Forever"; Brinton, R-NARA, March 13, 1986, 7.

312 *Until that moment:* Entries for January 20–27, 1986, Boisjoly, "Notebook #2," JSTC, Box Two, Roger Boisjoly Material.

312 *Now Boisjoly sat:* Boisjoly, *Challenger Book Material,* 388–89; Bill MacBeth, R-NARA, March 14, 1986, 5, 14; Ebeling, R-NARA, March 19, 1986, 13–14; Brinton, R-NARA, March 13, 1986, 7; Brian Russell, author interview, April 6, 2022; Jerry Burn, author interview, April 7, 2022.

313 *It was after 4:00 p.m.:* Ebeling, R-NARA, March 19, 1986, 14–15; McDonald with Hansen, *Truth, Lies, and O-Rings,* 94, 96–97; the full weather forecast is described by McConnell, *Challenger,* 171.

314 *"This has to be":* McDonald with Hansen, *Truth, Lies, and O-Rings,* 97.

314 *It now fell to McDonald:* Ibid., 97; Houston, R-NARA, April 10, 1986, 15, 20; Allan McDonald, "Typed version of handwritten notes of A. J. McDonald, Manager Shuttle Rocket Motor Project, Morton Thiokol, Inc.," reproduced in RCR, vol. IV, 740.

315 *When the call began:* Thiokol engineer Jack Kapp recalled that by this time his colleagues agreed that the launch should be delayed but did not explicitly say "don't fire" in the first teleconference; Kapp, R-NARA, March 19, 1986, 10–12.

315 *But the long-distance:* Houston, R-NARA, April 10, 1986, 20–23.

315 *the first time:* Ibid., 58; James Smith, R-NARA, March 13, 1986, 24.

315 *Over in the crew quarters:* "STS-51L Challenger—Launch Coverage," YouTube video, 34:34, https://www.youtube.com/watch?v=SdWlcmnFIuI; McConnell, *Challenger,* 175; author interviews with Jane Smith-Wolcott and Alison Balch, November 15, 2022, and June Scobee Rodgers, October 15, 2021; Marcia Jarvis, journal entry for January 27, 1986.

316 *As cheerful as ever:* Ibid.; Scobee Rodgers, author interview, October 15, 2021.

316 *McAuliffe spent an hour:* Hohler, *I Touch the Future,* 246.

316 *Judy Resnik called:* Sylvia Salinas Stottlemyer, author interview, April 7, 2021.

316 *And Scobee phoned:* Nelson, author interview, January 6, 2022.

317 *It was 6:30 p.m. on the East Coast:* McConnell, *Challenger,* 175; David Ignatius, "Did the Media Goad NASA into the *Challenger* Disaster?," WP, March 30, 1986, https://www.washingtonpost.com/archive/opinions/1986/03/30/did-the-media-goad-nasa-into-the-challenger-disaster/e0c8669d-a809-4c8d-a4f8-50652b892274/.

317 *In his motel room:* Mark Maier, " 'Condemned in the Court of My Own Conscience': Lessons in Leadership from NASA's Shuttle Disasters," *Ethikos,* January 1,

2022; Hamburg, "Jan. 28, 1986, Etched Forever"; Leslie Adams, R-NARA, March 12, 1986, 18.

318 *"Keep me informed"*: Lucas, "Press Conference of Dr. William Lucas, Director, Marshall Space Flight Center," R-NARA, n.d., 4.

318 *One hundred miles out*: McConnell, *Challenger*, 188–89; Aldrich testimony, RCR vol. V, 1019. In his later interview with Rogers Commission investigators, Thiokol's manager of operations at the Cape, Jack Buchanan, recognized from his own experience at sea during World War II that these were life-threatening conditions. Buchanan, R-NARA, March 25, 1986, 31.

319 *Back in Utah*: Boisjoly, *Challenger Book Material*, 393–407; Larry Sayer, R-NARA, March 20, 1986, 7, 11; Ebeling, R-NARA, March 19, 1986, 15; Arnie Thompson, R-NARA, April 4, 1986, 28; Kapp, R-NARA, March 19, 1986, 12–14.

319 *Meanwhile, the department specialists*: Joel Maw, R-NARA, March 20, 1986, 4–6.

319 *The conference was*: Jerry Mason testimony, RCR, vol. IV, 756; Kapp, R-NARA, March 19, 1986, 6.

319 *At fifty-nine*: Brian Russell, author interview, April 26, 2023; Jerald Mason obituary, *Pyramid* (Mount Pleasant, UT), May 20, 2004; Boisjoly, *Challenger Book Material*, 409–10.

320 *it was so late*: Boisjoly, *Challenger Book Material*, 390; McConnell, *Challenger*, 190.

320 *As Lund prepared*: Ebeling, R-NARA, March 19, 1986, 15; McConnell, *Challenger*, 190; chart of room layout and seating plan drawn by Roger Boisjoly; copy provided to the author by Brian Russell.

320 *By 6:00 p.m.*: Boisjoly, *Challenger Book Material*, 390; Kapp, R-NARA, March 19, 1986, 28.

320 *the unanimous recommendation*: Ebeling, R-NARA, March 19, 1986, 19–20.

CHAPTER TWENTY-TWO: MONDAY, JANUARY 27, 1986, 7:55 P.M.

321 *It was already freezing*: Hamburg, "Jan. 28, 1986, Etched Forever"; Cecil Houston, R-NARA, April 10, 1986, 23.

321 *Inside the office*: McConnell, *Challenger*, 191–93; Cecil Houston, R-NARA, April 10, 1986, 21–22; Michael Lafferty, "Winds May Blow Booster Retrieval," *Florida Today*, January 28, 1986.

322 *Conference Room 411*: Adams, R-NARA, March 12, 1986, 14–18.

322 *fifteen of them*: For full list of participants, see RCR, vol. I, 111; McConnell, *Challenger*, 191.

322 *one page at a time*: Sayer, R-NARA, March 20, 1986, 12.

322 *in an unusual breach*: McDonald with Hansen, *Truth, Lies, and O-Rings*, 111.

322 *It was 9:00 p.m.*: Ebeling, R-NARA, March 19, 1986, 15.

322 TEMPERATURE CONCERN: Boisjoly, *Challenger Book Material*, 392.

323 *During the first nine*: RCR, vol. I, 155.

323 *From the Utah*: McDonald with Hansen, *Truth, Lies, and O-Rings*, 98–100.

323 *"more like a brick"*: Ibid., 100.

324 *The numbers were bracing*: If launched at an ambient temperature of 26 degrees, the Thiokol engineers estimated that the temperature of the O-rings in the 51-L rockets would be 27 degrees to 29 degrees Fahrenheit. McDonald with Hansen, *Truth, Lies, and O-Rings*, 102.

324 *As he wrapped up*: Boisjoly, *Challenger Book Material*, 398–401. Boisjoly "cringed"

when Arnie Thompson displayed data from gas tests that seemed to show the O-rings would form a tight seal between metal surfaces down to temperatures below freezing. He thought the technicalities of the test made the data irrelevant to their discussion but knew it could be used to undermine their case: *Challenger Book Material*, 401.

324 *"the direction of 'goodness' "*: McDonald with Hansen, *Truth, Lies, and O-Rings*, 98.

324 *At last, Bob Lund*: Ibid., 103–4.

325 *For a moment*: Boisjoly, *Challenger Book Material*, 408.

325 *High above the steel*: Charlie Stevenson testimony, RCR, vol. V, 957–58; McConnell writes that the freeze protection work was complete by 8:00 p.m. (McConnell, *Challenger*, 187); KSC Ice Team communication net transcript reproduced in *Investigation of the Challenger Accident*, 233–36.

326 *"is that it, Thiokol?"*: Kapp, R-NARA, March 19, 1986, 24.

326 *simply inconclusive*: Boisjoly, *Challenger Book Material*, 408.

326 *Although the Thiokol*: Ibid., 411; Brian Russell, author interview, April 6, 2022.

327 *"My God, Thiokol"*: Ebeling, R-NARA, March 19, 1986, 22; McDonald with Hansen, *Truth, Lies, and O-Rings*, 104–5.

327 *"I'm appalled"*: Kapp, R-NARA, March 19, 1986, 25–27.

327 *"If they don't want to launch"*: Cecil Houston, R-NARA, April 10, 1986, 34.

327 *But the Thiokol team*: Kapp, R-NARA, March 19, 1986, 43–44; McDonald with Hansen, *Truth, Lies, and O-Rings*, 106.

327 *Boisjoly felt his stomach churn*: Boisjoly, *Challenger Book Material*, 408.

327 *But to others*: Cecil Houston, R-NARA, April 10, 1986, 32, 44.

327 *"ask them"*: Sayer, R-NARA, March 20, 1986, 17; Kapp, R-NARA, March 19, 1986, 28.

327 *"NASA, we need"*: Kapp, R-NARA, March 19, 1986, 29.

327 *It was 10:30 p.m.*: RCR, vol. I, 108. Shortly before the caucus began, Allan McDonald added a final comment to the debate, reminding the assembled engineers of an observation made earlier by George Hardy that, even if the primary O-ring in a field joint failed, the secondary would usually be in a position to seal; several of those present at Marshall chose to interpret this as an argument in favor of launching—although McDonald insisted afterward that this had not been his intention. (See, for example, James Smith, R-NARA, March 13, 1986, 32–33, and Wayne Littles, R-NARA, March 16, 1986, 5–6.)

327 *As a humming quiet*: McDonald with Hansen, *Truth, Lies, and O-Rings*, 107–8.

328 *Down in Alabama*: Ben Powers, R-NARA, March 12, 1986, 16–17.

328 *"We've got to make"*: Boisjoly, *Challenger Book Material*, 409.

328 *Nonetheless, for the next few minutes*: Kapp, R-NARA, March 19, 1986, 30–33, 36.

329 *Sensing that Mason*: Boisjoly, *Challenger Book Material*, 410; postflight examination photographs of STS-51-C SRB joints in the collection of Jerry Burn.

329 *almost half an hour*: Kapp, R-NARA, March 19, 1986, 35.

329 *"It's time"*: Ibid., 33.

329 *One at a time*: Maw, R-NARA, March 20, 1986, 14–15.

329 *A tormented look*: Kapp, R-NARA, March 19, 1986, 34–35; Sayer, R-NARA, March 20, 1986, 21.

330 *"Now, Bob"*: Ibid., 22.

330 *The agony of the choice*: Jack Kapp sensed this pressure himself but felt he could not protest; he had worked with Mulloy and Hardy for years and had the utmost respect for them as engineers. He knew that Lund was in an impossible position; the

decision came down to him alone. "I remember thinking that isn't fair, not for the company or the country," he told the investigators. Kapp, R-NARA, 40.

330 *Lund began:* Sayer, R-NARA, March 20, 1986, 21–23; Maw, R-NARA, March 20, 1986, 15.

330 *"Let's fire":* Kapp, R-NARA, 35.

330 *At around 11:00 p.m.:* RCR, vol. I, 108.

330 *"We have reconsidered":* Kapp, R-NARA, 35.

330 *"OK, Thiokol":* Ibid., 36. Opinions differ about who said this: Kapp, who was in Utah, recalled in an investigative interview that it was Larry Mulloy. But Al McDonald—who, unlike Kapp, was sitting in the room with Mulloy at the time—states that it was Hardy, on the line from Marshall; engineer Kyle Speas, who was in the room in Alabama with Hardy and Jud Lovingood, thought it was Lovingood: Speas, R-NARA, March 20, 1986, 18. Other sources agree it was Hardy.

330 *While Hardy spoke:* McDonald with Hansen, *Truth, Lies, and O-Rings,* 2–3, 110.

331 *"You can't feel bad":* Sayer, R-NARA, March 20, 1986, 24.

331 *"If anything happens":* McDonald with Hansen, *Truth, Lies, and O-Rings,* 113–14.

331 *approaching midnight:* "MTI Assessment of Temperature Concern on SRM-25 (51L) Launch, January 27, 1986," timed at 9:45 p.m. MST., RCR, vol. I, 97.

331 *McDonald walked:* McDonald with Hansen, *Truth, Lies, and O-Rings,* 114–16; Houston, R-NARA, April 10, 1986, 74–75, 77–78.

CHAPTER TWENTY-THREE: TUESDAY, JANUARY 28, 1986, 2:00 A.M.

333 *2:00 a.m.:* Charlie Stevenson, testimony, RCR, vol. V, 957. Stevenson arrived at around 1:30 a.m. and stayed for roughly ninety minutes.

333 *High up:* Ibid., 955–57; *Investigation of the Challenger Accident,* October 29, 1986, 232–51; photographs of gantry, RCR, vol. I, 113; McConnell, *Challenger,* 207.

334 *Firing Room 2:* Stevenson testimony, RCR, vol. V, 969.

334 *"Then what":* *Investigation of the Challenger Accident,* October 29, 1986, 235–36.

334 *Thomas had already been:* Thomas, *Some Trust in Chariots,* 182–83; *Investigation of the Challenger Accident,* October 29, 1986, 236–37.

334 *At 5:00 a.m.:* Hamburg, "Jan. 28, 1986, Etched Forever"; Ogawa and Grant, *Ellison S. Onizuka,* 157.

335 *At the Ocean Landings:* Author interview with Jane Smith-Wolcott and Alison Balch, November 15, 2022.

335 *It was still dark:* McConnell, *Challenger,* 212–15; Mulloy and Lucas testimony, RCR, vol. I, 100–101; list of managers in the Ops Support Center: RCR, vol. V, 1022; Mulloy, printed statement of events, Ref. 2/14-2, 7 of 7: RCR, vol. IV, 615.

336 *a single page:* "MTI Assessment of Temperature Concern on SRM-25 (51L) Launch, January 27, 1986."

336 *the matter was closed:* Lucas testimony, RCR, vol. I, 100–101.

336 *Over on the third:* McConnell, *Challenger,* 216; "STS-51L Challenger-Launch Coverage," 7:38; Rick Nygren, author interview, February 24, 2021; breakfast was at 6:48 a.m.: "Challenger Explosion, Live Audience Reaction, 25th Anniversary," ABC News video, January 28, 1986, https://www.youtube.com/watch?v=pUALwYsXSm8.

336 *A little before 8:00 a.m.:* The Staff of the *Washington Post, Challengers,* 182.

337 *The last member:* "STS-51L Challenger-Launch Coverage," 9:00; photo in the collection of Marcia Jarvis.

337 *Behind Jarvis:* National Archives Still Photograph Collection file 255-KSC-386C-4633, Frame 35.

337 *24 degrees:* McConnell, *Challenger,* 217.

337 *The seething:* Hamburg, "Jan. 28, 1986, Etched Forever"; Rockwell International KSC liaison John Tribe, quoted in *Investigation of the Challenger Accident,* October 29, 1986," 238; Tribe, author interview, July 12, 2021.

337 *Drowsy from lack of sleep:* Tribe, *My Story,* 392.

337 *In the meantime:* Testimony from Charlie Stevenson and B.K. Davis, RCR, vol. V, 958–61; "Chronology of Events Related to Temperature Concerns Prior to Launch of Challenger (STS 51-L)," RCR, vol. I, 110.

338 *Inside the Launch Control Center:* Communication loop transcript, *Investigation of the Challenger Accident,* October 29, 1986, 239–41.

338 *Led by a white:* Video of Astrovan: "The Challenger Disaster STS-51-L Pre Mission Reports," YouTube video, 1:13:39, https://www.youtube.com/watch?v=Q4Dz RKGKmQ8&list=PLwxFr1zAEfonxZagcNLPkTWqogem--Uoa&index=1; author drive from Operations and Checkout Building to Pad 39B, January 20, 2023; Rick Nygren, author interview, February 23, 2021; Cheryl L. Mansfield, "Catching a Ride to Destiny," July 15, 2008, archived at https://web.archive.org/web/20210308031008 /https://www.nasa.gov/mission_pages/shuttle/behindscenes/astrovan.html.

338 *As the high-speed elevator:* Hoot Gibson, OHI-J, by Jennifer Ross-Nazzal, January 22, 2016, 37; Steve Hawley, author interview, October 8, 2020; Harris, *Challenger,* 12; Hohler, *I Touch the Future,* 249; McConnell, *Challenger,* 222–24; "STS-51-L Challenger—Launch Coverage," YouTube video, from 9:34, https://www.youtube.com /watch?v=SdWlcmnFIuI.

340 *As he hunched over:* As described by Carter to John Devlin. Devlin, author interview, June 12, 2023.

340 *It was 9:07 a.m.:* William J. Broad, "Thousands Watch a Rain of Debris," *NYT,* January 29, 1986.

340 *Dawn was approaching:* Petrone testimony, RCR, vol. V, 1012–14; John Tribe, author interview, July 12, 2021.

341 *But the final decision:* Arnie Aldrich, "Challenger," August 27, 2008, 2; Aldrich, NASA bio sheet, AldrichAD_Bio.pdf; "Arnold Deane Aldrich," Money & King, https://www .moneyandking.com/obituaries/Arnold-Deane-Aldrich?obId=14837979.

341 *Since taking:* Arnie Aldrich testimony, RCR, vol. V, 1020–21.

341 *Soon after 9:00 a.m.:* The unexpected hold was announced to the crew at 9:08 a.m.: Hohler, *I Touch the Future,* 250; cockpit voice transcript in Lewis, *Challenger,* 11; although not mentioned specifically, 31 Fahrenheit was the overall shuttle system launch constraint temperature. Aldrich testimony, RCR, vol. V, 1022–25; Bob Glaysher testimony, RCR, vol. V, 1013.

342 *"What did you decide?":* Cassutt, *The Astronaut Maker,* 291.

342 *Standing on the aluminum:* Hamburg, "Jan. 28, 1986, Etched Forever"; McNair with Brewer, *In the Spirit of Ronald E. McNair,* 180; Hohler, *I Touch the Future,* 251.

343 *Up on the fourth floor:* McConnell, *Challenger,* 224–26; Hamburg, "Jan. 28, 1986, Etched Forever"; photographs from January 28, 1986, in the Smith family collection; Jane Smith-Wolcott and Alison Balch, author interview, November 15, 2022.

343 *"All of America":* "The Challenger Disaster STS-51L Pre Mission Reports," video, https://www.youtube.com/watch?v=Q4DzRKGKmQ8&list=PLwxFr1zAEfonx ZagcNLPkTWqogem--Uoa&index=2&t=4429s.

343   *In Washington, DC:* Reagan's daily diary for January 28, 1986: https://www.reagan
library.gov/public/digitallibrary/dailydiary/1986-01.pdf; McConnell, *Challenger,*
229–30; CNN morning news for January 28, 1986: "The Challenger Disaster
CNN Live Coverage 11:00 A.M.–12:00 P.M.," YouTube video, https://www.you
tube.com/watch?v=1rDg7S46ijM&list=PLwxFr1zAEfonxZagcNLPkTWqogem
--Uoa&index=2; Susan Page, "Reagan Conveys Nation's Sorrow," *Newsday,* January 29, 1986; "Space Coast Tunes Ears to State of Union," *Florida Today,* January
28, 1986.

344   *At 10:30 a.m.:* Stevenson testimony, RCR vol. V, 959; KSC communications net
transcript, January 28, 1986, 27.

344   *"I hope":* McConnell, *Challenger,* 230.

344   *Harnessed on:* Cockpit transcript in Lewis, *Challenger,* 10–14; author visit to Shuttle Fixed-Base Simulator, Lone Star Flight Museum, Houston, January 28, 2023;
photographs, taken inside the *Challenger* crew compartment during the January
Terminal Countdown Demonstration Test, in the Smith family collection; William
Harwood, "Voyage into History. Chapter Five: Launch," CBS News, https://www
.cbsnews.com/network/news/space/51L/51Lchap5launch.html.

345   *One hundred feet below:* Stevenson and Davis testimony, RCR, vol. V, 959–65.

345   *"Any problems?":* Polling dialogue with director of engineering Horace Lamberth,
KSC communications net transcript, January 28, 1986, 26–30.

345   *In the cockpit:* Lewis, *Challenger,* 11, 13.

345   *In the windowless vault:* McConnell, *Challenger,* 233; "Mission Control: The Heart
of US Manned Flight Operations," balettiedotcom, https://balettie.com/mcc/.

346   *"FIDO?":* McConnell, *Challenger,* 233; Flight Control roll call detailed in Seddon,
Go for Orbit, 169.

346   *In the top:* McConnell, *Major Malfunction,* 235.

346   *11:29 a.m.:* Time given in Hugh Harris commentary: "The Challenger Disaster:
STS-51-L Countdown and Launch," YouTube video, 17:00, https://www.youtube
.com/watch?v=WqDxYFzETCk.

346   *On the fourth floor:* Author visit to the roof of the Launch Control Center, January 20, 2023; photographs in the Smith family collection; Rick Nygren, author
interview, February 23, 2021.

347   *"The Ground Launch":* "The Challenger Disaster: STS-51-L Countdown and
Launch," 21:35.

347   *At the Morton Thiokol:* Hayhurst, "I Knew What Was about to Happen"; Boisjoly,
Challenger Book Material, 418; Jerry Burn, author interview, April 7, 2022.

348   *In the CNN studios:* "The Challenger Disaster CNN Live Coverage 11:00 A.M.–
12:00 P.M."

348   *"T-minus two minutes":* Harris commentary and video; NASA, "Transcript of the
Challenger Crew Comments from the Operational Recorder," https://www.nasa
.gov/missions/space-shuttle/sts-51l/challenger-crew-transcript/; Lewis, *Challenger:
The Final Voyage,* 15.

348   *In Firing Room 3:* McDonald with Hansen, *Truth, Lies, and O-Rings,* 120–21; Hamburg, "Jan. 28, 1986, Etched Forever."

349   *At his console:* McDonald with Hansen, *Truth, Lies, and O-Rings,* 120.

350   *"We just dodged":* Hayhurst, "I Knew What Was about to Happen"; Boisjoly, *Challenger Book Material,* 418.

350   *"Bye, Christa!":* "STS-51L—Barbara Morgan Watches the Launch and Explosion of

Space Shuttle Challenger," YouTube video, https://www.youtube.com/watch?v=aB
fZkQRWbIA.

350 *But in the bottom-most field joint:* RCR, vol. I, 19–20; "STS-51L Sequence of Major
Events," 37.

351 *Now the same forces of combustion:* Conclusions based on postaccident tests described in McDonald with Hansen, *Truth, Lies, and O-Rings,* 436; NASA analysis
suggested the initial leak stopped as early as 3.375 seconds after ignition; but senior
shuttle engineer Richard Kohrs at first reported it may have lasted as long as 12
seconds. "Accident Timeline, " RCR, vol. II, L-4 and Kohrs testimony, RCR, vol. IV,
433. Thiokol engineer Jerry Burn, who oversaw testing on the joint redesign after
the disaster, disagrees with the re-sealing theory, contending that the first O-ring
"burst" immediately, but enough of the second remained to seal the joint until later
in the flight when it, too, failed: Burn, author interview, December 6, 2023.

351 *In Houston:* Steve Nesbitt, author interview, April 7, 2021; author visit to JSC Build-
ing 30 with Nesbitt, January 27, 2023; Hamburg, "Jan. 28, 1986, Etched Forever";
"STS-51L—Launch Flight Directors Loop," January 28, 1986, YouTube video,
https://www.youtube.com/watch?v=H3SYYrKxHD8; "STS-51L Sequence of Major
Events"; RCR, vol. I, 19–20, 26–28; McDonald with Hansen, *Truth, Lies, and
O-Rings,* 142; NASA, "Transcript of the Challenger Crew Comments from the Operational Recorder"; Lynn Olson, "TV Brought the Trauma to Classroom Millions,"
Education Week, February 5, 1986, https://www.edweek.org/education/tv-brought
-the-trauma-to-classroom-millions/1986/02.

352 *The flame grew:* "STS-51L Sequence of Major Events"; RCR, vol. I, 19–20, 26–28,
and "Analysis of the Wreckage," 66–69; "STS-51L—Launch Flight Directors Loop";
Macidull and Blattner, *Challenger's Shadow,* 101–2.

352 *"So the twenty-fifth":* "The Challenger Disaster CNN Live Coverage 11:00 A.M.–
12:00 P.M."

353 *At seventy-two seconds:* McConnell, *Challenger,* 243–44; RCR, vol. I, 20–21; Ham-
burg, "Jan. 28, 1986, Etched Forever"; Steve Nesbitt, author interview, April 7, 2021;
"Challenger Lost While Leaving Region of Maximum Stress," *Aviation Week &
Space Technology,* February 3, 1986. For further explanation of the final moments
of the accident sequence, see RCR, vol. II, "Appendix L—NASA Accident Analysis
Team Report," L-6-16, and Macidull and Blattner, *Challenger's Shadow,* 102–3.

353 *On Jay Greene's console:* "NASA News Conference: Emphasis Is on Ground Activities," *NYT,* January 30, 1986.

CHAPTER TWENTY-FOUR: TUESDAY, JANUARY 28, 1986, 11:28 A.M.

354 *On the ground:* Jane Smith-Wolcott and Alison Balch, author interview, November
15, 2022; photographs in the Smith family collection.

354 *In the bleachers:* "STS-51L—Barbara Morgan Watches the Launch and Explosion of
Space Shuttle Challenger"; *"Challenger* Shuttle Disaster—RAW UNCUT Footage,"
YouTube video, https://www.youtube.com/watch?v=vd7dxmBLg48.

355 *Inside Firing Room 3:* Thomas, *Some Trust in Chariots,* 212.

355 *A heavy silence:* Bob Sieck, author interview, October 27, 2020; Cassutt, *The Astronaut Maker,* 291; Rick Nygren, author interview, February 23, 2021.

355 *"The fucking thing":* "CNN @ 20: The Challenger Disaster," YouTube video, https://
www.youtube.com/watch?v=VTEx-kBg5ZE.

355   *In Mission Control:* "STS-51L—Launch Flight Directors Loop"; Harwood, "Timeline of *Challenger's* Final Flight"; while the NASA transcript states that the Flight Dynamics Officer reported "discreting sources," a later interview with him suggests he said "disagreeing": Flight Dynamics Officer Brian Perry, quoted in Margaret Lazarus Dean, "The Oral History of the Space Shuttle *Challenger* Disaster," *Popular Mechanics,* January 28, 2019.

356   *Down at Cape Canaveral:* Harwood, "Timeline of *Challenger's* Final Flight"; Major Gerald F. Bieringer testimony, RCR, vol. I, 185.

356   *Seventeen miles:* Calculated from RCR, vol. III, Appendix O-18, Figure 1.

356   *On the roof:* Jane Smith-Wolcott and Alison Balch, author interview, November 15, 2022.

357   *In the crowd below:* "STS-51L—Barbara Morgan Watches the Launch and Explosion of Space Shuttle Challenger."

357   *At last, Steve Nesbitt's:* Nesbitt, author interview, April 7, 2021; Harwood, "Timeline of *Challenger's* Final Flight"; "STS-51L—Launch Flight Directors Loop."

357   *A hush settled:* Noel Houston, "We Watched, We Gasped, We Stayed Tuned for the Horrible Details," *Orlando Sentinel,* January 29, 1986.

358   *In Houston:* Brian Perry, quoted in Dean, "The Oral History of the Space Shuttle Challenger Disaster"; William J. Broad, "The Shuttle Explodes: 6 in Crew and High-School Teacher Are Killed 74 Seconds after Liftoff," *NYT,* January 29, 1986.

358   *"The vehicle has exploded":* "Stunned Relatives Watch," *Newsday,* January 29, 1986; *"Challenger* Shuttle Disaster—RAW UNCUT Footage"; James Fisher, "Challenger's Gone in Ball of Fire—Nation in shock," *Orlando Sentinel,* January 29, 1986.

358   *Back in Houston:* Harwood, "Timeline of *Challenger's* Final Flight"; Dick Covey, OHI-J, by Jennifer Ross-Nazzal, November 15, 2006, 26; Steve Nesbitt, author interview, April 7, 2021.

358   *In the conference room:* Boisjoly, *Challenger Book Material,* 418–19; Russell, author interview, April 6, 2022.

359   *It was just after 11:40 a.m.:* Broad, "The Shuttle Explodes"; Trento, *Prescription for Disaster,* 288, 292; Dean, "The Oral History of the Space Shuttle Challenger Disaster"; Houston, "We Watched, We Gasped, We Stayed Tuned for the Horrible Details"; "Covering the Awful Unexpected," *Newsweek,* February 10, 1986.

360   *"The Space Shuttle* Challenger": "The Challenger Disaster CBS News Special Report (Complete)," YouTube video, https://www.youtube.com/watch?v=N9kWG-lAOCA.

361   *On the fourth floor:* Rick Nygren, author interview, February 23, 2021; Rick Nygren, OHI-J, by Rebecca Wright, March 9, 2006, 55; Casper, *The Sky Above,* 145–46.

361   *George Abbey swept:* Jane Smith-Wolcott and Alison Balch, author interview, November 15, 2022; Staff of the *Washington Post, Challengers,* 184; June Scobee Rodgers, author interview, October 16, 2021; Accounts differ about how long the group remained inside the Launch Control Center after the accident. Marcia Jarvis, for instance, later recalled that the families remained in there "for a couple of hours" with Abbey checking in with them with updates about what had happened: Marcia Jarvis, author interview, March 14, 2022; others recall the transfer to crew quarters took place almost immediately.

361   *The crew quarters:* Author interviews with Rick Nygren, February 23, 2021, and Jane Smith-Wolcott and Alison Balch, November 15, 2022; John Wark, "Flood of Calls Shuts Down Brevard, KSC Phone Lines," *Orlando Sentinel,* January 29, 1986.

362   *Grace Corrigan kept repeating:* Staff of the *Washington Post, Challengers,* 184.

362   *In the kitchen:* Bagby, *The New Guys,* 267.

362   *It was an hour:* Marcia Jarvis, author interview, March 14, 2022.

363   *"We don't know":* Staff of the *Washington Post, Challengers,* 184; author interviews with George Abbey, September 25, 2020, and June Scobee Rodgers, October 16, 2021.

363   *Jane Smith went:* Smith-Wolcott and Balch, author interview, November 15, 2022; Scobee Rodgers, *Silver Linings,* 143–44; June Scobee Rodgers, author interview, October 16, 2021.

363   *Back in the Launch Control Center:* Thomas, *Some Trust in Chariots,* 212–14.

364   *many senior managers:* Author interviews with Jay Honeycutt, July 15, 2022, and Chet Vaughan, April 1, 2021; McDonald with Hansen, *Truth, Lies, and O-Rings,* 122–23, 130.

364   *Fortunately, a contingency plan:* William Safire, "Handling Bad News," *NYT,* January 30, 1986; John Noble Wilford, *"Challenger,* Disclosure and an 8th Casualty for NASA," *NYT,* February 14, 1986; Sam A Marshall, "NASA After Challenger: The Public Affairs Perspective," *Public Relations Journal* 42, No. 8, August 1986, 19–22; Harris, *Challenger,* 31; Chuck Biggs, OHI-J, by Rebecca Wright, August 1, 2002, 49; author interviews with Steve Nesbitt, April 7, 2021, and Ed Campion, May 11, 2021. A full simulation of the contingency plan had been scheduled for February 1986. (Marshall, "NASA after Challenger," 19.)

365   *At the White House:* "President Reagan's Remarks on Challenger Explosion to Network Anchors and Questions from Reporters, Roosevelt Room, January 28, 1986," YouTube video, https://www.youtube.com/watch?v=1mEiABsz5U4. The meeting took place "moments after" 11:45 a.m.: Bernard Weinraub, "The Shuttle Explosion; Reagan Postpones State of the Union Speech," *NYT,* January 29, 1986. Dan Rather did not attend the meeting, and Peter Jennings and Tom Brokaw left for their respective DC studios immediately after they heard the news. Tom Shales, "Horror of the Fire in the Sky: Cruel Visions, Over and Over, Bring the Nightmare Home," *WP,* January 29, 1986.

365   *In her office:* Noonan, *What I Saw at the Revolution,* 254–55; "Peggy Noonan on Reagan's *Challenger* Disaster Speech," C-SPAN, November 22, 2015, https://www.c-span.org/video/?c4562338/user-clip-peggy-noonan-reagans-challenger-disaster-speech.

367   *Down at Cape Canaveral:* Author interviews with Steve Nesbitt, April 7, 2021, and Ed Campion, May 11, 2021; Marshall, "NASA after Challenger," 18–19; Harris, *Challenger,* 31; Wilford, "Disclosure and an 8th Casualty for NASA."

367   *in shock and grief:* Ed Campion, author interview, May 11, 2021.

367   *In the growing:* Safire, "NASA Ill-Prepared to Handle News"; Harris, *Challenger,* 35; "STS-51L KSC Post-Accident Press Conference," YouTube video, https://www.youtube.com/watch?v=WkSx-xpgl-M.

368   *At 5:00 p.m. Eastern time:* "Address to the Nation on the Explosion of the Space Shuttle Challenger," Ronald Reagan Presidential Library & Museum, January 28, 1986, https://www.reaganlibrary.gov/archives/speech/address-nation-explosion-space-shuttle-challenger.

370   *It was late afternoon:* The landing was at 5:15 p.m.: Phil Williams, "Space Exploration Must Continue to Honor Astronauts, Bush Says," *Florida Today,* January 29, 1986.

370   *During a brief exchange:* Garland M. Boyd, "The Shuttle Explosion: Bush Flies to

Side of Crew's Families," *NYT,* January 29, 1986; Williams, "Space Exploration Must Continue to Honor Astronauts, Bush Says."

370 *By the time the politicians:* Alison Smith and Jane Smith-Wolcott, author interview, November 15, 2022.

371 *Numb with shock:* Author interviews with June Scobee Rodgers, October 16, 2021, Jane Smith-Wolcott and Alison Balch, November 15, 2022, and Marcia Jarvis, March 14, 2022; Scobee Rodgers, *Silver Linings,* 145. Accounts differ on Scobee's exact words to the delegation: this version was quoted in Boyd, "The Shuttle Explosion"; see also AP, "Scobee's Widow Says Tragedy Must Not Deter US from Leadership in Space," *Miami News,* January 29, 1986. Not everyone remained to meet the politicians: Marcia Jarvis was aghast to see Jake Garn—whom she held responsible for bumping Greg from his assigned flight in 1985—and left the room in disgust as soon as he began to speak. Jarvis, author interview, March 14, 2022.

371 Call us if you need us: Scobee Rodgers, *Silver Linings,* 145.

371 *Inside the Flight Control Room:* Steve Nesbitt, author interview, April 7, 2021; Jane Smith-Wolcott and Alison Balch, author interview, November 15, 2022; photographs in the Smith family collection.

372 *A steady rain:* Dick Covey, OHI-J, by Jennifer Ross-Nazzal, November 15, 2006, 30; Jane Smith-Wolcott and Alison Balch, author interview, November 15, 2022; Cassutt, *The Astronaut Maker,* 292–93.

CHAPTER TWENTY-FIVE: THE COMMISSION

373 *Surrendering slowly:* Ed Magnuson, "They Slipped the Surly Bonds of Earth to Touch the Face of God," *Time,* February 10, 1986; debris continued to fall for fifty-five minutes: Harris, *Challenger,* 32.

373 *By then, much of the nation:* Jon D. Miller, "The *Challenger* Accident and Public Opinion," *Space Policy,* May 1987, 126–28.

373 *To men and women:* See, for example, Lance Murrow, "A Nation Mourns," *Time,* February 10, 1986.

374 *Outside state capitols:* Magnuson, "They Slipped the Surly Bonds of Earth to Touch the Face of God"; Dan Zak, "Thirty Years Ago, a TV Critic Watched the *Challenger* Explosion. This Is What He Saw," *WP,* January 28, 2016, https://www.washingtonpost.com/news/arts-and-entertainment/wp/2016/01/28/thirty-years-ago-a-tv-critic-watched-the-challenger-explosion-this-is-what-he-saw/; Roberto Suro, " 'Deep Sorrow in My Soul,' Pope says," *NYT,* January 30, 1986; Judith Miller, "A Message from Qaddafi," *NYT,* January 30, 1986.

374 *Back in Houston:* Robert Reinhold, "The Mourning Families Return Home," *NYT,* January 30, 1986; Fred Gregory, OHI-J, by Jennifer Ross-Nazzal, March 14, 2015, 17; Scobee Rodgers, *Silver Linings,* 148; Steve Nesbitt, author interview, April 7, 2021; Pinky Nelson, OHI-J, by Jennifer Ross-Nazzal, May 6, 2004, 34; Scobee Rodgers, author interview, September 29, 2023.

375 *A few streets away:* Cheryl McNair, author interview, May 3, 2021.

375 *When Jane Smith:* Jane Smith-Wolcott and Alison Balch, author interview, November 15, 2022.

375 *Grace and Ed Corrigan:* Reinhold, "The Mourning Families Return Home"; John Grant, "Linda Long and the Challenger Disaster," *Impact Magazine,* Autumn 1986, vol. 32, no.1, 30.

376 *Three days after the accident:* "In Memory of the Challenger Seven, Johnson Space Center, January 31, 1986," YouTube video, https://www.youtube.com/watch ?v=SGrhmYxQpXY; Magnuson, "They Slipped the Surly Bonds of Earth to Touch the Face of God"; "Memorial Service for the Mission 51-L Crew (Edited)," NASA STI Program video, https://www.youtube.com/watch?v=bJfoJFFQvyA; Scobee Rodgers, *Silver Linings,* 149.

377 *That morning:* Jack Lule, "The Political Use of Victims: The Shaping of the Challenger Disaster," *Political Communication,* April 1990, 115–16; Miller, "The Challenger Accident and Public Opinion," 127–28.

377 *Initially, the proximate cause:* Susan Linden and Dorothy Nelkin, "Challenger: The High Cost of Hype," *Bulletin of the Atomic Scientists,* November 1986, 16–18; AP, "Shuttle Pieces Trickling In, Nation to Honor 7 'Heroes,' " *Omaha World-Herald,* January 29, 1986; William J. Broad, "Did Crew Die Instantly? Questions Re-Emerging," *NYT,* March 11, 1986.

378 *But in the photo lab:* Author interviews with Dick Covey, December 6, 2021, and Bob Crippen, November 18, 2021; "The Shuttle Inquiry: The Evidence Accumulates; Agency Releases Video of Plume," *NYT,* February 2, 1986; Bob Crippen, OHI-J, by Rebecca Wright, May 26, 2006, 78–79; Mike Coats, quoted in Bagby, *The New Guys,* 273; William Harwood, "Voyage into History. Chapter Seven: The First Month," CBS News, https://www.cbsnews.com/network/news/space/51L/51Lchap 7firstmonth.html.

378 *70-millimeter high-speed movie cameras:* "Agency Releases Photos of 'Plume.' "

378 *Down at the Marshall:* McDonald with Hansen, *Truth, Lies, and O-Rings,* 132–35; Harwood, "Voyage into History. Chapter Seven: The First Month."

379 *But now the agency's reticence:* Author interviews with David Sanger, May 13, 2021, William Broad, May 24, 2021, Ed Campion, May 11, 2021, and Steve Nesbitt, April 7, 2021; Alex S. Jones, "Journalists Say NASA's Reticence Forced Them to Gather Data Elsewhere," *NYT,* February 9, 1986.

380 *At around noon:* Jones, "Journalists Say NASA's Reticence Forced Them to Gather Data Elsewhere"; "The Challenger Disaster: 1-30-86 Evening News Reports," YouTube video, https://www.youtube.com/watch?v=GZGQEva2rIo. Barbree's first source was a recently retired official from Kennedy; the second source was KSC Director Dick Smith, according to Hugh Harris: Harris, *Challenger,* 49.

380 *Donald Regan:* Biddle, "Two Faces of Catastrophe"; Dunar and Waring, *Power to Explore,* 390–91.

380 *The President's team:* Biddle, "Two Faces of Catastrophe."

381 *That same day:* McDonald with Hansen, *Truth, Lies, and O-Rings,* 135–36. Details of cameras and smoke sequence: "STS 51L Incident Integrated Timeline," Figure 17, 1, RCR, vol. III, "Appendix N–NASA Photo and TV Support Team Report," N 20.

381 *In the meantime, the press:* Richard Witkin, "Data Said to Show Quick Power Drop Just before Blast," *NYT,* February 1, 1986; Barnaby J. Feder, "Booster Builder Is Joining Inquiry," *NYT,* February 2, 1986.

382 *That night:* "Agency Releases Photos of 'Plume,' " *NYT,* February 2, 1986.

382 *On Monday, February 3:* Gerald M. Boyd, "President Names 12-Member Panel in Shuttle Inquiry," *NYT,* February 4, 1986; Tom Scherberger, "8 on Inquiry Panel Have Links to NASA," *Orlando Sentinel,* February 16, 1986.

383 *While touting:* Boyd, "President Names 12-Member Panel in Shuttle Inquiry." According to the commission's executive director Alton Keel, at the start it was his and

Rogers's intention to root out any failures within NASA and correct them, without critically damaging the agency. The continued existence of NASA was too important for the nation "and the free world" to endanger its future, Keel said. Keel, author interview, October 21, 2020.

383　*Back in Utah*: Boisjoly, *Challenger Book Material*, 421–29.

384　*The first public*: RCR, vol. IV, 1–5; Harwood, "Voyage into History. Chapter Seven: The First Month."

385　*But whatever*: David Sanger, "Cold and Vibration in Rocket Studied; Questions about Solid-Fuel Booster Focus on Temperature and Changes in Pressure," *NYT*, February 5, 1986; RCR, vol. IV, 18, 57; Keel, author interview, October 21, 2020.

385　*Next, Judson Lovingood*: RCR, vol. IV, 83–95.

386　*Finally, as Lovingood*: Ibid., 97; "The Challenger Disaster 2-6-1986 CNN Coverage of 1st Rogers Commission Hearing," YouTube video, https://www.youtube.com/watch?v=-gw3kCUQZTg.

386　*But that weekend*: Philip Boffey, "NASA Had Warning of a Disaster Risk Posed by Booster," *NYT*, February 9, 1986.

387　*Rogers, embarrassed*: Alton Keel, author interview, October 21, 2020; McDonald with Hansen, *Truth, Lies, and O-Rings*, 149. Graham later said he was told about the story by his staff, including Mike Weeks, on Saturday afternoon, and called Rogers to warn him. Nonetheless, Keel says that Rogers phoned Graham on Sunday, angry at the revelations the story contained. Keel, author interview, October 21, 2020.

387　*That afternoon*: McDonald with Hansen, *Truth, Lies, and O-Rings*, 143–49.

387　*Chairman Rogers convened*: RCR, vol. IV, 243–45; Alton Keel, author interview, October 21, 2020.

388　*"This is not"*: RCR, vol. IV, 244.

388　*But as their testimony*: RCR, vol. IV, 297–302; McDonald with Hansen, *Truth, Lies, and O-Rings*, 158–63; Keel, author interview, October 21, 2020.

390　*"This is explosive"*: Ibid., October 21, 2020.

390　*In the offices of Morton Thiokol*: Leland Dribin, interview by Dr. Mark Maier, September 29, 1994 (transcript from the collection of Mark Maier).

CHAPTER TWENTY-SIX: THE TRUTH

391　*By the time he took the call*: Gleick, *Genius*, 417.

391　*At sixty-seven*: Ibid., 418–19; Feynman, *"Surely You're Joking, Mr. Feynman,"* 201, 236, 295–96, 301–2.

392　*Before flying*: Feynman, *"What Do You Care What Other People Think?,"* 133–37; page of notes reproduced in "Mr. Feynman Goes to Washington," *Engineering and Science*, Fall 1987.

392　*But once the meetings*: "Mr. Feynman Goes to Washington."

392　*But that night*: Ibid. Location of dinner: Alton Keel, author interview, October 12, 2023; "Presidential Commission on the Space Shuttle Challenger Accident," C-SPAN, February 11, 1986, https://www.c-span.org/video/?125993-1/presidential-commission-space-shuttle-challenger-accident.

393　*According to Chairman Rogers's*: Keel, author interview, October 21, 2020.

393　*And, once Mulloy began*: RCR, vol. IV, 352; Philip M. Boffey, "NASA Acknowledges Cold Affects Booster Seals," *NYT*, February 12, 1986; Mulloy testimony, RCR, vol. IV, 340–63; "Presidential Commission on the Space Shuttle Challenger Accident," C-SPAN, February 11, 1986.

393   *But Feynman had heard:* Although he gave the physicist permission, Rogers made his disapproval clear. "Do what you think is right," he told Feynman: Keel, author interview, October 21, 2020; RCR, vol. IV, 363; "Richard Feynman Debunks NASA," YouTube video, https://www.youtube.com/watch?v=raMmRKGkGD4.

394   *Two days after:* Time of flight given as 6:15 on February 13: Richard P. Feynman to Gweneth and Michelle Feynman, February 12, 1986, in Feynman and Feynman, *Perfectly Reasonable Deviations from the Beaten Track,* 401.

394   *By now, the causes of the accident:* McDonald with Hansen, *Truth, Lies, and O-Rings,* 174; David Sanger, "NASA Photos Hint Trouble Started Right at Liftoff," *NYT,* February 14, 1986; Boisjoly, *Challenger Book Material,* 431–2.

395   *Following his unexpected:* McDonald with Hansen, *Truth, Lies, and O-Rings,* 175–79.

395   *When the hearing began:* RCR, vol. IV, 609.

396   *Almost immediately:* Ibid., 634–35.

396   *The two engineers:* Ibid., 631; McDonald with Hansen, *Truth, Lies, and O-Rings,* 181–84.

397   *Yet the Thiokol Vice President:* RCR, vol. IV, 698–700.

397   *"Since I caused":* Ibid., 701.

397   *McDonald was torn:* McDonald with Hansen, *Truth, Lies, and O-Rings,* 184–86.

398   *When McDonald finished:* Ibid., 186–87, RCR, vol. IV, 707–8.

398   *That afternoon:* McDonald with Hansen, *Truth, Lies, and O-Rings,* 189.

398   *In the meantime:* Boisjoly, *Challenger Book Material,* 438; Rita Beamish, "Challenger Panel Questions Decision to Launch Shuttle," *Charlotte Observer,* February 16, 1986; Chairman's statement of February 15, quoted in RCR, vol. I, 206.

399   *It was another:* Philip M. Boffey, "NASA Taken Aback by Call to Screen Its Inquiry Panel," *NYT,* February 17, 1986; "NASA Administrator Steps Down," *Newsday,* February 26, 1986.

399   *Back on Merritt Island:* Avera, *The Truth about Challenger,* 71–76; "Shuttle Cabin Hit Ocean Intact," *Miami Herald,* April 9, 1986.

399   *Supervising the effort:* Crippen, author interviews, November 18, 2021, and October 4, 2023; Crippen, OHI-J, by Rebecca Wright, May 26, 2006, 80.

400   *"Crippen calls all the shots":* William J. Broad, "Astronauts Active in Quest for Answers about Disaster," *NYT,* February 18, 1986.

400   *Work to recover:* Dudley Clendinen, "Cutter Brings 600 Pounds of Debris of Shuttle to Port," *NYT,* January 30, 1986.

400   *It marked:* Commander, Naval Sea Systems Command, "Space Shuttle Challenger Salvage Report," April 29, 1988, 2.

400   *Joined by a growing:* William E. Schmidt, "Efforts Expanded to Find Wreckage," *NYT,* February 3, 1986.

401   *Carried by wind and tide:* Dudley Clendinen, "Search by Air and Sea Yields No Sign of the Shuttle Crew," *NYT,* January 29, 1986; Reston Jr., "The Astronauts After Challenger"; NASA photo of wreckage at dockside dated January 30, 1986, NASA ID 51L-s-020.

401   *In their Miami:* Dennis E. Powell, "Obviously, a Major Malfunction," *Tropic,* November 13, 1988; Commander, Naval Sea Systems Command, "Space Shuttle Challenger Salvage Report," 51.

401   *But NASA's official:* Powell, "Obviously, a Major Malfunction"; William J. Broad,

"Did Crew Die Instantly? Questions Re-Emerging," *NYT,* March 11, 1986; Crippen, author interviews, November 18, 2021, and October 4, 2023.

402 *"I don't want to hear":* Bagby, *The New Guys,* 277.

402 *By February 3:* "Recovery Priority" Ref. 3/7-7, RCR, vol. V, 1082; chronology in RCR, vol. III, Appendix O: "NASA Search Recovery and Reconstruction Task Force Team Report," O-152.

402 *By February 7:* Ibid.; Charles Fishman, "The Epic Search for the *Challenger,*" WP, May 28, 1986; Bartholomew, *Mud, Muscle, and Miracles,* 428, 430.

403 *On Sunday, February 16:* William J. Broad, "Latest Pictures from NASA Show Wider Fire on Booster," *NYT,* February 16, 1986; "STS-51-L Salvage Briefing, 2/19/86," video, National Archives Catalog, https://catalog.archives.gov/id/39655; "NASA Says Submarine May Have Found Rocket," *NYT,* February 17, 1986; William E. Schmidt, "Piece of Shuttle Rocket Identified on Sea Floor," *NYT,* February 20, 1986.

404 *Rogers reorganized:* RCR, vol. I, 206–7.

404 *In the meantime:* McDonald with Hansen, *Truth, Lies, and O-Rings,* 191–202; David Sanger, "Rocket Engineer Describes Arguing Against Launching," *NYT,* February 19, 1986.

404 *And, as the media focused:* Hanke Gratteau, "Townsfolk Ask: Why Punish Us for Shuttle Blast?," *Chicago Tribune,* February 25, 1986; Boisjoly, *Challenger Book Material,* 439–40.

405 *Back in Houston:* Author interviews with June Scobee Rodgers, October 16, 2021, and Jane Smith-Wolcott and Alison Balch, November 15, 2022; Scobee Rodgers, *Silver Linings,* 151–53.

405 *Each evening, the ships:* John Devlin, author interview, June 12, 2023; Harwood, "Voyage into History. Chapter Seven: The First Month."

405 *The crew families:* Author interviews with Scobee Rodgers, October 16, 2021, and Smith-Wolcott and Balch, November 16, 2022; Scobee Rodgers, *Silver Linings,* 151–53.

406 *By the beginning:* Alton Keel, author interview, October 21, 2020; McDonald with Hansen, *Truth, Lies, and O-Rings,* 206–10.

407 *"I'm not interested":* Leland Dribin, interview by Dr. Mark Maier, September 29, 1994.

407 *The culmination:* "Presidential Commission on the Space Shuttle Challenger Accident Hearing, February 25, 1986," Cape Canaveral Space Force Museum video, https://www.youtube.com/watch?v=1jPP7Ks6Rhk.

408 *That afternoon:* Boisjoly testimony, RCR, vol. IV, 781–806.

408 *Later, Thiokol:* "Presidential Commission on the Space Shuttle Challenger Accident," C-SPAN, February 26, 1986, https://www.c-span.org/video/?126041-1/presidential-commission-space-shuttle-challenger-accident.

409 *"if this were an airplane":* RCR, vol. V, 860.

409 *The three days of hearings:* David Sanger, *"Challenger's* Failures and NASA's Flaws," *NYT,* March 2, 1986.

410 *"Clearly flawed":* "Presidential Commission on the Space Shuttle Challenger Accident," February 27, 1986, C-SPAN video, 4:14: https://www.c-span.org/video/?126055-1/presidential-commission-space-shuttle-challenger-accident; Rogers's statement, RCR, vol. V, 1062.

410 *With their examination:* Broad, "Panel Shifting Shuttle Inquiry beyond Liftoff," *NYT,* March 2, 1986.

410 *At the same time:* RCR, vol. I, 207–8. Commission investigator Robert Thompson

explained to Larry Mulloy the plan to gather signed affidavits from all senior managers involved in the launch decision: Mulloy, R-NARA, April 2, 1986, 3.

410 *"It's like an artichoke":* Philip M. Boffey, "White House Aides Say Ex-Chief Leads Candidates for NASA Post," *NYT,* March 1, 1986.

CHAPTER TWENTY-SEVEN: APOCALYPSE

411 *It took almost exactly a month:* Mike McAllister, author interview, July 14, 2022; LCU Dive Log, provided to the author by Walter Hardman, March 7, 1986. Distance from Cape, sixteen nautical miles, given in "Space Shuttle Challenger Salvage Report," 51; *US Coast Pilot 4: Atlantic Coast: Cape Henry, VA, to Key West, FL, July 2023,* 317, https://nauticalcharts.noaa.gov/publications/coast-pilot/files/cp4/CPB4_C10_WEB.pdf.

411 *By the middle of the afternoon:* LCU Dive Log, March 7, 1986; McAllister, author interview, July 14, 2022. Details of 2A sonar set described by L. David Thompson, author interview, June 21, 2023.

411 *Bailey and McAllister:* Todd Halvorson, "They Found the Cabin and the Flag, They Lost Their Jobs," *Florida Today,* January 27, 1987; McAllister, author interview, July 14, 2022; "Sonar Operators Impressed with EG&G Sidescan Sonar Performance," *EG&G Monitor,* no. 5 (1986), 4–5. The ship's nickname was derived from its designation as a Landing Craft Utility, or LCU.

412 *But it was weeks:* "Sonar Operators Impressed"; LCU Dive Log, March 7, 1986; photographs in collection of Mike McAllister; the site was "a few hundred yards northeast" of the buoy: Powell, "Obviously, a Major Malfunction."

412 *The pinging sounds:* Author interviews with Mike McAllister, July 14, 2022, and Walter Hardman, August 18, 2022.

412 *Down on the bottom:* LCU Dive Log, March 7, 1986; McAllister, author interview, July 14, 2022.

412 *Wary of eavesdropping:* Walter Hardman, author interview, August 18, 2022; Devlin's rank given in "Space Shuttle Challenger Salvage Report," 51.

413 *At around eleven o'clock:* Devlin, author interview, June 12, 2023.

413 *Devlin and Bagian approached:* Devlin, author interview, June 12, 2023; Powell, "Obviously, a Major Malfunction."

413 *Devlin's men began:* William E. Schmidt, "Remains of Shuttle's Crew Are Apparently Identified," *NYT,* April 19, 1986; Powell, "Obviously, a Major Malfunction"; Author interviews with Devlin, June 12, 2023; David Belt, May 15, 2023; and L. David Thompson, June 21, 2023; Bagby, *The New Guys,* 282–85; William E. Schmidt, "Kin Given Hope Shuttle Bodies Are Identifiable," *NYT,* March 11, 1986.

414 *Jane Smith was in:* Cassutt, *The Astronaut Maker,* 297; Jay Hamburg, "Families Deal Again with Grief," *Orlando Sentinel,* March 10, 1986; "Remains of Shuttle Crew, Cabin, Found on Ocean Floor," *Newsday,* March 10, 1986; Jane Smith-Wolcott and Alison Balch, author interview, November 15, 2022.

414 *For others, the discovery:* Scobee Rodgers, *Silver Linings,* 152–53; Chip Warren, "Flights Must Continue, Mrs. Scobee Says at UA," *Arizona Daily Star,* April 10, 1986; June Scobee Rodgers, author interview, October 16, 2021.

415 *Although NASA waited:* "Remains to Help Solve How 7 Died," *Miami News,* March 10, 1986.

415 *Nevertheless, the discovery:* Michael Lafferty, "Explosion Did Not Destroy Cabin,"

Florida Today, April 10, 1986; Phil Long and Michael Merzer, "Shuttle Cabin Hit Ocean Intact," *Miami Herald,* April 10, 1986.

415 *he found the place in turmoil:* Truly, quoted in Cassutt, *The Astronaut Maker,* 297.

416 *Many of the astronauts:* Author interviews with Dick Covey, December 6, 2021, and Bob Crippen, November 18, 2021; Robert Reinhold, "Astronaut Assails NASA for Not Telling of Risk," *NYT,* March 4, 1986; Hank Hartsfield, OH-J, by Carol Butler, June 15, 2001, 25; Hartsfield's flight, STS-41-D, experienced primary O-ring erosion in both the right-hand forward field joint and erosion and blow-by in the left-hand nozzle joint: RCR, vol. II, "Appendix H: Flight Readiness Review Treatment of O-ring Problems," H-1-H-2.

416 *The message described:* Olive Talley and Jim Asker, "Astronaut Memo Tells of 'Awesome' Dangers," *Houston Post,* March 8, 1986.

416 *In Cape Canaveral:* William J. Broad, "NASA Aide Assails Methods of Investigators," *NYT,* March 16, 1986.

417 *"I'm very concerned":* Jim Asker, "Space Program's Future Bright, New JSC Head Says," *Houston Post,* March 1, 1986.

417 *In Huntsville:* Lawrence C. Levy, "NASA Aides Reject 'Flaw' Charge," *Newsday,* March 1, 1986.

417 *the complex seethed:* Lawrence C. Levy, "It's Not Fair to Lump It All on Us," *Newsday,* February 27, 1986; Philip Boffey, "Zeal and Fear Mingle at Vortex of Shuttle Inquiry," *NYT,* March 17, 1986.

417 *a three-page computer-printed letter:* "Apocalypse" letter, JSTC, Box Two, RB Writings. The copy of the letter among Roger Boisjoly's papers bears no visible date, but Dunar and Waring, in *Power to Explore,* date it to March 6: *Power to Explore,* 423, 48fn.

418 *Out in the Atlantic:* William Schmidt, "Effort Resumes to Recover Shuttle Cabin," *NYT,* March 12, 1986; John Devlin, author interview, June 12, 2023; AP, "Cabin Wreckage Brought Ashore," *NYT,* March 13, 1986; William J. Broad, "Examination of Remains Shifts to NASA Lab," *NYT,* March 14, 1986.

418 *In the end, the divers found:* "Official Flight Kit STS Mission 51-L, itemized description and recovered count," undated NASA document, and STS-51-L PPK and OFK manifests, January 23 and January 28, 1986, received in response to author FOIA request; Greg Jarvis, 51-L PPK List, from the collection of Marcia Jarvis.

419 *But amid the debris:* L. David Thompson, author interview, June 23, 2021; AP, "Shuttle Recorders and Computers Reported Found," *NYT,* March 14, 1986; William J. Broad, "Salt-Soaked Shuttle Tapes May Still Be Playable," *NYT,* March 15, 1986; David Sanger, "Obsolete Computers May Hold Key to Shuttle," *NYT,* March 15, 1986; William J. Broad, "Scientists Hopeful of Restoring Challenger Tapes," *NYT,* June 19, 1986; Avera, *The Truth about the* Challenger, 105.

419 *Work inside the submerged:* "Space Shuttle Challenger Salvage Report," 51, 53; L. David Thompson, author interview, June 23, 2021; Jay Barbree, "Chapter Six: Raising Heroes from the Sea," NBC News, January 25, 2004, https://www.nbc news.com/id/wbna3078060; UPI, "All 7 Astronauts Are Found," *Cincinnati Post,* April 18, 1986; author interviews with Marcia Jarvis, March 14, 2022; John Devlin, June 12, 2023; and David Belt, May 15, 2023; Todd Halvorson, "Scallop Ship Added to Salvors," *Florida Today,* April 6, 1986.

420 *Meanwhile, the dozen:* RCR, vol. I, 207–8; Feynman, *"What Do You Care What Other People Think?,"* 199.

420 *While continuing his work:* Gleick, *Genius,* 426.

420 *he scribbled notes on:* "Richard Feynman 4pp. Handwritten Document From the Challenger Investigation—Feynman's Detailed Notes for 4 Days Spanning 7–10 February 1986, Leading Up to & Including Discovery of O-Ring Failure," Nate D. Sanders Auctions, http://natedsanders.com/richard_feynman_4pp__handwritten _document_from_the-lot61780.aspx.

421 *To the growing irritation:* Keel, author interview, October 21, 2020.

421 *He chose:* Feynman, *"What Do You Care What Other People Think?,"* 130.

421 *When Feynman had first:* Feynman, *"What Do You Care What Other People Think?,"* 200–207; Otto K. Goetz, OHI-J, by Jennifer Ross-Nazzal, July 20, 2010, 10.

422 *"I believe we have eliminated":* Wayne Littles testimony, RCR, vol. V, 1398.

422 *In the deep water:* Colonel Edward O'Connor testimony, RCR, vol. V, 1242–47; "Search Update," chart, March 20, 1986, and "Right Solid Rocket Booster Components," chart, 1255.

423 *With time running out:* Fishman, "The Epic Search for the *Challenger*"; David Sanger, "Crucial Portion of Shuttle Joint Found in Ocean," *NYT,* April 15, 1986.

424 *In the second week of April:* John Noble Wilford, "Challenger Blast Left Cabin Intact," *NYT,* April 10, 1986; Phil Long and Martin Mercer, "Shuttle Cabin Hit Ocean Intact," *Miami Herald* (International Edition), April 10, 1986; John J. Glisch, "Challenger's Cabin Intact as It Smashed into Atlantic," *Orlando Sentinel,* April 10, 1986.

425 *Contact 131:* Fishman, "The Epic Search for the *Challenger*"; Sanger, "Crucial Portion of Shuttle Joint Found in Ocean"; Commander, Naval Sea Systems Command, "Space Shuttle Challenger Salvage Report," 42–49.

425 *They found:* Ibid., 47.

425 *The salvage operation:* Fishman, "The Epic Search for the *Challenger.*"

426 *Conducted by two military:* Powell, "Obviously, a Major Malfunction"; Seddon, *Go for Orbit,* 317, 322–24.

426 *The formal identification:* Seddon, *Go for Orbit,* 324; William Schmidt, "Remains of Shuttle's Crew Are Apparently Identified," *NYT,* April 19, 1986; Charlie Jean and John J. Glisch, "Agony Ends for Jarvis' Father," *Orlando Sentinel,* April 18, 1986.

426 *Joe Kerwin, the veteran:* Kerwin interview, in Hitt and Smith, *Bold They Rise,* 303.

426 *The pair of forensic pathologists:* Interview with Dr. Robert R. McMeekin by Charles Stuart Kennedy, Armed Forces Institute of Pathology Oral History Program, May 26, 1992, 6–7.

426 *By examining:* Seddon, *Go for Orbit,* 322.

427 *While the forensic work continued:* William Schmidt, "Astronauts' Families Expect to Get Remains Next Month," *NYT,* April 17, 1986; Schmidt, "Remains of Shuttle's Crew Are Apparently Identified"; William Schmidt, "All Shuttle Crew Remains Recovered, NASA Says," *NYT,* April 20, 1986.

427 *But inside the Astronaut Office:* Seddon, *Go for Orbit,* 315–17; "Challenger Crew Module Ejected from Fireball," *Aviation Week & Space Technology,* April 28, 1986, 23.

427 *Only after several:* Seddon, *Go for Orbit,* 315; David Sanger, "Shuttle Photos: Issue of Crew's Fate," *NYT,* April 25, 1986; John J. Glisch, "Photos Show Cabin Hurled Free of Fireball," *Orlando Sentinel,* April 24, 1986; Todd Halvorson, "Truly Praises Salvors," *Florida Today,* April 20, 1986.

427 *Sonny Carter told Jane Smith:* Jane Smith-Wolcott, author interview, November 15, 2022.

428 *At 9:30 a.m.:* John J. Glisch and Chris Reidy, "Jet Ferries Away Seven Heroes as Space Center Pays Silent, Somber Tribute," *Orlando Sentinel,* April 30, 1986; William E. Schmidt, "Bodies of Astronauts Flown to Delaware," *NYT,* April 30, 1986; "The Challenger Disaster 4-29-1986 CNN Coverage of Crew Transfer," YouTube video, https://www.youtube.com/watch?v=WZxz3Y-eGe8.

CHAPTER TWENTY-EIGHT: THE LONG FALL

429 *"We've suffered a tragedy":* "Remarks on Receiving the Final Report of the Presidential Commission on the Space Shuttle Challenger Accident," Ronald Reagan Presidential Library and Museum, June 9, 1986, https://www.reaganlibrary.gov/archives /speech/remarks-receiving-final-report-presidential-commission-space-shuttle -challenger; "President Reagan's Remarks on Receiving Final Report on the Space Shuttle Challenger on June 9, 1986," YouTube video, https://www.youtube.com /watch?v=alKD6WolnUA.

430 *"We know exactly":* H. Josef Hebert, "Some Sensitive Areas Left Alone," *Fresno Bee,* June 10, 1986.

430 *But the commissioners reserved:* RCR, vol. I, 85.

431 *The report concluded:* Ibid., 198–201.

431 *The position Graham had wanted:* Ian Fisher, "James Fletcher, 72, NASA Chief Who Urged Shuttle Program, Dies," *NYT,* December 24, 1991. Fletcher accepted the job in May 1986.

432 *Bob Ebeling:* Jerry Burn, author interview, April 7, 2022.

432 *The two men:* McDonald with Hansen, *Truth, Lies, and O-Rings,* 326; Boisjoly, *Challenger Book Material,* 499–500.

432 *Together with:* Ibid., 518–19.

433 *"Five Lepers":* Russell, author interview, April 6, 2022.

433 *When the commission's report:* Feynman, *"What Do You Care What Other People Think?,"* 226–28.

433 *Although it would not be:* Peter H. King, "Feynman Issues Report—More Critical Than Panel's," *LAT,* June 11, 1986; Mike Thomas, "NASA Safety Assurances Pure 'Fantasy,' " *Orlando Sentinel,* June 11, 1986; RCR, vol. II, Appendix F, F1.

433 *At the same time: Space Shuttle Accident: Hearings before the Subcommittee on Science, Technology, and Space of the Committee on Commerce, Science, and Transportation,* United States Senate, Ninety-Ninth Congress, Second Session, on Space Shuttle Accident and the Rogers Commission Report, February 18, June 10 and 17, 100–106; Chris Reidy, "Rogers Urges No Charges in Shuttle Loss," *Orlando Sentinel,* June 11, 1986; Jack Kelley, "Congress: Report Lacks Explanations," *Florida Today,* June 11, 1986; Philip M. Boffey, "Shuttle Panel Is Faulted for Not Naming Names," *NYT,* June 11, 1986.

434 *Like some members:* Kelley, "Congress: Report Lacks Explanations"; Don Kirkman, "NASA Firings Demanded as Shuttle Hearings Start," *San Francisco Examiner,* June 11, 1986.

434 *But, as a former prosecutor:* Reidy, "Rogers Urges No Charges in Shuttle Loss."

434 *Down in Huntsville:* Michael Mecham, "Mulloy Admits Launches Should Have Been Halted a Year Ago," *Florida Today,* June 12, 1986.

435 *In a series of public:* Richard March, "Graveside Rites Held for McAuliffe," *Tampa Bay Times,* May 2, 1986; "Pilot of Challenger Is Buried with Full Rites at Arlington,"

NYT, May 4, 1986; "Taps Sound Last Goodbye for Astronaut," Knight-Ridder, *Atlanta Journal,* May 4, 1986; Marcia Jarvis, author interview, March 14, 2022, and poem " 'To Greg,' 5/10/86," in collection of Jarvis; Peter Geiger, "Victims of Tragedy Urge Sensitive Reporting," *Akron Beacon-Journal,* May 18, 1986.

435   *On a hot Saturday morning:* Dudley Clendinen, "Astronaut Buried in Carolina: 35-Year 'Mission' Is Complete," *NYT,* May 18, 1986; Bruce Smith, "S.C. Citizens Pay Their Final Respects to Ronald McNair," *Times and Democrat,* May 18, 1986; Jay Hamburg, "McNair's Town Lays Its Hero of the Heavens to Earthly Rest," *Orlando Sentinel,* May 18, 1986.

435   *Two days later:* Scobee Rodgers, *Silver Linings,* 154; Lee Hockstader, "Astronaut Buried in Arlington Rite," *WP,* May 20, 1986; June Scobee Rodgers, author interview, September 29, 2023.

435   *It wasn't until June 2:* Kay Lynch, "Onizuka Returns to Isles," *Honolulu Advertiser,* June 1, 1986, AP, "Public Funeral Service Held for 7th Astronaut," *NYT,* June 4, 1986; "Puowaina (National Memorial Cemetery of the Pacific)," Historic Hawai'i Foundation, https://historichawaii.org/2014/02/19/puowaina-national-memorial-cemetery-of-the-pacific-2/.

436   *Even before the publication:* Deborah Mesce, "Families Mull Suing NASA," *Danville Register,* June 6, 1986.

436   *The families were among:* Scobee Rodgers, author interview, October 15, 2021; Cook, *The Burning Blue,* 225.

437   *Although drawn closer:* AP, "Astronauts' Families Tell of Dismay," *NYT,* June 10, 1986; Jane Smith-Wolcott and Alison Balch, author interview, November 15, 2022; Kathy Sawyer, "Pilot's Widow Castigates NASA," *WP,* June 17, 1986.

437   *In the meantime, NASA's doctors:* William J. Broad, "Scientists Hopeful of Restoring Challenger Tapes," *NYT,* June 19, 1986; date of late June given in John Noble Wilford, "Shuttle's Tapes Show Crew Had No Hint of Fate," *NYT,* July 18, 1986.

438   *The results of the forensic:* R. A. Zaldivar, "NASA Knew Challenger Crew Could Have Survived Blast," *Miami Herald,* August 2, 1986; Ben Evans, "The Fatal Flaws Which Doomed Challenger (Part 1)," https://www.americaspace.com/2014/01/25/missed-warnings-the-fatal-flaws-which-doomed-challenger-part-1/; Report, Joseph P. Kerwin to RADM Richard H. Truly, July 28, 1986, https://history.nasa.gov/kerwin.html; details of PEAP from Bob Crippen, author interview, October 4, 2023, and images and description in NASA Shuttle Operations Manual, JSC-12770, 128.

439   *After learning what Kerwin:* Jane Smith-Wolcott and Alison Balch, author interviews, November 15 and 16, 2022.

439   *On July 2, 1986:* "Jane Smith's 'Claim for Damage, Injury or Death' from the US Government, One Page," in Macidull and Blattner, *Challenger's Shadow,* 65; AP, "Shuttle Pilot's Widow Files Claim," *Herald and Review* (Decatur, IL), July 16, 1986.

440   *The decision:* Jane Smith-Wolcott and Alison Balch, author interview, November 16, 2022; Deborah Messe, "Lawyers Lay Groundwork for Possible Shuttle Suits," *Galion Inquirer,* June 6, 1986; UPI, "Shuttle Builder Lawsuit Target?," *Tyler Courier-Times,* June 5, 1986; Grissom and Still, *Starfall,* 222, 230–31; Katharine Q. Seelye, "Betty Grissom, Who Sued in Astronaut Husband's Death, Dies at 91," *NYT,* October 11, 2018.

440   *That suspicion:* Todd Halvorson and Michael Lafferty, "Shuttle's Last Word: Uh-Oh! Tape Adds to Sorrow of Families," and Michael Lafferty, "At Least 2 Likely Knew of Problem," *Florida Today,* July 29, 1986. In a letter to shuttle chief Dick Truly,

Kerwin summarized the conclusions of his investigation into the deaths of the crew, explaining that "the cause of death of the Challenger astronauts cannot be positively determined," that they probably survived the breakup of the spacecraft, and "possibly, but not certainly" lost consciousness soon afterward as a result of loss of cabin pressure: Report, Joseph P. Kerwin to RADM Richard H. Truly, July 28, 1986. Once the inquiry was complete, Dr. Kerwin chose to regard the subject as confidential, later explaining that "although the individuals involved were not my patients, I protect our interactions as though they were." Kerwin, text message to author, June 5, 2023.

440 *Eventually, all seven:* AP, "Challenger Settlements Disclosed," *LAT,* March 8, 1988; David E. Sanger, "How See-No-Evil Doomed Challenger," *NYT,* June 29, 1986.

441 *The company's treatment:* McDonald with Hansen, *Truth, Lies, and O-Rings,* 412–18.

442 *After Markey's letter:* Boisjoly, *Challenger Book Material,* 485–86, 500, 605, 787–89; McDonald with Hansen, *Truth, Lies, and O-Rings,* 403–5; 414–18; the later events ultimately took place during 1987. Boisjoly's medical report from Associated Behavior Consultants, October 15, 1986, quoted in Boisjoly, *Challenger Book Material,* 726–31.

442 *In July 1986, Boisjoly:* Ibid.; Boisjoly, *Challenger Book Material,* 751–52; " 'Whistle Blower' Sues Thiokol over Shuttle Tragedy," *Florida Today,* January 29, 1987; Philip M. Boffey, "Suits Against Maker of Space Rocket Thrown Out," *NYT,* September 3, 1988; "Judge Dismisses Two Lawsuits by Ex Thiokol Engineer," *Salt Lake Tribune,* September 3, 1988.

443 *In the end:* Martin Merzer and Stephen K. Doig, "Back to the Future," *Miami Herald,* September 30, 1988; "STS-26 Landing (Discovery)," YouTube video, https://www.youtube.com/watch?v=fYxyrFiBT0I.

443 *The seven families:* Scobee Rodgers, author interviews, October 15, 2021, and September 29, 2023.

443 *The changes:* William J. Broad, "Astronauts to Have Many New Ways to Escape," *NYT,* September 28, 1988; William J. Broad, "Changes Big and Small Aid Flight, Experts Say," *NYT,* October 1, 1988; William E. Leary, "Change of Command; A New NASA Guides Space Program into the Post-Challenger Era," *NYT,* October 2, 1988.

444 *But some of those:* Powell, "Obviously, a Major Malfunction."

444 *on Crippen's instructions:* Bob Crippen, author interview, October 4, 2023; "Smithsonian Considering Display of Fallen Shuttles Challenger and Columbia Debris," Collectspace, January 31, 2011, http://www.collectspace.com/news/news-013111a.html; Mike Cianelli, author interview, January 21, 2022.

444 *the diverse roles:* "Statement on the Building of a Fourth Shuttle Orbiter and the Future of the Space Program," Ronald Reagan Presidential Library & Museum, August 15, 1986; Alan Ladwig, personal communication with author, February 22, 2024; William J. Broad, "Pentagon Leaves the Shuttle Program," *NYT,* August 7, 1989.

445 *By the time it embarked:* John Uri, "20 Years Ago: Remembering *Columbia* and Her Crew," January 26, 2023, NASA, https://www.nasa.gov/history/20-years-ago-remembering-columbia-and-her-crew/; Jeffrey Kluger, "20 Years On, How the Columbia Shuttle Disaster Changed Space Travel," *Time,* February 1, 2023, https://time.com/6251161/columbia-shuttle-disaster-20th-anniversary/.

446 *On the morning of Saturday, February 1:* William Harwood, "Reconstructed Timeline of Columbia's Re-Entry," CBS News, https://www.cbsnews.com/network/news

/space/home/memorial/107.html; William Langeweische, "Columbia's Last Flight," *Atlantic Monthly*, November 2003.

447 *with the force of a rifle bullet:* McDonald with Hansen, *Truth, Lies, and O-Rings*, 561–62.

448 *At 9:12 a.m.:* CAIB Report, vol. I, 38–44; "Inside Mission Control during STS-107 Columbia's Failed Re-Entry and Disaster," YouTube video, https://www.youtube.com /watch?v=cbnT8Sf_LRs; Robert Lee Hotz, "Decoding Columbia: A Detective Story," *LAT*, December 21, 2003; Jason Whitely, "The Story Behind the Story of WFAA's Space Shuttle Columbia Coverage," WFAA-TV, January 31, 2018, https://www.wfaa .com/article/features/original/the-story-behind-the-story-of-wfaas-space-shuttle -columbia-coverage/287-513349113; AP, "Broadcast Feuds," *Lincoln Journal-Star*, February 13, 2003; Langeweische, "Columbia's Last Flight."

EPILOGUE

449 *many of the lessons:* "The parallels are striking," the report stated, explaining that "the acceptance of events that are not supposed to happen has been described by sociologist Diane Vaughan as the "normalization of deviance." CAIB Report, vol. I, 130.

450 *For a successful technology:* Feynman, RCR, vol. II, Appendix F, F-5.

450 *George Abbey was among those:* Cassutt, *The Astronaut Maker*, 310–13, 318, 366–87, 413–15, 420.

450 *Former NASA Administrator James Beggs:* Matt Schudel, "James M. Beggs, NASA administrator in the 1980s, dies at 94," WP, April 25, 2020; Sam Roberts, "James Beggs, 94, Is Dead; NASA Chief Championed Space Shuttle," *NYT*, April 28, 2020.

450 *In 1988, the American Association:* Douglas Martin, "Roger Boisjoly, 73, Dies; Warned of Shuttle Danger," *NYT*, February 12, 2012.

451 *Bob Crippen continued:* "Robert L. Crippen, Former NASA Astronaut and Kennedy Space Center Director (1992-1995)," https://www.nasa.gov/people/robert-l-crip pen/; Crippen, author interview, October 18, 2021; AP, "Thiokol signs contract extension for space shuttle rocket boosters," *Florida Today*, September 5, 1999.

451 *Astronaut Charlie Bolden:* "Charles F. Bolden, Jr.," https://www.nasa.gov/people /charles-f-bolden-jr/.

451 *Unable to face:* William Grimes, "Robert Ebeling, Challenger Engineer Who Warned of Disaster, Dies at 89," *NYT*, March 25, 2016; Sarah Kaplan, "Finally free from guilt over Challenger disaster, an engineer dies in peace," WP, March 22, 2016; Howard Berkes, "Your Letters Helped Challenger Shuttle Engineer Shed 30 Years of Guilt," NPR.org, February 25, 2016, https://www.npr.org/sections/thetwo -way/2016/02/25/466555217/your-letters-helped-challenger-shuttle-engineer -shed-30-years-of-guilt.

452 *After the disintegration:* Kirby, "The Wonder of it All"; Jonathan Hawkins, "The Tragic Triumph of the World's Largest Concert," CNN, April 5, 2016, https://edi tion.cnn.com/2016/04/05/us/challenger-astronaut-and-saxophone/index.html; Jarre, author interview, September 30, 2021; Craig Hlavaty, "Rendez-Vous Houston in 1986 brought Guinness record sound, light show to Houston," *Houston Chronicle*, April 3, 2018, https://www.chron.com/news/houston-texas/houston/article /Rendez-Vous-Houston-in-1986-brought-Guinness-7229060.php.

452 *Marcia Jarvis never returned:* Jarvis, author interviews, March 14, 2022 and January 17, 2024.

452   *After retiring from:* Charles Fishman, "The blame takers: New lives with little time for reflection," *Orlando Sentinel,* January 25, 1987; *Challenger: The Final Flight;* Dunar and Waring, *Power to Explore,* 624.

453   *In the months that followed:* June Scobee Rodgers, author interview, October 16, 2021; AP, " 'Teacher-in-space' widower weds," *Rapid City Journal,* March 6, 1992; Kathy McCormack, "McAuliffe's students followed example," and "Christa's goals have been largely accomplished," statement from Steven McAuliffe, *Concord Monitor,* January 28, 2016.

453   *After overseeing the:* McDonald with Hansen, *Truth, Lies and O-Rings,* 520–41, Clay Risen, "Allan McDonald Dies at 83; Tried to Stop the Challenger Launch," *NYT,* March 9, 2021; Brian Russell, author interview, April 6, 2022.

454   *Cheryl McNair stayed:* Cheryl McNair, author interview, May 3, 2021; AP, "McNair's mom concerned about launch," *Florence Morning News,* September 26, 1988; Reuters, "Challenger crew's children doing well ten years after explosion," *Leader-Post* (Canada), February 3, 1996; Loretta Fulton, "Program recognizes Challenger astronaut," *Abilene Reporter-News,* April 14, 2001; Otis R. Taylor, Junior, "Lake City builds on Ron McNair's legacy," *The State,* January 25, 2011; "Ronald E. McNair Postbaccalaureate Achievement Program," US Department of Education, https://www2.ed.gov/programs/triomcnair/awards.html.

454   *In July 1986, Larry Mulloy:* Fishman, "The blame takers: New lives with little time for reflection"; "Lawrence Benjamin Mulloy, Nashville Tennessee, April 13, 1934-October 2, 2020," Tribute Archive; Larry Mulloy to Allan McDonald, "Re:FYI," June 18, 2014, copy of email chain provided to the author by Mark Maier; Alton Keel, author interview, October 12, 2023.

455   *In 1990, Representative:* "Bill Nelson," https://www.nasa.gov/people/nasa-administrator-bill-nelson/; "Senator Bill Nelson," https://www.congress.gov/member/clarence-nelson/N000032.

455   *Steve Nesbitt withdrew:* Nesbitt, author interview, January 20, 2024.

455   *In the months following:* Beverly Creamer, "Hawai'i Hero," *Honolulu Advertiser,* January 28, 1996. Jane Smith-Walcott and Alison Balch, author interview, November 17 2022; Rich Phillips, "Challenger astronaut's widow reflects on the end of U.S. shuttle program," CNN, July 3 2011, http://www.cnn.com/2011/US/07/03/shuttle.challenger.widow/index.html; Cassutt, *The Astronaut Maker,* 413.

456   *Marvin Resnik:* "Marvin Resnik, father of Challenger astronaut Judith Resnik, was 90," *Cleveland Jewish News,* March 18, 2010 https://www.clevelandjewishnews.com/archives/marvin-resnik-father-of-challenger-astronaut-judith-resnik-was-90/article_2765d62d-8e7c-549c-a5f7-e252c7f048bd.html.

456   *Sally Ride remained:* Sherr, *Sally Ride,* 213–16; 248–49; 262–68; 281; Denise Grady, "American Woman Who Shattered Space Ceiling," *NYT,* July 23, 2012.

456   *As the chair:* June Scobee Rodgers, author interview, October 16, 2021; Scobee Rodgers, *Silver Linings,* 186–87; 203; 212; 230–31; Challenger Center Annual Report, 2021, https://challenger.org/news-insights/2021-annual-report/.

457   *Jane Smith remained:* Jane Smith Walcott and Alison Balch, author interview, November 17, 2022; Tamara Jones, "A Space in the Heart," *WP,* January 27, 1996; Balch, personal communication with author, February 6, 2024.

SELECTED
BIBLIOGRAPHY

Allen, Joseph. P., with R. Martin. *Entering Space: An Astronaut's Odyssey.* New York: Stewart Tabori & Chang, 1984.

Atkinson, Joseph D., and Jay M. Shafritz. *The Real Stuff: A History of NASA's Astronaut Recruitment Policy.* Westport, CT: Praeger, 1985.

Avera, Randy. *The Truth about Challenger.* Good Hope, GA: Randolph, 2002.

Bagby, Meredith. *The New Guys: The Historic Class of Astronauts That Broke Barriers and Changed the Face of Space Travel.* New York: William Morrow, 2023.

Barbree, Jay. *"Live from Cape Canaveral": Covering the Space Race, from Sputnik to Today,* Washington, DC: Smithsonian Books, 2007.

Bartholomew, C. A. *Mud, Muscle, and Miracles: Marine Salvage in the United States Navy.* Washington, DC: Department of the Navy, 1990.

Boisjoly, Roger. *Challenger Book Material.* Unpublished manuscript, 1999.

Brinkley, Douglas. *Cronkite.* New York: HarperCollins, 2012.

Burgess, Colin, and Francis French. *In the Shadow of the Moon: A Challenging Journey to Tranquility, 1965–1969.* Lincoln: University of Nebraska Press, 2007.

Cabbage, Michael, and William Harwood. *Comm Check . . . : The Final Flight of Shuttle Columbia.* New York: Free Press, 2004.

Casper, John H. *The Sky Above: An Astronaut's Memoir of Adventure, Persistence, and Faith.* West Lafayette, IN: Purdue University Press, 2022.

Cassutt, Michael. *The Astronaut Maker: How One Mysterious Engineer Ran Human Spaceflight for a Generation.* Chicago: Chicago Review Press, 2018.

Cernan, Eugene, and Don Davis. *The Last Man on the Moon: Astronaut Eugene Cernan and America's Race in Space,* 10th anniv. ed. New York: St. Martin's Griffin, 2009.

Chrysler, C. Donald, and Donald L. Chaffee. *On Course to the Stars: The Roger B. Chaffee Story.* Grand Rapids, MI: Kregel, 1968.

Collins, Michael. *Carrying the Fire: An Astronaut's Journeys.* New York: Cooper Square Press, 2001.

Cook, Kevin. *The Burning Blue: The Untold Story of Christa McAuliffe and NASA's Challenger Disaster.* New York: Henry Holt, 2021.

Cook, Richard. *Challenger Revealed: An Insider's Account of How the Reagan Administration Caused the Greatest Tragedy of the Space Age.* New York: Thunder's Mouth Press, 2006.

Cooper, Henry S. F., Jr. *Before Lift-Off: The Making of a Space Shuttle Crew.* Baltimore: Johns Hopkins University Press, 1987.

Corrigan, Grace George. *A Journal for Christa: Christa McAuliffe, Teacher in Space.* Lincoln: University of Nebraska Press, 1993.

Croft, Melvin, and John Youskauskas. *Come Fly with Us: NASA's Payload Specialist Program*. Lincoln: University of Nebraska Press, 2019.

Cunningham, Walter. *The All-American Boys: An Insider's Look at the U.S. Space Program*, rev. and updated ed. New York: Ipicturebooks, 2010.

Dethloff, Henry C. *Suddenly Tomorrow Came . . . A History of the Johnson Space Center*. Washington, DC: Lyndon B. Johnson Space Center, 1993.

Dunar, Andrew J., and Stephen P. Waring. *Power to Explore: A History of the Marshall Spaceflight Center, 1960–1990*. Washington, DC: NASA History Office, 1999.

Dwight, Ed. *Soaring on the Wings of a Dream: The Struggles & Adventures of the "First Black Astronaut" Candidate*. Denver: Ed Dwight Studios, 2009.

Eisele, Donn F. *Apollo Pilot: The Memoir of Astronaut Donn Eisele*. Lincoln: University of Nebraska Press, 2017.

Evans, Ben. *Tragedy and Triumph in Orbit: The Eighties and Early Nineties*. New York: Springer-Praxis, 2012.

Evans, Michelle. *The X-15 Rocket Plane: Flying the First Wings into Space*. Lincoln: University of Nebraska Press, 2013.

Feynman, Richard P. *Perfectly Reasonable Deviations from the Beaten Track: The Letters of Richard P. Feynman*. Edited by Michelle Feynman. New York: Basic Books, 2005.

Feynman, Richard P., and Ralph Leighton. *"Surely You're Joking, Mr. Feynman!": Adventures of a Curious Character*. Toronto: Bantam, 1986.

Feynman, Richard P., and Ralph Leighton. *"What Do You Care What Other People Think?": Further Adventures of a Curious Character*. New York: W. W. Norton, 1988.

Gleick, James: *Genius: The Life and Science of Richard Feynman*. New York: Pantheon, 1992.

Gorn, Michael J. *Expanding the Envelope: Flight Research at NACA and NASA*. Lexington: University Press of Kentucky, 2001.

Grissom, Betty, and Henry Still. *Starfall*. New York: Thomas Y. Crowell, 1974.

Hansen, James R. *Engineer in Charge: A History of the Langley Aeronautical Laboratory, 1917–1958*. Washington, DC: U.S. Government Printing Office, 1987.

Harris, Hugh. *Challenger: An American Tragedy: The Inside Story from Launch Control*. New York: Open Road Integrated Media, 2014.

Heppenheimer, T. A. *History of the Space Shuttle, Volume 1: The Space Shuttle Decision, 1965–1972*. Washington, DC: Smithsonian Institution Press, 2002.

———. *History of the Space Shuttle, Volume 2: Development of the Shuttle, 1972–1981*. Washington, DC: Smithsonian Institution Press, 2010.

Hitt, David, and Heather R. Smith. *Bold They Rise: The Space Shuttle Early Years, 1972–1986*. Lincoln: University of Nebraska Press, 2014.

Hohler, Robert T. *I Touch the Future: The Story of Christa McAuliffe*. New York: Random House, 1986.

Jenkins, Dennis R. *Space Shuttle: Developing an Icon, 1972–2013*, vols. I–III. Forest Lake, MN: Specialty Press, 2016.

Jenkins, Dennis R. *Space Shuttle: The History of the National Space Transportation System The First 100 Missions*, Cape Canaveral, Fla: D.R. Jenkins, 2001.

Johnston, Moira. *The Last Nine Minutes: The Story of Flight 981*. New York: William Morrow, 1976.

Kaplan, Marshall H. *Space Shuttle: America's Wings to the Future*: Fallbrook, CA: Aero Publishers, 1978.

Koppel, Lily. *The Astronaut Wives Club: A True Story.* New York, Grand Central Publishing, 2013.

Kraft, Chris. *Flight: My Life in Mission Control.* New York: Plume, 2002.

Kranz, Gene. *Failure Is Not an Option: Mission Control from Mercury to Apollo 13 and Beyond.* New York: Simon & Schuster, 2000.

Kuznetz, Lawrence. *Save the Shuttle: Memoirs of Life with the Space Shuttle and the Last Ditch Attempt to Save and Privatize It.* e-book, 2014.

Ladwig, Alan. *See You in Orbit?: Our Dream of Spaceflight.* Falls Church, VA: To Orbit Productions, 2019.

Lane, Helen, ed. *Wings in Orbit,* NASA SP-2010-3409.

Launius, Roger D., and Dennis R. Jenkins. *Coming Home: Reentry and Recovery from Space.* Washington, DC: U.S. National Aeronautics and Space Administration, 2012.

Leopold, George. *Calculated Risk: The Supersonic Life and Times of Gus Grissom.* West Lafayette, IN: Purdue University Press, 2016.

Lewis, Richard S. *Challenger: The Final Voyage.* New York: Columbia University Press, 1988.

Lipartito, Kenneth, and Orville R. Butler. *A History of the Kennedy Space Center.* Gainesville: University Press of Florida, 2007.

Logsdon, John M. *After Apollo?: Richard Nixon and the American Space Program.* New York: Palgrave Macmillan, 2015.

———. *Ronald Reagan and the Space Frontier.* New York: Palgrave Macmillan, 2019.

Macidull, John C., and Lester E. Blattner. *Challenger's Shadow: Did Government and Industry Management Kill Seven Astronauts?* Coral Springs, FL: Llumina Press, 2002.

Mahaffey, James. *Atomic Adventures: Secret Islands, Forgotten N-Rays, and Isotopic Murder—A Journey into the Wild World of Nuclear Science.* New York: Simon & Schuster, 2017.

Mark, Hans. *The Space Station: A Personal Journey.* Durham, NC: Duke University Press, 1987.

McCandless, Bruce, III. *Wonders All Around: The Incredible True Story of Astronaut Bruce McCandless II and the First Untethered Flight in Space.* Austin, TX: Greenleaf Book Group, 2021.

McConnell, Malcolm. *Challenger: A Major Malfunction.* Garden City, NY: Doubleday, 1987.

McDonald, Allan J., with James R. Hansen. *Truth, Lies, and O-Rings: Inside the Space Shuttle* Challenger *Disaster.* Gainesville: University Press of Florida, 2009.

McNair, Carl S., with H. Michael Brewer. *In the Spirit of Ronald E. McNair, Astronaut: An American Hero.* Atlanta: Publishing Associates, 2005.

Mullane, Mike. *Riding Rockets: The Outrageous Tales of a Space Shuttle Astronaut.* New York: Simon & Schuster, 2007.

Murray, Charles, and Catherine Bly Cox. *Apollo: The Race to the Moon.* New York: Simon & Schuster, 1989.

Nelson, Bill, with Jamie Buckingham. *Mission: An American Congressman's Voyage to Space.* San Diego: Harcourt Brace Jovanovich, 1988.

Neufeld, Michael J. *Von Braun: Dreamer of Space, Engineer of War.* New York: Alfred A. Knopf, 2007.

Noonan, Peggy. *What I Saw at the Revolution: A Political Life in the Reagan Era.* New York: Random House, 1990.

Ogawa, Dennis M., and Glen Grant. *Ellison S. Onizuka: A Remembrance.* Honolulu: Mutual, 1986.

Paul, Richard, and Steven Moss. *We Could Not Fail: The First African Americans in the Space Program.* Austin: University of Texas Press, 2016.

Perrow, Charles. *Normal Accidents: Living with High-Risk Technologies.* Princeton, NJ: Princeton University Press, 1999.

Petty, Chris. *Beyond Blue Skies: The Rocket Plane Programs That Led to the Space Age.* Lincoln: University of Nebraska Press, 2020.

Reichhardt, Tony, ed. *Space Shuttle: The First 20 Years.* London: DK Publishing, 2002.

Reuter, Claus. *The V2 and the German, Russian and American Rocket Program.* S. R. Research & Publishing, 2000.

Scobee Rodgers, June. *Silver Linings: My Life before and after* Challenger 7. Macon, GA: Smyth & Helwys, 2011.

Seddon, Rhea. *Go for Orbit: One of America's First Women Astronauts Finds Her Space.* Murfreesboro, TN: You Space Press, 2015.

Shayler, David J., and Colin Burgess. *NASA's First Space Shuttle Astronaut Selection: Redefining the Right Stuff.* New York: Springer Praxis, 2020.

Shayler, David J., and Ian A. Moule. *Women in Space: Following Valentina.* New York: Springer Praxis, 2005.

Sherr, Lynn. *Sally Ride: America's First Woman in Space.* New York: Simon & Schuster, 2014.

Staff of the *Washington Post.* *Challengers: The Inspiring Life Stories of the Seven Brave Astronauts of Shuttle Mission 51-L.* New York: Pocket Books, 1986.

Thomas, James A. (Gene). *Some Trust in Chariots: The Space Shuttle* Challenger *Experience.* Longwood, FL: Xulon Press, 2006.

Thompson, Milton O. *At the Edge of Space: The X-15 Flight Program.* Washington, DC: Smithsonian Books, 2003.

Trento, Joseph J. *Prescription for Disaster.* New York: Crown, 1987.

Tribe, John. *My Story: A Personal Account.* Unpublished memoir, 2016.

Vaughan, Diane. *The* Challenger *Launch Decision.* Chicago: University of Chicago Press, 2016.

Weitekamp, Margaret A. *Right Stuff Wrong Sex: America's First Women in Space Program.* Baltimore: Johns Hopkins University Press, 2004.

White, Rowland. *Into the Black: The Extraordinary Untold Story of the First Flight of the Space Shuttle* Columbia *and the Astronauts Who Flew Her.* New York: Touchstone, 2016.

Young, John W., and James R.Hansen. *Forever Young: A Life of Adventure in Air and Space.* Gainesville: University Press of Florida, 2012.

REPORTS AND PAPERS

The Columbia Accident Investigation Board Report, vols. I–VI. Washington, DC: NASA and U.S. Government Printing Office, 2003.

Investigation of the Challenger Accident: Hearings before the Committee on Science and Technology, U.S. House of Representatives, Ninety-ninth Congress, second session. Washington, DC: U.S. Government Printing Office, 1986.

NASA's Quality Assurance Program: Hearing before the Subcommittee on Space Science and Applications of the Committee on Science and Technology, U.S. House of Representatives, Ninety-Ninth Congress, Second Session, May 21, 1986, United States Congress

House Committee on Science and Technology. Washington, DC: U.S. Government Printing Office, 1986.

Space Shuttle Accident: Hearings before the Subcommittee on Science, Technology, and Space of the Committee on Commerce, Science, and Transportation. United States Senate, Ninety-Ninth Congress, Second Session, on Space Shuttle Accident and the Rogers Commission Report, February 18, June 10 and 17, 1986. Washington, DC: U.S. Government Printing Office, 1986.

Report of the Apollo 204 Review Board to the Administrator. National Aeronautics and Space Administration, 1967

Report of the Presidential Commission on the Space Shuttle Challenger Accident, vols. I–V. Washington, DC: Government Printing Office, 1986.

Turkish Airlines DC-10 TC-JAV Report on the Accident in the Ermenonville Forest, France on 3 March 1974. London: Department of Trade, Accidents Investigation Branch, HMSO, 1976.

TV AND FILM

Craig, Mark, dir. *The Last Man on the Moon.* 2014.

Leckart, Steven, and Daniel Junge, dir. *Challenger: The Final Flight,* 2020.

Watson, Paul, dir. *Lost in Space.* BBC TV, 1985.

Weil, Zachary, dir. *When We Were Shuttle,* 2022.

Engineering the Space Shuttle. MIT Open Learning Library Lectures.

"Presidential Commission on the Space Shuttle Challenger Hearings," C-SPAN, February 6–April 3, 1986.

The World about Us: The Ultimate Explorer. Michael Dean, reporter. BBC TV, 1982.

ARCHIVES

Christa McAuliffe Collection, Framingham State University

Joseph and Susan Trento NASA Safety Investigation Collection, Air and Space Archives, Udvar-Hazy Center, National Air and Space Museum, Virginia

NASA Headquarters History Collection

NASA Johnson Space Center Oral History Project

NASA Scientific and Technical Information Program Repository

National Archives at College Park, Maryland

Paley Center for Media Archive of Television and Radio, New York

Ronald Reagan Presidential Library & Museum

University of Alabama in Huntsville Archives, Special Collections and Digital Initiatives

AUTHOR INTERVIEWS

Abbey, George
Director of Flight Operations, Johnson Space Center
Telephone, September 25, 2020, and October 2, 2020

Allen, Joseph
Astronaut, STS-5, STS-51-A
Crawford, IN, June 9, 2021

Balch, Alison
Student; daughter of Mike Smith
Virginia Beach, VA, November 15 and 16, 2022

Belt, David
Lieutenant, operations and diving officer, USS Preserver
Telephone, May 15, 2023

Bluford, Guion
Astronaut, STS-8, STS-61-A, STS-39, STS-53
Telephone, January 8 and 13, 2021

Bolden, Charles
Astronaut, STS-61-C, STS-31, STS-45, STS-60
Washington, DC, August 11, 2021

Brasher, Warren
Propulsion design specialist, Reusable vehicle design, Engineering Directorate, Johnson
Space Center, 1969
Houston, TX, May 5, 2021

Broad, William
Reporter, New York Times
Telephone, May 24, 2021

Burn, Jerry
Design engineer, structures, Space Shuttle Solid Rocket Motor Project, Morton Thiokol
Wasatch Division
Wellsville, UT, February 7, 2022
Telephone, December 6, 2023

Buzzard, Frank
Director, Columbia Task Force, Johnson Space Center, 2003
Telephone, March 30, 2021

Campion, Edward
Public Affairs Officer, NASA Headquarters, Washington, DC
Annapolis, MD, May 11, 2021

Cianelli, Mike
Manager, Apollo, Challenger, Columbia Lessons Learned Program, 2023
Kennedy Space Center, FL, January 20, 2023
Merritt Island, FL, January 21, 2023

Covey, Dick
Astronaut, STS-51-I, STS-26, STS-38, STS-61
Scottsdale, AZ, December 6, 2021

Cox, Duane Morley
Program Manager for propellant, liner, and insulation during Space Shuttle Solid Rocket Motor Project design and development, Morton Thiokol Wasatch Division
Telephone, May 10, May 27, and August 5, 2021
Brigham City, UT, December 9, 2021

Crippen, Robert
Astronaut, STS-1, STS-7, STS-41-C, STS-41-G
West Palm Beach, FL, November 18, 2021, and October 4, 2023

Devlin, John
Commander, USS Preserver
Telephone, June 12, 2023

Fisher, Anna
Astronaut, STS-51-A
Houston, TX, January 13, 2022

Fishman, Charles
Reporter, Washington Post *and* Orlando Sentinel
Telephone, October 11, 2023

Frosch, Robert
NASA Administrator, 1977–81
Telephone, September 14, and 21, 2020

Gillette, Estella
Chief, Administrative Support Office, Astronaut Office/Flight Crew Operations Directorate, Johnson Space Center
Houston, TX, April 6, 2021

Hardman, Walter
Captain, LCU
Telephone, August 18, 2023

Hauck, Frederick
Astronaut, STS-7, STS-51-A, STS-26
Telephone, October 13, 2020, and December 15, 2021
Portland, ME, October 26, 2021

Hawley, Steve
Astronaut, STS-41-D, STS-61-C, STS-31, STS-82, STS-93
Telephone, October 1, 8, and 15, 2020

Heineman, Raymond
Chief, Aircraft Operations Division, Johnson Space Center, 2023
Ellington Field Joint Reserve Base, TX, January 27, 2023

Heisig, Kurt
Musician
Telephone, July 25, 2023

Henderson, Jeff
"A" Complex Test Director, Stennis Test Center, 2023
Stennis Test Center, MS, January 25, 2023

Honeycutt, Jay
Manager, Operations Integration, Space Shuttle Program Office
Cocoa Beach, FL, July 14, 2022

Huntoon, Carolyn
Member, Astronaut Group 8 Selection Board; Associate Director of Space and Life Sciences,
Johnson Space Center
Providence, RI, May 19, 2021

Hutchinson, Neil
Ascent Flight Director, Mission Control, STS-1
Telephone, July 20 and 21, 2021

Jarre, Jean-Michel
Musician
Zoom, September 30, 2021

Jarvis, Marcia
Dental assistant, wife of Greg Jarvis
Telephone, February 28, 2022, and January 17, 2024
St. James, NC, March 14, 2022

Keel, Alton
Executive Director, Rogers Commission
Stanardsville, VA, October 20, 2020
Telephone, October 12, 2023

Konrad, John
Engineer, Hughes Space and Communications Group; NASA Payload Specialist
Telephone, March 16, 2022

Kutyna, Don
General, USAF; member of Rogers Commission
Telephone, September 24 and November 2, 2020

Kuznetz, Lawrence
Engineer, Crew Systems Division, Johnson Space Center
New York, NY, April 21, 2022

Ladwig, Alan
Manager, Space Flight Participant Program
Washington, DC, August 12, 2021

McAllister, Mike
TPS mechanic, Kennedy Space Center, 1979–80, and Rockwell International,
Palmdale, CA, 1981–83; diver/deckhand, LCU, 1983-86
Melbourne, FL, July 14, 2022

McNair, Cheryl
Teacher; wife of Ron McNair
League City, TX, May 3, 2021

Morgan, Barbara
Teacher in Space backup; astronaut, STS-118
Telephone, December 12, 2023

Moser, Tom
Director of Engineering, Johnson Space Center
Kerrville, TX, February 23 and 24, 2022

Mullane, Mike
Astronaut, STS-41-D, STS-27, STS-36
Zoom, November 9, 2020

Musgrave, Story
Astronaut, STS-6, STS-51-F, STS-33, STS-44, STS-61, STS-80
Telephone, September 28, 2023

Nelson, George "Pinky"
Astronaut, STS-41-C, STS-61-C, STS-26
Telephone, January 5, 2022

Nesbitt, Steve
Public Affairs Officer, Johnson Space Center
Pine Brook, TX, April 7, 2021
Johnson Space Center, TX, January 27, 2023
Telephone, January 20, 2024

Nygren, Richard
Assistant to the Director of Flight Operations, Johnson Space Center
Telephone, January 6, 2021
La Grange, TX, February 23 and 24, 2021

Russell, Brian
Manager in the Solid Rocket Motor Project Office, Morton Thiokol Wasatch Division
Telephone, March 2 and April 26, 2022, and September 26, 2023
Ogden, UT, February 6, 2022

Salinas Stottlemyer, Sylvia
Administrative assistant, Astronaut Office/Flight Crew Operations Directorate, Johnson
Space Center
Clear Lake, TX, April 7, 2021

Sanger, David
Reporter, New York Times
Washington, DC, May 13, 2021

Scobee Rodgers, June
Teacher; wife of Francis "Dick" Scobee
Telephone, September 29, 2021 and September 29, 2023
Chattanooga, TN, October 15 and 16, 2021

Sieck, Robert
Director, Launch and Landing Operations, Kennedy Space Center
Telephone, October 26 and 27, 2020
Titusville, FL, July 12, 2021

Smith-Wolcott, Jane
Wife of Mike Smith
Virginia Beach, VA, November 15 and 16, 2022

Stein (Howard), Jenny
Flight Control Engineer, Booster, Johnson Space Center
Clear Lake, TX, May 5, 2021

Swain, Jerry
Shuttle Mission Simulator Instructor, team lead, STS-51-L, Johnson Space Center
Johnson Space Center, TX, January 27, 2023

Thompson, L. David
First Class Diver, USS Preserver
Telephone, June 21, 2023

Trento, Joseph
Correspondent, CNN
Telephone, September 25, 2020

Tribe, John
North American Aviation Lead Reaction Control Systems Engineer, Apollo Command
and Service Module, Kennedy Space Center, 1967; Rockwell International Director of
Engineering for Orbiter Launch Support Services, Kennedy Space Center,1986
Merritt Island, FL, July 12, 2021
Kennedy Space Center, FL, January 20, 2023

Vander, Maury
Chief of Test Operations, Stennis Test Center, 2023
Stennis Test Center, MS, January 25, 2023

Van Hoften, James "Ox"
Astronaut, STS-41-C and STS-51-I
Telephone, January 7 and 11, 2022

Vaughan, Chester
Deputy Chief, Propulsion and Power Division, Johnson Space Center
Telephone, April 1, 2021
Clear Lake, TX, May 4, 2021

Titles in early 1986 unless otherwise specified

INDEX

ABOUT THE AUTHOR

Adam Higginbotham has written for the *New Yorker*, the *New York Times Magazine*, *Wired*, *GQ*, and *Smithsonian*. His first book, *Midnight in Chernobyl*, was the winner of the William E. Colby Award and the Andrew Carnegie Medal for Excellence in Nonfiction, and a *New York Times* Top Ten Book of the Year. He lives with his family in New York City.